Honey Bee Medicine for the Veterinary Practitioner

Honey Bee Medicine for the Veterinary Practitioner

Edited by

Terry Ryan Kane, DVM, MS
A2 Bee Vet
Ann Arbor, MI, USA

Cynthia M. Faux, DVM, PhD, DACVIM-LA
The University of Arizona College of Veterinary Medicine
Oro Valley, AZ, USA

WILEY Blackwell

This edition first published 2021

Registered Office
John Wiley & Sons, Inc., 111 River Street, Hoboken, NJ 07030, USA

Editorial Office
111 River Street, Hoboken, NJ 07030, USA

For details of our global editorial offices, customer services, and more information about Wiley products visit us at www.wiley.com.

Wiley also publishes its books in a variety of electronic formats and by print-on-demand. Some content that appears in standard print versions of this book may not be available in other formats.

Library of Congress Cataloging-in-Publication Data

Names: Kane, Terry Ryan, editor. | Faux, Cynthia M., editor.
Title: Honey bee medicine for the veterinary practitioner / edited by Terry Ryan Kane, Cynthia M. Faux.
Description: Hoboken, NJ : Wiley-Blackwell, 2021. | Includes bibliographical references and index.
Identifiers: LCCN 2020026997 (print) | LCCN 2020026998 (ebook) | ISBN 9781119583370 (hardback) | ISBN 9781119583233 (adobe pdf) | ISBN 9781119583424 (epub)
Subjects: MESH: Bees | Veterinary Medicine
Classification: LCC QL568.A6 (print) | LCC QL568.A6 (ebook) | NLM QL 568.A6 | DDC 595.79/9–dc23
LC record available at https://lccn.loc.gov/2020026997
LC ebook record available at https://lccn.loc.gov/2020026998

Cover Design: Wiley
Cover Image: © przemeksuwalki/Getty Images

Set in 9.5/12.5pt STIXTwoText by SPi Global, Pondicherry, India

SMI007733_062425

This book is dedicated to the pollinators. These creatures enrich the world with their service to flowers and it is our sincerest hope to provide service to them.

Terry Ryan Kane and Cynthia M. Faux

Contents

List of Contributors

Editors

Terry Ryan Kane, DVM, MS
A2 Bee Vet
Ann Arbor, MI, USA

Cynthia M. Faux, DVM, PhD, ACVIM-LA
Professor of Veterinary Medicine
College of Veterinary Medicine
The University of Arizona
Oro Valley, AZ, USA

Contributors

Esmaeil Amiri, PhD
Department of Biology
University of North Carolina at Greensboro,
Greensboro, NC;
Department of Entomology & Plant Pathology
North Carolina State University
Raleigh, NC, USA

Jeffrey R. Applegate, Jr., DVM, DACZM
Nautilus Avian and Exotics Veterinary Specialist
Brick, NJ, USA;
Adjunct Clinical Professor
North Carolina State University
College of Veterinary Medicine
Raleigh, NC, USA

Dewey M. Caron, PhD
Emeritus Professor
Western Apiculture Society
University of Delaware
Affiliate Faculty Oregon State University
Portland, OR, USA

Yanping (Judy) Chen, PhD
Research Entomologist
USDA-ARS Bee Research Laboratory
Beltsville, MD, USA

Christopher J. Cripps, DVM
Betterbee
Greenwich, NY, USA

Gigi Davidson, BSP, DICVP
NC State College of Veterinary Medicine
Department of Clinical Pharmacy Services
Raleigh, NC, USA

Jay D. Evans, PhD
Research Leader
USDA-ARS Bee Research Laboratory
Beltsville, MD, USA

Krispn Given
Apiculture Specialist
Department of Entomology
Purdue University
West Lafayette, IN, USA

Jerry Hayes
Editor
Bee Culture Magazine
Medina, OH, USA

Brandon K. Hopkins, PhD
Department of Entomology
Washington State University
Pullman, WA, USA

Charlotte Hubbard
Schoolcraft, MI, USA
www.hubbardhive.com

Adam J. Ingrao, PhD
U.S. Army Veteran
Michigan Co-Coordinator
USDA North Central Region Sustainable Agriculture Research and Education Program (NCR-SARE);
Extension Specialist
Michigan State University Extension
MI, USA

Reed M. Johnson, PhD
Associate Professor
Department of Entomology
The Ohio State University
Wooster, OH, USA

Britteny Kyle, DVM
Department of Population Medicine
Ontario Veterinary College
University of Guelph
Ontario, Canada

Katie Lee, PhD
Department of Entomology
University of Minnesota
Bee Research Facility
St. Paul, MN, USA

Marcie Logsdon, DVM
Department of Veterinary Clinical Sciences
College of Veterinary Medicine
Pullman, WA, USA

Margarita M. López-Uribe, dPhD
Lorenzo L. Langstroth Early Career Professor
Assistant Professor of Entomology
Department of Entomology
Center for Pollinator Research
Penn State University
University Park, PA, USA

Scott McArt, PhD
Department of Entomology
College of Agriculture and Life Sciences
Cornell University
Ithaca, NY, USA

Meghan Milbrath, PhD
Michigan Pollinator Initiative
Department of Entomology
Michigan State University
East Lansing, MI, USA

Kristen K. Obbink, DVM, MPH, DACVPM
Center for Food Security and Public Health
Iowa State University College of Veterinary Medicine
Ames, IA, USA

Randy Oliver, MS
ScientificBeekeeping.com

David T. Peck, PhD
Department of Entomology
Cornell University
Ithaca, NY, USA

Robin W. Radcliffe, DVM, DACZM
Director, Cornell Conservation Medicine Program
Department of Clinical Sciences
College of Veterinary Medicine
Cornell University
Ithaca, NY, USA

Rolfe M. Radcliffe, DVM, DACVS, DACVECC
Lecturer, Large Animal Surgery and Emergency Critical Care
College of Veterinary Medicine
Cornell University
Ithaca, NY, USA

Kasie Raymann, PhD
Department of Biology
University of North Carolina at Greensboro
Greensboro, NC, USA

Gary S. Reuter
Apiculture Technician
Department of Entomology
University of Minnesota
Bee Research Facility
St. Paul, MN, USA

James A. Roth, DVM, MS, PhD, ACVM
Center for Food Security and Public Health
Iowa State University College of Veterinary Medicine
Ames, IA, SA

Olav Rueppell, PhD
Florence Schaeffer Professor of Science
Department of Biology
University of North Carolina at Greensboro
Greensboro, NC, USA;
Department of Biological Sciences
University of Alberta, Edmonton, Canada

Thomas D. Seeley, PhD
Professor Emeritus
Department of Neurobiology and Behavior
College of Arts and Life Sciences
Cornell University
Ithaca, NY, USA

David R. Tarpy, PhD
University Scholar Professor
Graduate Program in Ecology & Evolution
W. M. Keck Center for Behavioral Biology
Department of Entomology & Plant Pathology
North Carolina State University
Raleigh, NC, USA

Dan Wyns
Field Specialist – Bee Informed Partnership
Academic Specialist – Michigan State University
Department of Entomology
Michigan State University
East Lansing, MI, USA

Acknowledgments

This book would not be possible without the collaboration among veterinarians, entomologists, toxicologists, pharmacologists, and the beekeeping community. We thank all the authors who so generously donated their time and expertise. All of us that love and care for these important animals hope this book will be valuable to you even if you don't become a "bee doctor." We invite you to join the Honey Bee Veterinary Consortium (hbvc.org).

We would, in particular, like to thank Dr. Barrett Slenning, Charlotte Hubbard, and Christine King for their editorial expertise and Patrick D. Wilson for his numerous illustrations. Much thanks are also due to Dr. Gloria Degrandi-Hoffman, Henry Graham, and Emily Watkins de Jong of the USDA's Carl Hayden Bee Research Center, Tucson, AZ for their assistance and support in obtaining photographs and Dr. Jamie Perkins, University of Arizona College of Veterinary Medicine and Dr. Andrew Wessman of the University of Arizona for assistance in acquiring the high magnification photos. We would also like to thank Dr. Ryane Englar for her expertise.

Terry Ryan Kane DVM, MS
Cynthia M. Faux DVM, PhD, DACVIM-LA

I was very fortunate in my life to have benefited from many dedicated and inspirational science teachers. I want to thank Tom Poulson, my ecology professor, for opening the world of insects to me. Dr. Erv Small helped me launch a long and rewarding career in veterinary medicine. Thank you to all the caring and compassionate veterinarians I have had the privilege of working with over the years. The American Veterinary Medical Association and my colleagues on the Committee for Environmental Issues have been very supportive of our honey bee projects. Thank you to Gina Luke for your early encouragement. To my flight instructors, particularly Don Solms, who kept me doing my own waggle dancing in the sky. To all my beekeeping mentors and friends, a most sincere thank you for your wisdom. My love and appreciation to my ever-supportive funny family and friends. My sons and Tom Kane are always in my heart.

Terry Ryan Kane DVM, MS

I would like to thank my colleagues at the University of Arizona College of Veterinary Medicine and Washington State University for their patience and encouragement during this project. I owe great thanks and love to my friends and family who have put up with this bee craziness for quite a while. Love to my mom, Paula Anderson, who encouraged writing from when I was old enough to hold a pencil and to Jerry Anderson and DuWayne Marshall for being there when I needed them. And to my patient and resourceful spouse Randy Faux, who still buys me bee equipment for Christmas.

Cynthia M. Faux DVM, PhD, DACVIM-LA

Honey Bee Medicine: A One Health Challenge

Terry Ryan Kane

A2 Bee Vet, Ann Arbor, MI, USA

> *What a marvelous cooperative arrangement – plants and animals each inhaling each other's exhalations, a kind of planet-wide mutual mouth to stoma resuscitation, the entire elegant cycle powered by a star 150 million kilometers away.*
>
> Carl Sagan

More than 120 million years ago, when dinosaurs walked the earth and would-be mammals were no bigger than shrews, bees flew, and pollinated flowering plants. Bees coevolved with angiosperms over 100 million years, each contributing ingredients to this cooperative arrangement. This co-evolution was so successful that bees are found on every continent of the world where flowers grow.

We have much to thank the bees for. Beyond the critical role they play in securing our food supply, bees continue to provide a variety of hive products. We harvest the honey they make from nectar, the wax they produce for comb, the pollen they collect and pack into cells for stored protein to feed their young, the propolis they collect from tree resins to line and protect their hives, and even the royal jelly, the "bee milk," to feed larvae and produce their queen. We turn these into a variety of products: candles, salves, ointments, syrups, make-up, hair products, medicines, etc.

Bees are amazing and unique. Tens of millions of forager bees may travel up to 6 km to find a food resource before flying home to their hive, communicating in the dark on vertical surfaces to their sister foragers how far away the food is, its value, and how to find it. These foragers utilize the sun's position and polarized light to determine direction with an internal clock/odometer to tell her sisters how far she flew between the food resource and the hive. Humans have almost no innate ability to measure direction and distance, as our huge investments in maps, compasses and now Global Positioning Systems attest. Bees have had this innate capability for tens of millions of years. Kart Von Frisch won the Nobel Prize in Medicine in 1973 for his discovery of the "waggle dance" of the bee. Recent data analytics on the waggle dances have proven how accurate bee navigation really is.

One Health Issue: Planetary Health (Biodiversity and Climate Change)

The One Health concept is not new to veterinary medicine, but it is most timely now that we are facing multiple critical issues that involve our profession. Veterinary Medicine's greatest contributions to One Health have been in public health, particularly emerging zoonotic diseases, but environmental health has been largely neglected and requires our equal attention, now more than ever. Honey bees, native bees, bumble bees, and many other pollinators are the biosensors of our ecosystem health. Insects are the most diverse multicellular group of organisms on the planet – over one million species have been described, so far. And while the sheer biodiversity of insect species helps to ensure the group's survival, many of our pollinator species are in jeopardy. The decline of bees, as well as other animal pollinators, are in the public's consciousness, largely due to scientists' warning and media attention. Our ecosystems are out of balance. Habitat loss, pests and pesticide use, emerging diseases, and the extremes of global climate change all contribute to the instabilities we are experiencing. Veterinarians are trained problem solvers, but first we must recognize the problem. It is time our profession acknowledges and works to mitigate the challenges that climate change is having on animals and plant life, on agriculture, on zoonotic diseases, and on our environment. Mother Nature is relentlessly forcing us to face the threats of climate change and we must pursue all efforts to limit warming to 1.5°C.

One Health Issue: Food Security

Honey bee and pollinator health is crucial to our food supply. The pollination of flowering plants is an essential ecosystem service that produces the variety of vegetables, fruits, nuts, and seeds which, in turn, provide the necessary nutrients to sustain us, wildlife, and farm animals. The pollination services of birds, bats, butterflies, beetles, moths, ants, wasps, and the like, are vital to food systems – and to life itself. Without this variety of pollinators, we would not have the plant biodiversity that wildlife requires, or healthy soil and air. Without those things, we can never achieve global food security.

The public is increasingly aware that pollinators and honey bees are in trouble and people want to help. Hobbyist or "backyarder" and sideliner beekeeping has never been more popular, and veterinarians will be called upon more and more as we educate ourselves and the beekeepers learn our worth.

It is estimated that by 2050 there may be 9.8 billion people on earth and that global agriculture may need to increase by 30–70% in some areas. How will we feed a future population of 10 billion people? How will land and water resources be shared? How will we mitigate the increasing impact of global climate change on agriculture? Veterinarians will play an essential role in solving these issues. Food safety, food security, and public health are part of our jobs as veterinarians.

The honey bee is our top managed pollinator because it is the only bee that forms large colonies that can be transported in hive boxes. North America has the second largest commercial bee industry in the world. Today, millions of hives, the majority of the North American bee herd, are transported thousands of miles by truckloads around the United States and Canada to pollinate our food crops. The commercial beekeeper's life is a hard one – very labor intensive and with the new regulations, the spread of disease, increased fuel and transportation costs, and labor shortages, we are obliged to familiarize ourselves with their trade. Pollination services are a multi-billion-dollar industry.

Honey bees get the most "buzz" but actually some native bees are more efficient pollinators for some plants. Yes, honey bees are now considered livestock because we consume their products, but as far as getting pollen from one flower to another, honey bees are only one of a myriad of players. Native bees do not live in hives or colonies but in underground burrows. They come in all sizes and colors, and can be fuzzy, shiny, or metallic. They aren't as tidy, they don't pack pollen in little pouches, and they are messy. Farmers and producers have noted that when native bees are co-pollinating with the honey bees, production is even better. New management in Integrated Crop Pollination uses a combination of native bees and honey bees with farm practice tools, like no-till and cover crops, to increase production.

One Health Issue: The Global Epidemic of Antimicrobial Resistance

There is no doubt that antibiotic use improved the health of people and animals over the last 70 years. Antimicrobial resistance (AMR) is nothing new, it occurs in nature. Resistant genes are carried on plasmids (pieces of DNA) that are transferred between organisms. We now know that bacteria containing resistant genes can be transferred from livestock to humans via food. However, the misuse/overuse of antibiotics has led to the spread of resistant genes in medically important antibiotics and we now have diseases that are resistant to treatment. Multi-drug resistant bacteria are a threat to global health.

While in much of the world veterinarians have had a decades' long interaction with apiarists, veterinarians in the United States officially joined the honey bee's medical team as a result of the implementation of the 2017 US Food and Drug Administration regulations on the use of medically important antibiotics in livestock. Honey bees were officially defined as food-producing livestock in those regulations, putting their medical care into the hands of veterinarians. Writing Veterinary Feed Directives and prescriptions, however, should not be our profession's sole offering to honey bee medicine. Our expertise in herd health management will be an asset to the honey bee industry.

Antibiotic resistance has been documented in honey bees and we now know that there can be harmful effects on the honey bee microbiome. There is an increased effort to breed honey bees for hygienic behaviors to develop and enhance natural resistance.

Our Challenge

Just as you don't have to own pigs to be a swine veterinarian, you don't have to be a beekeeper to treat bee colonies. But you *do* have to know the biology, physiology, and behavior of these magnificent animals in order to forge a Veterinary Client Patient Relationship (VCPR) and feel confident in your handling, diagnosis, and treatment of this species, *Apis mellifera*, new to our profession.

All the authors in this book recommend experience beyond "book learning" – so join a local bee club, help a beekeeper in the field, or start a few hives of your own.

Learn about the types of beekeepers you may be working with; backyard hobbyists with a few hives, sideliners (whose apiary is a secondary source of income) and commercial beekeepers with many hundreds, or thousands, of hives. Sideliners are nothing new to veterinary medicine as most of our cow-calf and small ruminant clients have another primary source of income.

As with any other animal we work with, we need to know how to safely handle and manage bees. Will you get stung? Yes, you will. Know your response to bee venom in advance. Know the tools, equipment, and safety precautions you will be taking.

Once you have read the chapters on hive inspections and feel comfortable in a bee suit with insects flying all around you, quietly inspect the bee yard and hives. Observe the macro-environment for food sources and the activity around the hives. When you are ready to do an internal inspection, look for the different caste members, brood, and food. Get to know normal smells, sounds, and patterns. And don't be frustrated if you can't find the queen – that takes lots of practice!

In the first few chapters, you will learn about honey bees in nature, their arboreal homes, and behavior in the wild. It has been this unique eusocial behavior, division of labor, and adaptations that have allowed them to survive for many millennia. Proceed to the chapters on anatomy, physiology, behavior, colony organization, brood rearing, queen rearing, and swarming.

Bees, just like any other animal, get sick from a variety of diseases – bacterial, viral, fungal, and idiopathic. We provide specific chapters on these varied honey bee pathogens, including Colony Collapse Disorder (CCD). This disease syndrome, although not a major cause of bee mortality anymore, brought vital attention to the cause of global bee deaths, along with much needed research funding adding new knowledge on honey bee health and management.

Nutrition is a determining factor in honey bee colony health. The value and stability of an adequate food supply determine the homeostasis of the hive required for colony survival. Morphological changes within the caste members, be it queen development or forager phenotypes, depends on nutrition. We dedicated a large chapter of this text to honey bee nutrition for just these reasons.

Today, the top killer of our honey bees is an introduced pest, the Varroa mite. Many chapters will mention this pest and one is solely devoted to it. It is very difficult to kill an arachnid feeding upon an insect and that brings us to pesticide and pharmaceutical uses. This is an ever-evolving area of research, looking at the synergistic effects of diseases combined with chemicals in the environment.

As stewards of animal health and participants in One Health, this book is for all veterinarians, veterinary students, technicians, bee research scientists, state/provincial apiarists, and beekeepers. This book is an interdisciplinary collaboration among veterinarians, entomologists, and the beekeeping community. We have forged new relationships to protect honey bees and all pollinators.

For these reasons, there has never been a better or more important time for veterinarians to be involved in honey bee and pollinator health. The honey bee, *A. mellifera* (and subspecies), may be the most important animal that we veterinarians care for. It would be a different and more difficult world without honey bees and other pollinators. The mutual interdependence of humans, animals, and our environment are exemplified in this One Health challenge.

We end this book with a look to the future, for bee and pollinator research and for the role of veterinarians in this expanding field. Above all, this book intends to teach and amaze you. We should all be humbled by these remarkable animals.

Section I

Biology and Medical Foundations

1

Looking to Nature to Solve the Health Crisis of Honey Bees

Robin W. Radcliffe[1] *and Thomas D. Seeley*[2]

[1]*Department of Clinical Sciences, College of Veterinary Medicine, Cornell University, Ithaca, NY, USA*
[2]*Department of Neurobiology and Behavior, College of Arts and Life Sciences, Cornell University, Ithaca, NY, USA*
**Illustrations by Anna Connington*

Figure 1.1 *Gathering honey*, a beekeeping scene from the Tomb of Rekhmire. Egypt c. 1450 BCE (de Garis Davies 1930).

Prologue

Scientists recently discovered the lipid residues of ancient beeswax inside the earthen pottery vessels of Neolithic farmers, which suggests that the origin of domestication of honey bees dates back to the onset of agriculture (Roffet-Salque et al. 2015). The long association between humans and bees (Figure 1.1), with mankind harnessing honey bees for food, medicine, and spiritual wellness, can be summed up in a single word: beekeeper. In this book, we introduce a new term to the English language: bee doctor. Etymologists, who study word metamorphosis, follow how the use of particular words gradually evolve in our language – e.g. from bee keeper, to bee-keeper, and finally to beekeeper. Just as the "honey bee" is spelled as two separate words because it is a true bee, we will likewise separate "bee" and "doctor" since bee veterinarians are true doctors in every sense of the word. We work from single bee to whole colony, from individual cell to multicellular organism, and from microenvironment to ecosystem. Given the urgent call for modern *Homo sapiens* to reverse the anthropogenic impacts on pollinators everywhere, including our sacred *Apis mellifera*, we propose adoption of "bee doctor" without delay. Humans have been "keeping" bees for thousands of years, so we now have the word "beekeeper." Only by forging a close connection between human beings and honey bees in all matters relating to their health, do we stand a chance to save one of earth's most industrious species – the one who gives us food, health, and happiness, and was idolized on the walls of Egyptian tombs. Perhaps someday we will even have the word "beedoctor."

A Tenet of Medicine: Learn the Normal

Colonies of honey bees living in the wild are prospering in American forests even in the face of myriad stressors that are decimating the managed colonies living in apiaries. We know that both cohorts are exposed to the same parasites and pathogens. How then do wild colonies survive without beekeeper inputs, whereas managed colonies live just one to two seasons if humans do not intervene with various supplements or medicines? In examining this conundrum, we must ask ourselves as bee doctors, working hand-in-hand with beekeepers, how we should examine the health of the honey bee? A fundamental tenet of medicine is the need to learn what is *normal* (regarding anatomy, physiology, or the state of being known as health) before one can

Honey Bee Medicine for the Veterinary Practitioner, First Edition. Edited by Terry Ryan Kane and Cynthia M. Faux.

understand deviations from this baseline. We contend that the "normal" that bee veterinarians should be concerned about is the wonderfully adapted lifestyle of wild colonies of honey bees. In this chapter, we will highlight the important differences between wild and managed colonies of honey bees and we will suggest ways health professionals can make use of the marvelous tools for health and survival that evolution has bestowed upon *Apis mellifera* through adaptation and natural selection.

Declines of the world's pollinators are happening at an alarming rate, and it is predicted that these declines will have adverse impacts on pollinator-sensitive commodities worth billions of dollars (Morse and Calderone 2000). The threat to the honey bee is perhaps the best understood of the pollinator declines. Its causes are diverse: widespread use of agrochemicals, loss of plant and floral diversity, invasive species, migratory beekeeping practices, and monoculture pollen sources. Furthermore, the stresses created by these environmental stressors are intensified by the honey bee's pests, parasites, and pathogens. Although no single disease agent has been identified as *the* cause of honey bee colony collapse, pests and pathogens are recognized as the primary drivers of the massive deaths of managed bee colonies worldwide. Many of these agents of disease are vectored by an ectoparasitic mite introduced from Asia, *Varroa destructor* (Ellis et al. 2010; Ratnieks and Carreck 2010).

Investigations of honey bee declines have focused primarily on the pathogens themselves and their interactions, which are now understood to be multifactorial (vanEngelsdorp et al. 2009; Becher et al. 2013; Di Prisco et al. 2016). Besides the pathogens, the environments in which honey bees live also profoundly impact colony survival. In this chapter, we will examine honey bee health and the alarming levels of colony mortality from an ecological and evolutionary perspective. We will embrace the logic of natural selection and we will learn important lessons from long-term studies of honey bee colonies living in nature (Brosi et al. 2017; Seeley 2017b, 2019a; Neumann and Blacquière 2016).

Good Genes Versus Good Lifestyle: The Varroa Story

We will begin our account of the health and fitness of wild colonies by relating the story of the Varroa mite (*V. destructor*), a parasite that switched hosts from the Eastern honey bee (*Apis cerana*) to the Western honey bee (*A. mellifera*). In order to understand the resistance to Varroa mites that is found in wild honey bee colonies, we must examine more deeply their genes and their lifestyle.

Beekeepers today rely primarily on commercial queen producers for their bee stock. Most hobby beekeepers, for example, will start an apiary or add colonies to an apiary by purchasing either a "package" of bees shipped in a cage or a nucleus colony ("nuc") living in a small hive. In North America, packaged bees are shipped from various southern states in the U.S., as well as from California, and Hawaii, so they consist of stock that is not necessarily adapted to the beekeeper's local climate, temperatures, and agents of disease. Furthermore, even though queen bees are also produced and sold across North America – their genetics often traces to just a handful of colony lines. In many places, good colony health can be fostered by the use of locally-adapted bees.

From an evolutionary perspective, the observation that wild colonies have rapidly adapted to the Varroa mite, and to the diseases they vector, over a remarkably short timeframe (ca. 10 years), suggests that surviving wild colonies have either good genes (DNA), a good lifestyle, or both (Seeley 2017a).

Good Genes

The Varroa mite is the leading cause of honey bee health problems on all beekeeping-friendly continents except Australia. Beekeepers have always experienced colony losses, but it was not until the arrival of this parasitic mite that colony die-offs became severe in North America. The Varroa mite lies at the heart of poor colony health, because it acts both as a primary stressor (the adult mites feed on the "fat bodies" of adult bees and the immature mites feed on immature bees [pupae]) and as a vector for a myriad of the viral diseases of honey bees (vanEngelsdorp et al. 2009; Martin et al. 2012). If a managed colony of honey bees is left untreated, Varroa mites will kill it within two to three years (Rosenkranz et al. 2010). Remarkably, the wild colonies living in the forests of North America today, plus some notable examples of European honey bees living on islands, are resistant to the mite (De Jong and Soares 1997; Rinderer et al. 2001; Fries et al. 2006; Le Conte et al. 2007; Oddie et al. 2017). How did this resistance evolve? We know that wild colonies in the northeastern forests of North America went through a precipitous population decline in the 1990s, following the arrival of the mite (Seeley et al. 2015; Mikheyev et al. 2015; Locke 2016). Yet, studies show that these wild colonies recovered in the absence of mite treatments without appreciable loss of genetic diversity by evolving a stable host-parasite relationship with *V. destructor*.

The genetic bottleneck associated with a precipitous population decline would have devastated most species; cheetahs and Florida panthers, to name two prominent mammalian examples, exhibit extensive disease syndromes from low genetic variability. *A. mellifera*, however, came through its population decline with remarkable genetic variation intact because polyandry, a breeding strategy whereby the queen mates with 10–20 drones, helps maintain the genetic composition of a population. Polyandry also confers improved fitness through enhanced disease resistance (Seeley and Tarpy 2007); higher foraging rates, food storage, and population growth (Mattila and Seeley 2007); and possibly better queen physiology and lifespan in the colony (Richard et al. 2007). Fitness follows diversity and in honey bee colonies this comes through the multiple matings of the queen. In nature, there must be a trade-off between the optimal number of drone matings and the time that queens spend on their mating flights, which sometimes extend several miles from a queen bee's home. Delaplane and colleagues (2015) showed that queens artificially inseminated with sperm from 30 to 60 drones, rather than the 12 to 15 drones that are typical for the queens of wild colonies, produced more brood and had lower mite infestation rates relative to control colonies, supporting the idea that resistance to pathogens and parasites is a strong selection pressure favoring polyandry. One hypothesis to explain the high levels of polyandry of queen honey bees is that by mating with many males, the queen captures rare alleles that regulate resistance to pests and pathogens (Sherman et al. 1998; Delaplane et al. 2015). This has been confirmed in several studies in which colonies whose queens had either a high or a low number of mates were inoculated with the spores of chalkbrood (*Ascosphaera apis*) or American foulbrood (*Paenabacillus larvae*), and the levels of infection in their colonies were compared (Tarpy and Seeley 2006; Seeley and Tarpy 2007). The higher the number of mates, the lower the level of disease.

We know that Varroa mites initially killed off many wild colonies living in the forests of New York State, so maternal lines (mitochondrial DNA lineages) were lost (Mikheyev et al. 2015). Fortunately, the multiple mating by queen honey bees enabled the maintenance of the diversity of the bees' nuclear DNA despite the massive colony losses. Today, the density of wild colonies living in forests in the northeastern United States (c. 2.5 colonies per square mile, or 1 per square kilometer) is the same as it was prior to the invasion of the Varroa mites (Seeley et al. 2015; Radcliffe and Seeley 2018), and the survivor colonies possess resistance to these mites. In a comparison of the life history traits of wild colonies living in the forests around Ithaca, NY, between the 1970s (pre Varroa) and the 2010s (post Varroa), Seeley (2017b) found no differences, which implies that the wild colonies possess defenses against the mites that are not highly costly and so do not hinder colony reproduction.

Figure 1.2 Grooming, or mite-chewing, is a heritable trait in which honey bees remove and kill adult Varroa mites by chewing off parts of the mite's body, carapace, or legs.

Figure 1.3 Hygienic behavior or Varroa Sensitive Hygiene (VSH), is a form of social immunity in which honey bees selectively remove the varroa-infested larvae and pupae from beneath capped cells. The mites infecting these brood cells are killed along with the developing bee upon opening of the cell.

There exist multiple mechanisms of natural Varroa resistance, a form of behavioral social immunity, that have a genetic basis. These include grooming behavior, also known as "mite chewing," and hygienic behavior, also known as Varroa Sensitive Hygiene (VSH). Grooming behavior is the process whereby worker bees kill mites by deftly chewing off the carapace, ventral plate, or legs of a mite (Figure 1.2). The strength of a colony's ability to groom Varroa mites is indicated by the percentage of chewed mites among the mites that fall onto a sticky board placed beneath a screened bottom board in a hive (Rosenkranz et al. 1997). Hygienic behavior is the process whereby worker bees remove diseased (or dead) brood from the cells in which they are (or were) developing (Figure 1.3). VSH is measured by determining the percentage of sealed brood cells that contain Varroa mites shortly after cell capping and then again shortly before brood emergence (cell uncapping). Because this assay of a colony's VSH behavior is rather tricky to perform, people often use a different assessment of hygienic behavior: the freeze-killed brood (FKB) assay. Because the FKB assay does not involve Varroa infested brood, it is not a direct measure of VSH. The FKB assay works by freezing a c. 3 in. diameter circle of sealed brood cells, thereby killing the brood within, followed by calculating the percentage of the dead brood that have been removed, either 24 or 48 hours after the freezing of the brood (Spivak and Downey 1998).

In a long-term study in Norway, variation among colonies in their resistance to Varroa was found to be based on neither grooming behavior nor hygienic behavior, but on something else that was hindering mite reproduction. Oddie and colleagues (2017) examined managed honey bee colonies that had survived in the absence of Varroa control for >17 years alongside managed colonies that had received miticide treatments twice each year. Records were kept of daily mite drop counts, and of assays of the colonies' mite grooming and hygienic behaviors, for both survivor and control colonies. No difference was found in the proportion of damaged mites (~40% chewed in colonies of both groups) or in FKB removal rates (only ~5% brood removed). However, the average daily mite-drop counts (indicators of the mite populations in colonies) were 30% lower in surviving colonies compared to susceptible ones. Evidently, there were other colony factors (besides mite grooming and hygienic behaviors) responsible for reducing the reproductive success of the mites in these colonies of Norwegian honey bees. Since donor brood was used for the testing in both groups of colonies (mite susceptible and mite resistant), the possibility of protective traits of immature bees was eliminated. What Oddie et al. found is that in the mite-resistant colonies (but not in the mite-susceptible ones) the worker bees are uncapping brood cells and then recapping them several hours later, and that this reduces the mites' reproductive success to a level that protects the colony. An 80% reduction in mite reproductive success, together with a reduction in brood size, independent of grooming or hygienic behavior, was also described for populations of survivor (untreated) colonies of honey bees living on the island of Gotland in Sweden (Fries and Bommarco 2007; Locke and Fries 2011).

Good Lifestyle

To understand the survival of honey bee colonies living in the wild, we must look not only at their genetic makeup but also at their lifestyle. How do the ways in which wild colonies live combine with their genes to limit mite reproductive success and the virulence of mite-vectored pathogens? We know that modern beekeeping practices create living conditions for managed colonies that are far more stressful than the living conditions of colonies living in the wild (see Table 1.1). For example, we know that the artificial crowding of colonies in an apiary, the provision of large hives which foster Varroa reproduction, and the suppression of swarming behavior – are all apicultural manipulations that make large honey harvests possible for the beekeeper but are harmful to colony health (Seeley and Smith 2015; Loftus et al. 2016). Another important, but little understood, stressor experienced by managed colonies is the greater thermoregulation stresses experienced by colonies living in a standard hive compared to in a bee tree (Mitchell 2016). Our modern beekeeping practices – launched in 1852 with the invention of the movable frame hive, by Lorenzo L. Langstroth – have created new challenges for honey bee colonies, which are adapted for living without human management (interference). For the remainder of this chapter, we will explore the lifestyle features that help wild colonies of honey bees thrive despite their pests, parasites, and pathogens. We will also draw lessons that beekeepers and bee doctors can employ to help promote the health of the managed colonies living in apiaries.

Part 1: The Environment of a Wild Colony

Cavity Size

A good place to begin our exploration of wild honey bee health is understanding the home of a honey bee colony found in nature (Figure 1.4). Wild honey bees predominately make their homes inside the cavities of hollow trees, though any cavity of appropriate volume and specific characteristics will do, and this includes manmade structures, rock crevices, and other spaces. Wild colonies choose small cavities, with an average volume of just 45 l (range 30–60 l: Seeley and Morse 1976; Seeley 1977). When honey bee colonies choose their nesting sites, they seek cavities of this size, which is substantially smaller than the typical Langstroth hive in an apiary, with a volume of 120–160 l (Root and Root 1908; Loftus et al. 2016).

Nest cavity size has a major impact on honey bee health through its effect on mite population dynamics. A brief review of the Varroa life cycle will help us understand the role of nest cavity size on a colony's mite population. Varroa mites have two different life phases: the phoretic phase in which adult mites feed on the "fat bodies" of honey bees and the reproductive phase in which mites reproduce in the cells of sealed brood of workers and drones (Rosenkranz

Table 1.1 Characteristics of wild honey bees (*Apis mellifera*) that differ from managed honey bees and their impact on bee health.

Characteristic	Wild colonies	Reference	Managed colonies	Reference
Colony lifespan	Long-lived 5–6 yr once established	Seeley (2017b)	Short-lived; 2–3 yr without miticides	Rosenkranz et al. (2010)
Annual survival	High survivorship 84% (established) 20% (founder)	Seeley (2017b)	Low survivorship (0–50%)	Ellis et al. (2010)
Cavity size of home	Small cavity; 45 l (30–60 l)	Seeley and Morse (1976)	Large cavity; 120–160 l	Loftus et al. (2016)
Swarming frequency	87% annual queen turnover in established colonies	Seeley (2017b)	Swarming suppressed, so low queen turnover	Oliver (2015)
Propolis barrier	Complete barrier "propolis envelope"	Seeley and Morse (1976)	Incomplete barrier smooth hive walls	Hodges et al. (2018)
Colony spacing	Colonies far apart (~1 km)	Seeley and Smith (2015) Radcliffe and Seeley (2018)	Colonies close together (~1 km)	Root and Root (1908)
Virulence level	vertical transmission of mite-vectored pathogens, via swarming	Seeley and Smith (2015)	Virulence favored by horizontal transmission of mite-vectored pathogens, via drifting/robbing	Seeley and Smith (2015)
Nest insulation	Thick-walled (20 cm/8-in.) well insulated tree cavity	Seeley and Morse (1976)	Thin-walled (2.5 cm/1-in.) poorly insulated Langstroth	Root and Root (1908)
Immune Function	Strong social immunity, Immune genes downregulated	Simone et al. (2009)	Weak social immunity, Immune genes upregulated	Borba et al. (2015)

et al. 2010). Only adult female mites are phoretic; both the tiny males and the nymphal stage females remain within the capped brood cells. Honey bee larvae are essential for the mite because it has no free-living stage off the host – the mite is entirely dependent on honey bee brood for its own propagation. Honey bee colonies living in large hives hold more brood than those living in natural nest cavities, so colonies in large hives are especially favorable for mite reproduction.

All honey bee populations that have survived for more than a decade without miticide treatments share a common feature: their colonies are small (Locke 2016). Small colony size relates directly to the dynamics of brood development and swarming. Having relatively few brood has two significant impacts on mite reproduction. First, since Varroa mites only reproduce within the cells of sealed (pupal stage) brood, the reproduction of these mites is hampered by the relatively small brood nests of wild colonies. Second, a small nest cavity size shortens the time before the sealed brood fills a colony's brood nest, and this brood nest congestion is one of the primary cues for swarms and afterswarms (Winston 1980). When colonies living in large hives (two deep hive bodies plus two honey supers) were compared to colonies living in small hives (just one deep hive body, to mimic the nest cavity size in nature), it was found that the small-hive colonies had reduced mite loads and improved colony survival, as a result of more frequent swarming and lowered Varroa infestations (Loftus et al. 2016).

Wall Thickness and Thermoregulation

Seeley and Morse (1976) reported that the average wall thickness of natural nest cavities is approximately 20 cm (~8 in.). The wall thickness of a standard Langstroth hive is just 1.9 cm (0.75 in.), hence some 10 times thinner than the nest cavity wall of a bee tree. The reduced wall thickness in Langstroth hives creates a large reduction in nest insulation, possibly resulting in adverse effects on colony energetics. Large temperature fluctuations inside a hive exacerbate colony stress by increasing the demands on colony nutrition and hydration (more nectar and water foraging trips), by impairing a colony's ability to maintain thermal homeostasis (more fanning and "bearding" when it is hot, and more metabolic heat production when it is cold), and by hastening entry into a winter cluster – all of which increase the physiological demands on the colony (Mitchell 2016).

Coombs et al. (2010) found that natural tree cavities buffered environmental temperatures such that tree cavities were cooler than ambient during the day and warmer than ambient during the night. During the day, the tree diameter at breast height was the most important variable determining

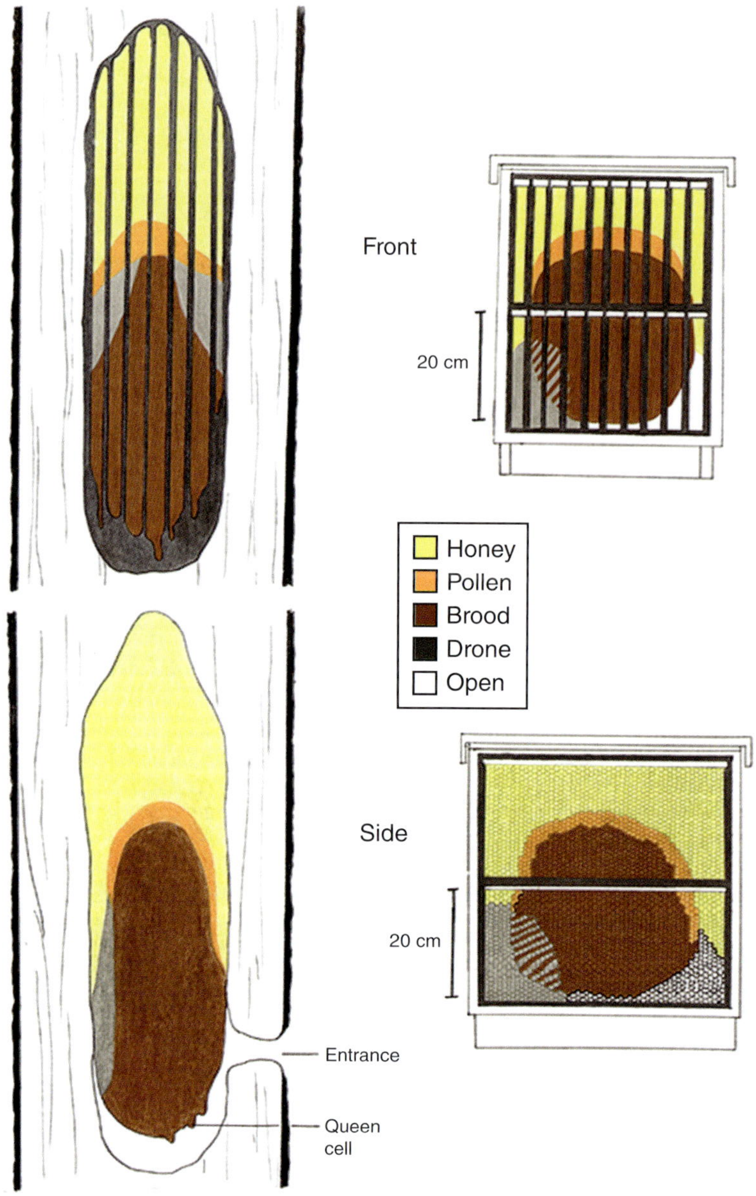

Figure 1.4 An illustration comparing the structure and organization of a honey bee nest as found in a bee tree (left) and a standard Langstroth hive made up of two deep hive bodies (right). The colors correspond to brood and hive products. A typical bee tree cavity has a volume averaging 40 l, whereas two deep hive bodies have a volume of 80 l. These differences in cavity volume are directly correlated with the size of a colony's brood nest and varroa reproductive success.

cavity temperature. At night, diameter and tree health were important with large *living* trees offering the most stable thermal environment. We compared the ambient temperatures inside two tall, man-made cavities; one was inside a rectangular wooden box (built of 1.9 cm thick pine boards, as used for Langstroth hives) and the other inside a living sugar maple tree (*Acer saccharum*) (Figure 1.5). These two cavities were built with the same dimensions (24 cm × 24 cm × 87 cm), which mimicked those of a typical tree cavity of a wild colony [see Tree Beekeeping by Powell (2015)]. Temperature recordings over a year revealed striking differences in interior temperature dynamics between the two cavities. In the poorly insulated box, the temperature closely followed the ambient temperature; the thin walls provided little or no temperature buffering. In the tree, though, the temperatures varied much less; they did not reach the extreme highs and lows found inside the uninsulated box (Seeley and Radcliffe unpublished data; see Figure 1.6a,b).

Mitchell (2016) found that heat is transferred four to seven times faster across the thin walls of a traditional hive relative to the walls of a natural (bee tree) enclosure. To maintain a colony's cluster core temperature of 35 °C (the set point of the brood nest), any energy lost through transfer from the hive walls must be replaced through the bees'

Figure 1.5 A research station beside the Shindagin Hollow State Forest in upstate New York. It was designed to test the environmental fluctuations – temperature (°C) and relative humidity (%) – inside two cavities of identical dimensions but with walls of different thicknesses, c. 2 cm vs. 20–30 cm. One (a) is a wooden box with walls like those of a Langstroth hive and the other (b) is a live sugar maple tree (*Acer saccharum*) in which a typical size bee cavity was cut using a chainsaw and adze. *Source:* Photo by Robin Radcliffe.

metabolic activity (bees isometrically contract their flight muscles to generate heat). Mitchell predicted that colonies living in hives (or trees) providing well-insulated cavities will not need to assemble into tight clusters until the ambient temperature is below 0°C. Mitchell concluded that the high thermal insulation of nests in bee trees results in increased relative humidity inside the cavity, decreased reproduction by Varroa mites, and enhanced survival of honey bee colonies.

Propolis Envelope

Propolis ("bee glue") is a resinous substance collected by honey bees from the buds and wounds of trees. When combined with beeswax, it makes a cement that bees use to fill the crevices and coat the walls of their nest cavities, often completely enshrouding their nests. This coating of the walls, floor, and ceiling of the nests of wild colonies with tree resins makes a "propolis envelope" that can be 2–3 mm thick (Seeley and Morse 1976). The propolis lining of the nest cavity probably serves several functions: creating a solid surface for comb attachment, reducing cavity draftiness, enhancing nest defense, waterproofing, and bolstering a colony's defense against microbial infections.

Ancient Greeks used propolis to treat abscesses, Assyrians put it on their wounds, and Egyptians used it for embalming their dead. Although humans have long recognized the health benefits of propolis for its antiseptic, anti-inflammatory, antibiotic, antifungal, anesthetic, and healing properties, only in the last century have humans discovered the specific compounds that give propolis its medicinal value – of the more than 180 compounds identified in propolis to date, one group (a class of plant-based polyphenols known as flavonoids) are of particular interest for their protective antioxidant properties. These same compounds that mankind values in propolis also confer health benefits to the honey bee colony through social immunity – a collective behavioral defense that produces

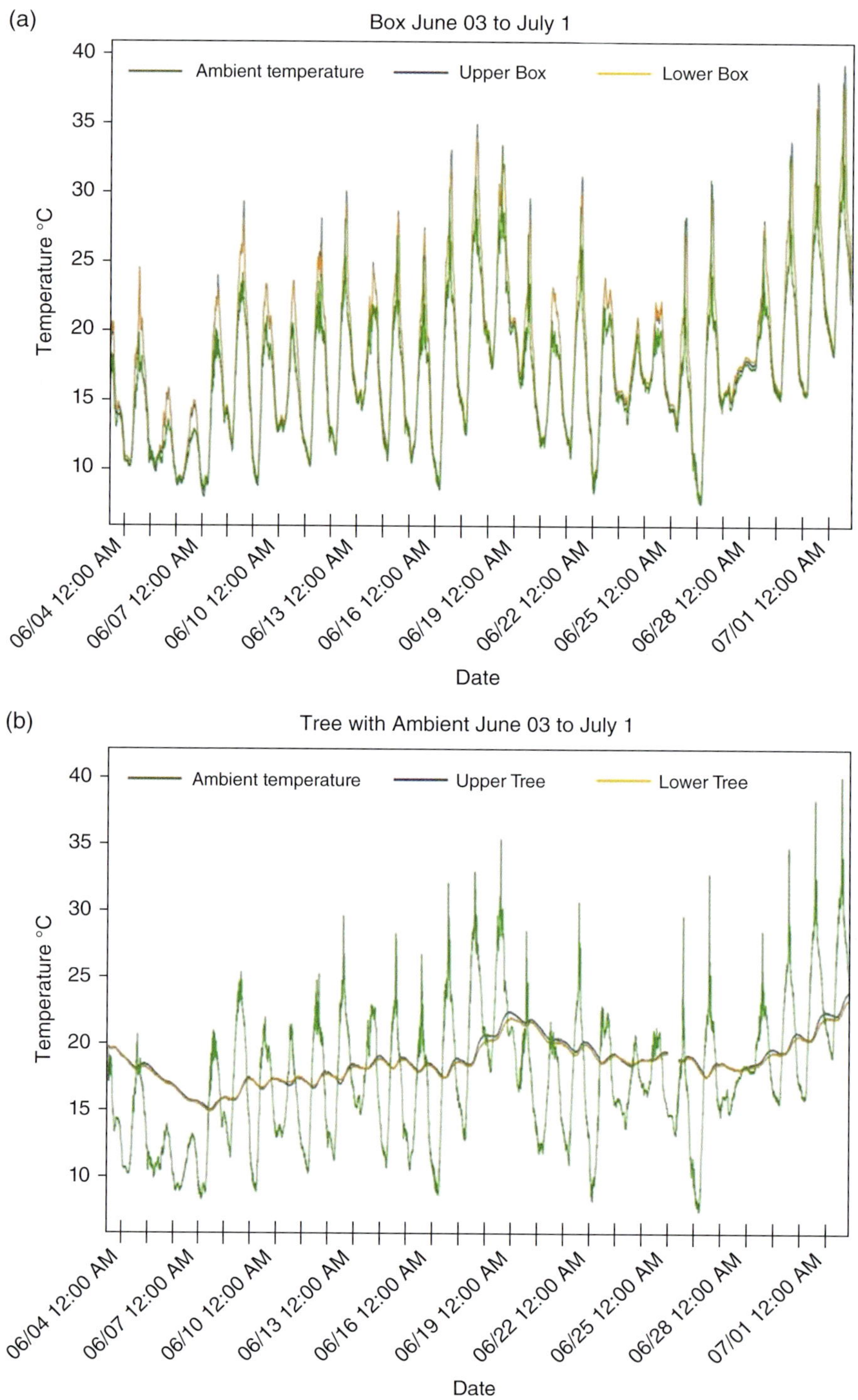

Figure 1.6 A month-long comparison of temperatures (°C) inside a thin-walled nest cavity made of 1.9-cm-thick lumber (a) and inside a thick-walled cavity made in a living sugar maple tree (*Acer saccharum*) having a wall thickness of 20–30 cm (b). Each cavity had two temperature probes, located c. 10 cm from either the floor or the ceiling of the cavity. In both figures the green line represents the ambient environmental temperature, while the orange and blue lines are the probes located within the respective cavities.

colony-wide immunity that in turn reduces the expression of immune genes in individual bees (Borba et al. 2015).

Curiously, the use of propolis for colony defense is limited to the temperate regions of the world. Neither the tropical honey bees in Asia (*A. cerana, Apis florea*, and *Apis dorsata*) nor those in Africa (the African subspecies of *A. mellifera*) make use of propolis other than for structural purposes (Simone et al. 2009; Kuropatnicki et al. 2013). It is the European honey bees living in nature for which the collection and use of propolis for its colony-level immunoprotective effects has reached its highest expression. Yet, rather than being viewed as a specific compound to be cultivated, propolis is more often than not regarded as an annoyance by modern beekeepers. Beekeepers are constantly scraping off propolis as they remove frames to manipulate their colonies. And the Langstroth hive bodies used by the vast majority of beekeepers today lack the rough inner surfaces of a bee tree or other natural cavity that stimulate propolis deposition by foragers. Colonies managed by beekeepers are not strongly stimulated to collect and use propolis. Indeed, it is the complex surface of the natural cavity that provides the tactile stimuli necessary for the deposition of propolis as a hive barrier by worker bees, something almost entirely lacking in modern hives made from smooth planed lumber (Hodges et al. 2018). Hodges and colleagues investigated three methods to increase the textural complexity of the interior surface of a standard hive body; these methods included using plastic propolis traps stapled to the inside wall surfaces, cutting horizontal parallel saw kerfs that were 7 cm apart and 0.3 cm deep, and roughening of the interior wall surface using a mechanical wire brush. The three interior hive wall types were compared to an unmodified, smooth-walled hive by measuring the bees' propolis application. Although the colonies were not challenged with specific pathogens, all three texturing methods induced significantly more propolis deposition compared to controls. The authors concluded that using unplaned, rough lumber for the interior hive surfaces would increase propolis deposition over standard hives built using lumber that is planed smooth on both sides.

A curious observation arising out of the mapping of the honey bee genome was the discovery that honey bees possess just one-third of the genes coding for immune function typically found in solitary insects (Evans et al. 2006; Honey Bee Gene Sequencing Consortium 2006). It was hypothesized that the weak capacity for an immune response in individual honey bees might be compensated by behavioral or colony-level defenses, or a form of social immunity. Indeed, as social insects, honey bees are steadfastly hygienic by removing alien organisms that gain entry to the nest, by feeding young bees antimicrobial products, by creating compounds that offer barriers to infection, and evolving complex interaction networks that serve to compartmentalize infections. The first indication that the bees' nest environment could influence immune expression in honey bees was discovered by Simone et al. (2009). Honey bees living in hives whose inner walls were coated with propolis extracts (derived from resins found in Minnesota and Brazil) invested less energy on immune function compared to bees living in hives without such coating. The colonies living in the propolis enriched hives also had lower bacterial loads. Scientists believe that individual bees are not immunocompromised, but rather that they conserve energy by not upregulating their immune genes except when a pathogen is encountered. This means that the defenses provided by social immunity (e.g. the collection of tree resins for propolis) allows individual bees to divert energy resources from immune function to other hive activities such as nursing, wax building, and foraging. This strategy likely maximizes the health and fitness of the entire colony.

Bee Microbiome

An oft-overlooked aspect of the bee environment that is essential to the good lifestyle of honey bees is their microbiome, that is, the community of specialized microbes (bacteria and yeasts) that have coevolved to live inside the bees and in their nests (e.g. in their pollen stores). We again return to the tenet of our chapter: the need to learn about the honey bee's natural biome to understand its biology, including its relationships with its pathogens. The honey bee microbiome is remarkable in that it is nearly consistent across thousands of individuals from hive to hive and even across continents. The honey bee's microbiome is similar to that of humans in that both feature specialized bacteria that have coevolved with their host and are socially transmitted (Engel et al. 2012; Zheng et al. 2018). Honey bees are first inoculated with bacteria in the larval stage, presumably through the food provided by nurse bees. However, during pupation, when bees undergo the final phase of metamorphosis, a bee's exoskeleton (including the gut lining and any associated bacteria) is shed in a process known as ecdysis. Therefore, honey bees emerge as young adults without a gut flora, except for those microorganisms they pick up when chewing through the wax cappings of their cells. The characteristic microflora of a worker bee is, therefore, developed mainly following emergence and through direct social interactions with conspecific worker bees. By four to six days of age, the population of a worker bee's gut flora stabilizes at 10^8–10^9 bacterial cells.

Although both wild honey bees and those living in apiaries possess complex microbiomes, some beekeeping practices – such as feeding pollen substitutes and treating with antibiotics – can alter the microflora of honey bees (Fleming et al. 2015; Maes et al. 2016). Dysbiosis, or unhealthy shifts in gut microflora, was observed in bees consuming aged pollen or pollen substitutes and was linked to impaired larval development, increased bee mortality and infection with pathogens such as *Nosema* and *Frischella*. Raymann et al. (2017) observed considerable changes in the gut microbial community composition and size following treatment with tetracycline, the most commonly used antibiotic in beekeeping operations globally. The authors concluded that decreased survival in honey bees was directly attributed to increased susceptibility to infection by opportunistic pathogens that colonized the gut after antibiotic use. The honey bee microbiome is thought to promote bee health and development in several ways. Gut microbes are required for normal bee weight gain, an effect which can be attributed to regulation of endocrine signaling of important bee hormones. The microbiome increases the levels of vitellogenin and juvenile hormone in worker bees, and these regulate the nutritional status and the development of their social behaviors, so it is likely that the state of the bees' microbiomes affects the health of the whole colony. Bee microbes are also implicated in modulating the worker bee's immune system (Zheng et al. 2018).

Alterations in the microbiota of the bee gut have been linked to disease and reduced fitness of the bee host. The use of tetracycline – an antibiotic commonly used to treat American foulbrood and European foulbrood, and often given prophylactically – reduces both the number and the composition of normal bacteria in the bee gut. Raymann and colleagues (2018) found that *Serratia marcescens*, a known pathogen of honey bees and other insects, normally inhabits the bee gut without eliciting a host immune response. However, bee disease occurs when this pathogen is inoculated into a bee's hemolymph through the bite of a Varroa mite or when the gut microbiome is disturbed with antibiotic use. Researchers studying Colony Collapse Disorder observed a shift in gut pathogen abundance and diversity, and proposed that such shifts within diseased honey bees may be a biomarker for collapsing colonies (Cornman et al. 2012). See Chapter 9 for more details on the bee microbiome.

Part 2: Epidemiology for Bee Health: How Lifestyle Impacts Disease Spread

The preceding comparison of the environments of honey bee colonies living in the wild versus in apiaries sets the stage for reviewing the host–parasite interactions that ultimately define colony health. Let us now compare the impacts of disease on colonies living in the differing settings in which honey bee colonies now find themselves. Compared to organisms that do not live in large and complex eusocial societies (i.e. ones with a reproductive division of labor and overlapping generations) honey bees have far greater complexities in their host–pathogen and host–parasite relationships.

Ecological Drivers of Disease

Living in crowded communities of thousands of individuals, honey bees interact closely through regular communication behaviors, grooming activities, and the trophallactic transfer of food and glandular secretions. This complex group living provides abundant opportunities for pathogens to spread and reproduce. Moreover, the high temperature and high humidity of a honey bee colony's home makes it a perfect environment for disease outbreaks. It comes as no surprise, then, that many of the protective mechanisms that honey bees have evolved to control the spread of disease operate at the level of the whole colony, the superorganism. The members of a colony work together closely to achieve a *social immunity*: they groom themselves and one another (allogroom); they work as undertakers to remove dead and diseased bees; they collect antibiotic enriched pollen and nectar; and they practice miticidal and hygienic behaviors by biting off the body parts of mites and by removing infected bee larvae and pupae from their nests (Fries and Camazine 2001). Relatively few mechanisms of disease resistance have evolved at the level of the individual bee. These include individual immune system functioning and filters in the proventriculus (the valve between esophagus and stomach) that remove spores of American foulbrood. Most of these protective mechanisms limit *intra*-colony transmission of disease agents, and they work well. What is probably the primary driver of disease problems for honey bees at present, however, is *inter*-colony disease transmission.

A Critical Distinction: Vertical vs. Horizontal Disease Transmission

The *method* by which a disease is transmitted from colony to colony is a fundamental determinant of pathogen virulence. Vertical transmission (the spread of disease from parent to offspring) favors the evolution of avirulence whereas horizontal transmission (the spread of disease among unrelated individuals) favors the evolution of virulence (Lipstich et al. 1996). This is because pathogens and parasites that spread vertically need their host to stay healthy to produce offspring, whereas those that spread horizontally do not have this need. Although numerous other host factors (i.e. host longevity, density, population structure, and novel hosts) and pathogen factors (i.e. vector availability and pathogen replication potential) also influence virulence, we will focus on how the mode of honey bee pathogen and parasite transmission within and among colonies impacts the evolution of the virulence of these agents of disease.

Vertical Transmission: Swarming

In honey bees, one way that a colony achieves reproductive success is by swarming: an established colony casts a swarm to produce a new colony. The other way that a colony achieves reproductive success is by producing drones; even though weak colonies can propagate their genes by producing drones, this does not create another colony. If a pathogen or parasite that is transmitted vertically (from parent to offspring) weakens its host and so hampers it from producing offspring (which for honey bee colonies equates to casting swarms) then it reduces its own reproductive success. In short, the natural mode of colony reproduction in honey bees favors the evolution of avirulence in most of its pathogens and parasites. The two exceptions to this generalization are American foulbrood and *Varroa destructor*, both of which are easily transmitted horizontally when one colony robs honey from another.

Swarming also helps inhibit the reproduction of Varroa mites (and other agents of brood diseases) by creating a natural break in brood production, which forces the mites to likewise suspend their reproduction (Seeley 2017b). Once a daughter queen emerges to replace the mother queen that has left in a swarm, this daughter queen must leave the hive to fly to a drone congregation area, where she will mate with multiple drones before returning to the hive to commence egg laying. This transition from mother queen to daughter queen creates a period without sealed brood (needed for mite reproduction) that can last from 7 to 14 days. This imposes a break in the reproduction of the Varroa mites. Furthermore, with each swarming event a sizable fraction (approximately a third) of the colony's mite population is exported with the departing workforce: the fraction of mites shed can be readily calculated since about half of the female breeding-age mites are on the workers in a colony at any given time, and nearly three-quarters of these workers depart in the prime swarm (Rangel and Seeley 2012). In a six-year study of the life-histories of wild honey bee colonies living in a forest in the northeast US, Seeley (2017b) found that most (~87%) swarmed each summer.

In contrast to the relatively small nest cavities of wild honey bee colonies, the colonies kept by beekeepers occupy large hives, and they are less apt to produce swarms (Oliver 2015). The swarm control methods of beekeepers include transferring sealed brood to the top of the hive and queen exclusion (the Demaree method), cutting out queen cells, preventing the filling of cells around the brood nest with nectar (possibly a cue for swarming) by providing empty combs above the brood nest, reversing the brood boxes and inserting empty combs in the brood nest, and reducing the worker populations of colonies by splitting them. All of these methods weaken the stimuli that trigger swarming, but only one helps control the Varroa mites: the removal of bees. We propose instead controlled colony fission by making "splits" to mimic the beneficial effects of swarming on mite control (Loftus et al. 2016).

Horizontal Transmission: Bee Drift, Robbing, Forager Contact, and Contamination

Fries and Camazine (2001) outline three distinct things that a pathogen must do to reproduce and disperse to a new honey bee colony. A pathogen must: (i) infect a single honey bee; (ii) infect multiple honey bees; and (iii) infect another colony. Of these, it is the spread to *another colony* that should most concern beekeepers and bee doctors:

> In terms of fitness, the successful transfer of a pathogen's offspring to a new colony is a critical step in its life history. If a parasite or pathogen fails to achieve a foothold in another host colony, the parasite will not increase its reproductive fitness, regardless of how prolific it has been within the original host colony. Thus, hurdles #1 and #2 (intra-individual and intra-colony transmission) are important aspects of pathogen fitness only to the extent that they contribute to more efficient inter-colony transmission (Fries and Camazine 2001).

The transfer of pathogens or parasites from one colony to another horizontally can occur by four main routes: drifting, robbing, contact while foraging, and shared use of a contaminated environment. Drifting occurs when a forager enters another colony by accident, something that is largely a byproduct of modern apiary management since the wide spacing of wild colonies largely precludes drifting (Seeley 2017b; Seeley and Smith 2015). Robbing occurs primarily during periods of a nectar dearth, when strong colonies attempt rob honey from weak ones. In this case, pathogen transfer is most likely to occur from the weak colony to the strong colony, though the opposite is also possible. The transfer of pathogens during contact while foraging has been described in both natural and experimental models, including video documentation of a Varroa mite jumping onto a foraging honey bee the instant the bee lands on a flower (Peck et al. 2016). Finally, diseases can be spread from one colony to another through sharing of contaminated water, as has been observed with infections of the microsporidium *Nosema apis* (L'Arrivée 1965).

Honey Bee Demographic Turnover

In the article entitled, *What epidemiology can teach us about honey bee health management*, Delaplane (2017) reviewed the ecological and evolutionary impacts of the host–parasite relationship and proposed that an important driver of virulence is the high rate of introduction of susceptible colonies into apiaries (i.e. the introduction of new individuals into existing populations). Epidemiologists recognize three distinct "compartments" for individuals in a population exposed to a disease: Susceptible (S), Infected (I), and Recovered (R) individuals. In the simplest SIR (Susceptible, Infected, and Recovered) model, once susceptible animals catch the disease they become members of the infected "compartment" and can spread the disease to susceptible individuals. The infected animals that survive then move into the recovered "compartment" and are considered immune for life (Kermack and McKendrick 1927). Delaplane argues that the beekeeping practice of restocking "dead-out" hives with nucleus colonies prolongs the epidemic by introducing new "S" individuals into the population of colonies in an apiary, a process that fosters the evolution of virulence (Fries and Camazine 2001). In a closed population, however, a disease epidemic is not artificially prolonged and the surviving individuals tend to have resistance, so there tends to be coevolution of the host–parasite or host–pathogen relationship. Given the high levels of colony losses experienced by beekeepers each year, the restocking of colonies with "nuc" replacements – thereby introducing a fresh batch of susceptible individuals to the apiary population – may represent one of the most noteworthy (and easy to address) management practices contributing to the collapse of honey bee colonies (Cornman et al. 2012).

Now let us return to those curious observations of populations of mite-surviving honey bee colonies in various places around the world. A common thread among these reports of populations of honey bee colonies surviving Varroa infestation for long periods without the use of miticides is the *isolation* of these populations of colonies from managed colonies. The colonies live on islands (Gotland Island in Sweden or the island of Fernando de Noronha off the coast of Brazil), in remote inaccessible regions (far-eastern Russia), or in an intact forest ecosystem (the Arnot Forest in the northeastern United States). The isolation from managed colonies found in all three of these scenarios must have favored the evolution of avirulence of Varroa and the multitude of viral diseases vectored by this mite. In essence, these populations all lack an important feature that drives virulence of infectious disease – a steady introduction of "S" individuals. With no new "Susceptible" colonies coming into these populations, in each case the mites and the bees have co-evolved a stable host–parasite relationship. In the case of the Arnot Forest bees, we know the Varroa invasion was associated with significant loss of genetic diversity in the bees (an indicator of heavy colony mortality caused by Varroa), but at the same time the surviving colonies of this population possessed effective defenses against the mites (Mikheyev et al. 2015; Seeley 2017b).

It is here that the "good lifestyle" of colonies occupying small nest cavities, living widely spaced, and swarming frequently meets the "good genes" of colonies that are living as an isolated "island" of colonies. Now that we have married the good genes and the good lifestyle aspects of health in our examination of honey bee management, where does the bee doctor fit into this picture? In the final section of our chapter, we will explore how we can use the knowledge garnered from a deep understanding of wild colonies to develop a new way of keeping healthy colonies in managed apiaries, an approach recently named Darwinian beekeeping (Seeley 2017a).

Lessons from the Wild Bees

Modern apiarists practice pest/disease control, close colony spacing, swarm control, queen rearing, mating control (sometimes), annual requeening of colonies, migratory beekeeping, queen imports, drone reduction, and various other alterations of the bee's natural biology. These apiculture practices tend to limit natural selection and to disrupt the hard-won adaptations of *A. mellifera*; they impact both the genes and the lifestyle of the honey bee (Neumann and Blacquière 2016). Now, what can be done from an animal husbandry and animal health perspective to reverse such trends?

The bee doctor must be prepared to examine honey bee health through a new lens that takes a holistic approach to medicine – one that features an understanding of and appreciation for the health of honey bees living in nature. In some parts of the world, beekeepers are already looking at beekeeping less as a process of domestication that forces the production of honey, wax, propolis, and pollination at great cost to colonies and more as the stewardship of a natural living system. The global decline in bee health is a direct consequence of man's disruption of this system: the

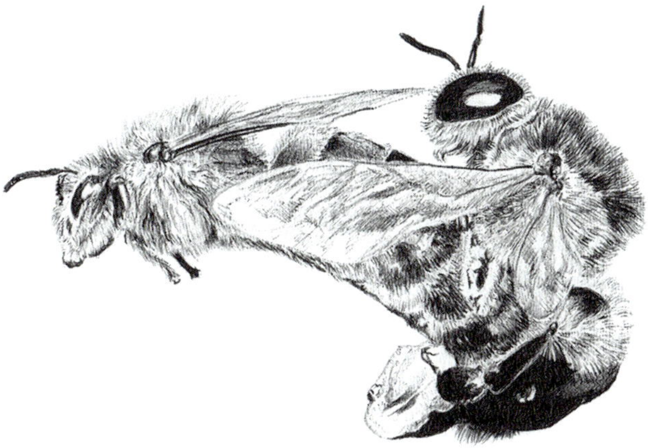

Figure 1.7 Polyandry, or the multiple matings of a queen with drones from different patrilines, has been associated with colony vigor and improved winter survival. The health benefits of polyandry are linked to improved foraging rates, greater brood production, lower mite infestations, and the possession of rare alleles important for control of infectious disease.

introduction of exotic parasites and pathogens, the rise in disease virulence driven by beekeeping practices, and the evolution of drug resistance caused by indiscriminate treatments of colonies. Indeed, it is the pharmaceutical-centric approach to preventative care for honey bees that is the fundamental reason behind the inclusion of honey bees among the food-producing animals in North America that now fall under FDA regulations requiring the services of a veterinarian for antibiotic use. A key feature of a healthy system is achieving a balance between the host and the pathogen that promotes host resistance and pathogen avirulence – we can find this balance by promoting good genes and a good lifestyle in the bees.

Promoting Good Genes

The idea that honey bees have been domesticated by mankind remains a matter of debate. What is clear is that across North America there are populations of wild colonies of *A. mellifera* that thrive independent of beekeeping activities and that do not require the regular input of new colonies from honey bee swarms arising from managed colonies (Oliver 2014; Seeley 2017b; Radcliffe and Seeley 2018). Furthermore, the wild colonies tend to be genetically distinct from those that queen breeders produce for commercial purposes; the former are both more diverse genetically and they show strong evidence of regional adaptation (Figure 1.7) (reviewed in Seeley 2019a,b). Evidently, the honey bee colonies managed by beekeepers are semi-domesticated, since their genes are influenced somewhat by queen breeders and their lifestyle is strongly influenced by their owners (Chapman et al. 2008; Oliver 2014).

An important lesson can be learned from the many animals that man has domesticated over the past thousand years: domestication carries with it a reliance on humans and generally a loss of the ability to survive in the wild. Here we can take some guidance from Charles Darwin:

> One of the most remarkable features in our domesticated races is that we see in them adaptation, not indeed to the animal's or plant's own good, but to man's use or fancy. Some variations have probably arisen suddenly, or by one step. However, we cannot suppose that all the breeds were suddenly produced as perfect and useful as we now see them. . . . The key is man's power of accumulative selection: Nature gives successive variations: man adds then up in certain directions useful to him. (Darwin 1868)

Among the honey bee traits that are known to have a genetic basis, resistance to disease has shown to be a strong component of colony fitness (Tarpy and Seeley 2006). With this in mind, we believe that both the beekeeper and the bee doctor will be wise to consider the following items when it comes to managing the genetics of honey bees.

Goal 1: Select Locally Adapted, Survivor Stock

Bait hives are a ready method for beekeepers to incorporate wild honey bees and their genes into their apiaries (Figure 1.8). A queen honey bee of local origin is well suited to an ecoregion or ecotype and has genes that provide a good fit with the local floral diversity, regional environmental conditions (including extremes of temperature, humidity, drought, etc.), and agents of disease.

A wonderful example of the adaptation of honey bees to their locale is the ecotype of *A. mellifera* that lives in the Landes heathlands of southwestern France (Louveaux 1973). The Landes bees have evolved to have a brood cycle with an unusual, second peak of brood production in August, just in time for the bloom of ling heather (*Calluna vulgaris*) in the Landes landscape. It is interesting to note that when Louveaux moved Paris honey bees to Landes, the Paris bees kept ahead (in colony weight gain, pollen collection, and brood production) of the Landes bees until the middle of July. Up to that point, the Landes bees had trailed behind the Paris bees because the Paris bees had reared more brood in May and June. In August, however, all the colonies of the Landes bees had a second burst of young bees emerging shortly before the heather bloom and by the end of summer these colonies had collected an astonishing 14 kg more honey than the colonies of Paris bees (Louveaux 1973).

Figure 1.8 Bait hives are small nest boxes that are filled with empty comb and sometimes lures or attractants (lemon grass oil or Nasanov pheromone). During the swarm season (May to June in the northeastern United States), scout bees search out and select bait hives during house hunting behaviors. Wild honey bees are well adapted to Varroa and often fare much better than managed bees. Therefore, bait hives are a simple means for the acquisition of locally adapted honey bee stock when they are used in places where there are few beekeepers.

Goal 2: Promote Drone Comb Building and Drone Mating in Congregation Areas

Modern apiarists work to limit the amount of drone comb produced by honey bees because beekeepers have learned that by preventing their colonies from producing drones, they can increase honey production (Seeley 2002). Furthermore, drone comb is the preferred site for *Varroa* reproduction. Limiting it, however, partially "castrates" a colony by reducing the ability of a colony to contribute to the population of drones in a region, an important driver of honey bee diversity and fitness (Seeley 2017b). It is now known that a colony's health and productivity is enhanced by its having high genetic diversity among its worker bees, which arises from the multiple mating (polyandry) strategy of queen honey bees (Tarpy and Seeley 2006; Seeley and Tarpy 2007; Mattila and Seeley 2007). On average, a queen honey bee mates with, and acquires sperm from 10 to 20 drones. Inhibiting drone production in colonies hinders the maintenance of genetic diversity within a region, including the genes that may hold resistance to mites (Rosenkranz et al. 2010).

Goal 3: Cull Failing Colonies Before Collapse

Some veterinarians with experience in honey bee disease and/or epidemiology have campaigned against the emergence of Treatment-Free Beekeeping or Natural Beekeeping because of the risk of spreading disease through the collapse of colonies. Perhaps most alarming is the phenomenon of "mite bomb" colonies (ones collapsing from high mite loads) that spread mites and virulent strains of the Deformed Wing Virus to neighboring colonies (Martin et al. 2012). When Varroa mites reached Hawaii, Martin and colleagues observed a drastic increase in the prevalence of DWV from 10% to 100% (the percentage of honey bee colonies infected with DWV virus), a millionfold increase in DWV viral copies in infected bees, and a reduction in DWV diversity to a single highly contagious strain. A collapse of 274 of 419 managed colonies on Oahu Island followed. The beekeeper should either treat Varroa-infested colonies once a critical mite infestation level is reached (typically c. three mites per hundred bees sampled) or cull (euthanize) highly infested colonies before they can spread their mites to neighboring colonies or surrounding apiaries.

Goal 4: Select Quality Queens and Let the Bees Requeen!

A vigorously laying queen is the most efficient promoter of good genes, so it is of utmost importance to keep colonies headed by highly fertile queens. If a hive must be requeened, it is better to allow the bees to choose their new queen (if age-appropriate larvae are present) than to replace her artificially since it has been shown that when bees are confronted with an emergency need for queen rearing, they do not select larvae at random for their queen cells (as a beekeeper might), but instead select larvae of certain patrilines (Moritz et al. 2005). In the future, beekeepers and bee doctors may be able to better assess queen quality through quantitative means; queen quality, judged in terms of body weight, is a good predictor of a queen's mating flight number, ovarian size, and overall mating success (Amiri et al. 2017).

Although insects lack the immunological memory provided by the antibodies of vertebrates, queen bees can recognize specific pathogens and prime their offspring against them (Salmela et al. 2015). The queen passes these immune signals to her future offspring via the egg-yolk vitellogenin, a protein that has been shown to bind harmful bacteria, including the *P. larvae* of American foulbrood. Queens of local origin will pass onto their larva the essential immune cells that are adapted to the pathogens she has encountered in her environment, giving her offspring the chance to build defenses against disease agents before they (the bees) emerge and become exposed to pathogens in the nest.

Promoting Good Lifestyle

The ways in which honey bee colonies live in the wild differ substantially from those experienced by colonies living in apiaries, where they are managed by beekeepers for honey production or crop pollination. Although there is debate about whether honey bees are truly domesticated (modified genetically to be more useful to humans), it is certain that humans have changed their living conditions through a variety of means. Just as domestic animals are manipulated by farmers in their housing, feed, and even medical care, so too are the colonies of honey bees that are managed by beekeepers. We suggest the following goals to help improve colony fitness through alterations of honey bee lifestyle.

Goal 1: Boost Rather than Disrupt Social Immunity of the Superorganism

In the next chapter we will learn that a honey bee colony is a superorganism. In other words, it is a highly integrated unit of function that has been shaped by natural selection to function as an integrated whole. One result of this high level of organization is that the immune system of a worker honey bee is relatively simple compared to those of non-social bees. With this in mind, we should note that there the beekeeper and bee doctor can inadvertantly weaken the social immunity of the colony. Perhaps the most damaging is breaking and reducing the propolis envelope, which will impair the colony's social immunity and compromise honey bee health. Therefore, the number of times a hive is opened for inspections or manipulations should be reduced to a minimum. The layers of propolis lining the walls and inner cover are playing an important role and should be left intact. The beekeeper can stimulate his/her bees living in a hive to build a complete propolis envelope by using hives whose inner walls have been roughened or by lining the interior surfaces with propolis collection screens.

Goal 2: Quarantine from Pests and Pathogens

Bee doctors should work closely with beekeepers to avoid bringing honey bee colonies from an outside location into an established apiary. The most important drivers of honey bee die-offs in North America have all been caused by emerging pests and pathogens that came from other parts of the world – Varroa mites from Asia, small hive beetles from Africa, and both chalkbrood fungus and acarine mites from Europe (Seeley 2017b). Returning to the SIR model, it follows that beekeepers should reduce as much as possible the introduction of new colonies that represent the "Susceptibles" into an apiary. If these introduced colonies are exposed to or are carrying a novel pathogen, then they can produce outbreaks. Specifically, Delaplane (2017) warns against bringing in outside bees to replace dead outs and recommends instead that these apiary losses should be replaced by splits made within the same apiary. Loftus et al. (2016) found in their study of the effects of colony size and frequent swarming on resistance to Varroa that 60 m was not a sufficient distance between apiaries to avoid spread of Varroa between apiaries during a nectar dearth. Three of the 12 small-hive colonies in this experiment suddenly acquired high mite loads when one of the large-hive colonies collapsed in the adjacent apiary. Evidently, robbers from these three small colonies brought home Varroa from the large colony that was collapsing, resulting in their own collapses several weeks later. It is therefore recommended that introducing new colonies to an apiary be done only after an appropriate period of quarantine in a separate location at least 1 km away.

Goal 3: Design Apiary as Close to Nature as Feasible

The idea that the "design" of an organism is a product of natural selection, which favors survival and reproduction, is the foundation for modern biology and is the basis for Darwinian beekeeping. The fitness of a honey bee colony is directly related to its ability to survive as a healthy unit and to cast viable swarms and produce fertile drones. It follows that we should aim to help our colonies survive and reproduce, if we want them to be part of a healthy population in the area. This viewpoint is perhaps the most challenging for the beekeeper to adopt because it is, in a sense, a break from managing colonies to maximize their production of goods (honey) and services (pollination). If, however, our goal as beekeepers and bee doctors is to sustain populations of healthy colonies of bees, then we should consider making changes in bee management practices that are in keeping with wild colony biology (Seeley 2017b):

First, keep the number of hives in an apiary to a small number to reduce crowding. High colony density promotes robbing and drifting, and thus the mixing of pathogens among host colonies. This mixing ("horizontal transmission") can favor the evolution of virulence in pathogens and eventually lead to the collapse of colonies.

Second, keep hive size small to avoid creating colonies with large brood chambers that support large, continually running "assembly lines" of mite reproduction. Seeley (2017b) suggests using one deep hive body for a brood nest and one shallow super over a queen excluder for harvesting some honey.

Third, perform colony splits (as a method to mimic swarming behavior) to initiate a broodless period that creates a break in reproduction by Varroa mites (Loftus et al. 2016). A beekeeper makes a split (a small, new colony) by removing from a colony its queen and some of its worker bees and brood, and putting them in a separate hive. The remainder of the colony, still living in the original hive, then rears a replacement queen.

Fourth, space colonies as widely as possible (>10 m) and face their hives in different directions to reduce the drifting of returning foragers into the hives of neighboring colonies (Seeley and Smith 2015). Artistic beekeepers can also color code their hives or add unique graphic designs (geometric shapes of color work well!) above the hive entrance to help the bees orient back to their own hives. The anatomy and physiology of the bee, which will be outlined in future chapters, will help guide the beekeeper in choosing colors and patterns most suitable to optimize color and shape recognition by returning bees. Honey bees discriminate colors across the range of green to ultraviolet. Hives painted red appear black to bees, and is a poor choice for hive color given that it is the color of a key predator – the black bear – therefore, hives painted in shades of yellows, greens, blues, or pastel colors are more easily distinguished by honey bees compared to ones painted red or purple.

Fifth, hives should provide the bees with a well-insulated nesting cavity, so that less of a colony's energy is expended on heating and cooling, to achieve thermal homeostasis. The health of a honey bee colony depends on keeping its brood nest at ca. 35 °C from spring to fall, and to keeping the outer layer of the winter cluster above about 10 °C throughout winter.

Finally, bee doctors should avoid treatment of pathogens without a clear diagnosis. A key component of the honey bee environment is the bee's microbiome, which is hidden from view to anyone without a microscope and culture plate. The social behaviors that produce the characteristic flora of the honey bee's gut serve important roles in prevention of disease; the indiscriminate use of antibiotic therapy is known to promote resistance as well as alter the symbiotic gut microbes that underlie the health of honey bee colonies.

Charles Darwin marveled at the honey bee organism and spent a great deal of time studying the organization and structure of their colonies, including the wonderous design of their hexagonal comb. Darwin could not have known the full extent of the threats that the world's honey bees would face in the twenty-first century – from climate change to mite-vectored pathogens. But perhaps he had the bees in mind when he wrote: *It is not the strongest of species that survives, nor the most intelligent, but the one most responsive to change.*

References

Amiri, E., Strand, M.K., Ruepell, O., and Tarpy, D.R. (2017). Queen quality and the impact of honey bee diseases on queen health: potential for interactions between two major threats to colony health. *Insects* 8 (48): 18.

Becher, M.A., Osborne, J.L., Thorbek, P. et al. (2013). Towards a systems approach for understanding honeybee decline: a stocktaking and synthesis of existing models. *Journal of Applied Ecology* 50: 868–880.

Borba, R.S., Klyczek, K.K., Mogen, K.L., and Spivak, M. (2015). Seasonal benefits of a natural propolis envelope to honey bee immunity and colony health. *Journal of Experimental Biology* 218: 3689–3699. https://doi.org/10.1242/jeb.127324.

Brosi, B.J., Delaplane, K.S., Boots, M., and de Roode, J.C. (2017). Ecological and evolutionary approaches to managing honey bee disease. *Nature Ecology and Evolution* 1 (9): 1250–1262.

Chapman, N.C., Lim, J., and Oldroyd, B.P. (2008). Population genetics of commercial and feral honey bees in Western Australia. *Journal of Economic Entomology* 101 (2): 272–277.

Coombs, A.B., Bowman, J., and Garroway, C.J. (2010). Thermal properties of tree cavities during winter in a northern hardwood forest. *Journal of Wildlife Management* 74 (8): 1875–1881.

Cornman, R.S., Tarpy, D.R., Chen, Y. et al. (2012). Pathogen webs in collapsing honey bee colonies. *PLoS One* 7 (8): 1–15.

Darwin, C. (1868). *The Variation of Animals and Plants Under Domestication,* vol. 2. London, UK: John Murray.

De Jong, D. and Soares, A.E.E. (1997). An isolated population of Italian bees that has survived *Varroa jacobsoni* infestation without treatment for over 12 years. *American Bee Journal* 137: 742–745.

Delaplane, K. (2017). What epidemiology can teach us about honey bee health management. *American Bee Journal* 157 (4): 419–421.

Delaplane, K.S., Pietravalle, S., Brown, M.A., and Budge, G.E. (2015). Honey bee colonies headed by hyperpolyandrous queens have improved brood rearing efficiency and lower infestation rates of parasitic Varroa mites. *PLoS One* 10 (12): e0142985. https://doi.org/10.1371/journal.pone.0142985.

Di Prisco, G., Annoscia, D., Margiotta, M. et al. (2016). A mutualistic symbiosis between a parasitic mite and a pathogenic virus undermines honey bee immunity and health. *Proceedings of the National Academy of Sciences* 113 (12): 3202–3208.

Ellis, J.D., Evans, J.D., and Pettis, J. (2010). Colony losses, managed colony population decline, and Colony Collapse Disorder in the United States. *Journal of Apicultural Research* 49 (1): 134–136.

Engel, P., Martinson, V.G., and Moran, N.A. (2012). Functional diversity within the simple gut microbiota of the honey bee. *Proceedings of the National Academy of Sciences* 109 (27): 11002–11007.

Evans, J.D., Aronstein, K., Chen, Y.P. et al. (2006). Immune pathways and defence mechanisms in honey bees *Apis mellifera*. *Insect Molecular Biology* 15 (5): 645–656.

Fleming, J.C., Schmehl, D.R., and Ellis, J.D. (2015). Characterizing the impact of commercial pollen substitute diets on the level of *Nosema* spp. in honey bees (*Apis mellifera* L.). *PLoS One* 10 (7): e0132014. https://doi.org/10.1371/journal.pone.0132014.

Fries, I. and Bommarco, R. (2007). Possible host-parasite adaptations in honey bees infested by *Varroa destructor* mites. *Apidologie* 38 (6): 525–533.

Fries, I. and Camazine, S. (2001). Implications of horizontal and vertical pathogen transmission for honey bee epidemiology. *Apidologie* 32: 199–214.

Fries, I., Imdorf, A., and Rosenkranz, P. (2006). Survival of mite infested (*Varroa destructor*) honey bee (*Apis mellifera*) colonies in a Nordic climate. *Apidologie* 37 (5): 564–570. https://doi.org/10.1051/apido:2006031.

de Garis Davies N. (1930). *Sculptors at Work, Tomb of Rekhmire (TT 100)*. Egypt, New Kingdom, Dynasty 18, Reign of Thutmose III–early Amenhotep II, ca. 1479–1425 B.C., Tempera on Paper. The Metropolitan Museum of Art, New York, USA.

Hodges, C.R., Delaplane, K.S., and Brosi, B.J. (2018). Textured hive interiors increase honey bee (Hymenoptera: Apidae) propolis-hoarding behavior. *Journal of Economic Entomology* 20 (10): 1–5. https://doi.org/10.1093/jee/toy363.

Honey Bee Gene Sequencing Consortium (2006). Insights into social insects from the genome of the honeybee *Apis mellifera*. *Nature* 443: 931–949.

L'Arrivée, J.C.M. (1965). Sources of Nosema infection. *American Bee Journal* 105: 246–248.

Kermack, W. and McKendrick, A. (1927). A contribution to the mathematical theory of epidemics. *Proceedings of the Royal Society of London Series A* 115: 700–721.

Kuropatnicki, A.K., Szliszka, E., and Krol, W. (2013). Historical aspects of propolis research in modern times. *Evidence-based Complementary and Alternative Medicine* 2013: 1–11.

Le Conte, Y., De Vaublanc, G., Crauser, D. et al. (2007). Honey bee colonies that have survived *Varroa destructor*. *Apidologie* 38 (6): 566–572.

Lipstich, M., Siller, S., and Nowak, M.A. (1996). The evolution of virulence in pathogens with vertical and horizontal transmission. *Evolution* 50 (5): 1729–1741.

Locke, B. (2016). Natural Varroa mite-surviving *Apis mellifera* honeybee populations. *Apidologie* 47 (3): 467–482. https://doi.org/10.1007/s13592-015-0412-8.

Locke, B. and Fries, I. (2011). Characteristics of honey bee colonies (*Apis mellifera*) in Sweden surviving *Varroa destructor* infestation. *Apidologie* 42 (4): 533–542. https://doi.org/10.1007/s13592-011-0029-5.

Loftus, J.C., Smith, M.L., and Seeley, T.D. (2016). How honey bee colonies survive in the wild: testing the importance of small nests and frequent swarming. *PLoS One* 11 (3): e0150362. https://doi.org/10.1371/journal.pone.0150362.

Louveaux, J. (1973). The acclimatization of bees to a heather region. *Bee World* 54 (3): 105–111. https://doi.org/10.1080/0005772X.1973.11097464.

Maes, P.W., Rodrigues, P.A.P., Oliver, R. et al. (2016). Diet-related gut bacterial dysbiosis correlates with impaired development, increased mortality and Nosema disease in the honeybee (*Apis mellifera*). *Molecular Ecology* 25: 5439–5450.

Martin, S.J., Highfield, A.C., Brettell, L. et al. (2012). Global honey bee viral landscape altered by a parasitic mite. *Science* 336: 1304–1306.

Mattila, H.R. and Seeley, T.D. (2007). Genetic diversity in honey bee colonies enhances productivity and fitness. *Science* 317: 362–364.

Mikheyev, A.S., Tin, M.M.Y., Arora, J., and Seeley, T.D. (2015). Museum samples reveal rapid evolution by wild honey bees exposed to a novel parasite. *Nature Communications* https://doi.org/10.1038/ncomms8991.

Mitchell, D. (2016). Ratios of colony mass to thermal conductance of tree and man-made nest enclosures of *Apis mellifera*: implications for survival, clustering, humidity regulation and *Varroa destructor*. *International Journal of Biometeorology* 60: 629–638.

Moritz, R.F.A., Lattorff, H.M.G., Neumann, P. et al. (2005). Rare royal families in honeybees, *Apis mellifera*. *Naturwissenschaften* 92: 488–491.

Morse, R.A. and Calderone, N.W. (2000). The value of honey bees as pollinators of US crops in 2000. *Bee Culture* 128 (3): 1–15.

Neumann, P. and Blacquière, T. (2016). The Darwin cure for apiculture? Natural selection and managed honeybee health. *Evolutionary Applications* 10: 226–230.

Oddie, M.A.Y., Dahle, B., and Neumann, P. (2017). Norwegian honey bees surviving *Varroa destructor* mite infestations by means of natural selection. *PeerJ* 5: e3956. https://doi.org/10.7717/peerj.3956.

Oliver, R. (2014). What's happening to the bees? Part 5: Is there a difference between domesticated and feral bees? *American Bee Journal* 154 (6): 679–682.

Oliver, R. (2015). Understanding colony buildup and decline: Part 7b. *American Bee Journal* 155 (9): 977–983.

Peck, D.T., Smith, M.L., and Seeley, T.D. (2016). *Varroa destructor* mites can nimbly climb from flowers onto foraging honey bees. *PLoS One* 11 (12): e0167798. https://doi.org/10.1371/journal.pone.0167798.

Powell, J. (2015). Tree beekeeping: reviving a lost tradition. *Permaculture* 83: 47–50.

Radcliffe, R.W. and Seeley, T.D. (2018). Deep forest bee hunting: a novel method for finding wild colonies of honey bees in old-growth forests. *American Bee Journal* 158 (8): 871–877.

Rangel, J. and Seeley, T.D. (2012). Colony fissioning in honey bees: size and significance of the swarm fraction. *Insectes Sociaux* 59: 453–462.

Ratnieks, F.L.W. and Carreck, N.L. (2010). Clarity on honey bee collapse? *Science* 327 (5926): 152–153.

Raymann, K., Shaffer, Z., and Moran, N.A. (2017). Antibiotic exposure perturbs the gut microbiota and elevates mortality in honeybees. *PLoS Biology* 15 (3): e2001861. https://doi.org/10.1371/journal.pbio.2001861.

Raymann, K., Coon, K.L., Shaffer, Z. et al. (2018). Pathogenicity of *Serratia marcescens* strains in honey bees. *MBio* 9: e01649–e01618.

Richard, F.-J., Tarpy, D.R., and Grozinger, C.M. (2007). Effects of insemination quantity on honey bee queen physiology. *PLoS One* 2 (10): e980. https://doi.org/10.1371/journal.pone.0000980.

Rinderer, T., De Guzman, L., Delatte, G. et al. (2001). Resistance to the parasitic mite *Varroa destructor* in honey bees from far-eastern Russia. *Apidologie* 32 (4): 381–394.

Roffet-Salque, M., Regert, M., Evershed, R. et al. (2015). Widespread exploitation of the honeybee by early Neolithic farmers. *Nature* 527 (7577): 226–230.

Root, A.I. and Root, E.R. (1908). ABC and XYZ of Bee Culture: A Cyclopedia of Everything Pertaining to the Care of the Honey-bee; Bees, Hives, Honey, Implements, Honey-plants, etc. Facts Gleaned from the Experience of Thousands of Bee-keepers, and Afterward Verified in Our Apiary. Medina, OH: AI Root Company.

Rosenkranz, P., Fries, I., Boecking, O., and Stürmer, M. (1997). Damaged Varroa mites in the debris of honey bee (*Apis mellifera* L.) colonies with and without hatching brood. *Apidologie* 28 (6): 427–437. https://doi.org/10.1051/apido:19970609.

Rosenkranz, P., Aumeier, P., and Ziegelmann, B. (2010). Biology and control of *Varroa destructor*. *Journal of Invertebrate Pathology* 103: S96–S119.

Salmela, H., Amdam, G.V., and Freitak, D. (2015). Transfer of immunity from mother to offspring is mediated via egg-yolk protein vitellogenin. *PLoS Pathogens* 11 (7): e1005015. https://doi.org/10.1371/journal.ppat.1005015.

Seeley, T.D. (1977). Measurement of nest cavity volume by the honey bee (*Apis mellifera*). *Behavioral Ecology and Sociobiology* 2: 201–227.

Seeley, T.D. (2002). The effect of drone comb on a honey bee colony's production of honey. *Apidologie* 33: 75–86.

Seeley, T.D. (2017a). Darwinian beekeeping: an evolutionary approach to apiculture. *American Bee Journal* 157: 277–282.

Seeley, T.D. (2017b). Life-history traits of wild honey bee colonies living in forests around Ithaca, NY, USA. *Apidologie* https://doi.org/10.1007/s13592-017-0519-1.

Seeley, T.D. (2019a). The Lives of Bees: The Untold Story of the Honey Bee in the Wild. Princeton, NJ: Princeton University Press.

Seeley, T.D. (2019b). The history of honey bees in North America. In: Phylogenetics of Bees (eds. R. Ilyasov and H.W. Kwon), 222–232. Boca Raton, FL: CRC Press.

Seeley, T.D. and Morse, R.A. (1976). The nest of the honey bee (*Apis mellifera* L.). *Insectes Sociaux* 23 (4): 495–512.

Seeley, T.D. and Smith, M. (2015). Crowding honey bee colonies in apiaries can increase their vulnerability to the deadly ectoparasite *Varroa destructor*. *Apidologie* 46: 716–727.

Seeley, T.D. and Tarpy, D.R. (2007). Queen promiscuity lowers disease within honeybee colonies. *Proceedings of the Royal Society of London Series B* 274: 67–72.

Seeley, T.D., Tarpy, D.R., Griffin, S.R. et al. (2015). A survivor population of wild colonies of European honeybees in the northeastern United States: investigating its genetic structure. *Apidologie* 46: 654–666.

Sherman, P.W., Seeley, T.D., and Reeve, H.K. (1998). Parasites, pathogens, and polyandry in honey bees. *The American Naturalist* 151 (4): 392–396.

Simone, M., Evans, J.D., and Spivak, M. (2009). Resin collection and social immunity in honey bees. *Evolution* 63 (11): 3106–3022.

Spivak, M. and Downey, D.L. (1998). Field assays for hygienic behavior in honey bees (Hymenoptera: Apidae). *Apiculture and Social Insects* 91 (1): 64–70.

Tarpy, D.R. and Seeley, T.D. (2006). Lower disease infections in honeybee (*Apis mellifera*) colonies headed by polyandrous vs monandrous queens. *Naturwissenschaften* 93: 195–199.

vanEngelsdorp, D., Evans, J.D., Saegerman, C. et al. (2009). Colony collapse disorder: a descriptive study. *PLoS One* 4 (8): e6481. https://doi.org/10.1371/journal.pone.0006481.

Winston, M.L. (1980). Swarming, afterswarming, and reproductive rate of unmanaged honey bee colonies (*Apis mellifera*). *Insectes Sociaux* 27 (4): 391–398.

Zheng, H., Steele, M.I., Leonard, S.P. et al. (2018). Honey bees as models for gut microbiota research. *Lab Animal* 47: 317–325.

2

The Superorganism and Herd Health for the Honey Bee

Robin W. Radcliffe

Department of Clinical Sciences, College of Veterinary Medicine, Cornell University, Ithaca, NY, USA
**Illustrations by Anna Connington*

Introduction

The honey bee colony is a magnificent product of evolution. Collective decision-making of thousands of individual bees, each with roles that change as they age, work seamlessly together to create a highly integrated system (the colony) that functions as a single organism (the superorganism). In this chapter we will explore the marvelous world of the honey bee with a focus on how the organization and structure of the colony allows honey bee societies to function as a single coordinated living entity. The superorganism must build new comb, make replacement bees, collect food, water, and hive materials all while protecting their home from pests and pathogens, and survive to reproduce by casting a swarm and sending off drones. Deviations in any one of these collective pathways can lead to disorders, disease, or colony failure.

We will follow the honey bee as it allocates tasks in sophisticated communication networks that help prevent the spread of pathogens, make and use organic compounds to fight disease, collect plant resins to make propolis, and manipulate the hive environment to prevent and even treat infections in the colony. In an extraordinary example of social behavior, we will also learn how honey bees can treat themselves and prevent disease by working as their own "doctors"! These novel methods of disease control and mitigation are just now becoming well understood. The marvels of resin and pollen collection and the myriad bioactive elements in these compounds, collected from nature itself, offers wonderful insights into the ways that honey bees protect themselves from harm. The health benefits of propolis to human health have been known since the days of the ancient Greeks, Romans, and Egyptians; the word itself comes from the Greek "pro" to defend and "opolis" the city, or in this case the beehive or wild nest. Here we will explore the value of propolis to the bees themselves, a topic deserving of more in-depth research. Honey bees can also control fundamental environmental conditions that are protective against disease, including the remarkable ability to regulate the "body temperature" of the superorganism. Used against large invaders such as a bumble bee that attempts to enter the colony, honey bees use heat to "bake" the invader in a ball of heater bees, while small invaders such as some bacteria and fungi that infect the brood are killed by small elevations in temperature (enough to kill the pathogen, but not the developing brood). Scientists call the latter a "social fever", and it is another example of how the colony can ward off infections through cooperative action.

Finally, the health and fitness of honey bees as a superorganism can be examined and evaluated in much the same way as a herd of livestock – *herd health* for honey bees offers a big picture "lens" through which serial monitoring of population level determinants of health are made. An understanding of how honey bees coordinate important hive processes (including collection of pollen, nectar and tree resins, coordination of bee caste populations, maintenance of biosecurity, and ensuring a healthy living environment) combined with the collection of relevant data will provide one of the most important tools for the bee doctor to help decipher health at the level of the colony working in concert with the beekeeper.

Honey Bee Medicine for the Veterinary Practitioner, First Edition. Edited by Terry Ryan Kane and Cynthia M. Faux.

Part 1: The Superorganism and Swarm Intelligence

At the peak of summer activity, an estimated 30 000–50 000 bees live in close proximity within the confines of the typical beehive, or a bee tree if a wild colony. The value of social living must exceed the disadvantages of being closely packed together since parasites and pathogens can exploit the high density of individuals and their network of interactions, predisposing the colony to disease outbreaks. Conditions inside the colony can likewise favor pathogen spread as the bees maintain strict control of the hive environment (brood nest $T = 34\text{–}36\,°\text{C}$; outer winter cluster $T > 10°\text{C}$; RH = 60–80%) (Avitabile 1978; Li et al. 2016). This combination of a stable temperature and humidity is essential to support key hive processes including brood development and rearing, yet these conditions also set up a perfect storm for infectious and parasitic disease to thrive. How then do bees work together to prevent infection and maintain colony health within the framework of this environment of potential harmful organisms? Before we delve into the wonderful colony-level adaptations that support health in the hive environment, we will begin by reviewing the structure of the superorganism.

The idea of the "superorganism" is more than a century old, having first been contemplated by the entomologist William Morten Wheeler (Wheeler 1911). Wheeler believed an ant colony is a system that possesses fundamentally the same properties of an individual organism – namely the complex system (the ant colony in Wheeler's studies) obtains and assimilates substances from the environment, produces offspring, and protects both the system and offspring from disruptions of the environment. Future researchers, including famed ecologists Edward O. Wilson and Bert Hölldobler, refined the definition of a superorganism (also known as eusocial insect societies) by outlining a fundamental couplet: division of labor whereby a small segment of a society produce offspring while the vast majority forgo reproduction to work for the hive – together with the overlap of generations (Wilson 1971; Hölldobler and Wilson 2009). The offspring of the reproductive queen must remain in the nest to help raise the next generation; it was the reproductive division of labor combined with the sibling care for younger siblings in the same nest that marked the rise of eusociality.

Honey bees represent the pinnacle of social evolution by taking this division of labor to the extreme with a single queen monopolizing the egg-laying role (a healthy queen can lay upward of 2000 eggs per day) while 50 000 or more worker bees toil in the hive. The workers still possess ovaries, but rarely lay eggs, and those that are laid are unfertilized and will only produce males. Worker bees cannot function as individual organisms – their only survival is linked inextricably to a promiscuous queen by way of a tightly choreographed system of communication among closely related sisters. The role of the male is simply as a conveyor of genes offering a mechanism to induce diversity (a maiden queen will mate with up to 20 drones) – such diversity is an essential ingredient for Darwin's recipe of natural selection. In their book, *The Superorganism*, the authors equate the male of these female-dominated societies as simply "sperm-guided missiles." While essential to the reproduction of the honey bee colony, the males (or "drones" in the honey bee society) do no work and offer no long-term value to the colony once the missile has launched (Hölldobler and Wilson 2009). Drones die during mating, while any survivors (those that have failed in their only life mission to pass on their genes) are cast aside as resources dwindle each autumn.

A short discourse on honey bee genetics is in order to make sense of how honey bees evolved a social system. Honey bees have a *haplodiploid* method of sex determination in which the queen bee dictates the sex of her own offspring by adding the drone's sperm, or withholding it, as each egg is laid. The worker bees also influence colony demographics when they make a cell: standard size comb cells and round queen cups receive a fertilized egg that become future female bees (workers or queens, respectively), while larger comb cells are fashioned for unfertilized eggs that become male bees (drones). The latter process is known as parthenogenesis – passing on just a single set of chromosomes (those of the mother) to the drone bee. Put simply, *haploid* gives half the number of chromosomes while *diploid* gives double the number of chromosomes. It was long thought that only female bees (workers and queens) were the outcome of a fertilized egg with the resulting bee receiving two sets of chromosomes, one from each parent. But it is not that simple in honey bee society.

Along came diploid drones from inbreeding studies. With their appearance, it was discovered that the number of chromosomes itself did not dictate the sex of honey bees, but rather a single sex determination locus (SDL) determines the sexual fate of honey bee offspring, a process known as complementary sex determination (Whiting 1933; Hasselmann and Beye 2004). Fertilized eggs are heterozygous at the SDL making females, unfertilized eggs are hemizygous and become fertile drones. And those peculiar diploid drones? They are homozygous at the SDL and never survive beyond their first days as a larva; eaten by workers who recognize such anomalous drones would never contribute to colony reproduction. In the curious world of honey bee gene flow, a drone has no father but does have a grandfather and is a parent to daughters, granddaughters and grandsons, but never to sons.

The superorganism exhibits both altruism and inequality, inconsistencies that Darwin himself struggled with in his unifying theory of evolution (Wilson 1971; Ratnieks and Helantera 2009). Darwin's ideas surrounding natural selection focused on how small heritable traits in the *individual* offer a survival advantage that is passed on to future generations (Figure 2.1). How then, could individuals that do not produce offspring (worker bees) evolve body shapes and functions far different from their fertile parents (the queen and drones)? Part of the answer can be found in the matter of kinship or a high level of relatedness among worker honey bees. More than a century after Darwin's *Origin of Species* (1859), Hamilton (1964) wrote that natural selection may favor altruism, but only among related individuals: worker bees are half-sisters having a single mother, and from a nonreproductive worker bee's perspective, success for her own sisters, and that of the queen, equates with success for herself and the colony. Yet, the kinship theory was cast aside with the subsequent discovery of other eusocial organisms having diplodiploidy (e.g. termites) as well as many haplodiploid species living in groups that failed to evolve a eusocial system (Hölldobler and Wilson 2009). Therefore, the extreme inequality and altruism observed in honey bee societies could not have emerged by close kinship alone. It must have been imposed on the sisterhood by the queen and other workers through coercion or "enforced altruism" – social pressures that essentially prevent workers from egg laying (Ratnieks and Helantera 2009). These pressures arise at both the larval stage by workers that control the level of feeding (phenotypically larger queens require more food) and at the adult stage through "policing," whereby the queen and other worker bees destroy worker-laid eggs.

Figure 2.1 Charles Darwin marveled at the superorganism. Recognizing the remarkable structure of honeycomb and the precision of its hexagonal shape, he nevertheless struggled to understand how it may have evolved. Darwin theorized that honeybees once had nests similar to bumblebees, with rough conglomerations of spherical cells. Honey bees, Darwin pondered, must have built circular cells closer and closer together over the generations until finally the cells became organized into the hexagon we see today.

It is fascinating to follow the evolution of the superorganism from solitary insect to primitive eusocial group living to the highly eusocial organism. But there is a difference between the evolution of, and maintenance of, eusociality. The multiple mating of queen honey bees and the resulting diversity of worker bees evolved after the formation of separate castes, a step in the journey to eusociality that Hölldobler and Wilson call "the point of no return." And it is this diversity in the honey bee that led to improved resistance to disease and the protective nature of colony living – in fact, on the path to eusociality the potency of protective defenses against disease in bee populations rises steeply with multiple matings (Stow et al. 2007). Diversity also brought about improvements in productivity and the regulation of hive temperatures, the latter made possible by a worker bee force having innately different thresholds of response to temperature cues that modulate hive ventilating behavior (Jones et al. 2004).

Whether ants or bees, the superorganism must have offered the society key advantages over life as an individual. Ultimately the concept can be viewed from the level of the gene. Seeley (1989) concluded that the emergence of the superorganism must have arisen through suppression of conflict over reproduction (and thereby geneflow) among its constituent parts. "It seems correct to classify a group of organisms as a superorganism when the organisms form a cooperative unit to propagate their genes, just as we classify a group of cells as an organism when the cells form a cooperative unit to propagate their genes" writes Tom Seeley (1989). Now let's turn our attention to the marvelous ways in which honey bees work together as a cooperative unit to maintain a healthy organism.

Part 2: Social Immunity: Bees as Their Own Doctors!

Group living in insects with its consequent division of labor, cooperative care of brood, and the overlap of more than a single generation in time and space are the hallmarks of the superorganism. Insects living within a coordinated framework, where tasks are divided among different bee castes and communication networks are compartmentalized in a confined space, are susceptible to the spread of disease from one individual to another. Likewise, their strict control of the nest cavity environment necessary to maintain the stable temperatures for brood care can be compared to a pathogen incubator. The group living of honey bees predisposes the individuals and the entire organism to epidemics. Fortunately, honey bees and other social insects have evolved highly adaptive behaviors that range from "constitutive" (aka prophylactic) to "inducible" (aka activated) responses that help prevent disease (Simone-Finstrom 2017). Behaviors that reduce or eliminate pathogen exposure or pest infestation at the level of the superorganism are collectively known as *social immunity*.

One of the advantages of a social (or group) response to preventing or actively eliminating an infection by a parasite or pathogen in honey bee(s) is a coordinated response from the colony. By doing so, the individual bee is able to conserve resources that it would otherwise expend on maintaining and delivering an individual response. The immune function of individual honey bees is costly and expressed to a lesser degree than in asocial insects; indeed, the mapping of the *Apis mellifera* genome revealed a surprising lack of immune specific genes (Evans and Pettis 2005; Simone et al. 2009). This does not mean that individual honey bees lack discrete methods for disease protection entirely. Like other insects, honey bees have a hard chitinous exoskeleton that protects against pathogen entry, possess hemocytes that can phagocytize foreign invaders (though they lack memory cells and any ability to produce protective antibodies like vertebrates), remove themselves from the colony when sick or dying, recruit specialized members to perform dangerous biosecurity tasks as guards and undertakers, and even mummify pests too large to carry out of the hive.

In his comprehensive review of social immunity in honey bees, Simone-Finstrom (2017) described the colony level adaptations for health in a continuum from prophylactic to activated: polyandry, task allocation, transfer of compounds and microbiota, resin use, allogrooming, hygienic behavior, social fever, and absconding. On the one extreme, diverse genes made possible by multiple matings and the compartmentalization of honey bee societies offer fixed preventative measures for health. The diversity that comes from numerous patrilines is linked closely to colony vigor and disease resistance and, once a queen mates, the colony's diversity (and thereby the protective alleles coding for disease protection) can only be changed by requeening. Likewise, the social structure of the honey bee colony, with its separation of castes, offers an important first line of defense against infectious disease since castes are separated in both time and space. Yet, the allocation of tasks is rarely altered by pathogen exposure.

On the other extreme, both social fever and absconding are actions taken by honey bees predominantly as a consequence of exposure to a pathogen and represent specific actions to combat the agent. Those social immune strategies located in-between on the continuum may offer both prophylactic and treatment modalities; for example, the collection of resins can be preventative when bees seal their nest cavity in a complete protective "propolis envelope" or resin gathering can be activated by a specific pathogen as a kind of "self-medication." In our overview of social immunity, we will focus on just three of these traits: allocation of tasks with compartmentalization, use of compounds with antimicrobial actions – both bee-derived and plant-derived, and social fever. The miticidal actions of grooming and hygienic behavior are covered in detail elsewhere in this book on chapters about wild colony health, the biology of the varroa mite, and queen breeding for mite resistant honey bees.

Task Allocation and Compartmentalization

Group living elevates the risk of disease transmission through the close intermingling of thousands of individuals, especially for pathogens that are spread by direct contact. In eusocial organisms like the honey bee, the homogeneity in closely-related individuals (all worker bees are daughters of the queen) together with the uniform physical environment both contribute to heightened risk of pathogen transmission. However, the complex social structure of honey bee colonies with its division of labor and allocation of tasks is one of the most important first levels of protection against disease (Cremer et al. 2007). In fact, the selection pressure of pathogens likely contributed to the evolution of social organization in honey bees (Naug and Camazine 2002; Stow et al. 2007). Modeling of honey bee societies depict a highly compartmentalized structure inside the hive with the core of the colony consisting of young bees surrounding a single queen with the foragers existing on the periphery. Even the dance stage of the foragers is located just inside the hive entrance so that the returning foragers – the bees most likely to bring novel parasites and pathogens from their travels outside the hive – are confined in a form of localized quarantine. The distribution of bees into castes with corresponding age classes, further serves to isolate potential spread of infection with young bees of the same age interacting regularly and overlapping spatially, while bees of different ages have limited direct contact (Baracchi and Cini 2014).

Naug and Camazine (2002) outline three key features of colony organization that may influence pathogen transmission in a group of social insects. Division of labor, interaction

Figure 2.2 Trophyllaxis, or the transfer of food from bee to bee, augments disease transmission. Yet, the allocation of tasks across different castes and ages of bees together with separation of entire groups of bees across both space and time represents a sophisticated strategy for biosecurity in a honey bee colony.

network, and colony demography collectively define the epidemiology of transmission. In division of labor, different groups of bees perform different tasks and these tasks are allocated based on the bee's morphology (physical polyethism) or their age (temporal polyethism). It is well known that honey bees conduct the safest jobs inside the hive first, followed by the riskiest jobs outside the hive (primarily defense, scouting and foraging) in the last part of their lives. While worker bees may skip tasks and perform more than a single task at each age, the general progression of tasks begins with cell cleaning, brood care, and tending of the queen, followed by comb building, handling of nectar and pollen, and finally guarding the entrance and foraging. This separation of duties serves to reduce the spread of pathogens in the colony. Honey bees that perform cell cleaning tasks or the hygienic behavior of removing infected brood are highly specialized and do not perform other tasks such as feeding larva that could transfer pathogens from infected to healthy larvae (Seeley 1982). The timing at which bees pass through the various age dependent work schedules in a honey bee colony could also profoundly impact the spread of pathogens and such timing can be altered by the honey bee itself! In infections with the protozoan *Nosema apis* and the sacbrood virus, honey bees have evolved a process known as precocious foraging. In these infections, honey bees move through the temporal schedule of working inside the hive in a fewer number of days and thus move more quickly to outside hive tasks. In this way, the spread of the protozoan and virus are slowed as the number of susceptible individual bees declines more quickly.

In their modeling of disease dynamics within the confines of the social organization of a honey bee colony, Naug and Camazine (2002) observed that the separation of duties through discrete caste division was insufficient by itself to limit the transmission of a pathogen. Both the interaction networks and colony demography were found to be essential elements in the protective mechanism provided by social organization. Networks of interaction define the nature and frequency of contact among hive members and therefore influence the rate of pathogen transmission in a group of social insects. Frequent contact occurs during the transfer of food between individuals by trophyllaxis, a common occurrence that can augment the spread of disease through a honey bee colony (Figure 2.2). Brood diseases such as American foulbrood are transmitted by the passage of spores from worker bee to worker bee during trophyllaxis with subsequent infection of the larva during feeding by nurse bees. The cleaning of foreign material from the cuticle of another bee, or allogrooming, is a form of social immunity that helps remove mites and other pathogens from the exterior of individual bees. However, some viruses such as Chronic Bee Paralysis Virus seem to benefit from the activity, or may even exploit it, all in an effort to help spread the virus to other bees in the colony. Disease transmission is also influenced by colony demography–the size and density of the honey bee population. Big colonies are more likely to contact pathogens because of a larger workforce of foragers working outside the hive, and once a pathogen gains entrance, pathogen spread will occur faster in a dense colony where bee-to-bee interactions are more frequent (Naug and Camazine 2002).

Antimicrobial Compounds Produced by Bees

The recent discovery that bee venom, the collection of vasoactive peptides injected by worker bees using their stinger in defense of the colony, is found on the cuticle of worker bees as well as on the wax surface of the nest comb suggests that venom may play an antiseptic role in social immunity (Baracchi et al. 2011). Bee venom consists of a mix of biogenic amines, peptides, and proteins with neurotoxic action while also breaking down mast cells and stimulating the release of vasoactive substances. More recently, bee venom has been shown to have antimicrobial properties as well. Since there is a complete lack of venom peptides on the cuticle of drones and newly emerged bees, one can surmise that the venom found on the cuticle of worker bees is placed there by grooming behavior from the venom gland itself. Therefore, the allogrooming behavior of bees to remove pests and pathogens may be augmented by the defense provided by the neurotoxic peptides of bee venom (Figure 2.3).

Honey bees synthesize a variety of antimicrobial compounds in response to infection with microorganisms (Simone-Finstrom 2017). One of the better-known peptides produced from the cells of both vertebrates and invertebrates is defensin. A role in innate immunity has been suggested for defensins given their diverse activity against bacterial, fungal, and viral pathogens (Raj and

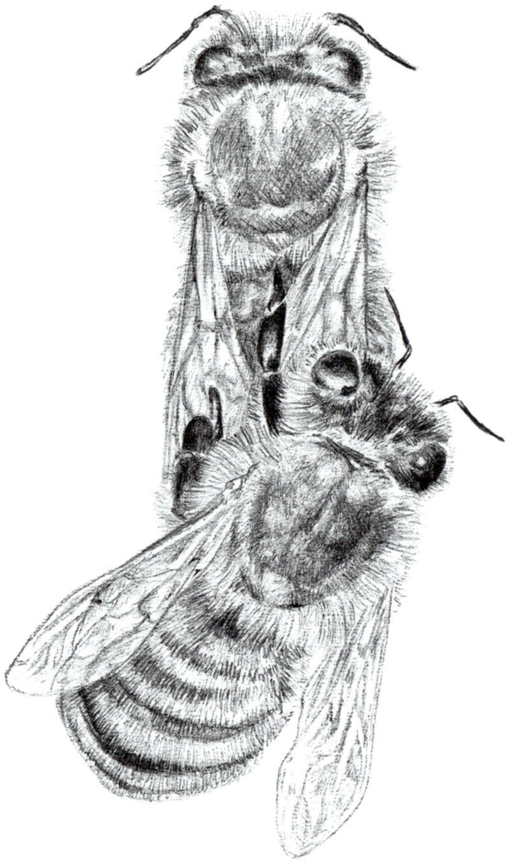

Figure 2.3 Allogrooming, or the grooming of one bee by another nestmate, is a form of social immunity that helps remove potential pathogens from the hive. Through the deliberate spread of bee venom, this regular grooming may also play a central protective role in biosecurity by acting as a cuticular form of antisepsis.

Dentino 2002). One such defensin compound is royalisin isolated from royal jelly – the nurse bee secretion fed to young worker larvae and developing queen larvae. Royalisin has broad antibacterial and antifungal properties, even possessing inhibitory growth against *Paenibacillus larvae*, the causative agent of American foulbrood (Bíliková et al. 2001). It is likely that these protective antimicrobial peptides can boost immunity at the level of the colony since the compounds are transferred widely during trophyllaxis.

The transmission of microorganisms from one bee to another does not always result in the spread of pathogens and disease. On the contrary, many beneficial bacteria are transferred throughout the colony during the complex interactions of honey bee society. Socially transmitted gut microbiota helps protect honey bees from infection by hive pathogens. Worker honey bees lack a bacterial microflora at the time of emergence and the future microbiome community of an individual bee is dictated by contact with

Figure 2.4 Tree saps or resins are collected from leaf buds and packed onto the worker bee's corbiculae for transfer back to the hive where other bees offload the resin, mix it with beeswax and enzymes, to make propolis or "bee glue."

nurse bees, the hive environment, and through trophyllaxis (Powell et al. 2014). In an experimental model of bumble bees (*Bombus terrestris*), contact with nestmate fecal material upon pupal emergence was required for protection against a virulent trypanosome gut parasite *Crithidia bombi* (Koch and Schmid-Hempel 2011). The community of microorganisms in colonies of honey bees and bumble bees is distinctive from that found in solitary bees and its role in providing a first line of defense against potential pathogens is an area that deserves more thorough investigation.

Resin Collection, Propolis, and Immune Modulation

Only a very small group of honey bee foragers (5–15 per day) in any given colony devote their time to the collection of tree and plant resins while 10 times as many foragers are off collecting nectar and pollen (Namakura and Seeley 2006). The bees do this laborious job (it may take 30 minutes to several hours to offload the sticky resins from a bee's pollen basket) without any apparent benefit to the individual bee (Figure 2.4). As in other forms of social immunity, the collective health benefits of resin collection to the colony may be significant by limiting the entrance of pathogens into the nest and reducing the cost for maintaining expensive immune functions for every single colony member (Simone et al. 2009). The latter function of modulating costly immune activity may represent the most important benefit to the superorganism – conserved energy at the level of the individual bee can be directed toward important colony functions of brood rearing and foraging that builds strong bees having adequate vitellogenin storage for overwintering and spring emergence. The collection of plant resins likely evolved as a colony-level adaptation for relieving workers of the need for sustaining an energetically costly immune response, especially when the colony is not being challenged by pathogens (Simone et al. 2017; Borba et al. 2015).

Trees and plants synthesize resins (flavonoids, monoterpenes, and many other biologically active compounds) to protect young leaf buds and injured tissues from infection with pathogens and to deter feeding by browsing herbivores. Honey bees and some other social insects utilize tree resins for their antimicrobial, antifungal, and antiviral properties. It is unknown whether bees select tree species for their resins based on simple availability or more purposely for (as yet unknown) pharmacologic actions (Simone-Finstrom and Spivak 2010). Bees are not known to ingest these compounds directly. Rather, the tree resins collected by honey bees are mixed with wax to make a sticky glue-like substance called "propolis" that is used to secure combs to the roof and walls of the bees' nest cavity as a kind of cement and to seal holes or spaces in the nest architecture. In their detailed portrayal of the nests of wild honey bees, Seeley and Morse (1976) describe a complete propolis envelope surrounding the bee's wild home, essentially sealing off the inner cavity from invading parasites and pathogens. The propolis barrier is incomplete inside the smooth-walled hives of modern managed apiaries, but the barrier can be augmented with commercial propolis traps or by roughening the inner hive wall surface to stimulate propolis deposition (Hodges et al. 2018; Simone-Finstrom et al. 2017).

Propolis production is a heritable trait and varies considerably among lines of honey bees. In Africanized bees, more eggs were produced and more brood survived from larva through pupal stages in colonies having queen-drone crosses with high-propolis production; likewise, the adult bees from such colonies lived longer than bees from crosses from low-propolis colonies (Nicodemo et al. 2014). Bees from high propolis colonies also collected more nectar and pollen than bees from low propolis colonies and showed greater hygienic behavior than low propolis colonies (Nicodemo et al. 2013; Borba et al. 2015). Although other factors may impact resin collection and signs of fitness, the health benefits of propolis may be far-reaching and offer the colony the critical advantage it needs to survive across seasons and reproduce in an environment increasingly dominated by harmful pests and pathogens.

In vertebrates, propolis enhances cellular immune function by increasing the cytotoxic effects of macrophages and the lytic activity of lymphocytes against invading microbes, but few studies have been done on honey bee immune responses with and without propolis. In their study of the health benefits of a complete propolis barrier, Borba et al. (2015) did not see a reduction in bacterial or viral loads in colonies having a natural propolis barrier (by use of propolis traps in Langstroth hives) compared to colonies not having such a barrier, although their methods did not distinguish between pathogenic and commensal organisms. The authors concluded that the reduced investment in immune expression at the level of the individual bee during periods of low pathogen challenge suggests a direct effect of propolis on the bee immune system. And these immune sparing effects were sustained over the entire summer and fall foraging season, only diminishing over the winter when the bees no longer collected resin until such effects were negligible by spring (Simone-Finstrom et al. 2017).

More is known about the benefits of propolis for the health of humans than for the honey bee with reported antiseptic, anti-inflammatory, antioxidant, antibacterial, antifungal, antiulcer, anticancer, and immunomodulatory properties (Pasupuleti et al. 2017). Propolis consists of resin (50%), wax (30%), essential oils (10%), pollen (5%), and other organic compounds (5%). The organic compounds from the plant resins are responsible for the health benefits and include primarily phenols, esters, flavonoids, and terpenes. Propolis also contains both vitamins (B1, B2, B6, C, and E) and minerals (Mg, Ca, K, Mn, and Fe) and a few enzymes from the bee saliva. The health benefits of propolis for honey bees has focused on the important bee pathogens causing American foulbrood (*P. larvae*) and the fungal agent of chalkbrood disease (*Ascosphaera apis*). Historical studies showed a positive correlation with feeding of propolis to bees in sugar solution, but since bees are not thought to ingest propolis, such models may not reflect the way resin compounds inhibit microbes – by way of a barrier defense at comb edges, colony walls and around the nest entrance – and the use of propolis in sugar solutions could harm beneficial gut microbes or lead to pathogen resistance (Simone-Finstrom et al. 2017). Colonies having a propolis barrier can still be infected with American foulbrood, but the severity of infection is reduced and the larval food from propolis rich colonies appears to inhibit *P. larvae*. Propolis may also help control hive parasites. Laboratory studies of propolis extracts demonstrated that propolis has narcosis and lethal effects on *Varroa destructor*. Yet, since both mites and propolis exist together in a bee colony without apparent varroacidal effects, it may be that the active compounds in propolis are insoluble and unavailable in adequate concentrations within the waxy glue that bees lay down (Garedew et al. 2002).

Self-medication in Honey Bees

The idea that non-human animals can self-medicate – that is, use organic compounds to clear an infection or reduce its symptoms – was long thought to be limited to vertebrates since it was presumed that it required learning (de Roode et al. 2013). However, we now know that self-medication or "zoopharmacognosy" is widespread in the insect world, in part because insects utilize a wide variety of organic compounds and have evolved methods to medicate their relatives, offspring, or even societal members. Given there are a variety of reasons why an animal might consume an organic substance independent of improving its own health or that of its kin, true self-medication has a strict definition: the organism must intentionally seek out the compound, the compound must harm the parasite, the compound must benefit the host, and finally, its use must come with a cost to the host if consumed in the absence of an infection (Abbott 2014).

Honey bees exhibit self-medication both as a way to prevent infection and to treat an acquired infection. While most insects consume organic compounds to protect their own health or that of their offspring, eusocial honey bees collect resins to treat the entire colony rather than the individual bee, a form of mass medication. Simone-Finstrom and Spivak (2012) observed that honey bees increased their resin foraging in response to exposure to the chalkbrood fungus, *A. apis*. In their study, rates of pollen collection declined while resin collection increased after honey bees were challenged with chalkbrood. Since chalkbrood is a disease of larvae and not adult bees, the increase in collection of resins in response to a fungal pathogen is a marvelous example of social immunity in which the colony, rather than the individual bee, is the beneficiary of the adaptive behavior (Simone-Finstrom and Spivak 2012). Curiously, the bacteria causing American foulbrood and another fungus, *Metarhizium*, failed to elicit increased resin foraging in their investigation.

Pollen plays a key role in brood rearing, worker bee lifespan, and bee resistance to pathogens. In particular, pollen and protein availability influence hypopharyngeal gland development in worker bees and an abundance is associated with lowered infection titers with deformed wing virus (DeGrandi-Hoffman et al. 2010). Although not a form of self-medication since bees do not increase pollen collection in response to infection, a pollen rich diet has been shown to provide protective benefits against a variety of pathogens, including the Varroa mite (Annoscia et al. 2017). In particular, the apolar fraction of pollen (that portion of pollen especially high in fatty acids, hydrocarbons, and sterols, and distinguishable in the laboratory from the polar fraction) appears to provide a dietary protective measure against disease. In the case of Varroa mites, the adults penetrate the bee cuticle and increase water loss, feed on the bee's fat body creating a negative energy balance, and vector viral diseases. Pollen is protective by providing a source of hydrocarbons for cuticle integrity, the unsaturated fatty acid component of pollen shows antibiotic activity, and pollens enhance immune function. The authors conclude that in bees infested with *V. destructor*, access to a pollen-rich diet increases lifespan and can compensate for the negative effects of the mite (Annoscia et al. 2017). Bumble bees (*Bombus impatiens*) are known to alter their foraging patterns based on the quality of the nutritional resource, with high Pollen:Lipid ratios of highest attraction (Vaudo et al. 2016). Likewise, the secondary metabolites in floral nectar (alkaloids, teropenoids, and glycosides) have been shown to reduce bumble bee parasite loads (Richardson et al. 2015). Such observations confirm suspicions that changes in bee forage, particularly in agricultural dominated landscapes or in migratory beekeeping practices, likely contributes to colony declines. The important message for the bee doctor from all this research is that colony nutrition is ultimately connected to colony health and that the role of the veterinarian in helping the beekeeper manage disease should always include a thorough evaluation of colony nutrition, including review of local bloom calendars, hive pollen stores, and the use of protein supplements.

Social Fever

It should come as no surprise that the honey bee superorganism can mount a biological "fever" as a direct preventative measure against a heat-sensitive pathogen. This fever is not mounted in the individual bee but rather in the heart of the colony in the developing brood, and is a remarkable example of convergent evolution between the organism and superorganism. In a fascinating experiment, Starks and colleagues (2000) measured brood comb temperatures in three colonies and one control colony in response to changes in ambient environmental temperatures and following the inoculation of an infective dose of the fungal pathogen for chalkbrood disease (*A. apis*). Chalkbrood is triggered by chilling of the brood; therefore, it is a seasonal condition most prevalent in the spring of the year or in small colonies that are unable to maintain homeostasis by way of thermoregulation. Normal brood comb temperatures are maintained within a very narrow range from 33 to 36 °C and only vary by small amounts in direct relation to ambient temperature – such a relationship allowed the authors to determine expected brood comb temperatures at each ambient temperature and measure variations from expected results. The brood comb temperature rose 0.56 °C after inoculation with an infectious dose of *A. apis* (Starks et al. 2000). The authors argue that this small elevation in temperature, representing 20% of the range in normal brood comb temperatures, is sufficient to provide protection against *A. apis* since only a slight cooling of the bee larvae is needed to cause disease. Of the three treatment-hives inoculated with *A. apis* and subjected to the biological fever of the superorganism, only one colony developed minor chalkbrood mummies. Furthermore, the social fever Starks observed in the experimental infection with chalkbrood appears to be preventative as the elevation in brood comb temperature happened before the larvae were killed.

Part 3: Herd Health for the Honey Bee

The bee doctor now understands that the complex interactions of honey bees are only achieved through a highly coordinated system of communication and feedback regulation within an environment that essentially offers an "incubator" for pathogens. Yet, we also know that honey bees have developed remarkable adaptations that promote colony health in the form of biosecurity, immunity at the colony level, thermoregulation and social fever, and even self-medication. The role then, for the bee doctor, is to become a proactive partner with the beekeeper, rather than a reactive harbinger of disease or colony failure. To do so, the bee doctor can use the tools of *herd health* that have been developed to manage populations of animals, principally those of dairy herds.

Health inspection of honey bee colonies is focused on colony factors (bee caste populations, brood size and pattern, eggs, honey and pollen stores, etc.) with less importance given to examination of individual bees (important exceptions include obvious organism level defects such as deformed wings, parasitic mites, seizure activity, etc.).

Herd health as it relates to dairy practice has been defined as "a method to optimize health, welfare, and production in a population of dairy cows through the systematic analysis of relevant data and through regular objective observations of the cows and their environment, such that informed, timely decisions are made to adjust and improve herd management over time" (Down et al. 2012). While the end goal for the bee doctor may differ from that of the herd health practitioner (not all beekeepers will be focused on production and profit margins), the lessons of a herd health approach can be applied across taxa. Perhaps most importantly, a thorough data-driven strategy to assess and manage bee health sets up a proactive dialog between beekeeper and bee doctor and serves to eliminate the reactive response to pathogens, disease, and colony loss. In short, a herd health approach re-orientates interactions with the beekeeper from one of disease and die-off investigations to one of health maintenance and preventative medicine.

Herd health is a continual process where the veterinarian and farmer (or beekeeper in our case) meet regularly to review colony-level data that informs decision-making in the apiary. The regular contact between bee doctor and beekeeper are essential in order to closely follow the success or failure of management interventions. These interactions also serve to develop a close Veterinary-Client-Patient-Relationship that will guide all aspects of the bee veterinarian's work with the beekeeper. As in the dairy farmer, beekeepers likely have wide variation in what motivates their actions. Kirstensen and Jakobsen (2011) observe:

> The practising cattle veterinarian's ability to translate knowledge into on-farm application requires a profound understanding of the dairy farm as an integrated system. Consequently, educating and motivating farmers are key issues. To achieve such insight the veterinarian needs to work with several scientific disciplines, especially epidemiology and (behavioural) economics. This trans-disciplinary approach offers new methodological possibilities and challenges to students of dairy herd health management.

Likewise, as we have already recognized in the beginnings of this chapter, the bee veterinarian must possess a deep understanding of honey bee biology and communication, collective intelligence of the superorganism, as well as appropriate measures of health and fitness at the level of the individual and the colony.

For example, a small backyard beekeeper may be perfectly happy with a management style that mimics natural colony biology (see Chapter 1) with moderate honey, pollen or propolis crops focused on value-added local markets. Such markets may include specialty varietal honeys, hive products such as propolis and royal jelly that offer health benefits to humans, or even small single farm pollination services for orchards, specialty crops, or hobby farms. Alternatively, a large apiary with thousands of hives may be focused on honey production for larger distribution or may travel with their colonies to fulfill lucrative pollination contracts. In the middle are producers that may share some of both motivations. In each case, the veterinarian must forge a working relationship that provides measurable benefits to the producer – the bee doctor must therefore understand honey bee terminology, demonstrate an in-depth knowledge of colony biology, be versed in the beekeeping industry, and be knowledgeable of the stages of the "factory" where honey is produced so timely interventions can be made.

In order to realize these shared goals, the veterinarian should establish several key components of a colony health program. These include developing a working relationship with the beekeeper and offer preventative medicine and care through regular site visits for the evaluation of bee health (i.e. review control measures, results of regular mite checks, progress of integrated pest management, colony nutrition and long-term data collection and record keeping). These services may not require regular visits to the apiary once a Veterinary–Client–Patient–Relationship is formed, but establishing a regular routine to check-in with the beekeeper will complement visits and may include the opportunity to prepare periodic reports or written summaries of data collected.

A good example of the valuable information to be gained from a herd health approach to beekeeping is the data-driven strategy to apiary health management that guides Randy Oliver's Scientific Beekeeping program. Randy devotes significant time and resources to examining problems using the scientific method to evaluate critical areas of bee health such as mite survival, queen longevity and genetics, colony loss, emerging pathogens, and the like. While not every beekeeper can take such a comprehensive data-intensive approach as Randy, the lesson is that colony health mandates that beekeepers collect information through careful observation and diligent record keeping. And, in the case of managing health, incorporating the skills of a competent bee doctor with knowledge of honey bee biology, medicine, and disease combined with an interest in epidemiology and bee science, will help improve management of bees in an apiary environment.

References

Abbott, J. (2014). Self-medication in insects: current evidence and future perspectives. *Ecological Entomology* 39: 273–280.

Annoscia, D., Zanni, V., Galbraith, D. et al. (2017). Elucidating the mechanisms underlying the beneficial health effects of dietary pollen on honey bees (*Apis mellifera*) infested by Varroa mite ectoparasites. *Scientific Reports* 7: 6258. https://doi.org/10.1038/s41598-017-06488-2.

Avitabile, A. (1978). Brood rearing in honeybee colonies from late autumn to early spring. *Journal of Apicultural Research* 17 (2): 69–73.

Baracchi, D. and Cini, A. (2014). A socio-spatial combined approach confirms a highly compartmentalized structure in honey bees. *Ethology* 120: 1167–1176.

Baracchi, D., Francese, S., and Turillazzi, S. (2011). Beyond the antipredator defence: honey bee venom function as a component of social immunity. *Toxicon* 58: 550–557.

Bíliková, K., Wu, G., and Šimúth, J. (2001). Isolation of a peptide fraction from honeybee royal jelly as a potential antifoulbrood factor. *Apidologie* 32: 275–283.

Borba, R.S., Klyczek, K.K., Mogen, K.L., and Spivak, M. (2015). Seasonal benefits of a natural propolis envelope to honey bee immunity and colony health. *Journal of Experimental Biology* 218: 3689–3699.

Cremer, S., Armitage, S.A., and Schmid-Hempel, P. (2007). Social immunity. *Current Biology* 17 (16): R693–R702. https://doi.org/10.1016/j.cub.2007.06.008.

Darwin, C. (1859). On the Origin of Species by Means of Natural Selection; or the Preservation of Favoured Races in the Struggle for Life. London, UK: John Murray.

DeGrandi-Hoffman, G., Chen, Y., Huang, E., and Huang, M.H. (2010). The effect of diet on protein concentration, hypopharyngeal gland development and virus load in worker honey bees (*Apis mellifera L.*). *Journal of Insect Physiology* 56: 1184–1191.

Down, P.M., Kerby, M., Hall, J. et al. (2012). Providing herd health management in practice – how does it work on a farm? *Cattle Practice* 20 (2): 112–119.

Evans, J.D. and Pettis, J.S. (2005). Colony-level impacts of immune responsiveness in honey bees, *Apis mellifera*. *Evolution* 59: 2270–2274.

Garedew, A., Lampretcht, I., Schmolz, E., and Schricker, B. (2002). The varroacidal action of propolis: a laboratory assay. *Apidologie* 33: 41–50.

Hamilton, W.D. (1964). Genetical evolution of social behaviour I. *Journal of Theoretical Biololgy* 7: 1–16. (doi:10.1016/0022-5193(64)90038-4)

Hasselmann, M. and Beye, M. (2004). Signatures of selection among sex-determining alleles of the honey bee. *Proceedings of the National Academy of Sciences* 101 (14): 4888–4893.

Hodges, C.R., Delaplane, K.S., and Brosi, B.J. (2018). Textured hive interiors increase honey bee (Hymenoptera: Apidae) propolis-hoarding behavior. *Journal of Economic Entomology* 20 (10): 1–5. https://doi.org/10.1093/jee/toy363.

Hölldobler, B. and Wilson, E.O. (2009). The Superorganism: The Beauty, Elegance and Strangeness of Insect Societies. New York, NY: WW Norton & Co.

Jones, J.C., Myerscough, M.R., Graham, S., and Oldroyd, B.P. (2004). Honey bee nest thermoregulation: diversity promotes stability. *Science* 305: 402–404.

Koch, H. and Schmid-Hempel, P. (2011). Socially transmitted gut microbiota protect bumble bees against an intestinal parasite. *Proceedings of the National Academy of Sciences of the United States of America* 108: 19288–19292.

Kristensen, E. and Jakobsen, E.B. (2011). Challenging the myth of the irrational dairy farmer; understanding decision-making related to herd health. *New Zealand Veterinary Journal* 59 (1): 1–7. https://doi.org/10.1080/00480169.2011.547162.

Li, Z., Huang, Z.Y., Sharma, D.B. et al. (2016). Drone and worker brood microclimates are regulated differentially in honey bees, *Apis mellifera*. *PLoS One* 11 (2): e0148740. https://doi.org/10.1371/journal.pone.0148740.

Namakura, J. and Seeley, T.D. (2006). The functional organization of resin work in honeybee colonies. *Behavioral Ecology and Sociobiology* 60: 339–349.

Naug, D. and Camazine, S. (2002). The role of colony organization on pathogen transmission in social insects. *Journal of Theoretical Biology* 215: 427–439.

Nicodemo, D., De Jong, D., Couto, R.H.N., and Malheiros, E.B. (2013). Honey bee lines selected for high propolis also have superior hygienic behavior and increased honey and pollen stores. *Genetics and Molecular Research* 12 (4): 6931–6938.

Nicodemo, D., Malheiros, E.B., De Jong, D., and Couto, R.H.N. (2014). Increased brood viability and longer lifespan of honeybees selected for propolis production. *Apidologie* 45: 269–275.

Pasupuleti, V.R., Sammugam, L., Ramesh, N., and Gan, S.H. (2017). Honey, propolis, and royal jelly: a comprehensive review of their biological actions and health benefits. *Oxidative Medicine and Cellular Longevity* 2017: 1–21.

Powell, E.J., Martinson, V.G., Urban-Mead, K., and Moran, N.A. (2014). Routes of acquisition of the gut microbiota of the honey bee *Apis mellifera*. *Applied and Environmental Microbiology* 80 (23): 7378–7387.

Raj, P.A. and Dentino, A.R. (2002). Current status of defensins and their role in innate and adaptive immunity. *FEMS Microbiology Letters* 206: 9–18.

Ratnieks, F.L.W. and Helantera, H. (2009). The evolution of extreme altruism and inequality in insect societies. *Philosophical Transactions of the Royal Society Series B* 364: 3169–3179.

Richardson, L.L., Adler, L.S., Leonard, A.S. et al. (2015). Secondary metabolites in floral nectar reduce parasite infections in bumblebees. *Proceedings Royal Society B* 282: 20142471.

de Roode, J.C., Lefèvre, T., and Hunter, M.D. (2013). Self-Medication in Animals. *Science* 340 (5129): 150–151.

Seeley, T.D. (1982). Adaptive significance of the age polyethism schedule in honeybee colonies. *Behavioral Ecology and Sociobiology* 11: 287–293.

Seeley, T.D. (1989). The honey bee colony as a superorganism. *American Scientist* 77 (6): 546–553.

Seeley, T.D. and Morse, R.A. (1976). The nest of the honey bee (Apis mellifera L.). *Insectes Sociaux* 23 (4): 495–512.

Simone, M., Evans, J.D., and Spivak, M. (2009). Resin collection and social immunity in honey bees. *Evolution* 63: 3016–3022.

Simone-Finstrom, M. (2017). Social immunity and the superorganism: behavioral defenses protecting honey bee colonies from pathogens and parasites. *Bee World* 94 (1): 21–29. https://doi.org/10.1080/0005772X.2017.1307800.

Simone-Finstrom, M. and Spivak, M. (2010). Propolis and bee health: the natural history and significance of resin use by honey bees. *Apidologie* 41: 295–311.

Simone-Finstrom, M.D. and Spivak, M. (2012). Increased resin collection after parasite challenge: a case of self-medication in honey bees? *PLoS One* 7: e34601.

Simone-Finstrom, M., Borba, R.S., Wilson, M., and Spivak, M. (2017). Propolis counteracts some threats to honey bee health. *Insects* 8: 46.

Starks, P.T., Blackie, C.A., and Seeley, T.D. (2000). Fever in honeybee colonies. *Naturwissenschaften* 87: 229–231.

Stow, A., Briscoe, D., Gillings, M. et al. (2007). Antimicrobial defenses increase with sociality in bees. *Biology Letters* 3 (4): 422–424.

Vaudo, A.D., Patch, H.M., Mortensen, D.A. et al. (2016). Macronutrient ratios in pollen shape bumble bee (*Bombus impatiens*) foraging strategies and floral preferences. *Proceedings of the National Academy of Sciences of the United States of America* 113: 4035–4042.

Wheeler, W.M. (1911). The ant colony as an organism. *Journal of Morphology* 22 (2): 307–325.

Whiting, P.W. (1933). Selective fertilization and sex-determination in Hymenoptera. *Science* 78 (2032): 537–538.

Wilson, E.O. (1971). The Insect Societies. Cambridge, MA: Harvard University Press.

3

Honey Bee Anatomy

Cynthia M. Faux

College of Veterinary Medicine, The University of Arizona, Oro Valley, AZ, USA
**Illustrations by Patrick D. Wilson*

Introduction

In veterinary school, insects are primarily studied as pests, parasites, and vectors of disease. However, in the honey bee we have both a patient and an agricultural partner. Veterinarians are used to dealing with differences among species, but with the honey bee there are both great similarities and vast contrasts between the bee and the more familiar veterinary patients. A working knowledge of honey bee anatomy and terminology will better equip the veterinarian to understand the clinically relevant physiology and pathologies of the honey bee and to communicate effectively with beekeepers.

Honey bees belong to the order Hymenoptera, a large clade that includes other bees as well as wasps, ants, and sawflies. Among other features, the group is distinguished by having tiny hooklets or "hamuli" on the leading edge of the back or hind wings which serve to secure the hind wings to the front wings so that the wing pairs function as a unit (Snodgrass et al. 2015; Vidal-Naquet 2015) (Figure 3.6 below).

Being invertebrates, honey bees have a rigid exoskeleton which defines their external shape (Figure 3.2). The bee body consists of three distinct sections: head, thorax, and abdomen. The head comprises the organs one would expect: brain, mouthparts, and sensory organs. All of the locomotory appendages are attached to the thorax and are paired (left and right); they include two pairs of wings (front and hind) and three pairs of legs. The abdomen contains the majority of the gastrointestinal tract and the reproductive organs. In the case of the female honey bee, the abdomen terminates in a sting apparatus.

Three distinct morphologies occur in honey bees: the drone (male), the worker (female), and the queen (female). Drones are distinguished by their rather large, plump contours, and large eyes. They are typically around 19 mm long. Drones do not possess a stinger. The queen's size depends upon if she is mated or not (average size of mated queen is ~20 mm long) and she is identified by her large, long abdomen. Her wings are short relative to her abdomen in contrast to the workers and drones, whose wings extend beyond the abdomen. Workers average 15 mm in length (Vidal-Naquet 2015). See Figure 3.1.

Unless otherwise noted, the following sections describe the features of the worker honey bee (Figure 3.2).

Head

Eyes

The honey bee has a pair of large, "compound" eyes that are readily visible on the lateral aspect of the head (left and right), and three miniscule "simple" eyes, called the ocelli,

Figure 3.1 A photograph showing the three members of the honey bee hive – drones, workers, and the queen. *Source:* Photo courtesy of Randy Oliver.

Honey Bee Medicine for the Veterinary Practitioner, First Edition. Edited by Terry Ryan Kane and Cynthia M. Faux.

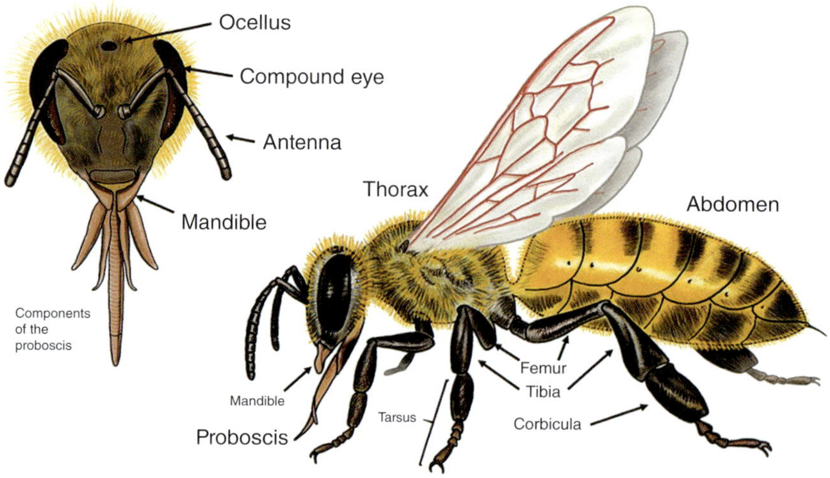

Figure 3.2 External anatomy of the honey bee. *Source:* Illustration by Patrick D. Wilson.

Figure 3.3 Magnified view of a honey bee compound eye. The hexagonal shape of the ommatidia are visible. *Source:* Photo courtesy of Jamie Perkins.

Figure 3.4 The three ocelli are indicated by the small arrows. *Source:* Photo courtesy of Zachary Y. Huang.

that are arranged in a triangular pattern on the top of the head. Honey bees are remarkably "hairy" under magnification, and these hairs include the compound eyes (Figure 3.3).

The compound eyes are composed of approximately 5500 hexagonal "ommatidia." An ommatidium can be thought of as an individual eye, each with a sensory (optic) nerve, that sends its own unique signals to the brain. Collectively, the input from these 5500 "eyes" compose what the honey bee "sees."

However, it is presumptuous to assume that the honey bee perceives its environment the same way we see or visually sense our environment. Although bees are considered to possess true color vision, the wavelengths to which a honey bee eye is sensitive ranges from 344 to 556 nm – a lower limit which supports the assumption that honey bee vision extends into the ultraviolet wavelengths (Kelber et al. 2003; Avarguès-Weber et al. 2012).

The three ocelli, or simple eyes, are difficult to see without magnification. Each has a single lens, but each eye contains approximately 800 photoreceptors. The ocelli are believed to function simply as light sensors; they do not form an image (Figure 3.4).

Antennae are segmented sensory structures that in honey bees contain chemo and other sensory receptors. The honey bee's antennae provide direct tactile, thermal, and humidity information as well as sensing vibrations and detecting pheromones in the surrounding air (Figure 3.5).

Mouthparts

The external mouthparts of interest in honey bees are the mandibles and the proboscis. Worker bees use their mandibles to manipulate such items as wax and pollen. Unlike

Figure 3.5 Closeup of a honey bee antenna, showing the segments. *Source:* Photo courtesy of Jamie Perkins.

the vertebrate mandible, honey bee mandibles move from lateral to medial and thus function as pincers.

The proboscis is a tubular structure used by the bee to suck in fluids such as nectar and water. In honey bees, the proboscis is formed by separate mouthpart components that combine to create a tube. In other nectar-feeding insects such as butterflies, the proboscis is a dedicated tubular structure (Snodgrass et al. 2015) (Figure 3.2).

Thorax

Wings

Honey bees have two pairs of wings: front and hind, left and right. As mentioned above, on each side of the thorax the front and hind wings are held together with tiny hooklets on the leading edge of the hind wing that catch on the caudal edge of the front wing. In this way, the front and hind wings function as a unit in flight (Figure 3.6).

The wings are powered by muscles within the thorax which act to compress or expand the shape of the thorax, and thus raise and lower the wings. Although the flight mechanism is relatively simple, the honey bee has great maneuverability, being able to hover and to fly forward, backward, and sideways.

Deformed wings may be observed in hives affected by deformed wing virus, so it behooves the veterinarian to make note of the wing structure of any honey bees on a frame, as bees with this condition may be unable to fly.

Figure 3.6 Wing hooklets, greatly magnified. *Source:* Photo courtesy of Jamie Perkins.

Legs

Like all insects, honey bees have three pairs of legs (front, middle, and back), all attached to the thorax (left and right). There are six segments per leg, thus six joints, including that which attaches the leg to the thorax (Figure 3.2). Each joint is restricted to motion in only one plane, but this constraint is overcome as each of the six joints in the leg move in different planes, thus enabling considerable dexterity and agility (Snodgrass et al. 2015). Each leg segment is named as shown in Figure 3.2, so the honey bee has six femurs, six tibias, etc.

In addition to locomotion, the honey bee uses her legs to carry pollen back to the hive. Specialized "pollen baskets" called corbiculae are found on the modified tibia of each hind leg. The bee uses her front and middle legs, as well as specialized brushes and combs on her hind legs, to remove pollen collected on hairs covering her body. She packs the pollen into her corbiculae for transport back to the hive. These so-called "pollen pants" can be seen on bees returning to the hive; they are an important observation when monitoring hive activity. Bees also carry propolis (a resinous mixture of bee saliva, wax, and botanical material) in their corbiculae (Snodgrass et al. 2015) (Figures 3.7 and 3.8).

The honey bee's antennae are critical information-gathering organs, so they require regular cleaning and maintenance. The bee's front legs each possess an antenna-cleaning notch that is lined by a tiny comb-like structure (Figure 3.5). The bee scrapes the ipsilateral antenna free of pollen and other debris using this tool (Figure 3.9).

Figure 3.7 (a, b) Magnified view of the corbicula (pollen basket) on the lateral side of the hind leg (yellow arrow). The honey bee uses her front and middle legs to place pollen on the bristles (pollen brush) of the medial side of the leg. Then, using the pollen rake and pollen press on the opposite leg, she packs the pollen into the corbicula (Snodgrass et al. 2015). *Source:* Photo courtesy of Cynthia Faux.

Figure 3.8 Honey bee foraging on prairie spiderwort *Tradescantia occidentalis* with pollen loaded into her corbiculae. *Source:* Photo courtesy of Zachary Y. Huang.

Figure 3.9 Antenna cleaner. *Source:* Photo courtesy of Jamie Perkins.

Abdomen

The abdomen of the queen is greatly elongated compared to that of the worker honey bee, in order to accommodate the sperm contained within her reproductive tract. The male honey bee, the drone, has a more rounded, plump appearance than the worker. The abdomen terminates in a cavity containing the opening of the rectum (anus) and, in female bees, the sting apparatus (Figure 3.10).

Figure 3.10 Stinger of a worker honey bee. The queen's is barbless. *Source:* Photo courtesy of Jamie Perkins.

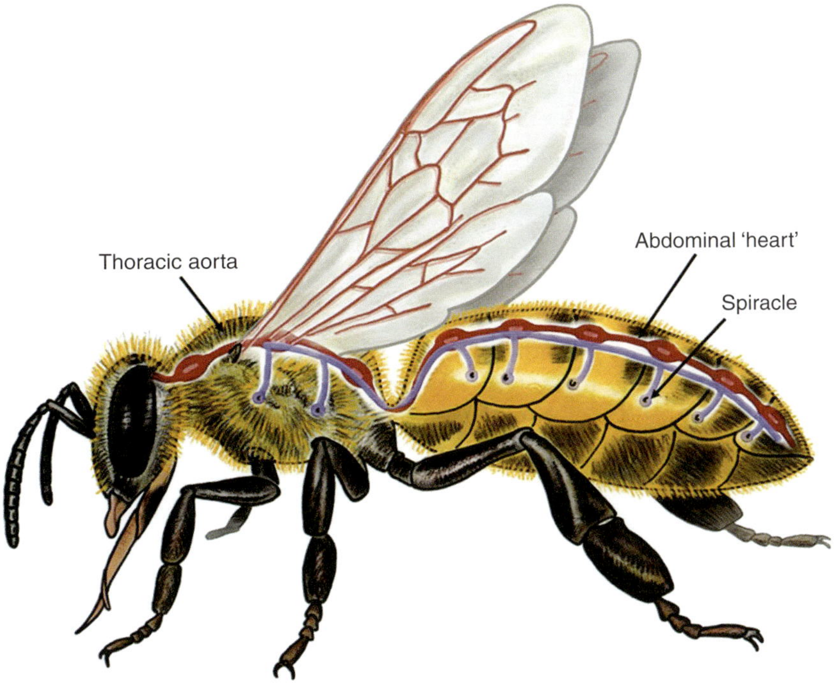

Figure 3.11 Circulatory and respiratory system. *Source:* Illustration by Patrick D. Wilson.

Circulatory and Respiratory Systems

Insect "blood" is called hemolymph, as it is devoid of oxygen-carrying cells. Delivery of oxygen and nutrients to, and collection of waste products from, the tissues and organs is accomplished by diffusion in an "open" circulatory system. (Figure 3.7) A single hemolymph vessel lies along the dorsal midline. Hemolymph flows into openings within this region, called the heart, and is pumped cranially through the thoracic region (aorta) toward the head. Hemolymph leaves the aorta and flows back toward the caudal thorax and abdomen, where it re-enters the circulation. Hemocytes within the hemolymph play a role in the insect's immune system (Snodgrass et al. 2015).

The respiratory system of the honey bee begins with a series of openings called spiracles along the lateral sides of the bee. These openings lead to thin-walled air sacs and then to a complex of tubes called the tracheal system. The air sacs allow air to be moved through the system by abdominal contractions. Oxygen is thus delivered to the tissues by diffusion (Snodgrass et al. 2015; Vidal-Naquet 2015) (Figure 3.11).

Nervous System

As in mammals, the brain lies within the bee's head; however, in bees a ventral nerve cord exits the head and passes caudally through the ventral aspect of the thorax and abdomen. The peripheral nervous system extends from the ventral nerve cord (Figure 3.8) (Snodgrass et al. 2015).

Honey bees have a remarkable capacity for memory and learning. The mere fact that foragers leave the hive daily and return to tell their hive-mates about their discoveries is amazing! (Figure 3.12).

Figure 3.12 Ventral nerve cord of the honey bee, shown in purple. *Source:* Illustration by Patrick D. Wilson.

Digestive and Excretory System

The digestive system of the honey bee begins with the mouth, including the proboscis. The esophagus is long, passing through the head and thorax before it empties into the crop, or "honey stomach," in the cranial portion of the abdomen. The abdomen of the bee visibly expands (see Figure 8.7) when the insect ingests fluids (water, nectar, or honey) and the crop fills (Snodgrass et al. 2015).

The crop is used for carrying resources to and from the hive. The proventricular valve between the crop and the next portion of the digestive tract, the ventriculus, prevents contents of the crop from passing further along the tract where it would otherwise be digested. The ventriculus or midgut is where digestion occurs. Distal to the midgut is the intestine, which has a short, narrow section and a wider, expandable portion, the rectum (Snodgrass et al. 2015). Normally, bees will not defecate in the hive, so an expandable rectum allows retention of fecal waste until the bee is able to fly from the hive; in winter, it may be several months before the bee leaves the hive. This nice-weather release of feces occurs in what beekeepers call "cleansing flights" (Figure 3.13).

In insects, nitrogenous waste excretion is managed by the Malpighian tubules – a series of long, meandering tubes that collect waste products from the hemolymph to be excreted with digestive waste via the rectum. Malpighian tubules are broadly equivalent in function to the vertebrate kidney.

The function of the vertebrate liver is provided in insects by the "fat bodies." Fat bodies are small organs that lie on the dorsal and ventral aspects of the abdomen and play a critical role in the synthesis of hemolymph proteins and the synthesis and storage of lipids. The fat bodies are critical for the survival and successful overwintering of honey bees. It was recently reported that the Varroa mite, which can so devastate hives, feeds not on hemolymph but on the fat bodies of infested bees (Ramsey et al. 2019).

Glands

Bees possess numerous specialty glands that secrete substances necessary for feeding larvae, defending the hive, building comb, recognizing the home hive, and functions (Bortolotti and Costa 2014). The glands illustrated in Figures 3.14 and 3.15 are more fully described in Chapter 4 – Physiology.

Reproductive System

The drone possesses the apparatus necessary for insemination of the queen – large eyes to spot a queen, excellent flying skills, and an endophallus. Spermatozoa are transferred to the queen with the endophallus, which breaks from the drone. The mating process kills the drone (Vidal-Naquet 2015).

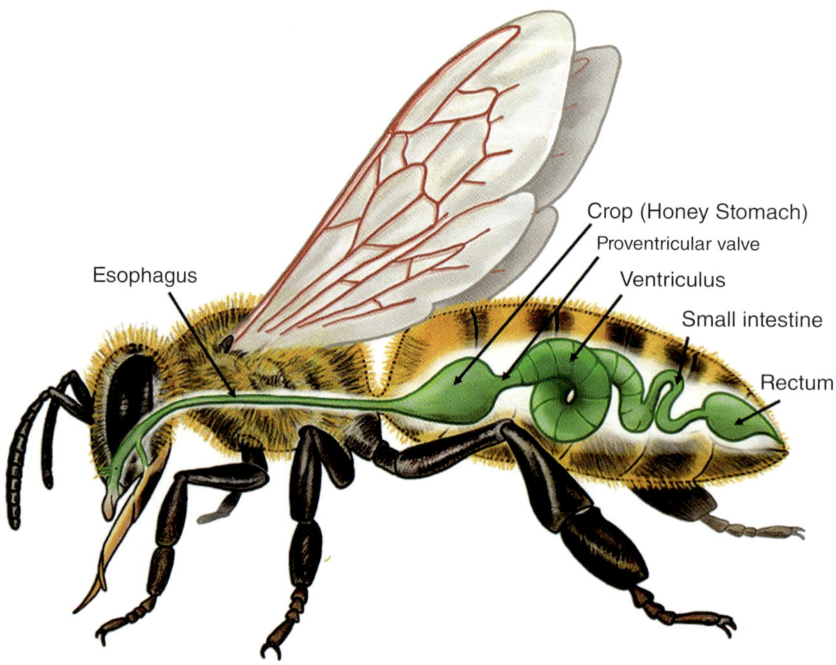

Figure 3.13 Digestive system. *Source:* Illustration by Patrick D. Wilson.

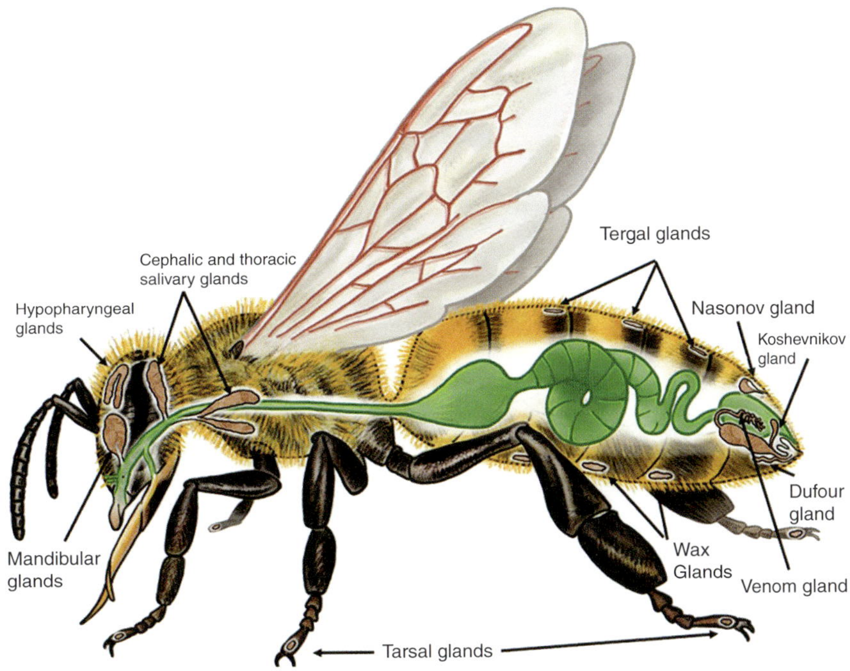

Figure 3.14 Overview of the glands of the honey bee. The presence or absence of particular glands at any given time in the individual's life depends on the age/caste of the worker (nurse bee, forager, etc.) versus a queen or a drone (Vidal-Naquet 2015). *Source:* Illustration by Patrick D. Wilson.

Figure 3.15 Wax scales emerging from the wax glands. *Source:* Photo courtesy of Zachary Y. Huang.

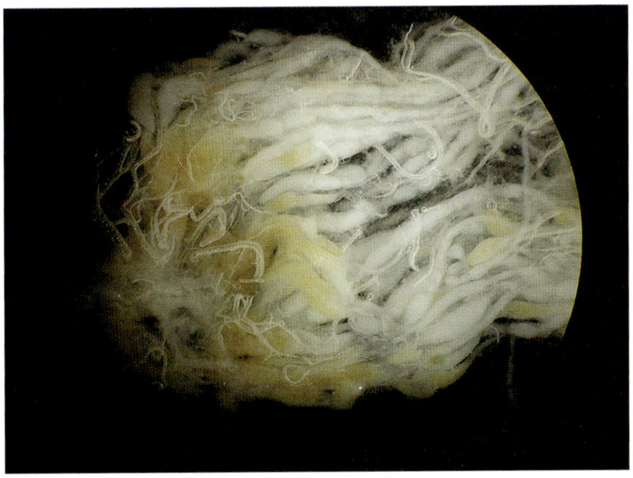

Figure 3.16 Ovary of a laying queen. Individual ovarioles can be observed with eggs. *Source:* Photo courtesy of Zachary Y. Huang.

The queen receives spermatozoa from multiple drones and stores it within a spermatotheca. The spermatotheca releases sperm into the vagina to fertilize the egg as it passes (Vidal-Naquet 2015) (Figure 3.16).

More detailed information and descriptions of honey bee anatomy can be found in the references.

References

Avarguès-Weber, A., Mota, T., and Giurfa, M. (2012). New vistas on honey bee vision. *Apidologie* 43: 244–268.

Bortolotti L, Costa C. Chemical Communication in the Honey Bee Society. In: Mucignat-Caretta C, editor. Neurobiology of Chemical Communication. Boca Raton (FL): CRC Press/Taylor & Francis; 2014. Chapter 5. Available from: https://www.ncbi.nlm.nih.gov/books/NBK200983/

Kelber, A., Vorobyev, M., and Osorio, D. (2003). Animal color vision – behavioral tests and physiological concepts. *Biological Reviews* 78: 81–118.

Ramsey, S.D., Ochoa, R., Bauchan, G. et al. (2019). *Varroa destructor* feeds primarily on honey bee fat body tissue and not hemolymph. *Proceedings of the National Academy of Sciences of the United States of America* 116: 1792–1801.

Snodgrass, R.E., Erickson, E.H., and Fahrbach, S.E. (2015). The Hive and the Honey Bee. Hamilton, IL: Dadant and Son.

Vidal-Naquet, N. (2015). Honeybee Veterinary Medicine: Apis Mellifera L. Sheffield, UK: 5M.

4

Physiology of the Honey Bee – Principles for the Beekeeper and Veterinarian

Rolfe M. Radcliffe

Large Animal Surgery and Emergency Critical Care, College of Veterinary Medicine, Cornell University, Ithaca, NY, USA
**Illustrations by Lauren D. Sawchyn*

Both beekeepers and veterinarians working with bees require an understanding of the honey bee individual and superorganism. It is imperative to be able to identify what is normal biology – anatomy, function, and behavior – for the individual honey bee and its colony and what is abnormal or unhealthy, to provide the appropriate management steps to control or eliminate a problem or disease. For example, understanding honey bee physiology and communication will provide the basis for making informed decisions about diseases like varroa mite that may affect many different features (e.g. bee weight, lifespan and bee numbers, deformities, behavior, reproduction) of a honey bee and their collective hive functions. Further, a full understanding of honey bee biology will help the beekeeper and veterinarian achieve their goal(s), whether it be for the production of honey, beeswax, pollination or other services.

The honey bee and its collective colony is a marvel of nature. The following passage from *The Superorganism: the beauty, elegance and strangeness of insect societies*, by Hölldobler and Wilson (2009), eloquently summaries the complexity of such a life:

> Consider a honey bee gathering nectar from a flower bed. Although simple in appearance, the act is a performance of high virtuosity. The forager was guided to this spot by dances of her nestmates that contained symbolic information about the direction, distance, and quality of the nectar source. To reach her destination, she traveled the equivalent of hundreds of human miles at bee-equivalent supersonic speed. She has arrived at an hour when the flowers are most likely to be richly productive. Now she closely inspects the willing blossoms by touch and smell and extracts the nectar with intricate movements of her legs and proboscis. Then she flies home in a straight line. All this she accomplishes with a brain the size of a grain of sand and with little or no prior experience.

Part 1: Comparing Vertebrates and Bees

Physiology of the Honey Bee Compared with Vertebrates

Veterinarians receive a broad understanding of vertebrate life yet little education concerning invertebrates because of this focus in their veterinary school training. The honey bee's recent recognition as a food producing animal is helping to reshape the future of our profession, and now veterinarians are becoming trained in this important capacity. The leap from vertebrate medicine to that of insects may not be that far-off. Even though these two groups of animals are distinct in many ways, they share similar bodily functions of a nervous system, respiration, blood circulation, digestion and excretion, metabolism, and reproduction (Figure 4.1) (Ritter 2014). In most respects the vertebrate body is considerably more complex compared to invertebrates, especially regarding the nervous system, body size, and structure. Interestingly however, compared with vertebrates, the honey bee colony – with its team of cooperating individuals working together as a superorganism – is one step above the organizational order of the vertebrates whose basic elements are composed of various cells and tissues (Hölldobler and Wilson 2009). This advanced organizational structure is dependent upon various anatomical and physiological adaptations.

Honey Bee Medicine for the Veterinary Practitioner, First Edition. Edited by Terry Ryan Kane and Cynthia M. Faux.

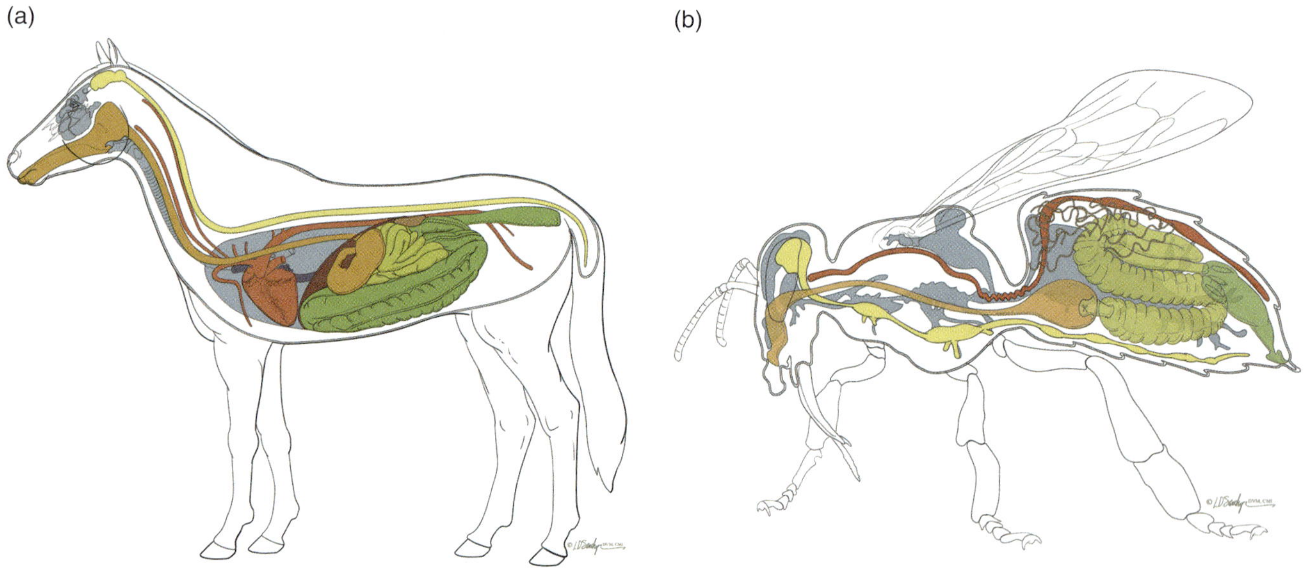

Figure 4.1 Comparative anatomy of the vertebrate equine (a) with the invertebrate honey bee (b). Despite their marked physical differences, these two groups of animals share similar functional body systems. Note the separate colors for the various body systems: GI System – Brown = Foregut, Light Green = Midgut, Dark Green = Hindgut; Blue = Respiratory System; Red = Cardiovascular System; Yellow = Nervous System. *Source:* © Lauren D. Sawchyn, DVM, CMI. Chapter: Physiology of the honey bee, authored by Rolfe M. Radcliffe and illustrated by Lauren D. Sawchyn.

According to R.E. Snodgrass, "An insect is a living machine; no other animal is provided with so many anatomical tools, gadgets or mechanisms for doing such a variety of things as is a winged insect."

Digestive System and Metabolism

Insects are able to feed on a great variety of organic material in nature and their digestive systems are modified to reflect their specific diet (Wigglesworth 1972). The more complex vertebrate systems have also evolved to complement their diet, whether it be monogastrics, ruminants or birds (Molnar and Gair 2015). The digestive function of the insects is similar to vertebrates both having an alimentary canal composed of foregut, midgut, and hindgut segments (Ellis 2015; Ritter 2014). The digestive system of the honey bee serves three main functions: the intake and absorption of nutrients, the elimination of waste, and a means of transport and storage of nectar and honey (Snodgrass 1956). Importantly, the three honey bee castes – drone, worker, and queen – display many physiologic, morphologic, and behavioral differences (Snodgrass 1956; Wigglesworth 1972; Hrassnigg and Crailsheim 2005).

The foregut of insects is composed of the mouth, esophagus, and crop, or honey stomach as it is known in the honeybee. The mouthparts of insects are adapted to fit their needs such as chewing solid materials like foliage, wood, or other creatures, as in many beetles or, as with the honeybee, to collect fluids such as the nectar of flowers (Wigglesworth 1972). The esophagus simply functions to transport food from the mouth through the head and thorax to the crop located in the abdomen, while the crop is the organ for storing nectar or water. The proventriculus functionally separates the crop and the ventriculus, and in the worker bee it regulates the food transfer between these two organs, retaining nectar in the crop for delivery to the hive (Snodgrass 1956). The proventriculus is comprised of four folds that act to separate the pollen from the nectar and control movement into the ventriculus. Because the proventriculus coordinates the nutrition between each honey bee and that of the hive, it represents the connection of the individual and social metabolic cycles of the honey bee colony (Ritter 2014).

The midgut or ventriculus of insects is the primary site for the digestion and absorption of nutrients. The ventriculus is the largest part of the alimentary tract and is the true stomach of the insect similar to the stomach of monogastric animals or the abomasum of the ruminant. The epithelium of the ventriculus functions to secrete digestive enzymes, absorb food materials, and excrete redundant products, such as calcium (Snodgrass 1956). In addition, cylindrical food envelopes, known as peritrophic membranes, are present the entire length of the ventricular epithelium; this layer is present in many insects and may improve digestion, act as a barrier to pathogens, and protect the epithelium from coarse pollen granules similar to the mucous boundary protecting the intestinal tract of vertebrates (Snodgrass 1956; Wigglesworth 1972; Vidal-Naquet 2015). However, because the midgut region of the honey bee is semi-permeable, this is also where many pathogens gain entrance into the insect, such as *Nosema* species and several viruses. The ventriculus opens into the anterior intestine – the beginning of the hindgut of insects, and this pyloric juncture is where the Malpighian tubules enter the intestinal lumen. These tubules are the excretory organs of the honey bee similar to the kidneys of vertebrates where waste products from the hemolymph are removed including nitrogen, urates, phosphates, and calcium (Wigglesworth 1972; Ritter 2014; Snodgrass 1956).

The hindgut of insects includes the anterior or small intestine and the posterior intestine or rectum; like that of the cecum and colon of vertebrates, its primary function is the absorption of water. The rectal glands, or more appropriately termed rectal pads, are thought to be the location of water conservation in terrestrial insects (Snodgrass 1956; Wigglesworth 1972). Importantly, in the honey bee the rectum also serves to store waste products during the extended time spent inside the hive during the winter months in temperate climates. In fact, in the over-winter worker bee, almost the entire abdomen may be distended with feces.

Both vertebrates and invertebrates utilize carbohydrates, proteins, and fats as energy for growth and work activities (Wigglesworth 1972). Carbohydrates are the most readily available energy source of insects, and the honey bee is dependent upon glucose and other sugars to maintain an almost constant supply of energy for flight and the various hive activities (Wigglesworth 1972). Blood glucose alone may be able to maintain flight in honey bees for 15 minutes or a distance of approximately 5.5 kms, while extended forays for hours are possible with a full honey crop (Wigglesworth 1972). Proteins are necessary as the building blocks of insect muscles, glands, and other tissues, while fat is the primary form of energy storage.

Unique to insects, the fat body is a large organ distributed throughout the body cavity and involved in metabolism and sustaining food reserves; this organ functions similar to the liver in vertebrate energy storage (Arrese and Soulages 2010; Wigglesworth 1972). Importantly, the fat body has many other functions including maintaining body homeostasis and supporting immunity. The insect fat body produces hemolymph proteins, circulating metabolites, antimicrobial peptides, assists nitrogen removal, collects toxic compounds and its function may be affected by disease and poor management (Arrese and Soulages 2010; Ritter 2014). In the honey bee the fat body stores fat, protein, and glycogen to be used for energy reserves during winter rest, summer colony growth, times of high energy demand such as flight or during times of food shortages (Snodgrass 1956; Wigglesworth 1972; Ritter 2014). Winter worker bees have a longer lifespan partly because of a less active lifestyle, reduced metabolic rate and increased fat bodies compared with summer worker bees.

Circulatory and Respiratory Systems

In contrast to vertebrates, insects have an open compared to a closed circulation system with a single blood vessel and simple heart (Wigglesworth 1972; Ritter 2014). This system fills most of the body of insects outside of the other organs and tissues (Snodgrass 1956). Blood flow is produced by the action of a pulsating dorsal vessel that has a heart-like function in the abdomen, directing the hemolymph forward through the aorta toward the brain; the blood returns via the body cavity to the abdomen to repeat the cycle (Wigglesworth 1972; Ritter 2014; Snodgrass 1956). Similar to vertebrates, hemolymph, or bee blood, contains many blood cells or hemocytes and functions to transport nutritional substances and waste products to and from the tissues, respectively; however, unlike vertebrates, insect hemolymph does not have hemoglobin and therefore cannot carry oxygen. In addition to their role in metabolism, insect hemocytes are also thought to have several other functions, including phagocytosis, resistance to microorganisms and parasites, blood coagulation and immunity (Wigglesworth 1972).

The vertebrate respiratory system is characterized by gills or lungs that act to exchange oxygen and carbon dioxide gases in concert with a complex circulatory system. A simpler method of gas exchange occurs in insects where the respiratory and circulatory systems are separate and blood has only a minor role in gas transport (Winston 1987). The respiratory system of most insects consists of a series finely branched trachea and tracheoles that allow for the direct diffusion of oxygen to their tissues and the removal of carbon dioxide (Wigglesworth 1972; Ritter 2014; Snodgrass 1956). Some insects, like the honey bee, also have saccular dilations of the trachea forming great air sacs. Air is brought into the insect through breathing apertures along the lateral thorax and abdomen known as spiracles (Snodgrass 1956). Ventilation of the air sacs and larger tracheal tree is thought to occur via specialized body wall movements that act to renew the tidal volume (Wigglesworth 1972). Most breathing insects have control over expiration alone, while inspiration is passive via the elasticity of the exoskeleton; however, honey bees have muscles that control both inspiration and expiration. The tracheal air sacs act largely as reservoirs for ventilation, and the respiratory movements of the abdomen in honey bees produce expansion and contraction of the air sacs similar to vertebrate lungs (Snodgrass 1956). Carbon dioxide is delivered via the hemolymph to the trachea for exhalation although elimination also occurs via direct diffusion through body tissues (Wigglesworth 1972; Ritter 2014).

Nervous System

Even though an insects' nervous system is relatively simple compared with a vertebrate's complex nervous system, it is well adapted to the environment (Vidal-Naquet 2015). The central nervous system of the honey bee consists of a primitive brain and a ventral cord that controls sensory perception, movement, navigation, defense, etc. . . whereas the vertebrate nervous system is comprised of a well-developed brain, medulla, and spinal cord. The brain of insects largely coordinates sensory perception of the environment (Vidal-Naquet 2015) and is comprised of three sections: the protocerebrum enables vision and forms two large optic lobes linking the compound eyes, the deutocerebrum controls olfaction via the sense organs of the antennae, and the tritocerebrum facilitates taste through the labrum. The ventral cord section of the central nervous systems – comprised of seven ganglia throughout the thorax and abdomen – innervates insect mouthparts, all of the legs and wings and the sting apparatus (Vidal-Naquet 2015). The peripheral nervous system supports the various sense organs supporting interactions among honey bees and perception of their environment.

Sense Organs

Sensory perception is quite advanced in insects, providing these animals a remarkable ability to sense and adjust to their environment. Insects, including honey bees, have several types of sense organs of their exoskeleton that may respond to pressure, odor, taste, sound, or light (Wigglesworth 1972; Snodgrass 1956). Seven different sense organs are described and even though they vary markedly in structure, they have in common a basic unit called the sensillum. The sensillum is composed of one or more sense cells connected to the central nervous system via a sensory axon, and a specific cuticular structure with accessory cells (Snodgrass 1956; Vidal-Naquet 2015; Wigglesworth 1972). Some of these organs have a small hair, peg, or plate, or a group of sensilla connected to the sense cell or cells and provide many sensory functions for insects (Wigglesworth 1972; Snodgrass 1956). The antennae of honey bees (faced with several types of sense organs) are a major center of communication with many sensory roles, including odor and chemoreception, detection of movement and vibration, as well as the perception of sound, temperature, and humidity (Snodgrass 1956; Vidal-Naquet 2015).

The Organ of Johnston, located on the pedicel of the antennae, is another sensory organ used in a variety of ways by insects including flight control, navigation, detection of gravitational and electromagnetic fields, mate identification, sound perception, and communication (Wigglesworth 1972). In many insects this organ indicates velocity and orientation in flight and other movements. In the honey bee the Johnston's Organ is also thought to be important for communication between foraging bees during the waggle dance (Tsujiuchi et al. 2007). This organ detects changes in the position of the antennae via mechanical stimulation – from abdomen waggling and wing vibration of a dancing bee – and together with other sensilla translates direction and distance communication to following forager bees during the waggle dance (Brockmann and Robinson 2007; Tsujiuchi et al. 2007).

In addition to the abundant sense organs of the antennae, mouth parts and other portions of the body, honey bees have two types of eyes – the paired compound eyes and three ocelli (Snodgrass 1956). The compound eyes are complex visual organs composed of thousands of hexagonal facets – known as ommatidia – that each function independently to receive, concentrate, and perceive light (Winston 1987). Specific groups of facets are specialized and work together for various functions including detecting light polarization, pattern recognition, color vision, and head movement (Winston 1987). The honey bee collects a mosaic of sensory input to the brain that is integrated to form an image; bees are good at identifying shapes and detecting movement, and may visualize shorter wavelengths (ultraviolet) of light compared with humans. Further, the compound eyes have sensory hairs near the facet junctions that perceive airflow and likely aid in navigation and orientation (Snodgrass 1956). The three ocelli or simple eyes of the honey bee likely do not form an image as with the compound eye, but rather are thought to be important for detecting variations of light intensity that may help diurnal navigation and orientation (Winston 1987).

Immune System

Social, compared with solitary organisms, are at an increased risk of disease because often, as with the honey bee colony, large numbers of individuals are living in a confined nest with stored resources; however, group living also imparts heightened infection control measures (Evans et al. 2006; Fefferman et al. 2007; Kurze et al. 2016). The social structure of the hive helps defend against disease in many ways (e.g. grooming, hygienic, and necrophoric behaviors, use of propolis with anti-microbial properties, social fever in response to disease, as well as nest hygiene and defense among others) (Evans et al. 2006; Vidal-Naquet 2015).

Individual bees also have physical properties that help prevent infection in addition to both cellular and humoral immunity for the recognition and removal of pathogens. Several morphologic characteristics of insects help combat infection (Evans et al. 2006). A layer of antimicrobial secretions covers the external surfaces of many types of insects and the intestinal tract with digestive enzymes is not friendly to pathogen survival, although the semipermeable midgut is documented as entrance site for several honey bee pathogens (Davidson 1973). In addition, the gastrointestinal tract of the adult worker honey bee is characterized by a core set of bacteria that are not only important in nutrition and metabolism, but also protective against pathogen infection (Raymann and Moran 2018). Next, the exoskeleton cuticle of the honey bee forms a physical barrier against pathogen invasion, as does the peritrophic membrane of the intestinal tract (Vidal-Naquet 2015; DeGrandi-Hoffman and Chen 2015). Further penetration of the honey bee by an infectious agent elicits an immune response at the level of the hemolymph and fat body. Such immune defenses of insects are similar to the innate immune system of vertebrates, both sharing many characteristics including the actions of phagocytosis, secretion of antimicrobial peptides, enzymatic degradation of pathogens, as well as similar architecture and orthologous components (Evans et al. 2006). Unlike vertebrates however, insects lack adaptive immunity and cannot produce antibodies; rather the honey bee immune response is characterized by non-specific reactions against pathogens via both cellular and humoral immunity (Vidal-Naquet 2015; DeGrandi-Hoffman and Chen 2015). Specifically, the binding of highly conserved structural motifs of pathogens by special receptors activate hemocyte-mediated cellular events such as phagocytosis or encapsulation of the pathogen, induction of hemolymph coagulation or melanization, or the synthesis of antimicrobial peptides (DeGrandi-Hoffman and Chen 2015). Further, RNA interference, a major antiviral immune response of insects, has also been identified in several honey bee studies (DeGrandi-Hoffman and Chen 2015).

Age Polyethism

An age-related division of labor among worker honey bees is well known, where younger bees perform the inside tasks (first two to three weeks of life), while older bees complete the outside jobs (last one to three weeks of life) (Winston 1987). However, because bees are sensitive to social changes within their colony, this division of labor is not firm and behavioral maturation of work activities may change, directly regulated by worker to worker interactions (Leoncini et al. 2004). For example, when a colony begins to lack older foraging bees, some bees commence foraging as young as five days of age, two weeks earlier than usual. Similarly, when a colony has an overabundance of older bees, younger bees delay their maturation to foraging (Leoncini et al. 2004).

The transition of a honey bee during aging from a life without flying and only inside hive tasks to one of outside defense and foraging, imparts unique functional demands and energy requirements (Elekonich and Roberts 2005). Vitellogenin, a storage protein produced in the fat body of insects and secreted into the hemolymph, plays a central role in such social organization of the honey bee colony, influencing social behavior, stress resilience, immunity, and longevity (Amdam et al. 2012). Besides guiding behavior and lifespan, vitellogenin also affects brood food production and worker specialization for pollen versus nectar collection (Nelson et al. 2007). Similarly, juvenile hormone acts to regulate the rate of behavioral development in honey bees, and guides many of the necessary conversions from a bee fostering brood to one collecting and processing nectar (Sullivan et al. 2000; Elekonich and Roberts 2005). Vitellogenin together with juvenile hormone are inversely correlated to the onset of foraging behavior in worker honey bees; younger nurse bees have low levels of juvenile hormone that increase with age and high vitellogenin, while older foraging bees exhibit the opposite relationship (Johnson 2010; Corona et al. 2007; Sullivan et al. 2000; Robinson 1987).

Other mechanisms are also involved in the behavioral transition of worker honey bees, including the queen and brood pheromones, which inhibit and accelerate worker maturation toward foraging, respectively (Doke et al. 2015; Bortolotti and Costa 2014). In addition, older foraging bees release a pheromone, ethyl oleate, that affects the behavioral maturation of young bees as outlined above, slowing down the transition of nurse bees into foragers (Leoncini et al. 2004). Complex models of division of labor in insect societies are proposed that integrate social, environmental, and nutritional factors, as well as both primer and releasing pheromonal mechanisms; the queen and brood pheromones, as well as vitellogenin and juvenile hormone are considered important parts of the story (Doke et al. 2015; Johnson 2010).

Overwintering Biology

Honey bees exhibit remarkable seasonal changes in their behavior and physiology in temperate climates with the changing seasons, as well as other tropical or arid climates following the cycle of flowering plants (Doke et al. 2015; Winston 1987). Winter conditions or a dearth in nectar and pollen production result in cessation of brood rearing and foraging, lifespans of the worker bee increase, and the active lifestyles of the worker and queen bees slow. Honey bees in temperate climates have stronger responses to seasonal changes compared to tropical areas (Winston 1987). Spring, summer, and fall worker bees (exhibiting an age-based division of labor) complete most of the hive tasks – except for reproduction – sequentially with only a short lifespan of around 30–45 days, while winter honey bees – also known as *diutinus* bees – become generalists working to maintain a thermoregulating cluster and may live up to eight months (Winston 1987; Doke et al. 2015; Johnson 2010). Levels of vitellogenin help shape such seasonal changes in honey bee behavior and physiology and have a positive impact on lifespan; the short-lived foraging worker bees produce less vitellogenin than nurse bees, wintering bees have the highest levels among workers, and the longest living queen bee has the highest levels of all castes (Amdam et al. 2012; Corona et al. 2007). Juvenile hormone levels (low in nurse and wintering bees) decrease in the fall and rise markedly in again in spring, suggesting that fall bees are already in their winter physiologic state, and overwintering bees return to a forager bee physiology in the spring (Doke et al. 2015; Fluri et al. 1982).

Temperature Regulation

Honey bees are heterothermic, meaning their individual temperature varies with the outside environment, yet they can also regulate body temperature via endothermic activity (Stabentheiner et al. 2010; Vidal-Naquet 2015). In addition, honey bee larvae and pupae have a low metabolic rate and cannot maintain thermal constancy in a changing environment; therefore, the brood are strongly dependent upon nest temperature regulation for development (Stabentheiner et al. 2010; Kronenberg and Heller 1982). Fortunately, the social organization of the honey bee hive facilitates colony level homeostasis including the migration activity of the worker bees within the nest and various bee behaviors. Other factors including the size and insulating properties of the nest also affect colony thermoregulation. During the winter, when cooling of the colony occurs, a portion of the worker bees (those older than two days) produce heat by movement of thoracic flight muscles while other nest mates (less than two days old) remain ectothermic (Stabentheiner et al. 2010). Such heat conduction among bees provides effective heat transfer and helps maintain the brood nest temperature in a precise range of 32–36 °C for normal development. Similarly, various strategies are used by social honey bees to cool the nest when the temperature becomes too high, and temperatures above 36 °C are reported to damage brood and negatively affect development (Vidal-Naquet 2015; Kronenberg and Heller 1982). Cooling of the nest cavity in warm conditions may occur via several mechanisms including dispersal of bees away from the brood, hive ventilation by worker bees fanning their wings or evaporative cooling using water at the hive entrance or within the colony (Winston 1987).

Part 2: Communication in Honey Bees

Renowned biologists, Hölldobler and Wilson (2009) stated, “the essence of social existence is reciprocal, cooperative communication.” Similarly, Seeley (1995) emphasized “all biologists are keenly aware of the amazing adaptive responses of cells and organisms, and are awed by the complexity of the underlying mechanisms of cellular and organismal physiology. But probably few biologists recognize that evolution has likewise endowed certain animal societies with impressive abilities and has fashioned elaborate mechanisms of communication and control inside these societies to produce their remarkable group-level skills.” Thus, advanced communication processes have evolved in honey bee colonies to coordinate interactions among its members and enable their intricate social organization. For example, honey bees have developed a complex system of communication strategies (including various dance behaviors, chemical messaging, vision and olfaction, and magnetic fields) to help them locate and store food, grow, build and maintain their colony, locate and defend their hive, attract other bees and for many other colony activities (Seeley 1995; Winston 1987).

Physical Communication

Waggle Dance

Of the diverse methods of communication known to humankind, the honey bee dance language is among the most intricate and well-studied in all of the animal kingdom (Hölldobler and Wilson 2009). The waggle dance of the honey bee is perhaps best known and permits foraging worker bees to precisely communicate the location and value of a food source. Through several ingenious experiments, the Austrian scientist Karl von Frisch first described this remarkable dance language where cooperative honey bees of a colony share food gathering information (von Frisch 1967). Once a honey bee locates a rich food source, say a patch of flowers, she returns to her hive and performs the waggle dance. Here she will stage her recent journey to the flowers, like an actor in a play providing her audience with the knowledge they will need to also find her valuable discovery. Bees following the dance will learn several important facts about the flower patch: the distance to the flowers, the direction they need to fly to locate them, the odor of the flowers, and their perceived value as a nectar or pollen food source (Seeley 1995). The dance is characterized by a small figure eight movement pattern on the vertical surface of the comb in the darkness of the colony often near the hive entrance (Figure 4.2). The honey bee first performs a straight waggle run followed by alternating right

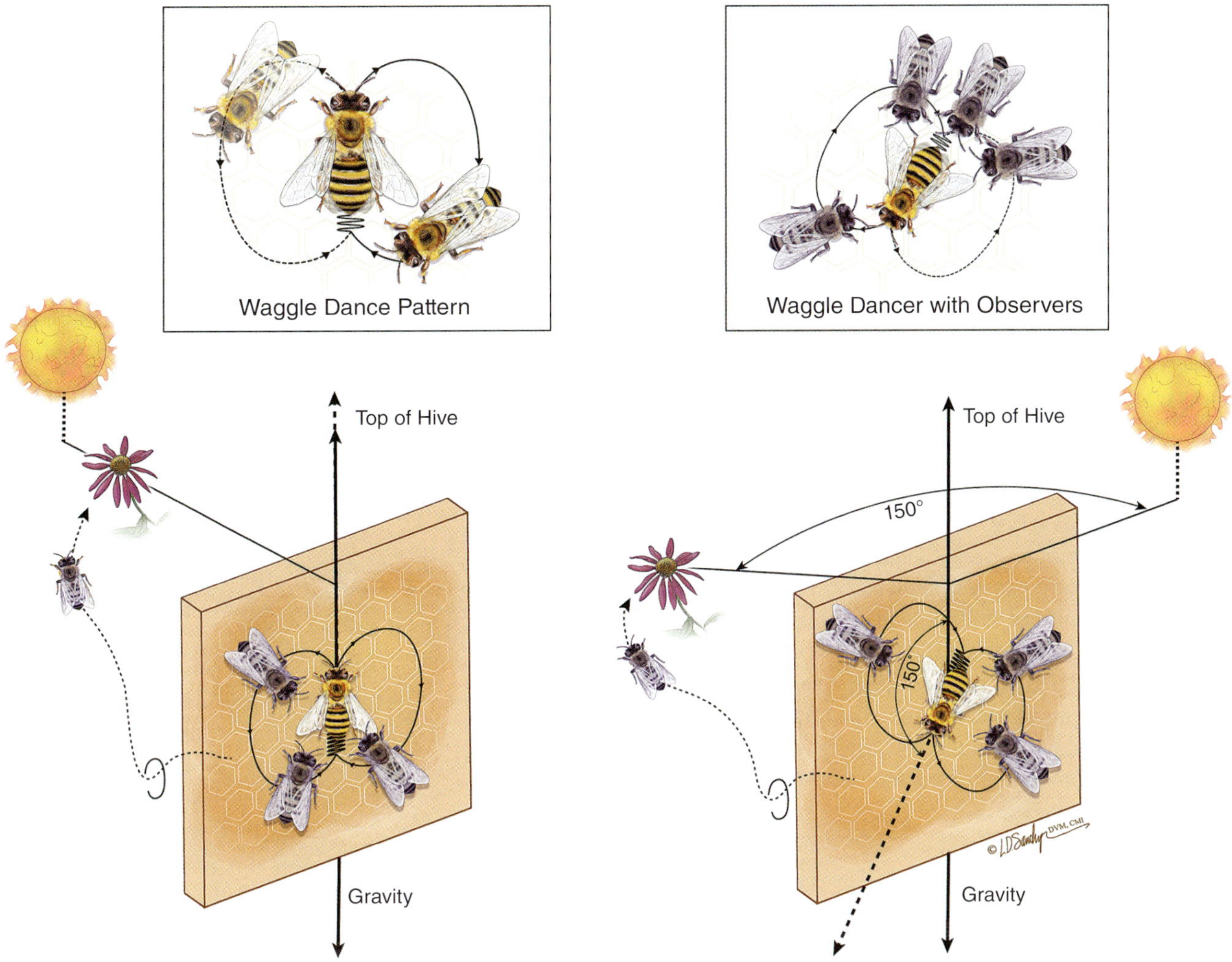

Figure 4.2 Physical communication in the honey bee using the Waggle Dance. The waggle dance occurs when a foraging bee returns to the hive and shares information on a food source; the bee performs a figure-8 dance on the vertical comb near the hive entrance. The center of the dance pattern signifies the *direction, distance,* and *quality* of the food source with reference to the sun (Winston 1987; Seeley 2010). *Direction:* on the left side of the diagram the flower food source is located directly in the path of the sun, and the bee waggles straight upward toward the top of the hive indicating the food source is directly in line with the sun. On the right side of the diagram, the food source is at an angle of 150 ° to the left of the vertical axis of the hive, and this tells the surrounding observing bees that the food source is located in a direction 150 ° to the left of the solar azimuth. *Distance:* the distance to the food source is specified by the duration of the waggle dance – approximately one second of body waggle symbolizes 1000 m of flight (Seeley 2010). *Quality:* the quality of the food source is indicated by the intensity and longevity of the waggle dance (Winston 1987). *Source:* © Lauren D. Sawchyn, DVM, CMI. Chapter: Physiology of the honey bee, authored by Rolfe M Radcliffe and illustrated by Lauren D. Sawchyn.

and left loops returning to the starting point, another waggle run, and so on (Seeley 1995). During the waggle run, the dancing bee shakes her abdomen back and forth while also vibrating her wings; the duration of the waggle run and dance tempo speaks for the distance to the flowers, and the direction of the waggle run on the vertical comb of the hive relative to gravity symbolizes the direction to the flowers with respect to relative position of the sun, or solar azimuth angle (Winston 1987; Seeley 1995; Tsujiuchi et al. 2007). Further, an increasing intensity of dancing behavior and length of dance performance within the hive communicates a higher quality food resource (Winston 1987). Other signals involving pheromones, tactile contact, dance sounds, comb vibrations, and temperature are also thought to be conveyed during the waggle dance. Even though much has been revealed about how this remarkable dance communication is achieved, mysteries still remain (Thom et al. 2007; Tsujiuchi et al. 2007).

Tremble Dance

Another dancing behavior also recognized by Karl von Frisch long ago was the tremble dance. He described the dance as a strange behavior – a neurosis – where the bees run about the combs making constant trembling movements of their bodies, similar to the disease known as St. Vitus' dance or chorea, but von Frisch was not able to identify its significance (Seeley 1995). The tremble dance is a long signal, persisting approximately 30 minutes, occurring throughout the broodnest portion of the hive. Seeley later deciphered its meaning, discovering that the tremble dance was used by foragers to recruit more nectar processing bees when a nectar flow results in increased nectar arriving at the hive without enough bees to help unload the food source (Seeley 1992). In a clever experiment Seeley switched the dancing behavior of foraging bees from performing the waggle to the tremble dance by only changing one variable: how long it takes for a foraging bee to unload its nectar resource. When a foraging bee experienced a search time of 20 seconds or less to find a food storing bee to unload its nectar, she performed the waggle dance, whereas a search time of 50 seconds or more resulted in the tremble dance (Seeley 1992). Through this work, Seeley and others determined that the tremble dance had multiple meanings: it stimulates a shift to processing nectar for bees working inside the hive, and to stop recruiting additional foragers for bees gathering nectar outside the hive. The waggle and tremble dances of foraging bees are now known to be complementary behaviors working in concert to keep a colony's rate of nectar collecting and processing balanced (Seeley 1995).

Shaking Dance

The shaking dance or signal is yet another method of communication performed by foraging worker bees and helps foragers increase the number of bees in a colony engaged in foraging during a rise in nectar supply or a high demand for food or both (Seeley 1995). A returning honey bee forager will perform this dance often in conjunction with the waggle dance as a way to entice resting bees to begin foraging, often following prolonged successful foraging or a period of nectar dearth. A honey bee transmitting this signal will literally shake a number of different bees in the hive (approximately 1 to 20 bees per minute) by vibrating her whole body in a dorso-ventral direction briefly for one to two seconds while holding the other bee tightly in her grasp (Seeley 1995). In contrast to the waggle dance enacted principally on the vertical hive comb near the colony entrance, the shaking signal is performed throughout the hive in an effort to persuade non-foraging bees involved in other hive activities to switch tasks and begin foraging.

Although the dancing behaviors of honey bees are most familiar, several other physical styles of communication are also used by this social insect to help convey messages both inside and outside of the colony (Winston 1987; Seeley 2010).

Chemical Communication

Complex social living necessitates a rich language, and the significant chemical language of social insects like the honey bee has been compared to the visual and auditory talents of the higher vertebrates (Bell and Carde' 1984; Slessor et al. 2005). Thomas Seeley (1995) reached the ensuing insight during his long summers spent working on the remarkable social physiology of the honey bee colony: ". . . the system of control devices found in a honey bee colony is extremely sophisticated and endows a colony with exquisite powers of adaptive response, both to internal changes and to external contingencies." Together with their notable dancing performances, chemical messages are fundamentally responsible for such extraordinary ability of the honey bee superorganism to adapt to changing conditions (Bortolotti and Costa 2014). However, it is the sociochemicals of the queen, adult worker bees, and brood that largely determine the complex social organization of the colony (Slessor et al. 2005; Jarriault and Mercer 2012).

Pheromones are chemical signals produced by a honey bee and released outside the body to effect a response (Free 1987; Slessor et al. 2005; Bortolotti and Costa 2014). These substances are used among colony members of all castes (queen, workers, drone, and brood) to coordinate hive activities. Such signals may be transferred via antennae contact, olfaction, food transfer, grooming, as well as trail marking or marking resources among others, and are involved in many functions including brood development, foraging, mating, defense, orientation, colony recognition, reproduction, swarming, and division of labor (Figure 4.3). Honey bees use two broad types of pheromones to communicate – *primer* and *releaser* pheromones (Free 1987; Slessor et al. 2005). A *primer* pheromone elicits a complex reaction in the receiver creating both behavioral and developmental changes. Such pheromones are important in the organization and cohesion of eusocial living and include the inhibition of reproduction; examples in the honey bee colony include the Queen Mandibular and Brood pheromones. Primer pheromones are well developed in social insects and act to maintain colony homeostasis. A *releaser* pheromone has a weaker effect and influences behavior only. Most worker bees utilize releaser pheromones for various hive functions including alarm and aggression, sex attraction, trail marking and recognition (Free 1987; Slessor et al. 2005). Importantly, pheromone communication within a honey bee colony is shaped by the complexity, synergy, context and dose of each signal and conveyed via both temporal and spatial distribution throughout the hive (Slessor et al. 2005).

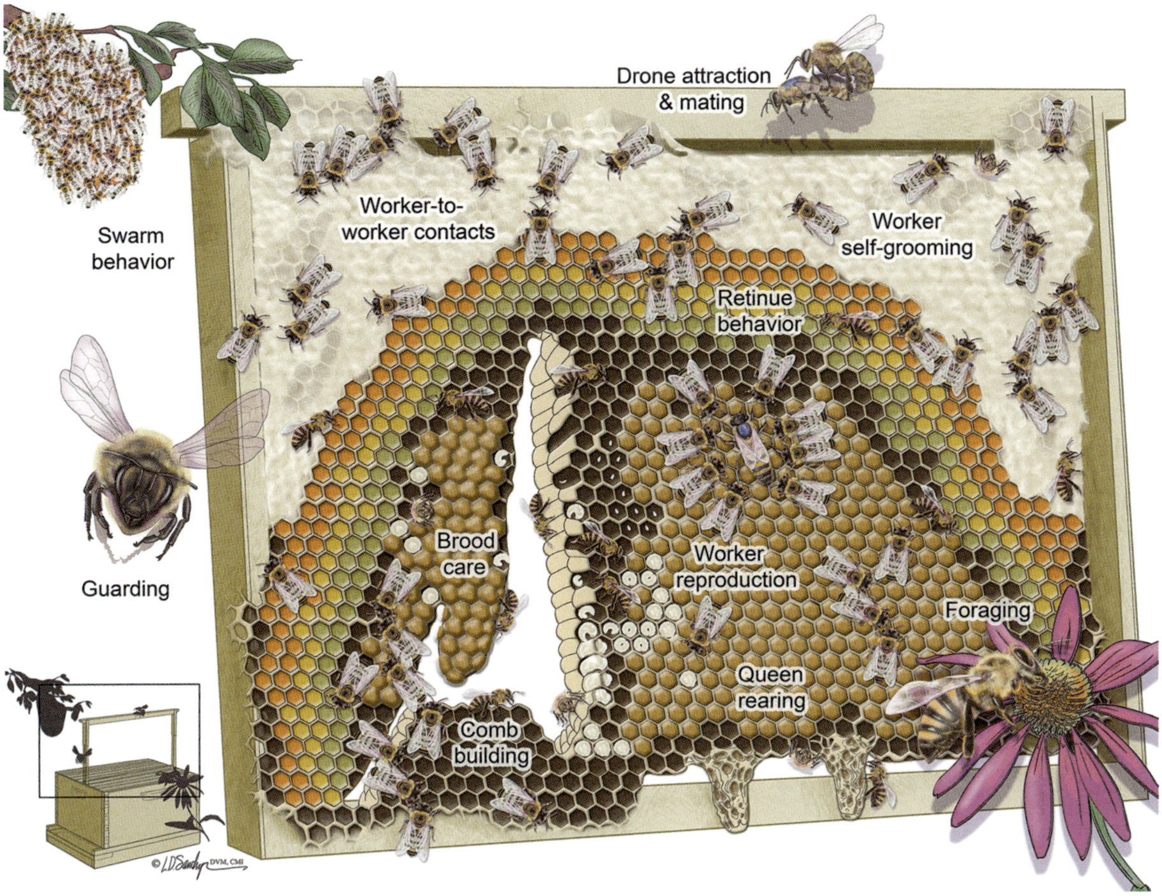

Figure 4.3 Life inside a honey bee hive is complex with many work activities, behaviors and functions occurring among the various worker, drone, and queen bees. Such social living demands effective communication for survival. Many of the hive activities and the ability of a colony to adapt to changing conditions are shaped by the chemical language of honey bees; pheromones may elicit behavioral or developmental changes among bees to maintain colony homeostasis and guide these hive functions. *Source:* © Lauren D. Sawchyn, DVM, CMI. Chapter: Physiology of the honey bee, authored by Rolfe M Radcliffe and illustrated by Lauren D. Sawchyn.

Queen Pheromones

The honey bee queen largely regulates colony activities, producing a variety of pheromones, known collectively as the queen signal, to support many hive functions such as cleaning, brood rearing, comb-building, guarding, and foraging (Free 1987; Slessor et al. 2005; Bortolotti and Costa 2014). Such pheromones also influence drone mating, swarm clustering and the queen retinue behavior, while also suppressing worker bee ovary function and egg laying, queen supersedure, and limiting the potential for queen rearing (Figure 4.4). When the queen is removed from a colony or dies, her absence is discovered with little delay and the colony swiftly initiates queen rearing because of the loss of queen pheromones (Free 1987; Slessor et al. 2005; Bortolotti and Costa 2014). After an extended time without a queen, the colony's worker bees stop completing their hive tasks, begin laying unfertilized eggs that mature into drone bees, and the colony declines toward collapse.

Queen mandibular pheromone, previously believed only secreted from the mandibular gland, has been most studied and is comprised of many different chemical compounds affecting many of the aforementioned hive behaviors (Free 1987; Slessor et al. 2005; Bortolotti and Costa 2014). Queen mandibular pheromone is now recognized as queen retinue pheromone because of its multiglandular origin (Slessor et al. 2005). This multicomponent queen pheromone attracts a continually shifting cohort of 6–10 young worker bees that feed and groom the queen (Free 1987). These young bees form the queen's retinue who, in addition to caring for the queen, collect and distribute her pheromones about the colony via antennal contacts and trophallaxis (Slessor et al. 2005; Jarriault and Mercer 2012). Queen retinue pheromone has many stimulatory properties beyond shaping the queen's court or retinue, including formation and steadiness of the swarm cluster, attracting drone bees and mating, influencing worker tasks and development, comb building, brood rearing, as well as foraging

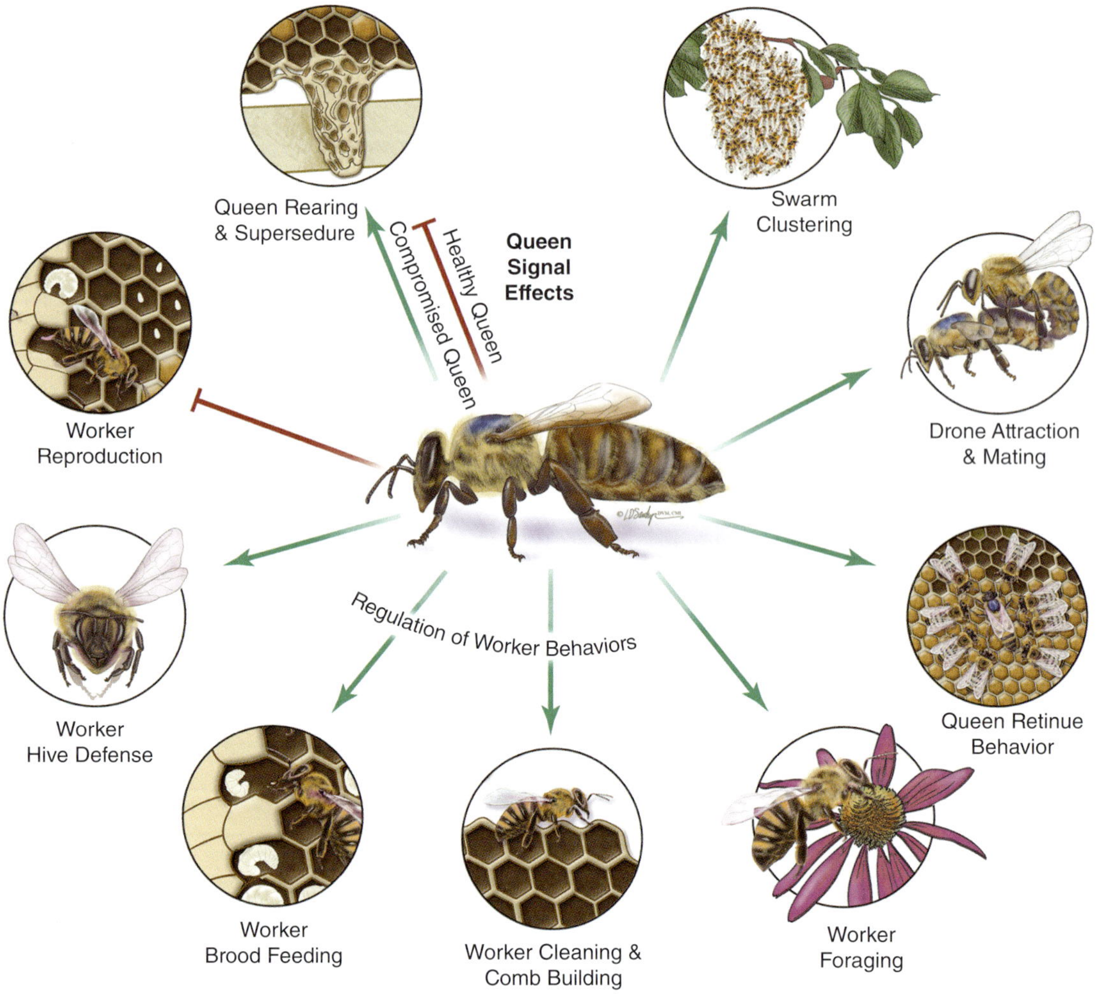

Figure 4.4 A colony of honey bees survives in large part because of the presence of a healthy queen. The queen honey bee produces several pheromones, known collectively as the queen signal, helping to regulate colony activities. Such pheromones support many important worker bee hive functions including cleaning, comb building, brood rearing, foraging behavior, and hive defense. Further, the queen signal is necessary for drone bee attraction and mating, maintaining the swarm cluster and queen retinue behaviors, as well as inhibiting worker bee reproduction and egg laying, queen supersedure, and queen rearing. *Source:* © Lauren D. Sawchyn, DVM, CMI. Chapter: Physiology of the honey bee, authored by Rolfe M. Radcliffe and illustrated by Lauren D. Sawchyn.

behavior. In addition, this queen pheromone is vital for the regulation of colony reproduction and worker physiology, inducing worker sterility when the structure of the hive – having a productive, fertile queen – favors the long-term genetic interests of the worker bees (Slessor et al. 2005; Princen et al. 2019).

Significant queen signal redundancy has been identified throughout the pheromone communication repertoire of the honey bee colony (Slessor et al. 2005; Princen et al. 2019). Besides the queen mandibular gland, other sources of the queen signal have been isolated from the tergal, labial, tarsal, Dufour's, and Koschevnikov glands (Slessor et al. 2005; Bortolotti and Costa 2014). Located beneath the abdominal tergites and more developed in the queen, the tergal glands secrete pheromones that support the function of the queen retinue pheromone. The tarsal glands, present in all three bee castes, release a footprint pheromone that in the queen inhibits queen cell construction by the worker bees (Bortolotti and Costa 2014). The Dufour's gland, closely associated with the sting apparatus in the female honey bee, is another part of the queen signal and also provides worker bees with information about queen fertility and reproductive potential (Dor et al. 2005; Bortolotti and Costa 2014). The Koschevnikov gland, located near the sting shaft of females, is yet another source of queen signal, and with gland degeneration beginning at one year of age contributes to the failing of aging queens (Bortolotti and Costa 2014).

Worker Pheromones

Worker honey bees also make use of a variety of pheromones, and most glands in both the queen and worker honey bees are fully developed yet secrete different pheromones (Free 1987; Ritter 2014; Bortolotti and Costa 2014). Further, gland maturity in worker bees follows a temporal pattern that parallels the changing tasks and activities of the worker bee within the colony. Worker bees employ several pheromones during their lifetime including alarm, Nasonov, footprint, and forager signals to help in colony defense, for orientation and marking sites, and for identifying the colony (Bortolotti and Costa 2014; Ritter 2014). Worker mandibular pheromone is also produced when needed in queenless colonies to suppress ovarian development in other workers when egg-laying workers or psuedoqueens emerge (Bortolotti and Costa 2014).

The defense behavior of a honey bee colony is well known and alarm pheromones released from worker bees help guide this response (Bortolotti and Costa 2014; Free 1987; Breed et al. 2004). Defensive behavior tasks occur prior to foraging in worker bees and two types of behaviors are recognized: guarding and defending (Breed et al. 2004). Guard bees watch and monitor the hive entrance, inspect all who wish to enter, and reject non-nestmates; soldier bees fly out in the face of danger, and chase, bite, or sting intruders. Honey bees are able to recognize nestmates of the same colony or non-nestmates of other colonies through chemical signals present in the lipid covering of the insect cuticle. Both the sting apparatus, including the Koschevnikov gland, as well as the mandibular gland produce pheromones that evoke defensive behavior through the recruitment and amplification of other worker bees (Bortolotti and Costa 2014).

Worker honey bees produce a pheromone from the Nasonov gland that is restricted only to this caste (Bortolotti and Costa 2014; Free 1987). The gland secretion is composed of several volatile compounds that function to attract other bees, and is released during three main contexts: during swarm clustering, marking their hive entrance, and marking of water and foraging sources (Bortolotti and Costa 2014; Free 1987). The worker footprint pheromone, secreted from the tarsal glands of worker honey bees, may share the role of marking the hive entrance and food sources. The Nasonov pheromone may also be used for recruiting other workers to help develop queen cups during the process of queen rearing.

A honey bee colony has a social organization characterized by a division of labor among the worker honey bees – worker polyethism – that changes over time (Winston 1987; Seeley 1995). During their early life (0–20 days) they work inside the colony cleaning cells and caring for the brood, receiving nectar and handling pollen, comb building, and tending the queen among other tasks; as they age (20–45 days) they begin working outside the nest ventilating the hive and guarding its entrance, as well as commencing foraging flights (Winston 1987). The control of worker polyethism has been a mystery, yet recent evidence suggests that a pheromone, produced by older foraging sister worker bees, delays the age of onset of foraging and other age-related tasks (Leoncini et al. 2004; Slessor et al. 2005).

Drone Pheromones

The male or drone honey bee differs significantly from his worker sisters, and with their singular reproductive role of sperm production and mating, drones produce very few pheromonal signals (Bortolotti and Costa 2014; Hrassnigg and Crailsheim 2005). The most important pheromone of drone bees is released from their mandibular gland, and functions to attract other drone bees to the male congregation areas in preparation for mating. Similar to female honey bees, drone bees also possess unique cuticular pheromone signals that allow worker bees to distinguish between the sexes and drone bees of different ages (Bortolotti and Costa 2014). In addition, the regulation of drone brood production and the rejection of adult drone bees from a colony are likely under pheromonal control (Free 1987).

Brood Pheromones

Larvae within the honey bee colony release pheromones that are important for the regulation of brood care and development, worker behavioral transformation, and worker reproduction (Bortolotti and Costa 2014; Free 1987). Different components of the brood pheromone are released as a function of the caste and larval age, and as such guide the nurse bees to provide the appropriate response during brood development (Bortolotti and Costa 2014). Brood pheromone affects the colony foraging behavior in a dose-dependent manner according to the age of the larvae. Young larvae (having little nursing needs) stimulate foraging and pollen collection, while older larvae (having greater nursing needs) delay foraging and instead promote increased brood care (Bortolotti and Costa 2014). Honey bee eggs, larvae, and pupae also stimulate pollen collection, and such brood pheromones are important modulators of colony growth (Bortolotti and Costa 2014; Free 1987). Further, brood pheromones work together with the queen signal to inhibit worker ovary development.

Acoustic Communication

Most of the social life of honey bees occurs within the darkness of the hive where vision plays a limited role (Kirchner 1993). Undeniably, pheromone communication provides the foundation for communication throughout this dark world of the honey bee (Slessor et al. 2005). However, honey bees can also detect and communicate via sound and vibrations that are transmitted throughout the beeswax structure of the hive (Hrncir et al. 2005; Kirchner 1993). Several types of acoustic communications are known within the honey bee colony and such signals may be transmitted via both substrate vibrations and airborne sound. These include the "tooting" and "quacking" signals among queens during the course of swarming, worker piping through dance language and swarming, as well as the hissing or shimmering behavior of honey bees during colony defense (Hrncir et al. 2005). Even though our knowledge of the complex interactions of the chemical, tactile, and acoustic sounds exploited by honey bees continues to advance, parts of their fascinating social lifestyle remain yet unknown.

Conclusion

When beekeepers and veterinarians understand the fascinating biology of the honey bee, including their anatomy and physiology, they will learn to appreciate the importance of the individual bee and the collective colony. They will both share a respect for the intricate and complex work, actions, and behaviors that allow thousands of relatives to share a common home and protect a cooperative future. And perhaps most importantly, the beekeeper and their veterinary colleague will build a foundation for effective communication, intervention and prevention of management problems and disease, helping to secure the future of this essential resource.

References

Amdam, G.V., Fennern, E., and Havukainen, H. (2012). Vitellogenin in honey bee behavior and lifespan. In: *Honeybee Neurobiology and Behavior: A Tribute to Randolf Menzel* (eds. C.G. Galizia, D. Eisenhardt and M. Giurfa). Switzerland: Springer Nature.

Arrese, E.L. and Soulages, J.L. (2010). Insect fat body: energy, metabolism, and regulation. *Annual Review of Entomology* 55: 207–225.

Bell, W.J. and Carde´, R.T. (1984). Preface, Page xiv. In: *Chemical Ecology of Insects* (eds. W.J. Bell and R.T. Carde´). London: Chapman and Hall.

Bortolotti, L. and Costa, C. (2014). Chemical communication in the honey bee society Chapter 5. In: *Neurobiology of Chemical Communication* (ed. C. Mucignat-Caretta). Boca Raton, FL: CRC Press/Taylor & Francis.

Breed, M.D., Guzman-Novoa, E., and Hunt, G.J. (2004). Defensive behavior of honey bees: organization, genetics and comparisons with other bees. *Annual Review of Entomology* 49: 271–298.

Brockmann, A. and Robinson, G.E. (2007). Central projections of sensory systems involved in honey bee dance language communication. *Brain Behav Evol* 70: 125–136.

Corona, M., Velarde, R.A., Remolina, S. et al. (2007). Vitellogenin, juvenile hormone, insulin signaling and queen bee longevity. *Proceedings of the National Academy of Sciences of the United States of America* 104 (17): 7128–7133.

Davidson, E.W. (1973). Ultrastructure of American foulbrood disease pathogenesis in larvae of the worker honey bee, *Apis mellifera*. *Journal of Invertebrate Pathology* 21: 53–61.

DeGrandi-Hoffman, G. and Chen, Y. (2015). Nutrition, immunity and viral infections in honey bees. *Current Opinion in Insect Science* 10: 170–176.

Döke, M.A., Frazier, M., and Grozinger, C.M. (2015). Overwintering honey bees: biology and management. *Current Opinion in Insect Science* 10: 185–193.

Dor, R., Katzav-Gozansky, T., and Hefetz, A. (2005). Dufour's gland pheromone as a reliable fertility signal among honeybee (*Apis mellifera*) workers. *Behavioral Ecology and Sociobiology* 58 (3): 270–276.

Elekonich, M.M. and Roberts, S.P. (2005). Honey bees as a model for understanding mechanisms of life history transitions. *Comparative Biochemistry and Physiology Part A: Molecular & Integrative Physiology* 141: 362–371.

Ellis, J.D. (2015). The internal anatomy of the honey bee. *American Bee Journal*: 971–974.

Evans, J.D., Aronstein, K., Chen, Y.P. et al. (2006). Immune pathways and defense mechanisms in honey bees *Apis mellifera*. *Insect Molecular Biology* 15 (5): 645–656.

Fefferman, N.H., Traniello, J.F.A., Rosengaus, R.B., and Calleri, D.V. (2007). Disease prevention and resistance in social insects: modeling the survival consequence of immunity, hygienic behavior and colony organization. *Behavioral Ecology and Sociobiology* 61 (4): 565–577.

Fluri, P., Lüscher, M., Wille, H., and Gerig, L. (1982). Changes in weight of the pharyngeal gland and haemolymph titers of juvenile hormone, protein and vitellogenin in worker honey bees. *Journal of Insect Physiology* 28 (1): 61–68.

Free, J.B. (1987). *Pheromones of Social Bees*. Ithaca, NY: Cornell University Press.

von Frisch, K. (1967). *The Dance Language and Orientation of Bees*. Cambridge, MA: Harvard University Press.

Hölldobler, B. and Wilson, E.O. (2009). *The Superorganism: The Beauty, Elegance and Strangeness of Insect Societies*. New York, NY: W. W. Norton & Company.

Hrassnigg, N. and Crailsheim, K. (2005). Differences in drone and worker physiology in honeybees (*Apis mellifera*). *Apidologie* 36 (2): 255–277.

Hrncir, M., Barth, F.G., and Tautz, J. (2005). Vibratory and airborne-sound signals in bee communication (hymenoptera). In: *Insect Sounds and Communication: Physiology, Behavior, Ecology and Evolution* (eds. S. Drosopoulos and M.F. Claridge). Boca Raton, FL: CRC Press/Taylor & Francis.

Jarriault, D. and Mercer, A.R. (2012). Queen mandibular pheromone: questions that remain to be resolved. *Apidologie* 43 (3): 292–307.

Johnson, B.R. (2010). Division of labor in honey bees: form, function and proximate mechanisms. *Behavioral Ecology and Sociobiology* 64 (3): 305–316.

Kirchner, W.H. (1993). Acoustical communication in honeybees. *Apidologie* 24 (3): 297–307.

Kronenberg, F. and Heller, H.C. (1982). Colonial thermoregulation in honey bees (*Apis mellifera*). *Journal of Comparative Physiology* 148: 65–76.

Kurze, C., Routtu, J., and Moritz, R.F.A. (2016). Parasite resistance and tolerance in honey bees at the individual and social level. *Zoology* 119: 290–297.

Leoncini, I., Le Conte, Y., Costaglioa, G. et al. (2004). Regulation of behavioral maturation by a primer pheromone produced by adult worker honey bees. *Proceedings of the National Academy of Sciences of the United States of America* 101 (50): 17559–17564.

Molnar, C. and Gair, J. (2015). Animal structure and function, digestive system Chapter 15.1. In: *Concepts of Biology – 1st Canadian Edition*. BCcampus, Victoria, B.C.

Nelson, C.M., Fondrk, M.K., Page, R.E. Jr., and Amdam, G.V. (2007). The gene vitellogenin has multiple coordinating effects on social organization. *PLoS Biology* 5 (3): e62.

Princen, S.A., Oliveira, R.C., Ernst, U.R. et al. (2019). Honeybees possess a structurally diverse and functionally redundant set of queen pheromones. *Proceedings of the Royal Society B: Biological Sciences* 286 (1905): 20190517.

Raymann, K. and Moran, N.A. (2018). The role of the gut microbiome in health and disease of adult honey bee workers. *Curr Opin Insect Sci* 26: 97–104.

Ritter, W. (2014). *Bee Health and Veterinarians. World Organization for Animal Health*. Paris: OIE.

Robinson, G.E. (1987). Regulation of honey bee age polyethism by juvenile hormone. *Behavioral Ecology and Sociobiology* 20: 329–338.

Seeley, T.D. (1992). The tremble dance of the honey bee: message and meanings. *Behavioral Ecology and Sociobiology* 31: 375–383.

Seeley, T.D. (1995). *The Wisdom of the Hive: The Social Physiology of Honey Bee Colonies*. Cambridge, MA: Harvard University Press.

Seeley, T.D. (2010). *Honeybee Democracy*. Princeton, NJ: Princeton University Press.

Slessor, K.N., Winston, M.L., and Le Conte, Y. (2005). Pheromone communication in the honeybee (*Apis mellifera L.*). *Journal of Chemical Ecology* 31 (11): 2731–2745.

Snodgrass, R.E. (1956). *Anatomy of the Honey Bee*. Ithaca, NY: Cornell University Press.

Stabentheiner, A., Kovac, H., and Brodschneider, R. (2010). Honeybee colony thermoregulation – regulatory mechanisms and contribution of individuals in dependence on age, location and thermal stress. *PLoS One* 5 (1): e8967.

Sullivan, J.P., Fahrbach, S.E., and Robinson, G.E. (2000). Juvenile hormone paces behavioral development in the adult worker honey bee. *Hormones and Behavior* 37 (1): 1–14.

Thom, C., Gilley, D.C., Hooper, J., and Esch, H.E. (2007). The scent of the waggle dance. *PLoS Biol* 5 (9): e228.

Tsujiuchi, S., Sivan-Loukianova, E., Eberl, D.F. et al. (2007). Dynamic range compression in the honey bee auditory system toward waggle dance sounds. *PLoS One* 2 (2): e234.

Vidal-Naquet, N. (2015). *Honeybee Veterinary Medicine: Apis mellifera L.* Sheffield: 5m Publishing.

Wigglesworth, V.B. (1972). *The Principles of Insect Physiology*. London: 7th ed. Chapman and Hall.

Winston, M.L. (1987). *The Biology of the Honey Bee*. Cambridge, MA: Harvard University Press.

5

The Honey Bee Queen

*Randy Oliver**

ScientificBeekeeping.com

The Queen's Function in the Hive

The queen bee holds a special mystique to beekeepers. Although she is treated as royalty, she is not responsible for much decision making in the colony. From a biological standpoint there is actually nothing particularly exceptional about her – she's a relatively normal reproductive female insect.

Practical application: The exceptional individuals in eusocial insect colonies are not their queens, but rather the *non-reproductive* female worker caste (and subcastes) – the queen being the sole member of the female reproductive caste. All the functionally-sterile worker caste members devote their energy and resources to support that singular egg-producing "queen," who acts not only as the "ovary" of the honey bee superorganism, but also as the pheromonal "gravitational center" of the colony (Figure 5.1).

The queen is indeed the heart of the hive – not only due to her being the mother of all the other members of the colony, but also by secreting pheromones that induce colony cohesiveness, suppress ovary development in the rest of the females, and perhaps most importantly, to provide an "honest signal" as to her reproductive status (as opposed to merely suppressing competitive egg laying) (Niño et al. 2013). Queen pheromone (actually a variable mixture of several pheromones) appears to also suppress the feeding of most female larvae, thus epigenetically resulting in them becoming functionally-sterile "workers" rather than reproductive queens.

A young, well-mated, healthy queen will attract a strong "retinue" of nearby nurse bees that offer her jelly, groom her, and transfer her pheromones to the rest of the members of the colony (Figure 5.2). An experienced beekeeper can immediately recognize, by colony organization and temperament, whether it has a queen that is producing abundant queen pheromone.

Practical application: Despite her royal treatment and being the mother of all the workers in the hive, the queen enjoys no particular fealty from her functionally-sterile daughters. She is entirely fungible, and the colony will replace her at the drop of a hat should it sense, via pheromonal or other cues, that either she or the colony is failing.

Queen Development and Performance

Queen Larval Development

A queen is reared in large, vertical cell as opposed to the smaller horizontal cell of a worker. All female larvae are fed essentially the same diet for the first 24 hours, up to which point any could potentially become a queen. After that, developmental paths switch, with larvae chosen by the nurses to be a queen being fed to excess, with much of the jelly not being consumed until after her cell is "sealed."

Although the common conception is that queen larvae are fed a special "royal jelly," whereas worker larvae are fed "worker jelly" and pollen, this explanation is controversial, especially considering how logistically difficult it would be for the nurse bees which are constantly moving.[1] The most

* All photographs were taken by the Author.

1 For example, the way that queens are mass reared is by placing 50 transferred worker larvae to a queenless group of nurse bees that were producing only "worker" jelly. It is difficult to understand how they could immediately shift to producing a jelly made up of components unique to "royal" jelly. And there is no evidence that nurse bees intentionally feed pollen to worker larvae, it more likely being an inadvertent contaminant from their mouthparts (Simpson 1955).

Honey Bee Medicine for the Veterinary Practitioner, First Edition. Edited by Terry Ryan Kane and Cynthia M. Faux.

Figure 5.1 The queen functions not only as the ovary of the honey bee superorganism, but also as the pheromonal "heart" of the hive, critical for colony cohesiveness. In a rapidly-growing colony, roughly a thousand of her daughters die each day from natural aging and mortality. Thus, in order to maintain colony growth, the queen must not only lay an adequate number of eggs to replace those lost workers, but also enough for population increase, as many as 1500 a day.

Figure 5.2 Whenever a "good" queen pauses on the comb, an ad hoc group of adjacent nurse bees will turn to face her, offering her food, and antennating her to pick up her pheromones (Collison 2017). This ring of attendants is called a "retinue." Since the advent of varroa, some are suggesting that queens these days don't seem to attract retinues the way they used to. It is possible that this may correlate with what appear to be reports of greater rates of queen failure.

Figure 5.3 A well-fed queen is an egg-laying machine, capable of producing an egg per minute, 24 hours a day (and up to double that rate in bursts). If she cannot locate an empty cell, she will just drop the egg (as this queen is doing), which will be consumed and recycled by a nurse bee.

parsimonious (but debatable) explanation is that the nurses adjust the proportions of the three components of jelly dependent upon the age and caste of the larva being fed. Those components are (i) a protein-rich secretion from the hypopharyngeal glands, (ii) a lipid-rich secretion from the mandibular glands, and (iii) the amount of nectar added. Queen larvae receive not only a far greater amount of jelly, but higher proportions of lipids and sugars (Winston 1987; Wang et al. 2016). Queen larvae thus grow more rapidly than do worker larvae, to a greater size, and emerge at an earlier age.

Queen Diet

Similar to workers and drones, upon emergence, a young queen may consume nectar or honey, but unlike them, she does not consume pollen as a protein source,[2] instead depending upon receiving a diet of jelly begged from nurse bees for her entire life. A queen can lay more than her body weight in eggs in a day, and thus requires an exceptionally nutritious diet. The jelly diet is so perfect that a queen's feces look like droplets of water. Due to this diet, queens carry a different endosymbiotic microbial community structure than do workers (Anderson et al. 2018), and typically live to a much longer age (Figure 5.3).

Queen Mating

A queen is not really a queen until she has successfully mated, and is barely recognized by the workers until then.

2 The author, unpublished; if any pollen is accidentally consumed, it passes through the gut undigested.

A few to several days after emergence, a "virgin queen" (Figure 5.4) flies out to mate with about 15 drones (often more) in what is known as a "drone congregation area," during a one to two day period when the temperature is above 70 °F (Figure 5.5). She returns to the hive exhibiting the torn endophallus of the last drone with which she mated, which must be removed by the workers (Figure 5.6). The workers then begin paying attention to her (Figure 5.2) (Richard et al. 2007), and she shifts her pheromonal output to signal that she is adequately mated. A few days later (typically 10–14 days after emergence, weather permitting), the queen begins to lay eggs (Figure 5.7).

Practical application: It's important to know that if a virgin is constrained from mating by weather, that the chance of her ever successfully being mated decreases greatly after three weeks.

After mating, the queen homogenizes the received semen, and discards roughly 95% of it, holding the remaining mixed spermatazoa in a clear sac called the spermatheca, in which the spermatozoa can remain viable for years.

Practical application: Temperature extremes, or certain insecticides and beekeeper-applied miticides, may diminish the viability of the spermatozoa, causing early failure of the queen. The seminal fluid received may confer some immunity to pathogens (as well as pathogen exposure), and affect the spermatozoa of other drones.

Figure 5.4 To the untrained eye, virgin queens are difficult to spot. Look for her longer legs, slightly more angular "hips," and rapid movement atypical of the rest of the bees on the comb.

Figure 5.6 A freshly-returned no-longer-a-virgin exhibiting "mating sign" – the torn endophallus of the last drone to successfully mate with her. This will soon be removed by the workers.

Figure 5.5 The leading edge of a "drone comet" chasing a "virgin" queen. At the top left you can see a drone starting to mount the queen. Slightly lower is what appears to be the previously-successful drone paralyzed and falling to his death after his explosive ejaculation.

Figure 5.7 Typically 10–14 days after emergence (dependent upon weather), a queen will be mated and commence laying the first of the half a million eggs that she may produce over her lifetime. Each cell in the above photo contains a single egg.

Practical application: Many novice beekeepers, resistant to put on their reading glasses, are unable to identify eggs in the combs. To see eggs, face toward your shadow, which will allow sunlight to come from behind you to illuminate the bottoms of the cells. Don't confuse white reflections with eggs – eggs look like white grains of rice standing on end.

Be aware that an egg may carry virions on its shell (notably Deformed Wing Virus), but inside carry proteins that confer transgenerational immune priming to her offspring (Salmela et al. 2015).

Queen Performance

A queen's performance is mainly measured by how eggs she lays each day.

Practical application: A queen's potential performance is often throttled by the number of prepared, thermoregulated, empty brood cells available in the hive. A good queen cannot exhibit her full laying capacity until the cluster covers at least 10 deep combs, and even then she would be limited if there were appreciable amounts of honey or beebread in those combs. The exceptional queen can nearly completely fill 10 deep combs with brood.

Young queens are typically more "exuberant" in their egg-laying than are older queens (newly-mated queens may even lay multiple eggs in a cell if there is not adequate room in the cluster). Young queens as a rule outperform older queens, although many queens are highly productive in their second year (which then often leads to swarming).

The performance of a queen is based upon a few main factors (listed in approximate order of importance):

1) **How well she was fed and cared for during her larval development, which determines her egg-laying capacity,**[3]
2) **How well she was mated – the number of viable spermatozoa in her spermatheca, and**
3) **Environmental factors, such as chilling and heating during shipment, or pathogen or pesticide exposure in the colony.**

A queen's potential performance is largely fixed by how lavishly she is fed jelly during the approximately four days between the emergence of her larva from its egg until her cell is sealed (she continues to feed and grow after sealing). And then she must get properly mated.

3 Miticide residues in her developmental cell may later affect queen performance (Rangel and Tarpy 2016).

Queen Performance vs. Colony Performance

The queen's contributions to colony "colony performance" are:

1) The number of eggs that she produces,
2) Her genetics, which then supply half the genetics to each of her daughters – the workers,
3) The genetics of the guys she mated with – the other half of the genetic equation, and
4) Some passing on of transgenerational epigenetic immunity.

After that, it's up to her offspring. Colony-level performance is mostly the result of how well the genetically-diverse patrilines of workers work as a "team."

Practical application: In general, a well-reared, well-mated queen of average genetics will outperform a poorly reared, poorly mated queen of the best genetics. After that, it's largely a matter of chance of how the patrilines of daughters happen to function together as a "team" (Figure 5.8), "performance" being more the result of the workers than the queen.

Practical application: Unlike other livestock, honey bee queens are polyandrous, resulting in a colony of bees consisting of a mix of patrilines of half-sisters, each fathered by one of the many drones with which their mother mated. Thus, there will always be genetic differences and diversity, even with colonies headed by sister queens. The performance of each colony as a whole is thus a matter of chance, similar to putting together a sports team of players (each patriline of worker bees) without knowing how well they will perform as a group (the colony). Bottom line: you might advise beekeepers to start in spring

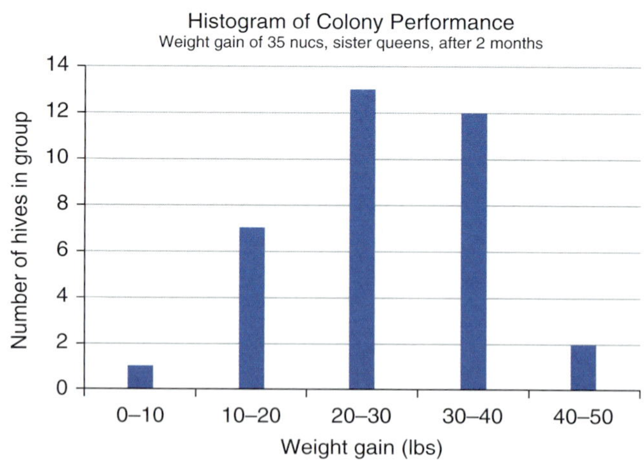

Figure 5.8 A histogram of colony performance (weight gain) of 35 colonies, started with sister queens mated and managed identically in one of the author's yards. Note the normal distribution of performance.

with twice as many colonies as they hope to take through the winter, and not waste their time on those that don't perform well. Some commercial beekeepers (the author included) simply cull any poor performers.

Practical application: On the other hand, many a poorly-performing colony can completely turn around once it clears itself of disease, or enjoys better nutrition. If honey production is not the beekeeper's main goal, they may find joy in helping a struggling colony to get back on its feet.

Queen Longevity

Beekeepers often refer to a queen's "age" in years. But biologically, chronology has little to do with it – a queen's "age" is a function of the number of viable spermatozoa remaining in her spermatheca, or more specifically, her ability to fertilize each egg laid in a worker cell (Baer et al. 2016).

Practical application: Thus, in a colony in a commercial operation in California or Florida, supplementally fed for near year-round broodrearing, a queen may begin to run out of spermatozoa in the late summer of her second year. On the other hand, a queen in a cold-winter area, in which she rests for five months of the year, may be productive for several seasons.

Queen Succession

A colony of bees is theoretically immortal, but not so its queen. In the natural state, with colonies living in tree cavities, the queen is replaced at least once a year (due to swarming). And prior to varroa it was not uncommon for a relatively unmanaged hive to live for many years, also replacing its queen without help from the beekeeper.

Practical application: There are three ways in which a colony can replace its queen, each due to different circumstances:

- **Supersedure – when the workers build a *supersedure cell* to smoothly replace a failing queen.**
- **Swarming – in which a colony in a crowded cavity builds *swarm cells*, and then divides by fission, with roughly half the workers flying off with their mother, leaving behind daughters in those cells to take her place.**
- **Emergency – when workers respond to the sudden loss of a queen by converting a young larva in a worker cell into an *emergency queen*.**

There is much overlap involved in the three methods above; the vet needs to be familiar with each.

Supersedure

A colony normally replaces an aging or failing queen by a process called "supersedure." The main factors that determine when a queen gets superseded are when she starts to run out of viable sperm with which to fertilize worker eggs, or by colony stress.

Practical application: Bees appear to "blame" their queen if the colony becomes seriously stressed by disease or parasites, which then triggers supersedure.

Worker bees often build "cell cups" on the face of a brood comb (Figure 5.9), but this does not necessarily mean that they are preparing to supersede or swarm. An "aging" queen will "willingly" lay an egg in a prepared cup, at which point the workers *may* turn it into a queen cell.

Practical application: The presence of cell cups does not indicate that a colony is about to supersede their queen. Even if supersedure larvae initially get fed, that doesn't mean that supersedure is inevitable, since the colony will often tear the cells down before emergence. Russian bees are noted for their continual starting of supersedure cells.

A mature supersedure cells looks similar to a peanut stuck to the side of the comb (Figure 5.10).

Practical application: beekeepers often ask whether they should destroy supersedure cells. I tell them that the bees likely have a better idea of the quality of

Figure 5.9 Preconstructed natural "queen cups." It is not unusual to see these in a colony. Their presence does not necessarily mean that the queen is about to be replaced. Even when there is an egg or a larva with jelly in a queen cell, they are often destroyed by the bees during that new queen's development.

Figure 5.10 A typical supersedure cell. Supersedure cells are generally found singly, as opposed to swarm or emergency cells, of which there are generally several. They are also generally found near the center of the brood area, rather than along the edges, as are swarm cells, or scattered, as are emergency cells.

Figure 5.11 A dissected supersedure cell after emergence, showing that the egg had been laid in a preconstructed cell cup (as for a swarm cell), as opposed to the cell being postconstructed from a worker cell (as for an emergency cell). We are not yet clear as to the proportion of supersedure cells that are produced by these two different routes. There is also some evidence that workers may be able to transfer an egg or young larva from a worker cell to a queen cup (Punnett and Winston 1983).

their queen than they do. I see no reason to remove supersedure cells.[4]

Not all supersedure cells come from prepared cell cups. Some that I have dissected originated from worker cells (Figure 5.11), so the line between supersedure and emergency cells may be fuzzy.

4 Other than when you need to prevent a colony from replacing an instrumentally-inseminated breeder queen.

Figure 5.12 Two laying queens on the same comb face (the lower one is a bit above the finger). On the other hand, I occasionally see a failing queen being "balled," suggesting that her own workers are killing her.

Although sister queens will invariably fight to the death, an emerged daughter queen and her mother often peacefully coexist.[5] It is not uncommon to observe mother and daughter laying eggs side by side for days or weeks, with the mother eventually disappearing from the hive (Figure 5.12).

Practical application: Although it is generally the case, never assume that there is only a single queen in a hive. The only way to be sure that *all* queens have been removed is to shake the bees through an excluder "sieve box" (Figures 5.13 and 5.14).

"Balling" of the Queen

In the process of supersedure, the workers often show no mercy toward a no-longer-wanted queen. Bees kill a queen in the same way that they kill invading hornets, by biting her and forming a tight, heads-in, walnut-sized ball of bees around her (Figure 5.15). The bees then produce heat and direct it to their heads to "cook" the queen. Thus, even if one manages to "save" a queen being seriously balled, she rarely survives.

Practical application: There is great demand for "early mated" queens. But the weather does not necessarily cooperate, resulting in a percentage of early queens being poorly mated. And during shipping, temperature stress can cause loss of viability of the spermatozoa. Whatever the reason, many beekeepers report that newly-introduced, or package bee queens, get superseded within the first month (this is not necessarily a problem).

5 There is strong evolutionary pressure for a queen to eliminate genetic competition from her sisters. There wouldn't be such pressure for a daughter to eliminate her mother, or vice versa.

Figure 5.13 Shaking bees through a sieve box to recover any queens. This is important when setting up a cell-builder colony, as above, in order to mass-produce emergency queens from selected mothers.

Figure 5.14 The thorax of a queen (center) or drone (lower right) is too broad to pass through a queen excluder.

Queen "Failure"

"Queen failure" is a nebulous and poorly-defined term, nowadays given in surveys as a choice for the cause of colony mortality. As such, it is often listed as the most

Figure 5.15 This aged queen (note the well-polished thorax) became a "drone layer." This photo shows her workers aggressively attempting to "ball" her to kill her.

common cause of colony loss by beekeepers (BIP 2019). This is somewhat surprising, since "back in the day," colonies tended to quietly and efficiently replace their queens via supersedure without any help from the beekeeper, and when I review older beekeeping textbooks, the term "queen failure" isn't mentioned.

Practical application: Infection by *Nosema apis* used to be strongly associated with early queen *supersedure* (Farrar 1947), but this does not appear to be the case with *N. ceranae,* which has largely supplanted its cousin. Thus, treatment with fumagillin against nosema may no longer be necessary to protect queens.

Since the arrival of varroa, the stress from greater virus exposure, as well as miticide residues in the combs may be having adverse effects upon queen survival. Similarly, residues of some agricultural insecticides may affect queen longevity.

Practical application: The question may be, why aren't colonies simply superseding failing queens, which would prevent a "colony loss." I wonder whether what we are now seeing is more "unsuccessful supersedure" rather than "queen failure." Some possible suspects would be genetics, pesticide residues, miticide residues, or something to do with varroa and its associated viruses.

Swarming

Given the right conditions (especially in springtime), a colony will typically divide itself and swarm, taking the old queen with them, and leaving behind "swarm cells" containing replacement daughters about to emerge.

Practical application: It is normal for a colony to reproduce by swarming once it's filled its cavity (whether a tree hollow or a hive). This normally occurs during the local "swarm season," and especially with queens that are over a year old. Some swarming may also occur in crowded colonies later in the season.

One can generally tell whether a colony is preparing to swarm by tipping up the brood chamber (or upper brood box in a double). Most (but not all) swarm cells will be built along the bottom bars of those brood frames (Figures 5.16–5.18).

Practical application: Queen cells are easily damaged if one attempts to cut them from a wooden frame. But frames containing queen cells can be carefully moved to a split.

Figure 5.16 Typical queen cells, some sealed, some in development, along the bottom bars of an upper brood chamber. This colony could be split, with each portion receiving at least one frame with a queen cell.

Figure 5.17 Typical swarm cells along the bottom bar of the upper brood chamber.

Figure 5.18 Bottom view of swarm cells prior to sealing. Note the huge amount of jelly beneath the young larvae, which will be consumed after sealing of the cell. Also note the typical presence of drone cells and drones. Of interest is that the jelly fed to larvae (including the queen) allows for transgenerational immune priming via dsRNA. Source: Maori et al. (2019).

Figure 5.19 An emerged queen cell, indicating that the colony has recently swarmed, and that it is too late to take preventative measures.

The colony swarms prior to the emergence of the queen cells. If the weather does not permit swarming, or if resources dry up, the colony will destroy the swarm cells without swarming.

Practical application: If the colony has recently swarmed, you will generally be able to see emerged cells, often with the round capping still attached (Figure 5.19). If, on the other hand, *all* the queen cells have been chewed out from the side, the colony may have "changed its mind."

Practical application: A colony first issues a "prime swarm" containing a mated, successful, but typically aging queen. If the remaining colony is still too large

for the cavity, one or more virgins may leave in "afterswarms." Young queens will not tolerate a rival sister (Figure 5.20). In the case of either a swarm or an afterswarm, only one queen will be left remaining.

Swarming is coordinated by a few "experienced" workers (Rangel and Seeley 2008, 2010) rather than being controlled by the queen. The queen is pushed to exit the hive. Not all queens (which may not have flown for a year or more) are able to do so (Figure 5.21).

Practical application: With time, the setae ("hairs") on a queen's body get worn off, especially on the top of her thorax. This is one way to tell how old a queen is.

After emerging from the hive, a swarm temporarily bivouacs on a tree limb or other resting place while the scouts decide upon a permanent nest site. Beekeepers use this opportunity to "hive" a swarm by shaking it into an empty (not containing bees) hive (Figure 5.22).

Practical application: A swarm may contain several queens, most being virgins. Workers may form protective walnut-sized "balls" around them, that can be picked up and used to start new colonies.

Once a swarm settles into a new cavity, the new colony is in a race against time, since the average longevity of a worker bee is about 35 days, and it takes around 20 days to complete a brood cycle. It is astounding how rapidly a swarm can build out a new hive and grow – provided that there is either a nectar flow on, or the colony is fed sugar syrup.

Figure 5.21 An aged or damaged queen may not be able to fly far, and may land on the ground and not be able to return to the hive, in which case the swarm will return to the hive, only to leave again once a new queen emerges from a swarm cell. You may spot a flight-impaired queen on the ground, sometimes surrounded by a few workers (note this aged queen's damaged wing).

Management to Minimize Swarming

Swarming is the natural reproductive process of the honey bee. It is challenging, but possible to manage colonies to

Figure 5.20 A sharp-eyed vet can really impress a beekeeper when they walk up to a hive, and with a glance at the ground pronounce that the colony has recently swarmed. The presence of a dead virgin queen in front of the hive, as in the photo above, is a sure sign of that occurrence.

Figure 5.22 A beekeeper about to shake a low-hanging swarm into an empty box placed above a hive full of frames. A swarm will readily move into a box full of beeswax combs, and immediately begin foraging and establish a broodnest.

minimize swarming. A colony swarms once it has grown to a certain size, or when it senses that it has grown too large for the cavity that it inhabits. The cues for those two factors are, respectively, the dilution of queen pheromone, and the lack of pheromones from young larvae.

Practical application: The short version is that the swarm impulse can be kept in check by:

- **Providing drawn comb directly in contact with the broodnest so that the queen doesn't run out of cells in which she can lay eggs, or**
- **"Shaking bees" to decrease the population size, or**
- **Splitting the colony to reduce its size.**
- **More details can be found at (Oliver 2015).**

Practical application: Springtime swarm prevention can often be combined with requeening and varroa management.

Usurpation Swarms

I would be remiss not to mention that a swarm of honey bees may invade another colony and replace its queen with their own. Such behavior is well-known for the Africanized Honey Bee (Schneider et al. 2004) but is also occasionally observed in European colonies (Mangum 2010; Oliver pers. obs.). This sort of aggressive usurpation allows a small swarm to take over the combs and valuable food stores of an established colony, thus giving the invading swarm a much better chance at their genetics surviving the winter. The invaded colony is often weak, queenless, or with a failing queen.

Practical application: If a swarm lands next to the entrance of an established hive, an inspection may find a queen being balled on the bottom board, indicating that a hostile takeover has taken place.

Emergency Queen Rearing

A queen may be unexpectedly lost without a supersedure cell in place. This may be due to sudden queen failure, disease, predation, mishap, or frequently in the case of inexperienced beekeepers, due to mishandling of the frames, resulting in injury to the queen (Figure 5.23). Or, a queen will occasionally drop off a frame to the ground, unnoticed, or sometimes even fly away (although she usually returns).

Practical application: a vet not highly practiced at handling frames of bees may wish to have the hive owner pull any frames for inspection, in order to avoid be blamed for killing the queen (Figure 5.23).

The colony does have a trick up its sleeve, however, to replace an unexpectedly lost queen: within hours it will begin making postconstructed "emergency queen cells."

Figure 5.23 The author has too many times observed a novice beekeeper inadvertently injuring a colony's queen, only to later see her body having been drug out the entrance. After any colony inspection, always check the ground and sides of the hive for small clusters of bees that may contain a queen.

Figure 5.24 Typical emergency queen cells at about a day after initiation, showing the process of floating worker larvae to the top of their horizontal cell on surplus jelly, and then adding additional wax to turn what is now a queen cell vertically downwards.

All female larvae are totipotent (capable of developing into either caste) until they begin differentiating due to their dietary change beginning roughly 24 hours into their development.

Practical application: Should a queen be unexpectedly lost, the workers have only a few days to save the colony – by converting at least one worker cell containing a female larva less than 24 hours old into an "emergency queen cell" (a "postconstructed" queen cell). This is done by floating chosen 1st-instar larvae to the top of their cell on jelly, and then building a vertical queen cell downward (Figure 5.24).

Identification of emergency cells: The scattered appearance and multiple number of the above emergency cells, surrounded by cells of equal-aged young larvae (not visible in this view) identifies them as emergency, rather than swarm or supersedure cells. These cells are only just started and will soon look similar to a supersedure or swarm cell. Before their emergence, however, the bees will cull some emergency cells, leaving only those with the best queens to emerge (Punnett and Winston 1983).

Practical application: The process of replacing a queen takes time. The first queens won't emerge until at least 12 days, then a virgin needs a few days to mature enough to take a mating flight (weather permitting), and then develop her ovaries prior to commencing egg laying (which typically commences 10–14 days after her emergence). Thus, it takes around 25–30 days after losing its queen before the colony will again exhibit a laying queen.

"Laying Worker" and "Hopelessly Queenless" Colonies

Should a colony be unsuccessful at producing at least one successfully mated replacement queen, it will become "hopelessly queenless." Without a queen's (and perhaps the brood's) pheromones suppressing ovary development in the workers, some workers will then commence to lay eggs, and are called "laying workers." Unfortunately, those eggs, being unfertilized, can develop only into drones (Figure 5.25).

Figure 5.25 Typical eggs from laying workers. Unlike those laid by a queen (singly and upright in the center of each cell), laying workers often place several eggs in a cell, not centered, and sometimes on the cell wall. Compare this egg pattern to that in Figure 5.7.

Practical application: Signs of laying workers normally don't appear until around three weeks after the loss of the queen.

A laying worker hive can survive for quite some time, but tends to become "pissy" and full of undersized drones.

Practical application: Once a colony goes "laying worker," it is difficult to requeen. Although there are any number of "folk remedies" for requeening a laying worker colony, it is generally easiest to temporarily combine it with a queenright hive, wait a few days for *that* queen's pheromones to again suppress the ovaries of the laying workers, and then to split the combined hive, giving the queenless portion frames of brood and an introduced queen or a queen cell.

Queen Status Assessment

Beginning beekeepers often wonder whether their colony has "gone queenless," or if their queen if "failing." The vet should be able to address these questions. The vet's first assessment would be to determine whether the colony is:

- "Queenright" – containing a functioning, laying queen,
- "Queenless" – with perhaps a queen in the works, or
- "Hopelessly queenless" – meaning that the colony has not only lost its queen but has passed the window of opportunity to replace her. A colony at this stage often exhibits the signs of having gone "laying worker."

Signs of Queenrightness, Failing Queens, and Queenlessness

It is generally much easier to find signs of a queen than to spot the queen herself, and in most cases there is no need to actually see the queen. If there are eggs and larvae present, that usually indicates that the colony is queenright, especially if the egg pattern is as below (Figure 5.26).

Scattered Drone Cells

Queens do "age," and in the process may exhibit poor brood patterns, suggesting that they be replaced. In general, few queens are good for more than a year and a half of egg-laying, by which time their spermatheca simply runs out of viable sperm. At that point some eggs laid in worker cells will develop into drones (Figure 5.27).

The workers apparently cue on unfertilized eggs as a signal that the queen is failing and will usually supersede a queen before this becomes noticeable, but if they wait too long, they may not have any female larvae from which to raise a replacement queen.

Figure 5.26 A nice egg-laying pattern by a queen – each egg centered in its cell, with no misses. The eggs are glued upright and typically slightly tilted (often the eggs are all tilted in the same direction). A pattern such as this indicates a high-quality queen.

Figure 5.27 Raised drone cells scattered among worker cells, as opposed to groups of drone cells at the edge of the comb, indicate that a queen has gone "drone layer" – meaning that she is running out of stored sperm with which to fertilize worker eggs.

Queenlessness Colonies

A colony that loses its queen while there are eggs and young larvae present will generally (weather and predators permitting) successfully requeen itself.

Field signs for a queenless colony:

- **No eggs or young larvae, although there may be other causes for this.**
- **The sound of a "queenless roar" when smoke is applied across the combs.**
- **The workers are agitated, exhibit jittery wings, and become defensive.**
- **The presence of emergency cells.**
- **Eventually, a "hopelessly queenless" colony will exhibit signs of "laying workers."**

The lack of eggs and young larvae is not a conclusive sign of queenlessness, since the colony may have simply ceased broodrearing due to dearth, or may have a new queen in progress (typically referred to as a "virgin" until she starts laying eggs). The other signs in the above list are more conclusive.

Signs of Being in the Process of Self Requeening

A colony in the process will have either queen cells of some stage present, or a virgin queen. A virgin is difficult for the untrained eye to spot.

Practical application: if a colony has recently lost its queen, emergency queen cells should be present. One way to test for queenlessness is to insert a frame of eggs and young larvae from another hive; if the colony is indeed queenless, it will start the construction of emergency cells on that frame by the next day.

Practical application: A colony in the process of requeening itself is generally calm, not jittery, does not "roar" in response to smoke, and prepares the cells in the center of the broodnest for the new queen to lay in, by keeping them polished and free of food stores. If the bees are storing nectar and pollen in the center combs, suspect queenlessness.

Signs of Being Hopelessly Queenless

Practical application: A colony that has been truly queenless for more than five days, and which does not have emergency cells in progress, will be unable to requeen itself, and is termed "hopelessly queenless."

The typical signs of a hopelessly queenless colony are:

- Lack of cluster formation,
- The "queenless roar" when smoked,
- Jitteriness of the workers' wings,
- Defensive behavior by the workers,
- Frames containing scattered or only drone brood (Figure 5.28),
- Multiple eggs in the cells from laying workers (Figure 5.25).

How to Tell a Drone-Layer from a Laying Worker Colony

Since in either case there will be scattered drone cells, many beekeepers have trouble telling the two conditions apart (Figure 5.29).

Figure 5.28 Drone cells are normally built in discrete patches, rather than scattered on the combs as above. If you observe scattered drone cells, look for multiple eggs within the cells, which would indicate that the colony lost its queen, and that some workers have activated their ovaries to become "laying workers." A laying worker colony is doomed to death unless it is requeened by the beekeeper, which may be difficult.

Figure 5.29 A brood frame with scattered drone cells. In this case, there were only single eggs in the cells, indicating that it was a case in which the queen had gone "drone layer" (unable to fertilize worker eggs). This diagnosis was confirmed by noticing the presence of a queen.

Practical application: A drone-laying queen and a colony with laying workers both exhibit scattered drone cells. But only in the case of laying workers are there are multiple eggs scattered in the cells.

Odd Problems

A virgin queen, due to weather, may be unable to take mating flights. After a few weeks, she may begin to lay eggs. Or a queen may have a problem with her internal plumbing

Figure 5.30 A case of a young queen commencing egg laying without having mated (likely due to confinement by weather). Note that she has attracted a retinue of attendants, but that *every cell* contains a drone larva too large for the worker cell in which the eggs were laid. Note also that the workers have extended the cell walls, and in a few days the cells would be capped with bullet-shaped drone cappings, rather than slightly-domed worker cappings.

Figure 5.31 A solid brood pattern, indicating an excellent queen, and a well-nourished, disease-free colony.

that prevents her from fertilizing her worker eggs. In either case, the queen may lay a *solid* pattern of drone brood in worker cells (Figure 5.30).

"Spotty Brood"

Beekeepers thrill to see a "solid brood pattern" (Figure 5.31).

Sometimes one will observe what beekeepers call "spotty brood" (Figure 5.32). This may or may not be a reflection of the queen. A queen mated in an inbred population, due to lack of diversity in the sex alleles of the drones that she

Figure 5.32 An example of "spotty brood." Although beekeepers often blame the queen for this condition, it may or may not have anything to do with her.

Figure 5.33 The queen is most often to be found on a brood frame, and not surprisingly, most often on one containing fresh eggs. But she will also sometimes be found on an outside frame, or even on the wall of the hive.

mated with may lay fertilized eggs that develop into "diploid drones," which are quickly consumed as larvae by the nurses, resulting is a spotty brood pattern. This is seldom the case, unless a beekeeper is rearing their own queens.

More often, spotty brood becomes normal as the season progresses, due to poor nutrition, disease, or parasitism by varroa (Figure 5.32).

Spotty brood is caused by larval mortality due to dearth or poor nutrition, European Foulbrood (EFB), Chalkbrood, a virus, toxic pollen, or pesticide or miticide contamination of the combs. Far too many queens are unnecessarily replaced due to problems not of their own fault. But if a queen is over a year old, she may be starting to run out of viable spermatozoa.

Practical application: There's not much that one can do about Chalkbrood, but EFB can be cleared up with oxytetracycline. Give the queen a fresh comb to lay on, in order to see whether comb contamination or disease is the problem. If the spotty brood persists, then replace the queen.

Queen Replacement and Introduction

Since queens tend to begin to fail in their second season, most professional beekeepers replace them annually, instead of taking the chance of successful supersedure. A colony going into the winter with an aged queen is a recipe for disaster.

Practical application: colonies tend to perform and survive better if they have young queens.

Identifying and Locating the Queen

Although beginning beekeepers always want to "see the queen" (Figure 5.33) in truth, it is rarely necessary to observe the queen directly in order to determine her status.

Tips for finding the queen:

- **It is difficult to see queens when you are wearing a veil (sorry, that is just a fact).**
- **Check the brood frames first.**
- **As you pull out the frame, look down the face at an oblique angle – the queen stands somewhat taller than the workers.**
- **Glance at the face of the next frame in the hive, I often spot the queen there.**
- **The queen is far easier to spot in the first 5 seconds than she is after 10 seconds after being disturbed, so first glance quickly at both sides of the frame.**
- **Hold the frame in front of you, with good lighting, hold your eyes steady, and *move the frame* in a rectangular motion in front of your stationary eyes (rather than moving your eyes).**
- **Train your eye to recognize the queen's abdomen, more angular "hips," a possible retinue of attendants, and her longer (and often light-colored) legs.**

Handling the Queen

A queen will only sting another queen – they can be safely handled with bare fingers. They should never be handled through gloves, due to the loss of delicate touch. To prevent damaging her delicate ovaries, avoid ever touching her abdomen. To pick up a queen, approach her from the rear, and gently pinch her wings together (Figure 5.34).

There is likely little reason for a vet to handle a client's queens. But you may get a call about a queen suddenly dropping dead while being handled. It is a relatively

Figure 5.34 Queens are typically held by *both* wings. If you only get one wing, release her before she wrenches around and hurts herself.

Figure 5.35 Even when gently handled, some queens may play possum. Do not let this scare you – simply return her to her bees on a frame held horizontally, and she will soon "come to."

common behavior for a queen to play possum when handled, but very alarming to the novice (Figure 5.35).

If a queen is disturbed, she may start "running," causing her own workers to attack her. To prevent this, if I am going to be disturbing a colony greatly, I will often temporarily place the queen in a queen cage, returning her to the colony when I'm finished (Figure 5.36).

Introduction of Queens

There are a million suggestions for introducing queens. In general:

- Remove the old queen and wait a day.
- If the replacement queen is closely related, and in laying condition, she can be successfully introduced at the entrance with a few puffs of white smoke.

Figure 5.36 Placing a queen headfirst into a queen cage for her own protection, to be used to release her back into the colony after it has been put back together. I will plug the end with a miniature marshmallow or piece of green leaf, which then allows the colony to calm down by the time the bees have chewed their way through the plug.

Figure 5.37 Bee behavior indicating acceptance of a caged queen. The bees will be moving over the screen, offering the queen food, and can be easily brushed away with a finger. Once you observe this behavior, it is safe to remove the cover from the candy plug, in order to allow the workers to release the queen some time after you have closed the hive back up. Then replace the still-plugged cage into the hive with the screen exposed to the workers.

- Requeening via the insertion of queen cells into a queenright hive has been shown again and again not to be successful.
- For extremely valuable queens, use a push-in cage, or better yet, introduce her first into a nuc containing only brood and nurse bees.
- Otherwise, introduce her in a queen cage with a candy plug, pushed into a brood frame (but not into honey, which may drown her).
- Releasing any attendants from the queen cage will greatly improve success at introduction. Do so in an

Figure 5.38 Bees that *are not* accepting a queen. These workers are attempting to bite and "ball" the queen. They will be biting the screen and cannot be easily moved away with a finger. Such behavior generally means that there is already another queen in the hive, or that the queen in the cage has a very different odor than the "colony scent."

Figure 5.39 The round opening at the bottom of the inserted queen cell indicates successful emergence. Any chewing on the side of the cell would indicate that the bees killed the queen.

enclosed area to avoid having a queen fly off (although if you hold still, she may return to the cage).

- For best results, do not remove the candy plug cover until after inspecting the caged queen the next day (Figures 5.37 and 5.38)

Practical application: It is often difficult to replace a queen with a queen of another race.

Introduction via Queen Cell

When splitting a colony in the spring, a ripe queen cell (a swarm cell from another hive, or a grafted queen cell) can be introduced successfully after a queenless split has been sitting for a day (Figure 5.39).

Wrap Up

A well-reared queen of good stock can make beekeeping a pleasure. There is no reason to keep bees that are not gentle, productive, and disease and parasite resistant. In general, colonies should enter the winter with a queen that is not more than a year old. Every beekeeper should learn how to rear a few queens (Oliver 2014), and then keep queenright nucs on hand. All beekeepers should demand that queen producers start selecting for varroa-resistant stock.

References

Anderson, K. et al. (2018). The queen's gut refines with age: longevity phenotypes in a social insect model. *Microbiome* 6: 108.

Baer, B. et al. (2016). Sperm use economy of honeybee (*Apis mellifera*) queens. *Ecology and Evolution* 6 (9): 2877–2885.

Bee Informed Partnership (2019). Sentinel Apiary Program Final Report 2019.

Collison, C. (2017). A Closer Look. https://www.beeculture.com/a-closer-look-8.

Farrar, C. (1947). Nosema losses in package bees as related to queen supersedure and honey yields. *Journal of Economic Entomology* 40: 333–338.

Mangum, W. (2010). The usurpations (takeover) of established colonies by summer swarms in Virginia. *American Bee Journal* 150 (12): 1139–1144.

Maori, E. et al. (2019). A transmissible RNA pathway in honey bees. *Cell Reports* 27: 1–11.

Niño, E. et al. (2013). Chemical profiles of two pheromone glands are differentially regulated by distinct mating factors in honey bee queens (*Apis mellifera* L.). *PLoS One* 8 (11): e78637.

Oliver, R. (2014). Queens for Pennies. http://scientificbeekeeping.com/queens-for-pennies.

Oliver, R. (2015). Minimizing Swarming http://scientificbeekeeping.com/understanding-colony-buildup-and-decline-part-7b.

Park, O. (1946). The queen. In: *The Hive and the Honey Bee*, 84. Dadant and Sons.

Punnett, E. and Winston, M. (1983). Events following queen removal in colonies of European-derived honey bee races (*Apis mellifera*). *Insectes Sociaux* 30 (4): 376–383.

Rangel, J. and Seeley, T. (2008). The signals initiating the mass exodus of a honey bee swarm from its nest. *Animal Behaviour* 76: 1943–1952.

Rangel, J. and Seeley, T. (2010). An oligarchy of nest-site scouts triggers a honey bee swarm's departure from the hive. *Behavioral Ecology and Sociobiology* 64: 979–987.

Rangel, J. and Tarpy, D.R. (2016). In-hive miticides and their effect on queen supersedure and colony growth in the honey bee (*Apis mellifera*). *Journal of Environmental & Analytical Toxicology* 6: 377. https://doi.org/10.4172/2161-0525.1000377.

Richard, F.-J. et al. (2007). Effects of insemination quantity on honey bee queen physiology. *PLoS One* 2 (10): e980.

Salmela, H. et al. (2015). Transfer of immunity from mother to offspring is mediated via egg-yolk protein vitellogenin. *PLoS Pathogens* 11 (7): e1005015.

Schneider, S. et al. (2004). Seasonal nest usurpation of European colonies by African swarms in Arizona, USA. *Insectes Sociaux* 51: 359–364.

Simpson, J. (1955). The significance of the presence of pollen in the food of worker larvae of the honey-bee. *Quarterly Journal of Microscopical Science* 96 (1): 117–120.

Wang, Y. et al. (2016). Comparison of the nutrient composition of royal jelly and worker jelly of honey bees (*Apis mellifera*). *Apidologie* 47: 48–56.

Winston, M. (1987). *The Biology of the Honey Bee*. Harvard University Press.

6

Honey Bee Strains

Dewey M. Caron

University of Delaware, Affiliate Faculty Oregon State University, Portland, OR, USA

The familiar honey bee has many modifier names. The common name, "honey bee," may be used for all 10 species (Sheppard 2005). In North America there is but a single introduced honey bee, the western (sometimes termed European) honey bee, *Apis mellifera*. The European bee has been carried by man to all continents except Antarctica.

The native range of the European honey bee is Africa, Europe, and the Middle East. There are nine other honey bees, i.e. *Apis* species, found exclusively in Asia, in addition to introduced *A. mellifera*. Virtually all Asian *Apis* species, like *A. mellifera*, are either kept by humans or hunted for their immature brood, their stored foods (honey and bee bread), and/or their beeswax combs. The greatest impact of the honey bee, however, is through their pollination of flowering plants.

Honey bees, originally believed to have originated in Africa but evidence exists for an Asian origin (Han et al. 2012), are globally diverse. This great diversity is thought to have arisen due to regional adaptation to the distinctive ecological conditions in various geographic regions. A global analysis of the modern diversity of *A. mellifera* using morphology, coloration, behavior, and MtDNA/DNA, suggests that there are approximately 26 major varieties, classified as subspecies, grouped into at least five different evolutionary groups: (A) African subspecies, (M) northern and western European subspecies, (C) north Mediterranean subspecies, (O) Middle Eastern subspecies, and (Y) Ethiopian subspecies. (See Meixner et al. 2013, original work by Ruttner 1988) (Figure 6.1).

Subspecies designations are not widely used today: *A. mellifera* diversity is commonly called races, but which more properly might be considered ecotypes. With extensive movement of bees by humans, the race designations are less accurate today. Bee differences can be distinguished on how they look (morphological features), biological differences and behavioral variation. One behavioral distinctiveness is defensive (inappropriately called aggressive) behavior, frequently cited for subspecies *Apis mellifera scutellata* transported from Africa to South America in mid 1950s (and renamed Africanized bees in the Americas.)

Several honey bee populations (primarily the M and C subspecies) have been introduced into North America. Four major, initially (more or less), distinct bee populations comprise the bulk of successful introductions. The primary stock adaptable most widely is the Italian bee; thus, the honey bees of North America are commonly called the Italian bee; – some suggest calling the honey bees in North America an "American bee," (Borst 2019).

Distinctions between the major bees introduced into North, Central, and South America and developed selections are as follows (see Caron and Connor 2017 for details).

German or Black Bee

The original bee carried from Europe to the American colonies (first documented arrival 1622) remains the common bee of northern Europe. It was usually housed in the skep hive. The bee prospered in the Americas and rapidly spread via swarming.

The German or Black bee has the desirable traits of: Overwintering well in severe climates, tendency to maintain strong colonies while conservative in consumption of honey stores, forages in cooler weather, and is desirable for surplus honey production.

Undesirable traits of this bee include: Colonies can be defensive, susceptibility to disease, especially EFB bacterial disease, and it is generally thought to be a bee that is slow in spring population buildup.

Carniolan bee, (sometimes called the carney bee or carnies). This bee was introduced in 1883 from present-day Slovenia. They have dark bodies (like German bees), especially evident in drones, and a low robbing tendency.

Honey Bee Medicine for the Veterinary Practitioner, First Edition. Edited by Terry Ryan Kane and Cynthia M. Faux.

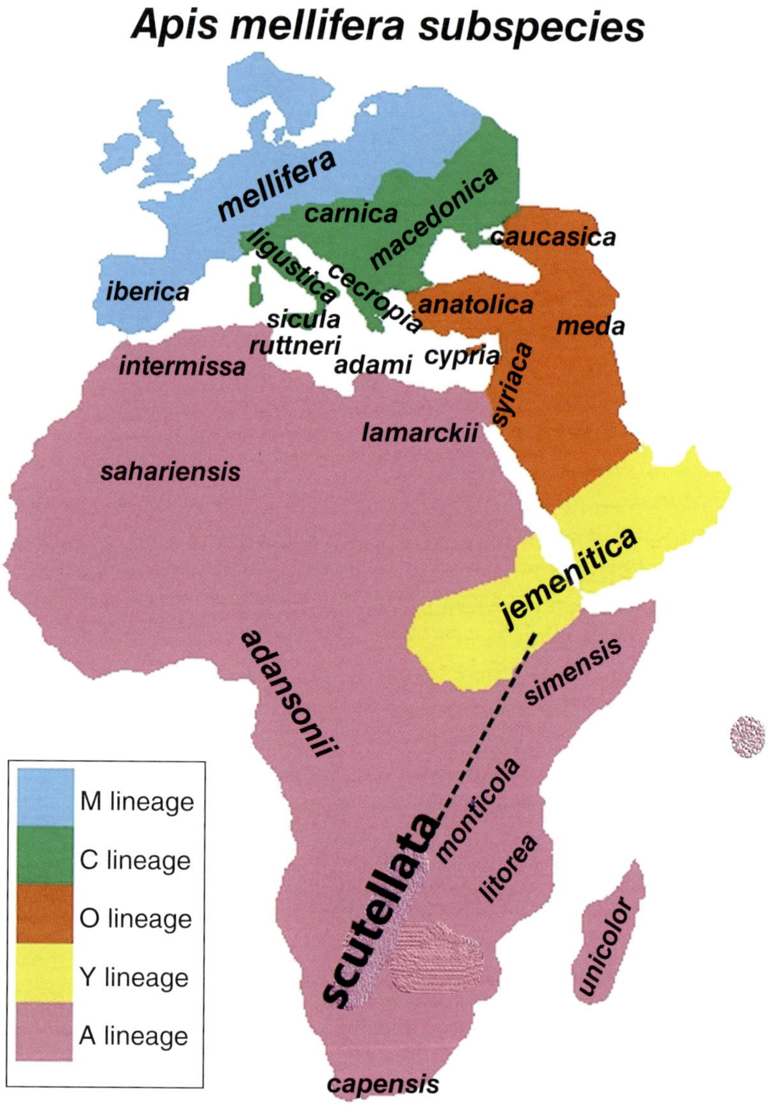

Figure 6.1 Lineages of bees: *Apis mellifera scutellata* Lepeletier, the African honey bee, is a subspecies that has a large distribution in Africa and ranges from South Africa northward along the eastern half of the continent to about Ethiopia. *Apis mellifera capensis* Escholtz (1822), the Cape honey bee, exclusively occurs in the Cape Region of South Africa. *Apis mellifera monticola* Smith (1961), the East African Mountain honey bee, is a subspecies that inhabits within the mountains of Eastern Africa (Kenya and Tanzania). *Apis mellifera unicolor* Latreille (1804), the Malagasy honey bee, is exclusively distributed in Madagascar. *Apis mellifera litorea* Smith (1961), the East African honey bee, is distributed along the eastern coast of tropical Africa occurring from Southern Kenya (perhaps even the southernmost portions of Somalia) to Mozambique. *Apis mellifera ruttneri* Sheppard et al. (1997), the Maltese honey bee, is only distributed on the island of Malta in the Mediterranean sea. *Apis mellifera simensis* Meixner et al. (2013), the Ethiopian Mountain honey bee, is distributed in the mountain systems of Ethiopia. *Source:* Modified from Galarza (2016).

The Carniolan bee has the additional desirable traits of: rapid spring buildup, will forage at lower temperatures, good fall close-down, tend to overwinter with compact clusters, very gentle, relatively less defensive, and demonstrates a resistance to many brood diseases.

Undesirable traits include a propensity to swarm, high requirement of pollen to sustain the population, and the dark queen can be difficult to locate on darker combs.

There is an active effort to bring in drone semen and improve the carniolan bee stock in the US This bee stock is slowly being distributed, primarily in the Pacific Northwest, and is often referred to as the New World Carniolan. A monk, Brother Adam, in Southern England sought to improve the dark bee of the UK with importation of primarily carniolan genetics. These bees are termed Buckfast. Although Brother Adam has died, the program continues today (Figure 6.2).

Figure 6.2 Examples of the Carniolan Bee. *Source:* Photo courtesy of (a) Kathy Garvey; (b and c) Sue Cobey.

Caucasian (or Gray) Bee

Introduced in late 1800s from eastern Europe (Caucasus Mountains), this bee is more of a gray color with shorter body hairs. The genetics of North American populations of Caucasian bees is degraded.

The Caucasian bee has the desirable traits of: a long proboscis, is generally considered gentle and calm, builds up into strong colonies with good overwintering, has a low swarming tendency and forages under marginal conditions.

Undesirable traits include a high susceptibility to Nosema, a heightened propensity to use propolis, a tendency to build excessive burr comb, and a slow spring buildup. In addition, this bee is defensive with persistent bees that may leave the apiary to defend, workers tend to drift to other colonies, and they may display heightened robbing of neighboring colonies (Figure 6.3).

Figure 6.3 An example of a Caucasian Queen. *Source:* Photo courtesy of Rob Snyder

Italian bees were introduced via importation of queens from northern Italy beginning in 1859. This bee has been selected by US breeders to be larger in body size and for light-colored queens (golden Italians) and drones.

Italian bees have the desirable traits of: prolific queens, a light colored queen which makes them easier to spot during colony inspections, a general resistance to European foulbrood (EFB), and they are considered relatively gentle and calm (although a bit less so than carniolan).

Undesirable traits of the Italian bee includes a high propensity to build burr comb, they readily drift to neighboring colonies, their tendency to rob, and their susceptibility to many parasites and pathogens. In addition, these bees are slow in spring buildup and remain large colonies in summer, that may consume the food stores needed for overwintering; they are slow to prepare during fall for overwintering (Figure 6.4).

Russian Bees are of uncertain origin (perhaps a mixture of German and/or Carniolan bee stock). The Russian bee was imported into the US from the Primorski region of far eastern Russia by USDA. This stock has demonstrated ability to keep both Varroa and tracheal mite populations reduced (Rinderer et al. 1999). Additionally, they are good honey producers, rear brood only with pollen availability (so often good in spring buildup and fall shut down) and low robbing tendencies. Undesirable characteristics include the need to requeen with Russian bees (hybrids are not as effective in keeping mite numbers lower), requeening is more difficult compared to Italian colonies and they are a dark colored queen. The stock has been released to a Russian Bee Breeders group where they continue to maintain and conduct stock selections.

Hygienic Bees

There are an increasing number of selections within the available North American bee stock for "hygienic bees." The USDA bee breeding program has identified a stock (initially termed SMR – suppressed mite resistance) but since renamed to VSH (Varroa Sensitive Hygiene) that demonstrate higher hygienic behaviors. This selection has greater ability to detect, and then remove, reproducing female Varroa mites within capped brood cells (see Glenn 2016). A number of northern plains beekeepers have a selection (Minnesota hybrid) and a group of higher elevation beekeepers in Colorado and New Mexico also

Figure 6.4 An Italian bee. *Source:* Photo courtesy of Dewey Caron.

have a local selection of bees demonstrating this hygienic behavior. Another selection, termed "mite or ankle biters" or simply "biters," developed in a breeding program at Purdue University exhibits improved grooming behaviors, physically removing mites from the infested bees (Hunt et al. 2016).

Selections have been made for bees with higher alfalfa pollen collection activity, hoarding behavior, American (AFB) and European (EFB) Foulbrood resistance, increased honey production, longer tongue length and other heritable characteristics. Few of such stock are currently maintained, although one line, the Saskatraz bee, is a selection, from northern Canada, for better honey production (under their very unique environmental conditions of that region) and mite resistance (see www.saskatraz.com).

Africanized Bee

In 1957 African honey bees (subspecies group A) were introduced into the western hemisphere. The aim was to breed a hybrid bee better suited for Brazil. A Brazilian geneticist considered Asian and African bees, eventually selecting queens from Tanganyika and South African bee breeders for importation.

The Africanized bee desirable traits include extremely vigorous and rapid colony buildup, flight activity at low light levels, and they are considered highly resistant to pests and diseases.

Undesirable traits, included the well known vigorous defense of their hive and apiary, as well as a tendency to abscond (migrate) when resources are reduced, heightened swarming and this is not a good overwintering bee (Caron 1995; Winston 1992).

Before the Brazilian breeding program developed, the Africanized stock was accidently released, and then widely distributed, through sale of virgin queens. There was surprisingly little hybridization with existing European bee genetic material in the region and a feral population, retaining much of its original behaviors including heightened colony and apiary defense and poor honey production, established.

Quite remarkably, the bee began to spread, reaching Paraguay (1964), Argentina (1965), and Bolivia (1967) within 10 years of introduction to Brazil. It crossed the isthmus into Panama in 1982, into Mexico in 1986 and was detected in Brownsville, Texas in 1990. It spread along the Pacific into Arizona and was found in California in the early 1990s. Arrival of the stock was often noted by human stinging incidences and the deaths of small animals and corralled or tethered farm animals (Caron 1995).

Why are they Called "Killer Bees?"

The Africanized bee is also called the "killer bee," a term applied by Time magazine to the African bees imported into Brazil. The Brazilian press often used "assassin bee" in newspaper accounts of human stinging incidents; Brazilian beekeepers often used "foreign bee." Was the Time Magazine term justified?

Prior to importation of African bees, the European honey bees maintained in Brazil were isolated in agricultural areas, cast few swarms and were only likely to sting beekeepers. The native social bee of Brazil is stingless. As more individuals became beekeepers, they captured the increasingly abundant swarms and absconds of the foreign bee which meant more human–bee interactions and increase in stinging incidences. Some incidences included individuals allergic to bee stings, humans with poor health issues, and people and animals receiving too many stings in too short a time period. Such incidences were subsequently reported in local newspapers. When stings involved humans or killed animals the local newspapers began to use term "assassin" bee.

Animal and human stinging incidences are often used as an early warning of the arrival of Africanized bees into a new area. When bee samples were taken it was almost always the introduced bee. Reportedly the editors of Time thought assassin bee an inappropriate word so they renamed it "killer bee."

Consult University of Florida website http://sfyl.ifas.ufl.edu/natural-resources/africanized-honey-bees for information on how to avoid being a sting victim of Africanized bees.

Initially Africanization was determined by colony defensiveness. To aid in identification of the invading bee, the US Department of Agriculture (USDA) developed a morphometric means of identifying the bee termed Fast Africanization Bee Identification System (FABIS) (Sylvester and Rinderer 1987). Currently the most common ID method combines morphometric with mitochondrial marker (Lin et al. 2018) although a DNA analysis using DNA barcode COI might be as precise. (Kono and Kohn 2015).

Cargo shipping resulted in separate introductions onto the Caribbean islands and the Tampa Bay area of South Florida as swarms hitchhiked on the shipping containers themselves. In some ports, local beekeepers captured such swarms and sought to manage them. However, feral colonies, (non-managed colonies living in cavities of trees/ human structures/ human discards and in bird and squirrel nests) became common and these in turn produced large numbers of swarms. By vigorous trapping and removal of feral swarms, Puerto Rico and Cuba have developed a hybrid bee considered better adapted to those respective islands (Avalos et al. 2017).

Today this bee inhabits all countries of South and Central America (Chile is an exception) along with the southern tier of US states (California to Florida). There have been reports of Africanized bee captures in the second tier of most states from Nevada to Georgia. Isolated captures have been reported in a number of more northern states. Where they exist, the Africanized bees are the predominant bee, and via several isolating mating behaviors, they maintain their identity. European stock is shortly converted to Africanized via queen replacement and usurpation. Therefore, beekeepers prefer the Africanized bees because they will prosper while colonies with European bee stock struggle and eventually die out.

Several states initially had detection programs to map the movement of Africanized bees (via swarm capture) but none do so currently except for California, where ongoing research to detect and map movement northward (Lin et al. 2018) continues. Otherwise, the extent of the current northward spread is not precisely known. The California work suggests the northern spread of Africanized honey bees may not have stopped; they appear to still be advancing northward but at a slow rate. In the southern hemisphere (Argentina/Uruguay) there is a latitudinal demarcation line above which Africanized bees dominate and below which European bees predominate.

The Africanized bee colony remains highly defensive and more difficult to manage for commercial/hobbyist purposes than the European honey bees. In rural areas of the Americas, however, an entire swarm trapping program has been developed. Where Africanized bees exist in numbers, keeping European bees alongside the feral Africanized bees is impossible without frequent (two times a year minimum) requeening with European queen stock. Beekeepers in southern regions of the US seek to requeen with gentle stock and have adopted measures to keep European bees. It is generally considered a poorer honey producer than European bees but is an excellent bee for trapping pollen or harvesting propolis. They are, like European bees, a good pollinating bee, though with their defensiveness and absconding/swarming behaviors, provide more of a challenge to move and manage within the pollination setting (Figure 6.5 and 6.6).

Figure 6.5 An Africanized honey bee apiary in Bolivia. Note the stone walled "corral" to help minimize disturbance of the hives by passing people and animals. *Source:* Photo courtesy of Dewey Caron.

Figure 6.6 A frame of brood from an Africanized hive. Note the extensive brood pattern as the bees have filled the frame with brood. *Source:* Photo courtesy of Dewey Caron.

Defensive Bees

Are bees aggressive? It is preferable to describe their behavior defensive. Social honey bees have a sense of home and territory, with some races as previously described (German or Black bees, African bee *A. m. scutellata,* and Africanized bees of the Americas) more extreme in such behaviors.

Alerting a bee colony to a potential threat occurs with vibratory disturbance, movement (in the apiary), color, smells (pheromones and others) and environmental changes. Cessation of a nectar flow as well as increasing or increased honey stores in beeswax combs may lead to heightened defensive behavior.

Bee defense has been described as a four-stage response: alerting, activating, attracting, and culminating in stinging or fleeing behavior. Alerted bees initially bounce into veils, burrow into body hair, bite and pull body hair. Activating is physically searching for a source of disturbance, while attracting is orienting to disturbance. These two middle stages are closely linked and usually not individually discernable (Collins et al. 1980).

Stinging, or fleeing, is the ultimate defense. Stinging is also a recruitment of others to respond via release of a pheromone, which remains with the excised stinger in certain types of materials (cotton clothing, bee suits, gloves, human skin) but may also be released by exposure of the sting apparatus.

As discussed, different bee races/populations have different temperaments. Additionally, temperament of a honey bee colony changes with the seasons, size of colony, resource availability to foragers, different times of day, and the weather. Events external to the colony can influence temperament, such as robbing behavior or movements of animals, humans, or other objects that are unfamiliar in color or smell to the bees. Ground disturbances leading to transmission of vibrations to colonies may lead to great stinging behavior, such as weed "whacking" or lawn mowers.

How beekeepers handle their bees when inspecting a colony is a significant factor in bee defensiveness. Alerting the colony by bumping the hive, sudden jerky motions that bees can observe, or retaining stings in clothing may initiate stinging behavior. Strong natural or supplemented body odors may likewise do so. Beekeeper lore suggests bees might not sting their keeper because they "remember" him or her. In all likelihood, a simpler explanation is the beekeeper who gets fewer stings has developed handling skills that result in reduced colony defense (Figure 6.7).

Removal of Defensive (Africanized) Bees

Prevention of stinging incidences in areas where Africanized bees are common has become a growth industry and a new bee business. Traps baited with old comb or pheromone lures are positioned around governmental property, schools, businesses, etc. and then monitored for capture of swarms. Any colonies that establish in trees or structures/debris on such properties are likewise destroyed and removed. Fees are collected for the trap installations and removals. Often educational efforts are developed to alert the general public, especially youngsters and older individuals, to the potential dangers of Africanized bees. See University of Florida extension website for a good example (http://sfyl.ifas.ufl.edu/natural-resources/africanized-honey-bees).

The bees in traps are either killed or moved to a remote area. Bees are killed with soapy water, or by pesticides by licensed pesticide applicators. A vacuum device is preferred by some to remove the bees from structures not otherwise easily accessed. Few of the removed colonies are suitable for beekeeping because they are defensive, so most are destroyed. Swarms initially might be relatively easy to manage but become defensive as they expand.

Figure 6.7 Working defensive bees in Bolivia. *Source:* Photo courtesy of Dewey Caron.

Recommendations for handling defensive bees:

- Don't – unless absolutely necessary and never by individuals unfamiliar with honey bees.
- Do not walk in front of colonies or manipulate colonies from the front (where the entrance is).
- Avoid ground vibration transmission to colonies.
- Keep colonies on individual hive stands to avoid having activity/vibrations with one colony being transmitted to (and alerting) neighboring colonies.
- Be properly outfitted with sturdy protective equipment. Prepare for colony inspection/manipulations prior to entering an apiary – it is far easier to remove clothing if not necessary than having to stop and put on such clothing after the bees are alerted.
- Manipulate potentially defensive colonies in pairs or with an experienced mentor. Keep a cell phone on your person in case of an emergency.
- Open and inspect colonies only under favorable weather conditions (70 °F or higher, little wind, sunny conditions) and preferably in midday (10 a.m. to 2 p.m.) when the majority of foragers will be outside their hive foraging. With Africanized beekeeping, some prefer to open and inspect toward the end of day, so the bees have a chance overnight to regain a calmer demeanor.
- Keep colonies open for a short time period and avoid excessive time-consuming manipulations.
- If possible, if one is inexperienced, learn to handle defensive bees by working with weaker/smaller colonies rather than open and manipulate full-sized colonies.
- If unable to perform necessary activities, split the defensive colony into 2–3 units and return another day to perform manipulations. If finding the queen is required (such as when requeening is desired), shake the bees from frames and boxes into an empty box fitted with a queen excluder to enable worker bees to escape below into a capture box. Queen and drones will remain in the shaker box.
- Do not tolerate excessively defensive colonies (as judged on three or more colony visits). Kill such colonies with soapy water and restock with gentler stock.

References

Avalos, A., Pan, A.H., Li, C. et al. (2017). A soft selective sweep during rapid evolution of gentle behaviour in an Africanized honeybee. *Nature Communications* 8 Art. No: 1550.

Borst, P.L. (2019). Locally adapted honey bees. *The American Bee Journal* 159 (4): 423–426.

Caron, D.M. (1995). *Africanized Honey Bees in the Americas*. A.I. Root Co.

Caron, D.M. and Connor, L.J. (2017). *Honey Bee Biology and Beekeeping*. Wicwas Press.

Collins, A.M., Rinderer, T.E., Tucker, K.W. et al. (1980). A model of honeybee defensive behaviours. *Journal of Apicultural Research* 19 (4): 224–231.

Escholtz, J.F. 1822. Entomographien, I. Reimer, Berlin.

Galarza, Julio César Chávez. (2016). Population genomics and landscape genetics of the Iberian honey bee (*Apis mellifera* iberiensis). Tese de Doutoramento em Biologia Molecular e Ambiental. Universidade do Minho.

Glenn, T. (2016). Varroa Sensitive Hygiene VSH. http://www.glenn-apiaries.com/vsh.html.

Han, F., Wallberg, A., and Webster, M.T. (2012). From where did the Western honeybee (*Apis mellifera*) originate? *Ecology and Evolution* 2 (8): 1949–1957. https://doi.org/10.1002/ece3.312.

Hunt, G., Krispn Given, J., Tsuruda Jennifer, M., and Andino Gladys, K. (2016). Breeding Mite-Biting Bees to Control Varroa. Bee Culture March 2016.

Kono, Yoshiaki and Joshua R. Kohn. (2015). Range and Frequency of Africanized Honey Bees in California (USA). PLoS One https://doi.org/10.1371/journal.pone.0137407

Latrielle, A. 10804. Notice de espècies d'abeilles vivant en grande societè, et formant de celliles hexagonales, ou des abeilles proprement dites. Annales du Museum D'Historire Naturelle Paris 5:161–178.

Lin, W., McBroome, J., Rehman, M., and Johnson, B. (2018). Africanized bees extend their distribution in California. *PLoS One* https://doi.org/10.1371/journal.pone.0190604.

Meixner, M.D., Pinto, M.A., Bouga, M. et al. (2013). Standard methods for characterising subspecies and ecotypes of *Apis mellifera*. *Journal of Apicultural Research* 52 (4): 1–28. https://doi.org/10.3896/IBRA.1.52.4.05.

Rinderer, T.E., Delatte, G.T., De Guzman, L.I. et al. (1999). Evaluations of the *Varroa*-resistance of honey bees imported from Far-Eastern Russia. *American Bee Journal* 139: 287–290.

Ruttner, F. (1988). *Biogeography and Taxonomy of Honeybees*, vol. 284. Heidelberg, Berlin, New York: Springer-Verlag.

Sheppard, W. (2005). Honey bee diversity – races, ecotypes and strains. In: *The Hive and the Honey Bee* (ed. J. Graham), 53–67. Hamilton, IL: Dadant & Sons.

Sheppard, W.S., M.C. Arias, A. Grech and M.D. Meixner. (1997). *Apis mellifera ruttneri,* a new honey bee subspecies from Malta. Apidologie 28: 287–293. DOI:10.1051/apido:19970505

Smith, F. G. 1961. Races of bees in East Africa. Bee World 42:255–260.

Sylvester, H. Allen and Thomas.E. Rinderer. 1986. Fast Africanized Bee Identification System (FABIS) Manual. American Bee Journal 127(7):511-516.

Winston, M. (1992). *Killer Bees: Africanized Honey Bees in the Americas*. Harvard University Press.

7

Wild Bees: Diversity, Ecology, and Stressors of Non-Apis Bees

Margarita M. López-Uribe

Department of Entomology, Center for Pollinator Research, Penn State University, University Park, PA, USA

The western honey bee (*Apis mellifera*) was one of the first animals domesticated in Anatolia and the Fertile Crescent region, about 10 000 years ago. Managed honey bees have now been introduced to all tropical and temperate ecosystems around the world and have been domesticated for honey production and pollination services. Because of its fascinating and complex social behavior, *A. mellifera* is also a model system for studies of chemical communication, social behavior and genomics, and it is by far the best studied species among bees.[1] While people may assume all bees are social species that live in large groups, nest in cavities and produce honey, honey bees are a unique group among the over 20 000 described species of bees worldwide. Indeed, about 80% of the bees live solitary lifestyles (i.e., there is only one female in each nest), they do not produce honey, and many show remarkable species-specific specialization with flowers. Even among social bee lineages, most social bees have modest colony sizes that vary between a handful to a few hundred individuals, unlike honey bees that live in large colonies that can reach tens of thousands of individuals in a single nest. As the basic biology of the non-*Apis* bees is vastly different from the biology of honey bees; an understanding of the life-history traits of these other bees is critical to better characterize how non-*Apis* bees respond to biotic and abiotic stressors. Below is a description of the diversity, nesting habits, variation in social behavior and life cycles of non-*Apis* bees and the known responses of non-*Apis* bees to environmental stressors with an emphasis on the fitness effects and prevalence of known bee pathogens.

Bee Diversity and Distribution

Bees (Hymenoptera: Apoidea: Anthophila) are a monophyletic group – meaning they have a single evolutionary origin – that comprises over 20 000 species distributed in all continents except for Antarctica (Ascher and Pickering 2020). One of the biological traits that defines bees is the use of pollen and nectar to feed their brood. With the exception of masarine wasps that also feed pollen and nectar to their brood (Murray et al. 2018), bees are the only group of insects that rely on floral resources as food. Bees are closely related to sphecid wasps, which are mostly predatory and feed their brood with insects and other arthropods (Figure 7.1). From an evolutionary perspective, bees are sphecid wasps that transitioned from an insect diet to a plant-based diet (Danforth et al. 2019). Because they depend on floral resources as a source of food, adult bees constantly visit flowers and in this process serve an important role as pollination agents of over 80% of flowering plants (Ollerton et al. 2011). It has been hypothesized that the interrelationship between insect pollination and plants has facilitated the major diversification of certain groups of Angiosperms (Cardinal and Danforth 2013). Indeed, the number of species in the group of flowering plants (Angiosperms = 300 000 spp.) greatly exceeds the diversity of species in their close relatives that lack flowers (e.g. Gymnosperms = 10 000 spp.). From the bee's side, the exclusive use of floral resources for food does not correlate with a higher diversification of bee species (Anthophila = ~20 000 spp.) compared to their close relatives, the sphecid wasps (10 000 spp.) (Murray et al. 2018).

1 A search in Google Scholar [on January 10th 2020] using the term "*Apis mellifera*" resulted in ~87 200 articles published from the years 1950 to 2020. Using the same filter, the number of articles returned for non-*Apis* bees such as "*Bombus impatiens*" (a North American bumble bee species commercially produced) was ~5320 and for "*Osmia cornifrons*" (a managed solitary bee introduced to North America from Asia) was ~923.

Honey Bee Medicine for the Veterinary Practitioner, First Edition. Edited by Terry Ryan Kane and Cynthia M. Faux.

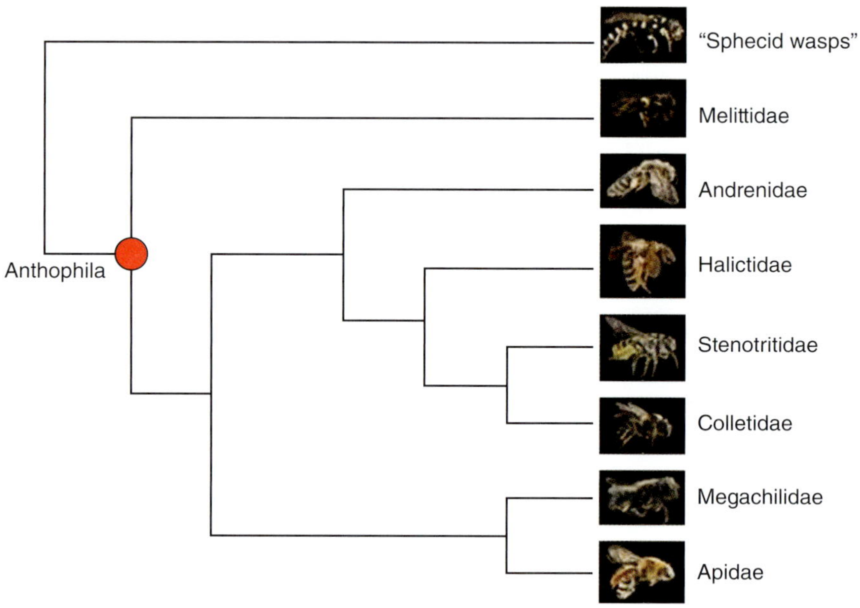

Figure 7.1 Cladogram depicting the phylogenetic relationships between "sphecid wasps" (photo *Cerceris hatuey*) and the seven extant bee families (Anthophila): Melittidae (photo *Macropis nuda*), Andrenidae (photo *Andrena cressonii*), Halictidae (photo *Halictus ligatus*), Stenotritidae (photo *Stenotritus pubescens*), Colletidae (photo *Colletes thoracicus*), Megachilidae (photo *Osmia lignaria*), and Apidae (photo *Eucera pruinosa*). Tree topology from (Danforth et al. 2013). *Source:* Photo courtesy of Sam Droege, USGS.

Table 7.1 Summary of the species richness, distribution, and life-history traits of the seven bee families (Anthophila lineage).

Family	Number of species	Distribution	Nesting habitat	Diet specialization	Levels of sociality	Notes
Andrenidae	~3000	Worldwide except for Australia	Ground	Many specialists	Solitary	Some species nest in large aggregations
Apidae	~6000	Worldwide	Ground and pre-existing cavities	Varies between narrow specialists to generalists	Varies between solitary to highly eusocial	Many species have parasitic lifestyles (clepto- and social parasites)
Colletidae	~2700	Worldwide (highest diversity in South America and Australia)	Ground, hollow stems, and cavities	Mostly generalists	Solitary	Some species nest in large aggregations
Halictidae	~4500	Worldwide (highest diversity in temperate areas)	Ground, and decaying wood	Varies between narrow specialists to generalists	Varies between solitary to eusocial	Small in size. Commonly known as "sweat bees"
Megachilidae	~4100	Worldwide	Hollow stems, wood, and cavities	Mostly generalists	Solitary	The majority of non-native species worldwide belong to this family
Melittidae	~200	Old World and Nearctic	Ground	Specialists	Solitary	Some species collect oil from flowers
Stenotritidae	21	Australia	Ground	Specialists	Solitary	Large and fuzzy bees.

Phylogenetically, bees are grouped in seven families: Andrenidae, Apidae, Colletidae, Halictidae, Megachilidae, Melittidae, and Stenotritidae (Danforth et al. 2013) (Figure 7.1). Bee families vary significantly in the numbers of species, geographic distribution, nesting habitats, diet specialization, and types of social or solitary behavior (Table 7.1). It is worth emphasizing that only 10% of all bee species exhibit some degree of social behavior, 80%

are solitary, and the remaining 10% have a parasitic lifestyle.

The biological details of parasitic bees (a.k.a. cuckoo bees) are outside of the scope of this chapter, but it is a common lifestyle among bees as it is present in four out of the seven bee families. Parasitic bees are characterized by the lack of nest building and pollen collection activity by the females. Instead, cuckoo bees usurp nests of other species and use the pollen and nectar already provisioned by the host to lay their eggs. Both females and males of parasitic species still visit flowers for nectar collection and can play a role in the transfer of pollen between flowers.

The close relationship between bees and flowering plants has led to the evolution of specialized anatomical structures that allow bees to collect and transport pollen (Michener 2007). Bees have statically charged feather-like branched hairs that facilitate the attachment of pollen grains to their bodies. For pollen transport, honey bees, bumble bees, stingless bees, and orchid bees possess a "basket" on their hind legs called corbicula (pl corbiculae) where wet pollen is placed after it is gathered from flowers (Figure 7.2a). Other bee species – like large carpenter bees or squash bees – have long hairs on their hind legs where they transport loose dry pollen. This mass of elongated hairs, called the scopa, can be located in the posterior tibia (Figure 7.2b), femur/trochanter (Figure 7.2c) or beneath the abdomen (Figure 7.2d). Parasitic bees and male bees lack these structures because they do not actively forage for pollen (Figure 7.3a).

Figure 7.2 Structures for pollen transport in different bee species. (a) The common eastern bumble bee (*Bombus impatiens*) has pollen baskets, also known as corbiculae, on their hind legs. (b) Burrowing bees (genus *Anthophora* spp.) have a brush like tuft of long hairs, called scopa, on their hind legs. (c) The sweat bee *Agapostemon virescens* transports pollen in the long hairs located in the femur of their hind legs. (d) Leafcutter bees (genus *Megachile* spp.) transport loose pollen grains in the long hairs located under their abdomens. *Source:* Photo courtesy of Anthony Vaudo.

(a)

(b)

Figure 7.3 Diversity of natural history traits among solitary non-*Apis* bee species. (a) An individual of a cleptoparasitic species of *Nomada* spp. visiting an apple flower. (b) Developing larva of the squash bee *Eucera* (*Peponapis*) *pruinosa* in a provisioned cell underground. *Source:* Photo courtesy of Henry Kindervatter and Margarita López-Uribe.

Diversity of Social Behavior

Bees are model systems for the study of the evolution of social behavior because they exhibit a great range of variation in behaviors from solitary to advanced eusocial. Among solitary bee species, the solitary female builds her nest, finds floral resources, and brings back pollen and nectar to provision cells[2] (Figure 7.3b). In most cases, solitary females die before their brood emerges. However, in some species, called subsocial, the mother provides extended parental care for the developing larvae and may still be in the nest when the first generation of daughters emerges (e.g., genus *Ceratina*, [Mikát et al. 2016]). Extended parental care is one of the first steps on the ladder of social evolution (Rehan and Toth 2015).

Other bee species live in colonies comprised of two or more adult bees with varying degrees of interactions. In incipiently social species, mother and daughters, or two or more sisters, can coexist in the same nest and develop cooperative brood care. Even though all adult females have fully developed ovaries and can produce eggs, some females (usually the mother or older sisters) will exclusively lay eggs while others will take the roles of foraging and nest guarding behavior (e.g. some species of *Xylocopa*; [Richards 2011]).

In primitively social species, a solitary female initiates the nest which then transitions into a cooperative colony after the first generation of daughters emerges. Bumble bees (genus *Bombus*) and many social sweat bees (genera *Halictus* and *Lasioglossum*) exhibit this type of primitively social behavior. In these colonies, the queen is usually larger in size than the workers and has more signs of age (e.g. worn-out wings). Primitively social colonies do not last more than a year because the colony breaks down after the production of new reproductive individuals – the future queens who initiate the cycle again the following season.

Advanced eusocial species exhibit morphologically and behaviorally distinct queen and worker castes, in addition to overlapping generations and cooperative care. Workers lose their ability to reproduce and function as nurses of developing larvae, guards that defend the nest, and foragers of the colony, while the queen never leaves the nest and monopolizes reproduction. Among bees, this type of advanced eusocial behavior is observed in honey bees (~12 spp.) and stingless bees (~500 spp.).

Diversity of Life Cycles

As holometabolous insects, bees undergo a complete metamorphosis from egg to larva to pupa and finally the adult stage. When adult females mate, they store the sperm in a spermatheca that will supply sperm for the rest of her life. Females lay eggs inside brood cells present in the nest. If the eggs are not fertilized, the individual will develop into a male while fertilized eggs develop into females. The number of eggs that a female can lay during her lifetime varies significantly from a handful in many solitary species to millions in advanced eusocial species (e.g. *A. mellifera*).

2 A cell is a cavity inside the nest where females store food and lay an egg that later develop into adult bees.

One of the most important differences in food provisioning between solitary and social bees is that solitary females lay their eggs on top of the food provision while social bees lay eggs on little or no food. In most social species, the food is actively provisioned to the developing bees by nurse bees. The time of development from first instar larva to pupa varies widely among bee species (from a few weeks to years; Danforth et al. 2019).

While the most visible life stage of bees is the adult stage, the majority of the developmental life cycle stages of bees occur inside the nest during the change from egg to pupa (Michener 2007). In solitary bees, the prepupa (aka last larval instar or defecating larva) is often the stage which undergoes the most unfavorable or harsh environmental conditions to which the bee will be subject to during its lifetime. This differs from honey bees that have perennial colonies with overwintering adults, or bumble bees where queens overwinter in a solitary stage. Many solitary bees that overwinter underground spend the winter months in the prepupal stage. In desert bees, individuals may stay as prepupae for many years until the environmental conditions are appropriate for emergence (Minckley et al. 2000). The life cycles of most non-*Apis* bees have only one generation per year (a.k.a. univoltine) with adults active for only four – eight weeks. However, species with more than one generation per year are not uncommon especially in tropical regions (Roubik 1989).

Abiotic Stressors to Wild Bees

Bees have been historically viewed as ecologically successful insects because of their abundance and widespread distributions. However, increasing evidence suggests that the abundance of bee species has decreased by 50% since the 1990s (Zattara and Aizen 2019) with ~10%–20% of the species already classified as critically endangered in some geographic regions (Cameron and Sadd 2020; Nieto et al. 2017). The rapid loss of bee biodiversity and the threat that these declines pose to pollination of wild plants and crops have precipitated huge efforts to (i) characterize the health status of pollinator populations (Brown and Paxton 2009), (ii) monitor how populations and communities are changing over time (Lebuhn et al. 2013), and (iii) identify biomarkers that provide information about the health status of individuals, colonies, and populations (López-Uribe et al. 2019). A summary of the main abiotic stressors to bee populations is presented below.

Reduced availability of floral resources – Currently, there is consensus about the drivers of bee population declines involving the interactions between multiple abiotic and biotic stressors linked to anthropogenic change (Goulson et al. 2015). One of the most important environmental changes linked to the decline of bee populations is the conversion of natural areas to productive lands for agricultural purposes (Otto et al. 2016). With approximately 40% of the terrestrial land cover being used for crop production and pastures (McDonald and Stukenbrock 2016), land use change has reduced the total diversity – both abundance and richness – of available floral resources, leading to poor nutritional landscapes for bees (Ogilvie and Forrest 2017). Even though the mass floral resources provided by crops and weeds are sources of pollen and nectar to a variety of insects, the overall quality of these floral resources may not be optimal for bees that require specific ratios of proteins, lipids, and carbohydrates in their diets (Vaudo et al 2016). Unlike honey bees, most non-*Apis* species are diet specialists to some extent – meaning they use pollen from a restricted number of plant species. Changes in the availability of floral resources is expected to impact specialist bee species more strongly than generalists, such as honey bees (Wood et al. 2019). As the proportion of pollen specialists varies from zero to up to 66% in different communities across the globe (Danforth et al. 2019), bee community responses to the loss of floral resources may vary geographically.

Pesticides – The massive conversion of natural areas into areas for agricultural production has precipitated an increase in the proportion of land exposed to pesticides. Even though pesticides are meant to target and control plant pests and pathogens, they can be highly toxic to nontarget species such as natural enemies of insect pests and also pollinators. Recent changes in the use patterns of insecticides and fungicides are a major stressor to bee populations across the globe (McArt et al. 2017). Specifically, the adoption of highly insect-toxic neonicotinoid insecticides has been directly linked to bee decline (DiBartolomeis et al. 2019). Because of the potency of neonicotinoids, the total amount of applied pesticides has reduced since their introduction to the pest control markets but their toxicity levels have increased up to ninefold (Douglas et al. 2020). Some of the more toxic neonicotinoids (e.g. imidacloprid) not only have acute lethal effects on bees but also have a myriad of sublethal effects including reductions in individual and colony survival (Whitehorn et al. 2012), foraging ability (Gill et al. 2012), and immunocompetence (Aufauvre et al. 2014). In addition to neonicotinoid use, increasing fungicide use has also been directly linked to increased pathogen prevalence and range contractions of bumble bees in North America (McArt et al. 2017). Fungicides disrupt the microbiota associated with bee bread – pollen stored in bee cells – and the digestive tract of developing larvae, which increases bee susceptibility to pathogens (Lozo et al. 2015). The synergistic effects between insecticides and fungicides are still poorly understood but can potentially amplify the negative effects of xenobiotics on bees (Raimets et al. 2018). Given the persistence of pesticides in soil, the risk of acute and chronic pesticide exposure for ground nesting bees via soil could be a significant stressor for the majority of non-*Apis* bees (Chan et al. 2019).

Climate Change – Rapid changes in abiotic factors resulting from climate change are also an imminent, but less understood, stressor to bee populations. For example, warming temperatures and changes in precipitation patterns are driving changes in the geographic distribution of pollinating species. Climate change is also changing plant phenology[3] and the quality of floral resources, which has an indirect impact on bees and other pollinators that use flowers as food sources. The effect of climate change on population stability and distribution has been better studied in bumble bees, which are cold-adapted species expected to respond negatively to global warming. The geographic ranges of bumble bees have shrunk in the past 50 years with species moving uphill on mountain ranges, and reductions on the southern limits of species in the northern hemisphere (Kerr et al. 2015). Ecological niche models based on future climate data predict further reductions in geographic ranges, with alpine species being the most critically threatened by global warming. Recent studies have demonstrated a direct link between thermal tolerance and the geographic distributions of bees (Hamblin et al. 2017). In urban areas where the increased proportion of impervious surface has significantly increased ambient temperatures (Rogan et al. 2013), it has been demonstrated that the most heat-tolerant bees are the species found in more urbanized sites that experience higher ambient temperatures (Hamblin et al. 2017). The limited number of studies on the impacts of climate change on bees indicate that warming is driving rapid shifts in bee distribution at large and small geographic scales. Cold-tolerant groups of non-*Apis* bees, such as bumble bees, already show strong negative responses to global warming (Kerr et al. 2015; Soroye et al. 2020). A remaining gap of knowledge in this context is how thermal stress may interact with other stressors such as poor nutrition, exposure to xenobiotic compounds and diseases.

3 Phenology refers to the cyclic and seasonal patterns of plant and animal life cycles. For example, time of emergence, flowering, migration, etc.

Biotic Stressors: Pathogens and Pests

Malnourished bees that consume pollen and nectar with toxic levels of pesticide residues can be more susceptible to bacterial, fungal, protozoan, and viral diseases (O'Neal et al. 2018). It is now well-established that flowers play an important role in the epidemiology of bee diseases by facilitating the horizontal transmission of pathogens within and between species (Graystock et al. 2015; Figueroa et al. 2019). The role of flowers as hubs for pathogen transmission most likely predates bee decline but it has aided the recent spread of exotic pests and pathogens between native and introduced bee species (Wilfert et al. 2016). Indeed, the introduction of *A. mellifera* and another ~80 bee species to areas outside of their native range has led to the transmission of novel pathogens (Russo 2016; Hedtke et al. 2015). The spillover of pathogens and parasites has been demonstrated between honey bee species (e.g. *Varroa* mites; [vanEngelsdorp and Meixner 2010]), between bumble bee species (e.g. *Nosema ceranae*; [Cameron et al. 2016; Cameron et al. 2011]) and between honey bees and bumble bees (e.g. Deformed Wing Virus [DWV]; [Fürst et al. 2014]). While the spillover of pathogens from managed to wild bees has been demonstrated, the role of wild bees as passive carriers or reservoirs of honey bee pathogens is not well understood. It is plausible that wild bees can lead to the amplification of pathogen abundance in active hosts and this is a future research direction of pathogen dynamics in bees.

Although many honey bee pathogens have been detected in wild social and solitary bees, evidence for virulence of most honey bee pathogens in wild bees is still weak (Dolezal et al. 2016; Ngor et al. 2020). More studies characterizing the pathosphere[4] of wild bees are necessary to better understand the degree of threat that pathogens pose on non-*Apis* bee health. Studies indicate that solitary insect species have stronger immune systems than social ones, suggesting that solitary bees could be less susceptible to pathogens than social bees like honey bees (López-Uribe et al. 2016). As with any other aspect of bee biology, the diversity, fitness effects and epidemiology of bee pathogens have been better described and characterized for honey bees (Evans and Schwarz 2011; Schwarz et al. 2015). The general biology of the most common honey bee pathogens and their prevalence in non- *Apis* bees is described below.

4 The pathosphere refers to all the pathogens and their gene pool in a certain environment or organisms.

Fungi – The genus *Ascosphaera* comprises ~30 species of fungi that require bee hosts to complete their entire life cycle (Wynns et al. 2013). In honey bees, *Ascosphaera apis* causes a larval disease called chalkbrood in which larvae turn into "mummies" (hard and covered by white fungal spores). See Chapter 23 for more information regarding chalkbrood in honey bees. *Ascosphaera apis* is found in *A. mellifera*, *A. cerana* (the Asian honey bee) and other bees in the family Apidae. However, other species of *Ascosphaera* are common across multiple bee lineages, particularly among species of the family Megachilidae (Evison et al. 2012). For example, *Ascosphaera aggregata* has a widespread geographic distribution and it has been reported in several species of the genera *Megachile* and *Osmia*. The widespread distribution of *A. aggregata* is probably the result of the transport and introduction of *Megachile rotundata* for alfalfa pollination around the world (Wynns et al. 2013). The introduction of the *Osmia cornifrons* from Japan to North America for apple pollination also facilitated the introduction of a non-native *Ascosphaera* species to the New World (Hedtke et al. 2015).

Microsporidia – The lineage Microsporidia comprises a large group of intracellular parasites that includes the genus *Nosema,* a parasite that specifically attacks insect hosts. *Apis mellifera* can be parasitized by two *Nosema* species that invade the epithelial cells of the midgut and increase the nutritional requirement and mortality of the individuals (Paris et al. 2018). *Nosema apis* was the historical causal agent of nosemosis in *A. mellifera*, a disease associated with diarrhea and transmitted via oral-fecal and oral-oral routes. In contrast, *N. ceranae* is a more recent pathogen of *A. mellifera*, which was originally reported in Asian honey bees (*A. cerana*) but it is now widespread in Western honey bees, bumble bees, and other solitary species (Paxton et al. 2007; Fürst et al. 2014; Vaudo et al. 2018). For more discussion of *Nosema*, see Chapter 23. Bumble bees can be parasitized by both *N. ceranae* and *N. bombi,* and both pathogens can be transmitted from parasitized individuals from commercially reared colonies to individuals in the wild (Colla et al. 2006). The degree of pathogenicity of *Nosema* infections has been attributed to several factors such as pathogen genetics (Branchiccela et al. 2017), climate (Martín-Hernández et al. 2012), and host exposure to pesticides (Whittington and Winston 2003). The increased prevalence and pathogenicity of *N. ceranae* around the globe is one of the major threats to bumble bee species in North and South America (Cameron and Sadd 2020).

Trypanosomes – Several trypanosome species have been documented in bees. Trypanosomes occur primarily in the hindgut of bees in two forms: a motile flagellated form and a non-flagellated, rounded stage form. The latter form produces a layer on the gut epithelial surface. *Crithidia mellificae* and *Lotmaria passim* are the primary trypanosomatid parasites attacking honey bees, which have been correlated with higher colony winter losses (Williams et al. 2019; Ravoet et al. 2013). A close relative species, *C. bombi*, is a prevalent and highly pathogenic species among bumble bees. *Crithidia bombi* spillover from introduced *Bombus terrestris* to wild species has been implicated in the declines of multiple South American wild bumble bees (Schmid-Hempel et al. 2014).

Apicomplexa – *Apicystis bombi* is a neogregarine parasite that has been reported infecting a wide range of bee hosts – including both honey bees and wild bees – and has a widespread geographic distribution (Colla et al. 2006; Gamboa et al. 2015). These parasites are transmitted via the oral-fecal route, replicate in the midgut (but reside in fat body tissue) and thus can have implications on overwintering survival, immune response and fecundity. Even though it is a relatively understudied pathogen, *A. bombi* is highly prevalent among bees (Ravoet et al. 2014) and it warrants more attention as an important biotic stressor of non-*Apis* species.

Bacteria – Species belonging to five genera of bacteria have been reported as pathogenic in bees. Two of those species have been the focus of a multitude of studies in honey bees: *Paenibacillus larvae* (the causal agent of American Foulbrood) and *Melissococcus plutonius* (the causal agent of European Foulbrood). While these two pathogenic bacteria are specific to honey bees, two other bacterial species in the genus *Spiroplasma* (*S. apis* and *S. melliferum*) are pathogens of honey bees (Schwarz et al. 2014), bumble bees (Meeus et al. 2012), mason bees (Clark et al. 1985), and squash bees (López-Uribe et al. unpublished data). Both species of *Spiroplasma* can be found in the digestive tract or hemolymph of infected individuals. Infected honey bees exhibit symptoms of a disease commonly referred to as "May disease," which is characterized by a worker's inability to fly that can lead to bees being found crawling outside near the colony. Recent studies in honey bees have identified two other bacterial species, *Serratia marcescens* and *Lysinibacillus sphaericus*, as pathogenic to bees (Fünfhaus et al. 2018).

Viruses – Bee viruses can affect the lifespan, morphology, physiology, and behavior of bees and have been implicated in the massive declines of honey bees in North America and Europe. Unlike several eukaryotic pathogens such a *Ascosphaera* spp. or *Nosema* spp., most of these viruses are less host specific and can be detected across a wide range of hosts from phylogenetically diverse lineages including bees, ants, wasps, and cockroaches (Singh et al. 2010; Tehel et al. 2016). The two most widespread and prevalent viruses among bees and relatives are DWV and Black Queen Cell Virus (BQCV). Deformed Wing Virus is currently found in most managed honey bee colonies in areas where the ectoparasitic mite *Varroa destructor* is present. Although *V. destructor* cannot parasitize non-*Apis* bee species, the synergistic interaction between mites and DWV has led to a significant rise in DWV prevalence and titers in both honey bees and wild bees (Fürst et al. 2014). In contrast, the epidemiology of BQCV has not been directly linked to the distribution and dynamics of *V. destructor* and honey bees. However, BQCV has been found to be highly prevalent in wild bees and it is spatially correlated with its prevalence in honey bee colonies (Murray et al. 2019). There is no empirical evidence suggesting direct negative fitness effects of BQCV in wild bees (Dolezal et al. 2016). While many of these honey bee viruses are present in wild bees, the impact of coinfections of viruses and other pathogens on wild bee health is a less understood biological interaction (Graystock et al. 2016).

Conclusion

Knowledge about the diverse ecology of non-*Apis* bees can help develop predictive models about the responses of wild bee populations to environmental change. With the increasing demands for pollination in agriculture, more non-*Apis* bee species are being managed for pollination services (Aizen and Harder 2009). The domestication and management of more bee species will certainly lead to the transport and introduction of these species into areas where they are not native. Intentional introductions of bees into novel environments may lead to ecological problems such as resource competition and pathogen spillover (Wojcik et al. 2018; Vaudo et al. 2020). The practical implications of resource competition between managed honey bees and wild bees means that more research is necessary to predict the role of multiple interacting stressors on wild non-*Apis* bee species (Meeus et al. 2018). The ability to detect novel pathogens and prevent their spillover will be critical to the long-term persistence of wild bees in their native areas as agriculture continues to intensify and demands for pollination services increase. Using biomarkers and implementing monitoring programs to assess pollinator health of non-*Apis* bees is critical for the development of effective plans that can mitigate stressors of wild bee populations across the globe.

Acknowledgements

I would like to thank Kristen Brochu for comments on an earlier version of this chapter, and to Terry Kane and Cynthia Faux for their insightful comments and edits. This work was supported by the USDA NIFA Appropriations under Project PEN04716 Accession no. 1020527.

References

Aizen, M.A. and Harder, L.D. (2009). The global stock of domesticated honey bees is growing slower than agricultural demand for pollination. *Current Biology* 19 (11): 915–918.

Ascher, J.S., and Pickering, J. (2020). Discover Life bee species guide and world checklist (Hymenoptera: Apoidea: Anthophila). http://www.discoverlife.org/mp/20q?guide=Apoidea_species (accessed 11 January 2020)

Aufauvre, J., Misme-Aucouturier, B., Viguès, B. et al. (2014). Transcriptome analyses of the honeybee response to *Nosema ceranae* and insecticides. *PLoS One* 9 (3): e91686.

Branchiccela, B., Arredondo, D., Higes, M. et al. (2017). Characterization of *Nosema ceranae* genetic variants from different geographic origins. *Microbial Ecology* 73 (4): 978–987.

Brown, M.J.F. and Paxton, R.J. (2009). The conservation of bees: a global perspective. *Apidologie* 40 (3): 410–416.

Cameron, S.A. and Sadd, B.M. (2020). Global trends in bumble bee health. *Annual Review of Entomology* 65: 209–232.

Cameron, S.A., Lozier, J.D., Strange, J.P. et al. (2011). Patterns of widespread decline in North American bumble bees. *Proceedings of the National Academy of Sciences of the United States of America* 108 (2): 662–667.

Cameron, S.A., Lim, H.C., Lozier, J.D. et al. (2016). Test of the invasive pathogen hypothesis of bumble bee decline in North America. *Proceedings of the National Academy of Sciences of the United States of America* 113 (16): 4386–4391.

Cardinal, S. and Danforth, B.N. (2013). Bees diversified in the age of eudicots. *Proceedings of the Royal Society B: Biological Sciences* 280 (1755): 20122686.

Chan, D.S., Prosser, R.S., Rodríguez-Gil, J.L., and Raine, N.E. (2019). Assessment of risk to hoary squash bees (*Peponapis pruinosa*) and other ground-nesting bees from systemic insecticides in agricultural soil. *Scientific Reports* 9 (1): 11870.

Clark, T.B., Whitcomb, R.F., Tully, J.G. et al. (1985). Spiroplasma melliferum, a new species from the honeybee (*Apis mellifera*). *International Journal of Systematic Bacteriology* 35 (3): 296–308.

Colla, S.R., Otterstatter, M.C., Gegear, R.J., and Thomson, J.D. (2006). Plight of the bumble bee: pathogen spillover from commercial to wild populations. *Biological Conservation* 129 (4): 461–467.

Danforth, B.N., Cardinal, S., Praz, C. et al. (2013). The impact of molecular data on our understanding of bee phylogeny and evolution. *Annual Review of Entomology* 58 (1): 57–78.

Danforth, B.N., Minckley, R.L., and Neff, J.L. (2019). *The Solitary Bees: Biology, Evolution, Conservation*. Princeton University Press.

DiBartolomeis, M., Kegley, S., Mineau, P. et al. (2019). An assessment of acute insecticide toxicity loading (AITL) of chemical pesticides used on agricultural land in the United States. *PLoS One* 14 (8): e0220029.

Dolezal, A.G., Hendrix, S.D., Scavo, N.A. et al. (2016). Honey bee viruses in wild bees: viral prevalence, loads, and experimental inoculation. *PLoS One* 11 (11): e0166190.

Douglas, M.R., Sponsler, D.B., Lonsdorf, E.V., and Grozinger, C.M. (2020). County-level analysis reveals a rapidly shifting landscape of insecticide hazard to honey bees (*Apis mellifera*) on US farmland. *Scientific Reports* 10 (1): 797.

Evans, J.D. and Schwarz, R.S. (2011). Bees brought to their knees: microbes affecting honey bee health. *Trends in Microbiology* 19 (12): 614–620.

Evison, S.E.F., Roberts, K.E., Laurenson, L. et al. (2012). Pervasiveness of parasites in pollinators. *PLoS One* 7 (1): e30641.

Figueroa, L.L., Blinder, M., Grincavitch, C. et al. (2019). Bee pathogen transmission dynamics: deposition, persistence and acquisition on flowers. *Proceedings of the Royal Society B: Biological Sciences* 286 (1903): 20190603.

Fünfhaus, A., Ebeling, J., and Genersch, E. (2018). Bacterial pathogens of bees. *Current Opinion in Insect Science* 26: 89–96.

Fürst, M.A., McMahon, D.P., Osborne, J.L. et al. (2014). Disease associations between honeybees and bumblebees as a threat to wild pollinators. *Nature* 506 (7488): 364–366.

Gamboa, V., Ravoet, J., Brunain, M. et al. (2015). Bee pathogens found in Bombus atratus from Colombia: a case study. *Journal of Invertebrate Pathology* 129: 36–39.

Gill, R.J., Ramos-Rodriguez, O., and Raine, N.E. (2012). Combined pesticide exposure severely affects individual- and colony-level traits in bees. *Nature* 491 (7422): 105–108.

Goulson, D., Nicholls, E., Botías, C., and Rotheray, E.L. (2015). Bee declines driven by combined stress from parasites, pesticides, and lack of flowers. *Science* 347 (6229): 1255957.

Graystock, P., Goulson, D., and Hughes, W.O.H. (2015). Parasites in bloom: flowers aid dispersal and transmission of pollinator parasites within and between bee species. *Proceedings of the Royal Society B: Biological Sciences* 282 (1813): 20151371.

Graystock, P., Meeus, I., Smagghe, G. et al. (2016). The effects of single and mixed infections of Apicystis bombi and deformed wing virus in Bombus terrestris. *Parasitology* 143 (3): 358–365.

Hamblin, A.L., Youngsteadt, E., López-Uribe, M.M., and Frank, S.D. (2017). Physiological thermal limits predict differential responses of bees to urban heat-island effects. *Biology Letters*.

Hedtke, S.M., Blitzer, E.J., Montgomery, G.A., and Danforth, B.N. (2015). Introduction of non-native pollinators can lead to trans-continental movement of bee-associated fungi. *PLoS One* 10 (6): e0130560.

Kerr, J.T., Pindar, A., Galpern, P. et al. (2015). Climate change impacts on bumblebees converge across continents. *Science* 349 (6244): 177–180.

Lebuhn, G., Droege, S., Connor, E.F. et al. (2013). Detecting insect pollinator declines on regional and global scales. *Conservation Biology* 27 (1): 113–120.

López-Uribe, M.M., Sconiers, W.B., Frank, S.D. et al. (2016). Reduced cellular immune response in social insect lineages. *Biology Letters* 12 (3): 20150984.

López-Uribe, M.M., Ricigliano, V.A., and Simone-Finstrom, M. (2019). Defining pollinator health: assessing bee ecological, genetic, and physiological factors at the individual, colony, and population levels. *Annual Review of Animal Biosciences*.

Lozo, J., Berić, T., Terzić-Vidojević, A. et al. (2015). Microbiota associated with pollen, bee bread, larvae and adults of solitary bee *Osmia cornuta* (Hymenoptera: Megachilidae). *Bulletin of Entomological Research* 105 (4): 470–476.

Martín-Hernández, R., Botías, C., Bailón, E.G. et al. (2012). Microsporidia infecting *Apis mellifera*: coexistence or competition. Is *Nosema ceranae* replacing *Nosema apis*? *Environmental Microbiology* 14 (8): 2127–2138.

McArt, S.H., Urbanowicz, C., McCoshum, S. et al. (2017). Landscape predictors of pathogen prevalence and range contractions in US bumblebees. *Proceedings of the Royal Society B: Biological Sciences* 284 (1867).

McDonald, B.A. and Stukenbrock, E.H. (2016). Rapid emergence of pathogens in agro-ecosystems: global threats to agricultural sustainability and food security.

Philosophical Transactions of the Royal Society of London. Series B, Biological Sciences 371 (1709).

Meeus, I., Vercruysse, V., and Smagghe, G. (2012). Molecular detection of Spiroplasma apis and Spiroplasma melliferum in bees. *Journal of Invertebrate Pathology* 109 (1): 172–174.

Meeus, I., Pisman, M., Smagghe, G., and Piot, N. (2018). Interaction effects of different drivers of wild bee decline and their influence on host-pathogen dynamics. *Current Opinion in Insect Science* 26: 136–141.

Michener, C.D. (2007). *The Bees of the World*, 595. Baltimore: Johns Hopkins University Press.

Mikát, M., Černá, K., and Straka, J. (2016). Major benefits of guarding behavior in subsocial bees: implications for social evolution. *Ecology and Evolution* 6 (19): 6784–6797.

Minckley, R.L., Cane, J.H., and Kervin, L. (2000). Origins and ecological consequences of pollen specialization among desert bees. *Proceedings of the Royal Society B: Biological Sciences* 267 (1440): 265–271.

Murray, E.A., Bossert, S., and Danforth, B.N. (2018). Pollinivory and the diversification dynamics of bees. *Biology Letters* 14 (11).

Murray, E.A., Burand, J., Trikoz, N. et al. (2019). Viral transmission in honey bees and native bees, supported by a global black queen cell virus phylogeny. *Environmental Microbiology* 21 (3): 972–983.

Ngor, L., Palmer-Young, E.C., Nevarez, R.B. et al. (2020). Cross-infectivity of honey and bumble bee-associated parasites across three bee families. *Parasitology* 18: 1–5.

Nieto, A., Roberts, S.P., Kemp, J. et al. (2017). *European Red List of Bees*. European Commission.

Ogilvie, J.E. and Forrest, J.R. (2017). Interactions between bee foraging and floral resource phenology shape bee populations and communities. *Current Opinion in Insect Science* 21: 75–82.

Ollerton, J., Winfree, R., and Tarrant, S. (2011). How many flowering plants are pollinated by animals? *Oikos* 120 (3): 321–326.

O'Neal, S.T., Anderson, T.D., and Wu-Smart, J.Y. (2018). Interactions between pesticides and pathogen susceptibility in honey bees. *Current Opinion in Insect Science* 26: 57–62.

Otto, C.R.V., Roth, C.L., Carlson, B.L., and Smart, M.D. (2016). Land-use change reduces habitat suitability for supporting managed honey bee colonies in the Northern Great Plains. *Proceedings of the National Academy of Sciences of the United States of America* 113 (37): 10430–10435.

Paris, L., El Alaoui, H., Delbac, F., and Diogon, M. (2018). Effects of the gut parasite *Nosema ceranae* on honey bee physiology and behavior. *Current Opinion in Insect Science* 26: 149–154.

Paxton, R.J., Klee, J., Korpela, S., and Fries, I. (2007). *Nosema ceranae* has infected *Apis mellifera* in Europe since at least 1998 and may be more virulent than *Nosema apis*. *Apidologie* 38 (6): 558–565.

Raimets, R., Karise, R., Mänd, M. et al. (2018). Synergistic interactions between a variety of insecticides and an ergosterol biosynthesis inhibitor fungicide in dietary exposures of bumble bees (*Bombus terrestris* L.). *Pest Management Science* 74 (3): 541–546.

Ravoet, J., Maharramov, J., Meeus, I. et al. (2013). Comprehensive bee pathogen screening in Belgium reveals *Crithidia mellificae* as a new contributory factor to winter mortality. *PLoS One* 8 (8): e72443.

Ravoet, J., De Smet, L., Meeus, I. et al. (2014). Widespread occurrence of honey bee pathogens in solitary bees. *Journal of Invertebrate Pathology* 122: 55–58.

Rehan, S.M. and Toth, A.L. (2015). Climbing the social ladder: the molecular evolution of sociality. *Trends in Ecology & Evolution* 30 (7): 426–433.

Richards, M.H. (2011). Colony social organisation and alternative social strategies in the eastern carpenter bee, *Xylocopa virginica*. *Journal of Insect Behavior* 24 (5): 399–411.

Rogan, J., Ziemer, M., Martin, D. et al. (2013). The impact of tree cover loss on land surface temperature: a case study of Central Massachusetts using Landsat thematic mapper thermal data. *Applied Geography* 45: 49–57.

Roubik, D.W. (ed.) (1989). *Ecology and Natural History of Tropical Bees*. Cambridge University Press.

Russo, L. (2016). Positive and negative impacts of non-native bee species around the world. *Insects* 7 (4).

Schmid-Hempel, R., Eckhardt, M., Goulson, D. et al. (2014). The invasion of southern South America by imported bumblebees and associated parasites. *The Journal of Animal Ecology* 83 (4): 823–837.

Schwarz, R.S., Teixeira, É.W., Tauber, J.P. et al. (2014). Honey bee colonies act as reservoirs for two Spiroplasma facultative symbionts and incur complex, multiyear infection dynamics. *MicrobiologyOpen* 3 (3): 341–355.

Schwarz, R.S., Huang, Q., and Evans, J.D. (2015). Hologenome theory and the honey bee pathosphere. *Current Opinion in Insect Science* 10: 1–7.

Singh, R., Levitt, A.L., Rajotte, E.G. et al. (2010). RNA viruses in hymenopteran pollinators: evidence of inter-taxa virus transmission via pollen and potential impact on non-Apis hymenopteran species. *PLoS One* 5 (12): e14357.

Soroye, P., Newbold, T., and Kerr, J. (2020). Climate change contributes to widespread declines among bumble bees across continents. *Science* 367 (6478): 685–688.

Tehel, A., Brown, M.J., and Paxton, R.J. (2016). Impact of managed honey bee viruses on wild bees. *Current Opinion in Virology* 19: 16–22.

vanEngelsdorp, D. and Meixner, M.D. (2010). A historical review of managed honey bee populations in Europe and the United States and the factors that may affect them. *Journal of Invertebrate Pathology* 103: S80–S95.

Vaudo AD, Patch HM, Mortensen DA, Tooker JF, Grozinger CM. Macronutrient ratios in pollen shape bumble bee (Bombus impatiens) foraging strategies and floral preferences. *Proceedings of the National Academy of Sciences*. 2016 Jul 12;113(28):E4035-42.

Vaudo, A.D., Fritz, M.L., and López-Uribe, M.M. (2018). Opening the door to the past: accessing phylogenetic, pathogen, and population data from museum curated bees. *Insect Systematics and Diversity* 2 (5): 1–14.

Vaudo, A.D., Biddinger, D.J., Sickel, W. et al. (2020). Introduced bees (*Osmia cornifrons*) collect pollen from both coevolved and novel host-plant species within their family-level phylogenetic preferences. *Royal Society Open Science* 7 (7): 200225.

Whitehorn, P.R., O'Connor, S., Wackers, F.L., and Goulson, D. (2012). Neonicotinoid pesticide reduces bumble bee colony growth and queen production. *Science* 336 (6079): 351–352.

Whittington, R. and Winston, M.L. (2003). Effects of Nosema bombi and its treatment fumagillin on bumble bee (Bombus occidentalis) colonies. *Journal of Invertebrate Pathology* 84 (1): 54–58.

Wilfert, L., Long, G., Leggett, H.C. et al. (2016). Deformed wing virus is a recent global epidemic in honeybees driven by Varroa mites. *Science* 351 (6273): 594–597.

Williams, M.-K.F., Tripodi, A.D., and Szalanski, A.L. (2019). Molecular survey for the honey bee (*Apis mellifera* L.) trypanosome parasites *Crithidia mellificae* and *Lotmaria passim*. *Journal of Apicultural Research*: 1–6.

Wojcik, V.A., Morandin, L.A., Davies Adams, L., and Rourke, K.E. (2018). Floral resource competition between honey bees and wild bees: is there clear evidence and can we guide management and conservation? *Environmental Entomology* 47 (4): 822–833.

Wood, T.J., Gibbs, J., Graham, K.K., and Isaacs, R. (2019). Narrow pollen diets are associated with declining Midwestern bumble bee species. *Ecology* 100 (6): e02697.

Wynns, A.A., Jensen, A.B., and Eilenberg, J. (2013). Ascosphaera callicarpa, a new species of bee-loving fungus, with a key to the genus for Europe. *PLoS One* 8 (9): e73419.

Zattara, E.E. and Aizen, M.A. Global Bee Decline. *bioRxiv*. 2019 Jan 1:869784.

8

Honey Bee Nutrition

Randy Oliver

ScientificBeekeeping.com

Success with beekeeping largely depends on two factors: management of varroa, and provision of good nutrition to the colony – either from local flora, or by supplemental feeding, As a veterinarian, it is important to understand the nutritional needs of a colony, the flow of nutrients within the hive, the effects of nutrition upon colony population dynamics and immunocompetence, and how to recognize the signs of nutritional deficiency, In addition, you will likely be asked questions about supplemental feeding.

Understanding the Honey Bee Superorganism

In order to understand honey bee nutrition, it is important to appreciate that it is not about the individual worker bee, but rather about the well-organized *colony as a whole* in which some members gather food, but other members process or digest it and then distribute it via the "communal stomach" to all members of the colony. When food is abundant, it is shared equally by all members, and the colony thrives; when food is scarce, it is restricted according to a hierarchy determining which members starve – saving only the queen and a cohort of physiologically-specialized "diutinus" bees to hold the fort until food resources again become available.

Think of a honey bee colony as a warm-blooded superorganism the size of a small dog, but with an ability to "dissolve off" a quarter of its cells each day to forage over several square miles, and then to coalesce back into a cluster each evening. Those foragers seek only two food resources – nectar (a carbohydrate source) and pollen (providing protein, lipids, vitamins, minerals, sterols, and plant secondary metabolites). The only other things gathered by foragers are water (providing some minerals), and plant resins used to create propolis.

The warm-blooded honey bee superorganism grows, divides, ***or shrinks*** in response to food availability. Thus, a deep understanding of how the colony responds to nutritional inputs from the environment (or beekeeper) is critical for making recommendations on management or disease control.

Practical application: from a management perspective, the beekeeper may not always want the colony to be in growth mode. This is especially the case with regard to varroa management, since varroa reproduces most rapidly when a colony is growing.

Carbohydrates vs. Other Nutrients

The honey bee is all about its namesake honey – sugar stored as honey allows the *colony* to survive periods of dearth and cold winters. European honey bees, by forming a homeothermic cluster, are the only insects able to maintain elevated body temperatures throughout high latitude winters (creating a "tropical" in-cluster environment sheltered within a tree or hive cavity). A colony's stored food reserves also allow it to maintain a standing (but largely inactive) field force over winter, and to then build up in early springtime ahead of competing pollinator species.

Practical application: a colony can survive for months on a diet consisting solely of sugar. But in order to rear new bees, it requires the protein, lipids, sterols, vitamins, and minerals obtained from pollen.

Seasonal colony population dynamics thus follow the availability of pollen and nectar, with pollen generally being the most critical component. As illustrated in Figure 8.1, a colony passes through four phases each year:

- Survival through the dearth period, living on stored food reserves.
- Population buildup, sustained as long as pollen is abundant. In some climates, pollen availability may be

Honey Bee Medicine for the Veterinary Practitioner, First Edition. Edited by Terry Ryan Kane and Cynthia M. Faux.

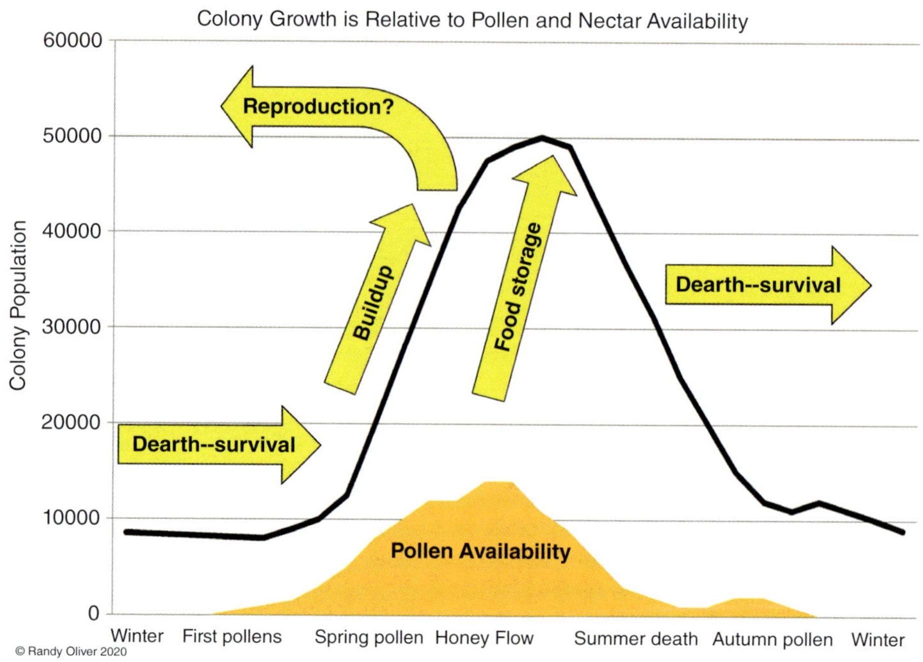

Figure 8.1 Relationship between pollen availability and colony population dynamics. Dates have been left out of the above chart, since colony population dynamics are driven by food availability, not calendar date or photoperiod. *Source*: Illustration by Randy Oliver.

bimodal or trimodal, so broodrearing (and thus colony population) will follow those dynamics (Rashad and Parker 1958).

- Once a colony has filled its living space (hive boxes, tree hole, etc.) and has adequate sealed brood ready to emerge, it will attempt reproduction by swarming.
- After reproduction, a colony's sole goal is to store enough honey to carry it through the next dearth period (which could span months of temperatures below freezing), until pollen again becomes available.

Practical application: The above image is to illustrate what occurs with *unmanaged* colonies. A veterinarian's input will generally be asked for regarding *managed* colonies, which are often supplementally fed in order to shift the curve for various reasons. The vet should be able to make informed recommendations regarding supplemental feeding – including whether it is even indicated.

Flowers depending on insect pollination bloom when it is normally warm enough for insect activity – typically above 50 °F (10 °C). So, in higher latitudes, the dearth period is associated with short days and cold winter weather, with pollen availability mainly during spring, early summer, and again in autumn. In lower latitude areas, however, pollen, and nectar availability are often a function of rainfall, so in such regions in the northern hemisphere (e.g. much of the California coast), "spring" (as far as bees are concerned) may begin with rainfall in November, and the "winter" dearth period is between July through November.

Practical application: colony population dynamics are not based upon calendar dates, but rather track the availability of nectar and pollen. Sugar syrup and pollen supplement are therefore important management tools. The beekeeper, by supplemental feeding of protein and sugar syrup, can encourage colonies to rear brood when they otherwise would not, and to store "honey" for the future even when no flowers are in bloom.

It is not just the overall colony population that changes with food availability – the entire age structure and physiology of a colony changes. (Figure 8.2).[1]

Practical Application: Figure 8.2 reflects a colony in a long-winter climate with intense nectar and pollen flow in spring and summer. In regions closer to the equator, plants flower all year, so the graph could be flat – with broodrearing through the year, and no development of "winter bees." An important point from a management perspective is that the age structure of the adult bee population, as well as the prevalence of host pupae for varroa reproduction, reflect colony nutrition – a colony receiving high protein input will consist of a younger workforce, but

1 http://scientificbeekeeping.com/understanding-colony-buildup-and-decline-part-2

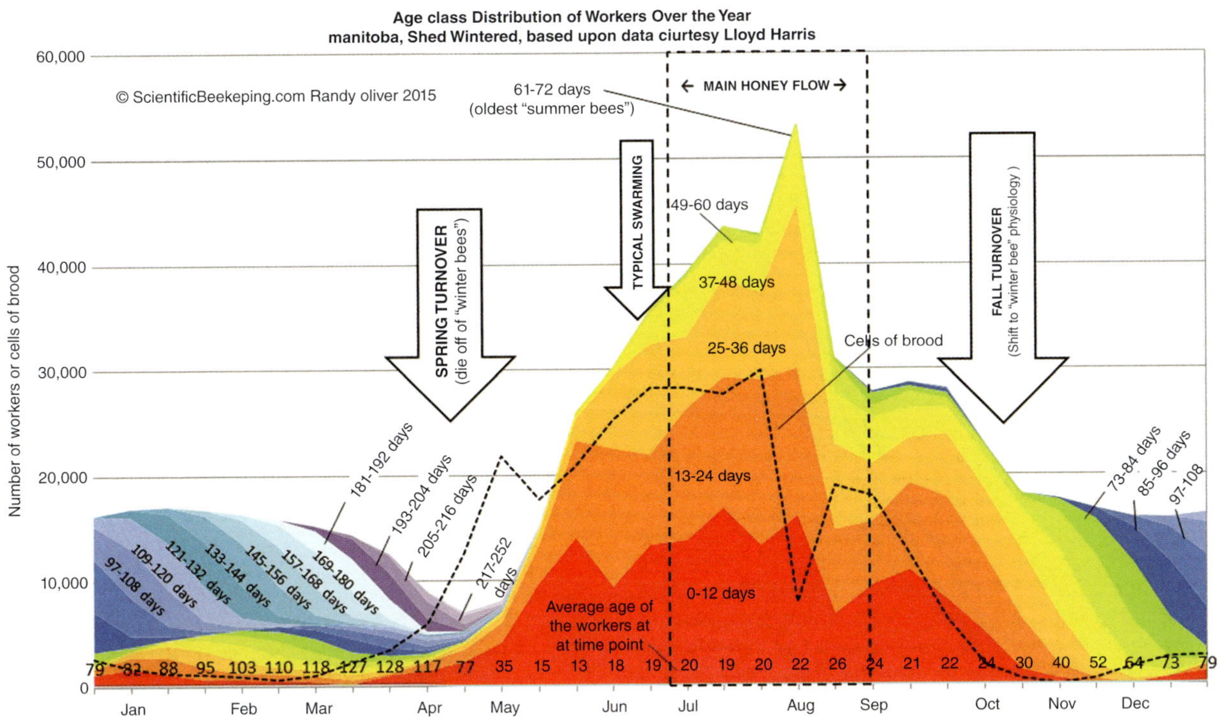

Figure 8.2 Age structure of a colony over a year in Manitoba, Canada. Number ranges and colors represent age-profiles of groups of workers. Dashed line indicates total amount of brood (tracking pollen availability) Note that in this case, there were decreases in broodrearing during a cold May rainstorm, and again in August when bees had filled brood combs with nectar. Immediately above the months are figures representing the average age of the workers at that time, showing how rapid the population turnover is when pollen is abundant. *Source*: Chart created by the author from data courtesy of Lloyd Harris. © Scientific Beekeping.com Randy Oliver 2015.

also be rearing more mites. Note also that in early spring that there are considerably more bees in the larval and pupal stages than as adults.

Nutritional Needs of the Colony Over Seasons

- During dearth (Oct-Mar in the region in Figure 8.2), the worker population consists primarily of long-lived "diutinus" or "winter" bees having physiology to store as much protein as possible in their fat bodies. ***During this phase, the colony has little demand for protein, ideally utilizing autumn pollen reserves preserved as beebread*** for the minimal broodrearing that takes place.
- In spring (early March above), stimulated by the first tree pollens, the colony initiates serious brood rearing, requiring the aged "winter" bees to convert to short-lived nursing/foraging physiology and behavior, resulting in most of them dying within a month. Therefore, the colony undergoes a "spring turnover" – a race in which the rapidly-aging "winter bees" must rear an abundance of new workers to replace them before they die. ***In this phase, demand for sugar and protein increases weekly.*** Adverse weather at this time can result in nutritional stress and colony losses due to starvation. *Beekeeper intervention with sugar or pollen substitute can mitigate those problems.*
- As the colony cluster expands over the next six to eight weeks, its protein requirement increases dramatically, until the queen is egg laying at her maximum (for a good queen this results in 10 deep frames containing brood (70% average coverage). ***This is the period of maximal protein demand.***
- The population continues to grow linearly (limited by its queen's egg laying rate and nutrition, mid-April through June above), until adult bee emergence equilibrates with worker death (mean worker survivorship during spring/summer is roughly 35 days). This growth may be interrupted by swarming.
- After (or during) the main honey flow, brood rearing decreases (as does the nonproductive workforce), ***and colony nutritional needs drop.***
- **For successful wintering, beekeepers must ensure that winter bees are reared in autumn under low stress (especially regarding varroa), *and high protein availability*, so the "winter bees" can fully develop their critical fat bodies.**
- Once broodrearing ceases (due to lack of incoming pollen), the Autumn Turnover occurs, during which most

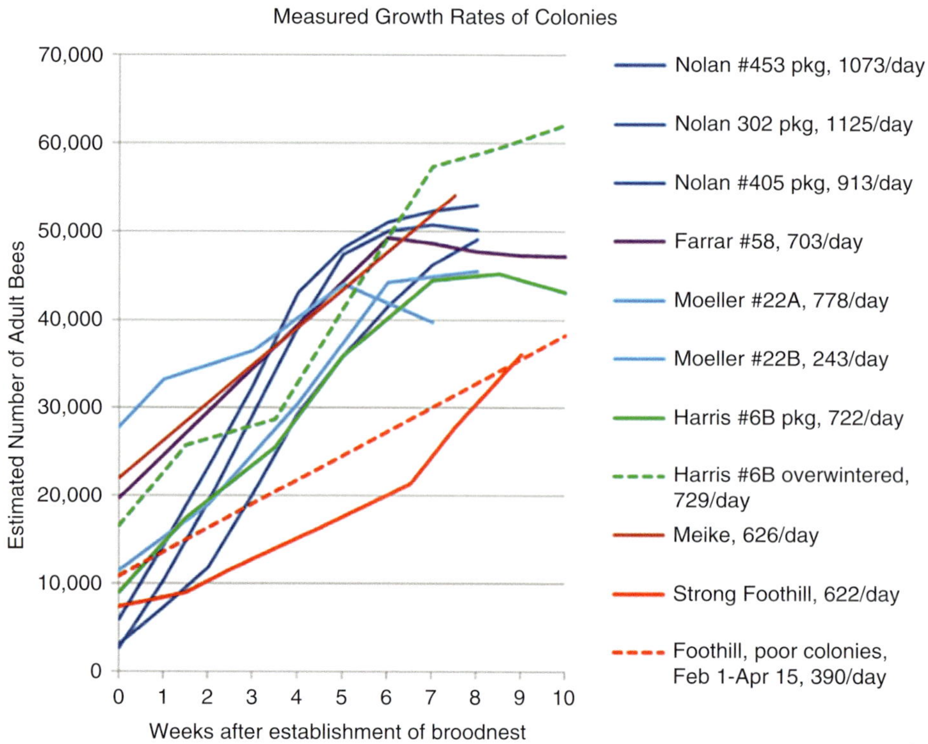

Figure 8.3 Growth rates of both package and overwintered colonies, generally starting with small populations. The green and blue plots are mostly historical data for exceptionally fast-growing colonies in the Northern Plains, the red for some of the author's in the Sierra Foothills. The steeper the slope, the more rapid the growth. The number following each legend entry is the net gain in bees per day for the most rapid period of growth. *Source*: Illustration by Randy Oliver.

bees previously engaged in broodrearing "disappear" via natural attrition. Only bees with diutinus physiology (and guts full of pollen) remain in the winter cluster.

Colony Growth Is Not Exponential

Although colony growth at first glance appears to be sigmoid (Figure 8.1), in fact, from when the cluster establishes its broodnest until it reaches maximum population, growth rate is essentially linear (Nolan 1932) – limited by the number of eggs the queen lays per day. It is instructive to compare the measured growth rates, as far as additional numbers of adult bees per day, from various studies (Figure 8.3).[2]

Practical application: a colony of bees, during its essentially linear growth phase in springtime, grows at a rate of 500–700 additional workers per day. This works out to a sustained total colony body weight gain of over a pound per week for 8–10 weeks. The requirement for pollen is therefore also linear. However, while strong colonies will outcompete weak colonies in the same yard, in early spring the strongest colonies are often the first to starve during inclement weather, due to their more-rapid consumption of their stores.

Colony Growth Rate Compared to Vertebrates

If we compare that growth rate to the fastest-growing land animal in agriculture today – the hybrid broiler chicken (which instead of taking the 19 weeks to grow to slaughter in the '50s, now takes only six with optimal feeding). Since the chicken's weight crosses the 2 lb line, we can handily compare its growth rate to that of a small colony of bees – a freshly-hived 2 lb package. The following chart compares honey bee colony body weight gain to that of a modern broiler chicken (Figure 8.4).[3]

Clearly, a bee colony grows at an amazing rate. But that isn't the half of it! The chicken has the immense

2 I arrived at rough estimates of my populations by counting the number of frame interstices filled with bees, and then multiplying that number by 1820 (the approx. number of bees on a covered frame, as determined by a large data set of computer-generated counts vs. our visual gradings).

3 http://scientificbeekeeping.com/sick-bees-17-nosema-the-smoldering-epidemic

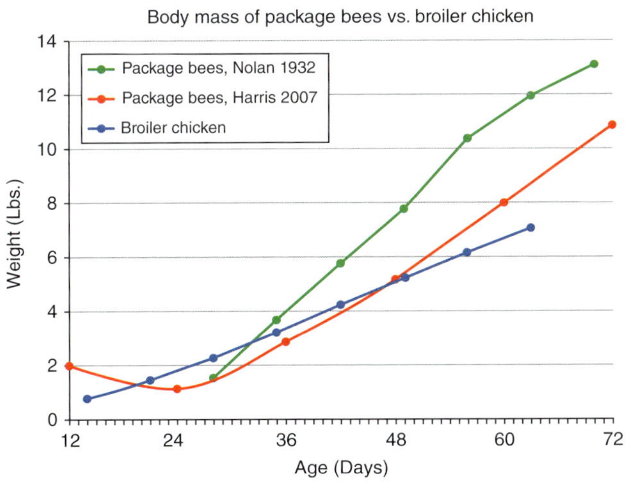

Figure 8.4 Comparison between the weight gain of package bees vs. that of a broiler chicken. The red curve shows how a package loses population until the first brood emerges. After that point, honey bee colonies grow considerably faster than even the fastest-growing chicken! For additional comparison, if a 7 lb human baby were to grow at the same rate, it would weigh 56 lbs two months following its birth (as opposed an expected 11 lbs)! *Source*: Illustration by Randy Oliver.

advantages of being penned in a warm room and provided with optimally-formulated chow, and maintains a compact body size, insulated by feathers. On the other hand, the industrious bee colony has to forage over a dozen square miles, spending a tremendous amount of energy in the process, as well as wasting a vast amount of body heat to the environment. Plus, the honey bee colony does this while shedding around a thousand bees every day due to natural attrition.

Practical application: Beekeepers expect our bees to perform a feat of rapid growth beyond the capability of perhaps any other animal! The bee superorganism can pull off this prodigious feat only by being extremely efficient at digesting and utilizing its food. That is why honey bees are completely dependent upon two of the richest foods in nature – pollen and nectar. A colony would starve to death on the sorts of diets that most organisms are adapted to.

Nutrition as a Limiting Factor to Colony Growth

The intrinsic limiting factors for colony growth are:

1) The potential egg laying rate of the queen.
2) The amount of cluster-warmed comb available for brood rearing.
3) Disease issues.
4) Adequate nutrition.

The incredible linear growth rate of the colony, under optimal conditions, comes into equilibrium with the natural survivorship rate of the workers at about 40 times the queen's daily egg laying rate. Thus, colony population typically tops out at 40 000–60 000 adult bees.

The first three factors can be largely controlled by the beekeeper and their vet. The last factor – nutrition – depends on the local floral environment, timing, and weather, but the beekeeper has the option of providing supplemental nutrition.

Nutritional Requirements of a Hive

Unfortunately, the specific nutritional requirements of honey bees are not yet as well defined as for other livestock. Over the course of a year, a colony of bees consumes – for maintenance – an estimated 250–400 pounds of sugars and 50–75 pounds of pollen (Standifer et al. 1977). A "productive" colony can, in addition, store well over a 100 pounds of sugars in the form of surplus honey.

Carbohydrates

While individual honey bees may exhibit very high metabolic rates, the colony as a whole is extremely energy efficient, consuming as little as a pound of sugar (as honey) per week during winter. However, once a colony begins broodrearing in late winter, its energy consumption climbs rapidly to around 5 lbs/week (as honey) during early broodrearing, increasing to at least 10 lbs/week during linear growth. Later in the season, as broodrearing tapers off, energy needs again decrease.

Practical application: Generally, it is not *necessary* to feed a colony sugar, provided that it has enough honey reserves. However, beekeepers routinely feed syrup (or other forms of sugar) to provide colony energy requirements (i.e. to avert starvation), to stimulate broodrearing or the drawing of comb, or to allow its conversion into "honey" as an energy reserve ahead of times of dearth.

Protein and Amino Acids

An individual worker bee consumes pollen (essentially its sole protein source) during the first two weeks after their emergence (approximately 4 mg/d), then has little need for protein for the rest of its life (which it then obtains by begging jelly from the nurse bees). The colony during dearth or winter, when it is populated by bees exhibiting diutinus physiology, requires little protein until it begins broodrearing. Body reserves of protein in the diutinus "winter bees" can initiate, but not sustain, broodrearing. Thus late-winter and spring broodrearing depend on reserves of beebread, or fresh pollen gathered by foragers. In order to realize optimal growth, a strong growing colony during spring or early summer must consume and efficiently process a minimum of 2–3 pounds of high-protein pollen per week.

Practical application: Beekeepers can feed pollen substitutes to colonies when nature does not provide adequate nutrition. For example, in order to supply hives for almond pollination in late winter, commercial beekeepers often feed 10–15 lbs (in total) of high-quality pollen substitute to a hive, divided between late summer and the six weeks before bloom (an in-depth discussion of protein substitutes occurs later in this chapter).

Honey bees, like other animals, depend on protein quality – typically determined by the ratio of the essential amino acids in the protein. Research on honey bee amino acid requirements by de Groot (1953) determined optimal ratios relative to tryptophan. Unfortunately, tryptophan analysis requires a different test than other amino acids and is thus often not determined. The author has since compared DeGroot's ratios to those of royal jelly, known nutritious pollens (Oliver 2020a), and that of other animals, to determine optimal amino acid ratios relative to the often-limiting amino acid in pollen: histidine (Table 8.1). These ratios can be used to evaluate protein qualities of substitute pollen sources.

Table 8.1 A suggested essential amino acid ratio (relative to Histidine content) for the bee diet. This optimal ratio is nearly identical to that of vertebrate animals.

Essential amino acid	Ideal ratio
Histidine	1
Arginine	2
Isoleucine	2
Leucine	3
Lysine	2.5
Methionine + Cysteine	1
Phenylalanine + Tyrosine	2
Threonine	2
Tryptophan	0.5
Valine	2.5

Serine, glycine, and proline, though not essential for growth, exert a stimulating effect at suboptimal growth levels. *Source*: deGroot, AP (1953).

Vitamins and Minerals

Although certain vitamins and minerals are essential to the honey bee diet, bees' requirements are not yet well enough known to merit review here and are likely only an issue for colonies subsisting solely upon artificial diets, unless the soil in a region is deficient in a specific trace element (Smith et al. 2019) (Figure 8.5).

On the other hand, as in the case of California's San Joaquin Valley, the high selenium content of the soil and water may cause toxicity to pollinators.

Practical application: Some minerals may also be obtained from water sources, since water foragers appear to prefer water containing the scent of manure, mud, chlorine, or salts of sodium or magnesium.

Lipids

According to Wright et al. (2018), palmitic, linoleic (omega-6), and alpha-linolenic (omega-3) acids, comprise an average of 60–80% of pollen fatty acids. These three, in addition to oleic and stearic acids, are the primary fatty acids in honey bee bodies. It is unclear, however, which lipids are nutritionally important.

Sterols

Insects lack the ability to synthesize cholesterol, making plant sterols essential nutrients. The major tissue sterol of brood is 24-methylenecholesterol (24-mCh), which is also the main sterol found in many pollens (Smith et al. 2019; Villette, 2015). Not surprisingly, it also is an important component of "major royal jelly protein 1," produced by nurse bees to feed other colony members (Tian et al. 2018; Ferioli et al. 2014; Kodai et al. 2007). Nurse bees appear to concentrate, accumulate, and then provide larvae – via jelly – with more 24-mCh than other sterols (Svoboda et al. 1981). However, the presence of fairly constant, but lower, levels of sitosterol, isofucosterol, and campesterol present in prepupae suggest a need of developing larvae for other sterols, which bees may convert into 24-mCh (Svoboda et al. 1986, Oliver 2020b).

Other Phytochemicals

There are potential effects of plant secondary metabolites on pollinators. To date, their importance in honey bee nutrition is largely unknown, although evidence suggests that some may affect bee behavior, longevity, and immunity to pathogens, or may act as

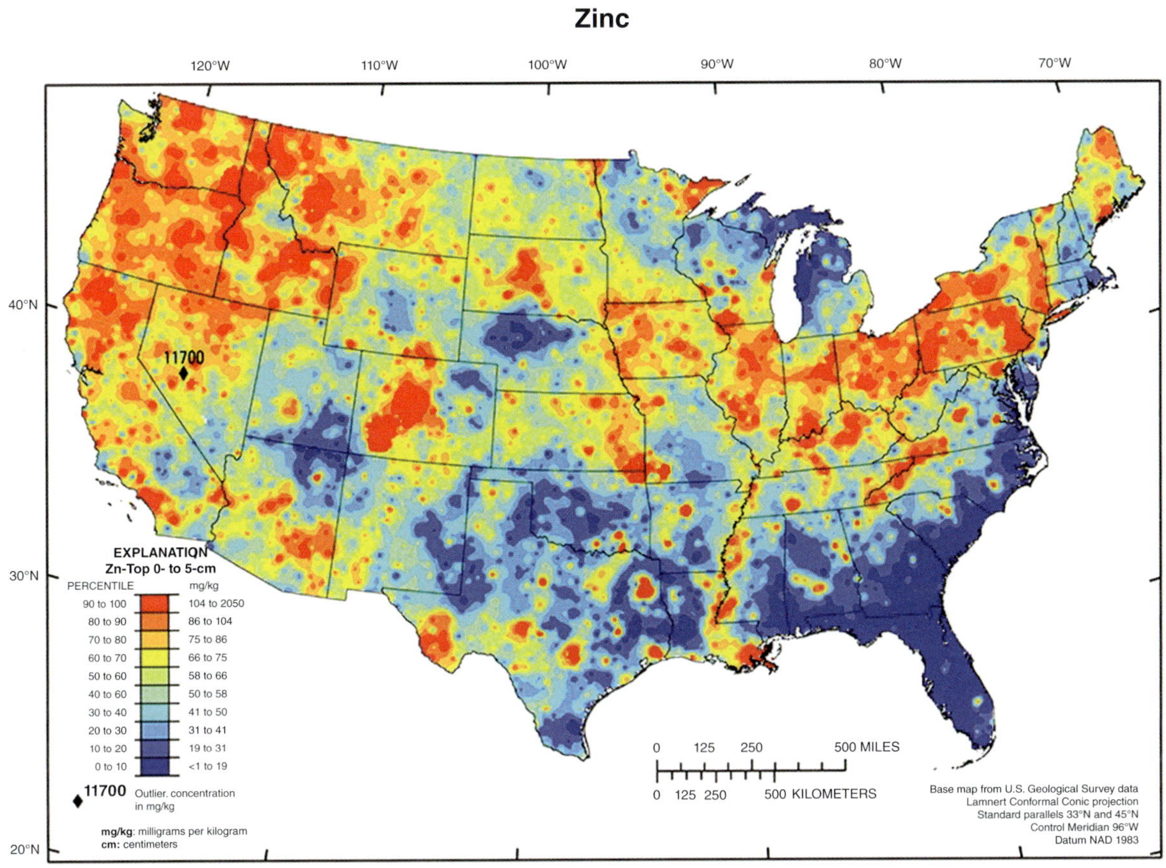

Figure 8.5 The availability of trace elements in the soil varies greatly across the country. Zinc is an indispensable trace element for honey bees (Zhang et al. 2015). The map above indicates the concentrations of zinc in the upper soil level, with red indicating the highest concentrations, blue a deficiency. Based upon this map, it would be reasonable to conclude that it would be more likely for honey bees to suffer zinc deficiency in Florida than in Oregon. *Source*: Maps such as the one above can be downloaded from (Smith et al. 2019).

"neutraceuticals" to regulate detoxification processes (Mao et al. 2013).

Pollen Consumption

Estimates of pollen requirements per hive per year differ, since pollens vary in protein content and overall nutritional profile. Nutritional benefits from pollen are constrained by its first limiting nutrient (Liebig's Law), since nurse bees appear to cease production of jelly rather than produce a nutritionally incomplete product. In both natural pollens and pollen substitutes, researchers have focused on crude protein or certain amino acids as being the limiting nutrients (Somerville 2005), but it may well be other, as yet unrecognized, trace elements or phytochemicals that are actually the limiting factor.

Practical application: Although the focus of researchers has typically been on high-protein pollens (Somerville 2005), colonies are also able to grow using low-protein pollen, if enough is available. With artificial diets, then, some high-protein pollen substitutes may be out-performed by an equal weight of lower-protein natural pollen, due perhaps to it having a better balance of essential nutrients.[4]

Food Handling and Processing in the Hive

Both nectar and pollen are gathered by bees in their last stage of behavioral development – when they undertake foraging. Foragers may consume nectar for their own needs, but are unable to digest pollen, and do not normally consume it. In the hive, nectar, and pollen are handled,

4 http://scientificbeekeeping.com/a-comparative-test-of-the-pollen-sub

processed, and stored very differently (Johnson 2009). In brief, pollen (and some incoming nectar) is stored near the brood (for easy access by nurse bees), whereas honey is stored above and to the sides of brood (allowing bees to utilize the warmth of the cluster to access it in cold weather). Brood frames, therefore, typically exhibit a dynamic ring of any surplus pollen and beebread surrounding the brood, with honey above and to the sides (Figure 8.6). Brood frames showing other distributions can be useful in diagnosing colony disorders (see Section 8.6).

Nectar

Nectar is the main carbohydrate source for a colony. Depending on source plant species and environmental conditions, the sugar concentration of nectar typically ranges from 20% to 50% (Shaw 1956). The primary sugars in honey are sucrose, glucose, and fructose, however, the proportions of these sugars vary by source plant species. Nectar may also contain amino acids such as glycine and proline, and secondary metabolites which may make the nectar more or less attractive to honey bees (or even toxic).

A forager carries nectar back to the hive in its foregut ("crop"), in which it is filtered to remove pollen, dust, bacterial spores, or other solid contaminants. It is also slightly acidified, and its sucrose largely converted to glucose and fructose by the enzyme invertase.

At the hive, a nectar forager must find a mid-aged nectar-processing bee to accept its load – foragers do not unload nectar directly into cells. Nectar receivers then share the nectar with other workers, creating a "communal stomach" for the colony.

Practical application: The shared "communal stomach" created by the transfer of nectar and jelly among the workers, informs every bee in the hive about the current nutritional status of the colony. In general, colonies thrive when all the workers are enjoying plenty of nectar, *provided that the colony also has nutritious pollen available* (Figure 8.7).

Surplus nectar not immediately needed by the colony is dehydrated by nectar processing bees until it reaches a sugar concentration of approximately 83%, at which point it is too hyperosmotic for yeasts to grow. This now "ripened" honey is then capped with beeswax for long-term preservation. Freshly-capped honey has an air space between the beeswax cap and the honey surface, and has a white appearance. As the honey ages, the air diffuses out through the cap, resulting in the cap eventually touching the honey surface, then giving the cappings a darker appearance.

Practical application: Bees preferentially consume low-sugar nectar (which does not require dilution with water to utilize), leaving honey as a food reserve. Thus, dilute syrup (1 : 1 or less w:w sugar: water) is more "stimulatory" to a colony (and more rapidly taken from a feeder), than is concentrated (2 : 1) syrup, which is typically used by bees for winter

Figure 8.6 Pollen foragers pack their loads into empty cells near young larvae. Such fresh pollen is preferentially consumed by the nurses. In times of pollen surplus, unconsumed pollen remains in a typical band arcing above the brood, where it undergoes lactic acid fermentation for preservation until needed. There is also a dynamic interchange of fresh nectar in the broodnest, with surplus being converted to low-moisture honey, and stored above and to the sides of the brood area. *Source*: Photo courtesy of Randy Oliver.

Figure 8.7 The degree of distension of the bees' abdomens is a quick field sign to tell whether the colony is sharing abundant nectar. In this photo, note the two bees in the center engaging in trophallaxis, and the wide spacing between the anterior sclerites (hardened abdominal plates) of the other bees, indicating that the crops of those bees contains a nectar reserve. Bees in a nectar-hungry hive will exhibit shorter abdomens, with the pigmented bands closer together. *Source*: Photo courtesy of Randy Oliver.

reserve storage. The bees in the photo below, due to the presence of abundant nectar and pollen, will be highly stimulated to rear brood.

Practical application: One can learn a great deal about a colony's condition by inspection of the brood-to-honey interface. From Figure 8.8, we can infer that this colony had until recently been consuming its way upwards into its honey reserves, as evidenced by the sunken cappings on the honey above, indicating (due to lack of air space below the cappings) that the honey has been capped for some time, with no recently-capped honey below it (which would have white cappings). A nectar/pollen flow has been going on for at least eight days, as evidenced by the presence of sealed brood right up to the pollen band. This colony is experiencing "backfilling" of nectar, and unless provided with additional drawn comb, may soon initiate swarming.

When there is a surplus of incoming nectar, the mid-age bees process it as rapidly as possible into honey and store it above, and around, the broodnest. Should they run out of open comb in which to store the honey, they will activate their wax glands and begin to produce beeswax with which to construct new comb (Figure 8.9)

Practical application: The freshly-deposited "white wax" shown Figure 8.10 is an indicator to the beekeeper of four things: (i) that there is a strong nectar flow on, (ii) that the colony lacks enough *drawn* comb for the storage of that nectar, (iii) that the colony may soon "plug out" and either crowd out brood rearing, miss a potential honey crop, or swarm, and (iv) that conditions are right for the drawing of foundation or the adding of honey supers.

Figure 8.8 This colony is enjoying a surplus of nectar and pollen, as evidenced by the bees' distended abdomens, and the presence of unconsumed beebread and cells full of ripening nectar. Note that the cells with nectar butt up against dark well-aged honey with sunken cappings, indicating that this nectar flow is following a dearth period. *Source*: Photo courtesy of Randy Oliver.

Problems with Nectar and Honey

Ideally, honey will remain liquid and stable, due to its high fructose content. However, some honeys with a high proportion of glucose will crystallize in the combs, making it difficult for bees to consume in winter (partly because they would need to add scarce water) or, under some conditions, the more-dilute supernatant may ferment or result in

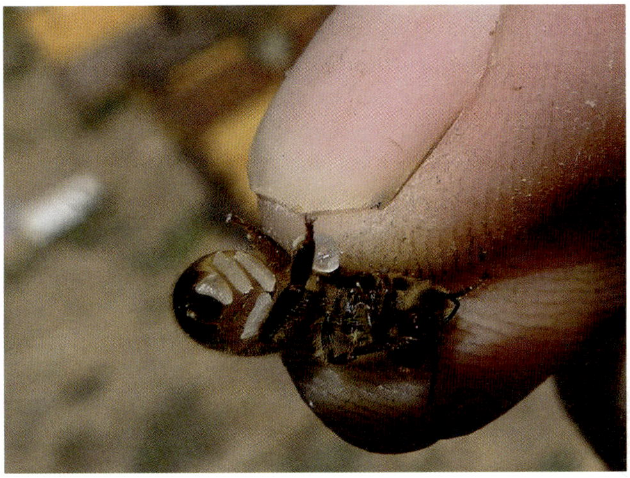

Figure 8.9 Wax flakes produced by the eight wax glands on a worker's abdomen. The production of beeswax is metabolically expensive for bees, and only occurs when a colony has run out of enough storage space for incoming nectar. When this occurs, the beekeeper will observe "white wax" being deposited around the hive. *Source*: Photo courtesy of Randy Oliver.

Figure 8.10 The presence of "white wax" indicates to the beekeeper that a strong nectar flow is on, and that the colony needs more room in which to store the incoming nectar for processing into honey. *Source*: Photo courtesy of Randy Oliver.

causing dysentery,[5] which can rapidly spread nosema spores, degrading overall colony health.

Honey that has been stored for some time at an elevated temperature may darken, and the fructose may convert to hydroxymethylfurfural (HMF), which may reach toxic levels (Shapla et al. 2018). This also may occur in sugar syrup stored in warm conditions or in metal containers.

Bees may also gather the sugary excrement of sap-sucking insects and convert it to "non-floral honey" called "honeydew" or "forest honey." This is typically darker and more strongly-flavored than floral honey, and may contain indigestible complex carbohydrates, a high mineral content, and plant secondary metabolites. Beekeepers report that colonies winter poorly on some honeydews.

Practical application: Beekeepers, often thinking that because honey is "natural," it is the best food upon which to overwinter colonies. In practice, colonies may fare better on sugar syrup. Some honeys, such as canola, alfalfa, or ivy, may crystallize, and make poor winter stores; the same with some honeydews. In this case, the beekeeper may take that honey for extraction, and then feed sugar syrup back to the hive to be processed into better-quality winter stores.

The vet will likely get calls from beekeepers concerned about colonies exhibiting signs of "dysentery" (Figure 8.11). It is a common misconception that dysentery indicates that a colony is suffering from infection by nosema. In fact, the author has been unable to find a single reference to support that claim, and several to the contrary,[7] nor do his examinations of the fecal material support the nosema claim(Figure 8.12)[8].

Practical application: Dysentery appears to be most commonly caused by diet, induced perhaps most commonly by the issue of water imbalance (from consumption of high-moisture stores). Other causes may be from gut dysbiosis (again perhaps due to poorly digested foodstuffs), or infection by pathogens other than nosema.

Practical application: Expect nosema levels to spike during periods of high pollen consumption, but unless the *prevalence* of infection exceeds more than 20% of the workers in the hive, it is unlikely to be problematic. However, if dysentery occurs due to poor diet, nosema spores will be rapidly transmitted

Figure 8.11 The above is an example of a colony suffering from dysentery, resulting in the explosive discharge of the rectal contents of desperate bees as they exit the hive. Such defecation, should it occur *within* the hive, can result in the rapid transmission of nosema spores, thus engendering the common misconception that dysentery is a sign of (as opposed to a fomite for) nosema infection. Infection by nosema cannot be diagnosed by field signs; the vet would need to use a microscope[6] to detect the presence of spores or prevalence of infection. *Source*: Photo courtesy of Monique Vescia.

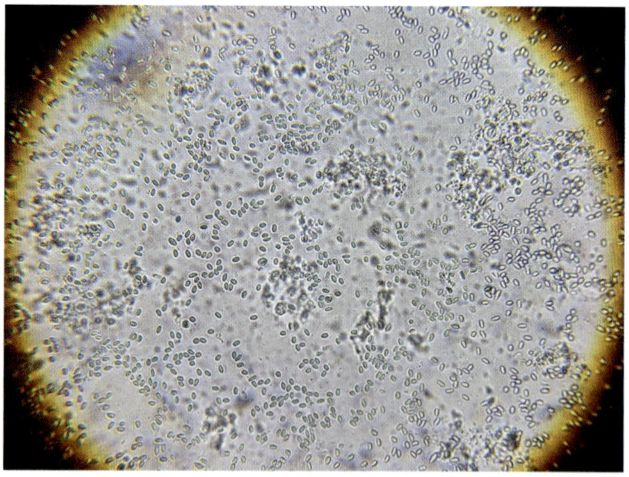

Figure 8.12 It requires microscopic examination of a bee's diluted gut contents at 400x magnification, to determine whether or not nosema is present. This bee exhibits a very high level of infection. The nosema spores (the numerous small oval objects) in this slide happened to rest at two different levels, thus exhibiting how a slight adjustment of the fine focus will cause their centers to glow (at the right) or dim (at left) with all spores exhibiting nearly identical smooth, regular outlines (as opposed to irregular yeast cells). *Source*: Photo courtesy of Randy Oliver.

5 http://scientificbeekeeping.com/the-causes-of-dysentery-in-honey-bees-part-2/
6 http://scientificbeekeeping.com/sick-bees-part-13-simplemicroscopy-of-nosema
7 http://scientificbeekeeping.com/its-common-knowledge-that-nosema-causes-dysentery-but-is-it-actually-true/
8 http://scientificbeekeeping.com/the-causes-of-dysentery-in-honey-bees-part-2.

within the hive. So, it's important to first address diet-caused dysentery. Provide the colony with a low-moisture sugar source.

Toxic Nectars and Pollens

Not all flowering plants are friendly to honey bees. Some plants, in order to favor specific pollinators or repel predators, produce toxic allelochemicals in their nectar or pollen (Rhoades and Bergdahl 1981). In their native ranges, locally-adapted bees avoid poisonous plants, but managed bees imported to the New World face plants to which they are not adapted. Some of these plants can poison a colony (Figure 8.13).

Below is a list of plants that may exhibit bee toxicity (Table 8.2).

Practical application: Established beekeepers are often aware of such toxicity and avoid areas with such plants, but novice beekeepers may be unaware of the dangers. In California, Buckeye poisoning is common, but may be mitigated by the amount of alternative attractive forage during its flowering period.[9] Brood poisoning may be difficult to identify by the veterinarian – they may need to consult with local experienced beekeepers.

Honey bees, as generalists, encounter those natural toxins when they forage, and in many cases are able to detoxify them. Indeed, some plants add potentially toxic attractants to nectar – the stimulants caffeine and nicotine being two examples. Bees may preferentially forage on these and other potentially toxic nectars, even in the case of nectar contaminated with systemic neonicotinoid insecticides (also stimulants) which may, at high enough levels, exhibit adverse effects upon the bee and its colony.

Colony Checkpoints for Avoiding Toxins

In the case of nectar, a multi-step screening process minimizes intake of toxins into a hive. Evolutionary pressure results in honey bees coevolving to adapt to endemic plants with toxic nectar, by instinctually avoiding them. Individual foragers are often able to recognize unacceptable levels of toxins in a nectar (Liao et al. 2017) since many phytotoxins are bitter or irritating. On the flight home, the forager may become sickened by its nectar, die, or intentionally sacrifice itself through "altruistic self-removal." In-hive nectar receiver bees would likely refuse any nectar that had previously bothered them.

9 Personal experience in California. If blackberry boom does not coincide with that of Buckeye, entire apiaries may be killed. Corn Lily poisoning is also of concern to beekeepers who move hives to mountain meadows.

Figure 8.13 Honey bees are not native to California, and thus did not evolve with California Buckeye, the attractive blossoms shown above. Buckeye's toxic pollen, if the predominant food to the colony, can cause serious brood mortality or even colony death. The signs are wholesale dying of brood, a distinctive foul odor in the hive, and snowy-white, recently-emerged bees walking on the combs. *Source*: Photo courtesy of Randy Oliver.

Table 8.2 Known toxic nectars and pollens.

Common name	Genus
California Buckeye, Horse-chestnut	*Aesculus*
Locoweed, Milk vetch	*Astragalus*
Dodder	*Cuscuta*
Summer titi	*Cyrilla*
Yellow jessamine	*Gelsemium*
Mountain laurel	*Kalmia*
Black nightshade	*Solanum*
Corn lily	*Veratrum*
Death camas	*Zigadenus*
New Zealand laurel	*Carynocarpus*
Angelica	*Angelica*
Tea tree	*Camellia*
Balsa tree	*Ochrama*
Kōwhai tree	*Sophora*

Figure 8.14 Bees collecting sawdust on a sunny morning in January, when there was little natural pollen to be had. *Source*: Photo courtesy of Randy Oliver.

Figure 8.15 Almond pollen and nectar both contain bee-toxic amygdalin, yet not only do honey bee colonies thrive on almond pollen and nectar, but foragers are preferentially attracted to the taste of amygdalin. As with all plant toxic metabolites and agricultural pesticides, it is the dose that makes something "toxic." *Source*: Photo courtesy of Randy Oliver.

So, the colony in most, *but not all,* cases has mechanisms in place to protect it from toxic nectars.

Surprisingly, no such screening appears occur with pollen. Pernal and Currie (2002) demonstrated that foragers discriminate between pollens by particle size and aroma, not by nutritional content.[10] As a result, foragers, especially during times of pollen dearth, may collect material of similar size to pollen, such as dust from animal feed troughs, or even sawdust (Figure 8.14). Since pollen foragers do not ingest the pollen they gather, they do not detect if it is toxic, and may collect pollens noxious enough to kill the colony (e.g. California Buckeye).

Another issue of toxic pollen may arise in springtime when systemic insecticides in "treated" seed is carried to the hive and stored in beebread. Since the dry insecticide does not penetrate a forager's cuticle, the forager cannot sense its toxicity. However, tainted pollen can then kill nurse bees that consume it once it's in the hive. It is common in the Corn Belt to see dead, recently emerged bees at the colony entrance during the planting of seed-treated corn; luckily, the colony generally rapidly recovers once the contaminated pollen has been consumed.

Practical application: Colonies may quickly recover from fast-acting insecticides that kill the foragers before they can return to the hive. Slower-acting insecticides, fungicides, and adjuvants can contaminate the beebread. This may adversely affect the nurse bees, and occasionally the brood. In most cases, however, the nurses do not transfer the pesticide to the jelly that they feed the queen and larvae, thus protecting those critical colony members from outside toxins.

The Case of Almond Pollen

One demonstrably toxic pollen of great interest is that of the almond tree (Figure 8.15), since roughly two-thirds of all managed hives in the US are taken to almond pollination in California each winter. Almond pollen and nectar may contain high concentrations of amygdalin (normally metabolically degraded to benzaldehyde and cyanide – both also toxic)(London-Shafir, L, et al. (2003)). Yet bees *thrive* on almond pollen and nectar, despite the fact that in some cultivars the concentration of amygdalin approaches that of acute bee toxicity (Detzel and Wink 1993).

Although each *individual* pollen forager exhibits strong species fidelity when in the field (down to the plant strain cultivar), the forager force, *as a whole*, appears to bring back a mixture of pollen types, packing them in layers in the cells (Figure 8.16). Since pollen from different plant species varies in its constituents, this behavior may help the colony experience a more complete overall nutritional profile.

Practical application: In some regions, or when surrounded by monocrops, colonies may be limited to pollen from a single species of plant that may be nutritionally deficient. Thus, despite a single type of pollen being abundant, the nurse bees may be unable to produce adequate amounts of jelly to share with the rest of the colony. Beekeepers are happy when

10 Other researches make the case that foragers can differentiate the nutritional value of pollens, but the evidence is weak.

Figure 8.16 Typical layers of pollen from different returning foragers. The pollen has been removed from the cells to demonstrate the layering. *Source*: Photo courtesy of Randy Oliver.

Figure 8.17 Beebread exhibiting a range of colors generally guarantees a colony good nutrition. On the other hand, even an abundance of pollen from some plant species does not supply a complete nutritional profile. *Source*: Photo courtesy of Randy Oliver.

they see a rainbow of colors of pollen being brought in (Figures 8.17–8.19)[11].

In some areas, foragers, tricked by sugary secretions, may collect the spores of rust fungi[12] and store them as beebread (Figures 8.20 and 8.21; Table 8.3).

Practical application: A colony foraging mainly on rust fungus will quickly go downhill. In the author's experience, providing such colonies with high-quality pollen substitute will correct the problem within days.

11 Other researchers make the case that foragers can differentiate the nutritional value of pollens, but to this author, the evidence to date is unconvincing.

12 Commonly in the California foothills.

Figure 8.18 Plants of the sunflower family, such as Dandelion, although very attractive to bees, are deficient in certain amino acids (Nicolson and Human 2013; Frias et al. 2016). The colony must gather additional types of pollen in order for the nurses to produce jelly. *Source*: Photo courtesy of Randy Oliver.

Figure 8.19 The owner of this prune orchard seeded mustard, knowing that beekeepers would be willing to place hives for pollination free of charge, due to the high nutritional value of the brassica pollen and nectar being produced before the trees come into bloom. *Source*: Photo courtesy of Randy Oliver.

Pollen Processing in the Hive

Unlike nectar foragers, pollen foragers deliver their loads directly, adding a little nectar, and use their heads to pack it into cells adjacent to larvae. Fresh pollen is preferentially consumed by nurse bees (although the ethanol from the initial fermentation appears to be phagostimulatory) (Carroll et al. 2017). Stored pollen undergoes yeast and bacterial fermentation over several hours, until lactic acid bacteria lower the pH and stop microbial degradation (Anderson et al. 2014a). Excess pollen collected in autumn

Figure 8.20 Although is colony is packed with what looks like pollen, the orange beebread is largely composed of rust fungus spores. Note the spotty brood pattern and dying larvae, Indicating the noxious effect of this food source. *Source*: Photo courtesy of Randy Oliver.

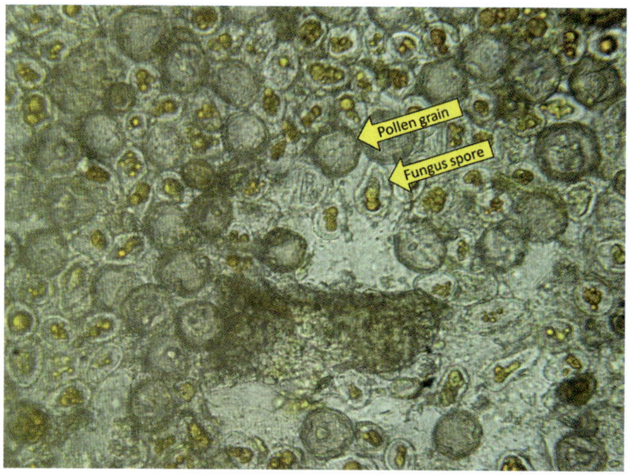

Figure 8.21 The above beebread at 400x magnification. The roundish grains are plant pollen; the irregular objects that look like "fried eggs" are spores from a rust fungus. *Source*: Photo courtesy of Randy Oliver.

Table 8.3 Pollens, by plant family, known to have overall good nutritional profiles for honey bees, taking into account amino acids, lipids, and sterols, and based upon lab feeding studies and field experience.

Rosaceae – apples, almonds, pears, loquats, almonds, peaches, apricots, plums, cherries, strawberries, blackberries, raspberries.
Brassicaceae – the mustard family.
Solanacea – the nightshade family, although not normally a major pollen source.
Papaveraceae – the poppy family.

Examples of nutritionally incomplete, not preferred, or poorly utilized pollens would be dandelion and other aster/sunflowers, alfalfa, field corn, squash, or melon, cranberry, blueberry, grasses, some eucalyptus, and conifers.

is fermented into beebread for long-term storage for winter broodrearing prior to early pollen flows. The fermentation of pollen into beebread appears to be performed for preservation, rather than to improve its nutritional value (Anderson et al. 2014a).

Practical application: look at the width of the "pollen band" around the brood. A band wider than a few cells indicates that the foragers are keeping ahead of pollen demand. If no pollen reserve is evident, the colony may benefit from supplemental protein feeding.

Digestion of Pollen

The only adult bees able to derive substantial nutrition from pollen are nurse bees and recently emerged drones; pollen consumed by foragers or the queen quickly passes through the gut undigested. Although larvae can digest pollen, it is an unnecessary addition to their diet of jelly.

Bees digest pollen surprisingly quickly – it takes only three hours to pass through the "stomach" (the ventriculus or midgut). Unlike most vertebrates, bees do not secrete digestive juices into the foregut – instead they shed the epithelial cells lining the gut, which burst and release their contents, including digestive enzymes. Pollen contents are digested through the pores of the tough pollen coat (the exine), leaving collapsed shells to be easily observed with magnification in hind gut contents.

Some pathogens may compromise the gut. Deformed Wing Virus, *Malipighamoeba apis,* and *Nosema* all likely affect digestive function.

Practical application: Nosema propagates most efficiently in the midguts of nurse bees during times of intense pollen consumption, damaging their intestinal villi and adversely affecting their digestion and ability to produce jelly. Nosema also causes infected workers to transition to foraging behavior earlier in their lives. Nonetheless, unless over about a fifth of the workers are infected, the overall affect upon a colony may be relatively minor and is typically self-correcting.

Consumption of Pollen

The honey bee digestive tract contains a well-defined and critical microbiome of "core" endosymbiotic bacteria. Newly-emerged bees inoculate themselves when they chew their way through their bacteria-charged cell capping, and then by cleaning cells. After a day or so, they start consuming pollen, with maximum consumption occurring approximately five days after emergence. By that time, they have a well-developed gut community (Anderson et al. 2015), which appears to be important for colony health.

When engaged in nursing behavior, workers convert extra protein into an egg-yolk precursor, vitellogenin (a zinc-binding glycolipoprotein), which is stored in the bees' fat bodies and hypopharyngeal glands. Worker bees, being nonreproductive, have no need for egg production and have evolved to co-opt vitellogenin in several critical aspects of individual and colony-level biology (Nelson et al. 2007). Vitellogenin serves as a protein reservoir in the bees' all-important fat bodies (critical for a range of functions), as a precursor for "major royal jelly protein," for transgenerational immune priming (Salmela 2015), and as an antagonist to juvenile hormone – important in the behavioral development and longevity of adult bees, especially in queens and long-lived diutinus "winter bees" (Corona et al. 2007).

Vitellogenin-rich fat bodies (Figure 8.22) are critical for worker longevity: During nutritional stress, fat body reservoirs allow workers to switch to diutinus physiology to wait out the dearth (potentially for months at a time). This reservoir in "winter bees" also allows some brood rearing even in the absence of pollen income.

As their hypopharyngeal ("brood food") glands develop, nurses begin producing jelly to feed larvae and other adults. Once they reach around two weeks of age, they shift into other duties, and cease producing the enzymes necessary to digest pollen, and live mainly on sugars. Upon transitioning to foraging, a bee's fat bodies atrophy, likely to save weight.

Practical application: a colony rich in vitellogenin reserves can better handle nutritional, pathogen, weather, or pesticide stress. Supplemental feeding of pollen substitute can help to prepare colonies for dearth periods, or allow them to commence or sustain broodrearing during a natural pollen dearth.

Jelly as the Currency of Protein in the Hive

Perhaps the most important aspect of bee nutrition for veterinarians to understand is that the colony uses the conversion of pollen into jelly to effect the distribution of nutrients in the hive – one that informs every bee in the hive to the nutritional status of the colony as a whole (Schmickl and Crailsheim 2004) (Figure 8.23). As explained by Crailsheim (1992).

> Since [nurses] are the protein-digesting temporal caste within the colony they should not have to depend on jelly from another nurse. However, if poor pollen supply causes a reduced willingness to exchange or to spend jelly, then these trophallactic contacts might provide them with information about availability of protein in the hive.

Practical application: The queen does not run the hive – she serves only as the ovary of the colony. It is the nurse bees who control the currency of protein – the jelly. They provide the queen with her sole diet (at perhaps twice her body weight per day), they decide whether larvae get fed, and they supply foragers with protein needed for body maintenance.

Figure 8.22 Fat bodies in the abdomen of a forager (left) and nurse or winter bee (right). Bees with fully-developed fat bodies are more resistant to stress, and can have extended lifespans. *Source*: Irene Keller et al. (2005). Reprinted by permission of Taylor & Francis Ltd.

Figure 8.23 Nurse bees sharing jelly via trophallaxis. Only colonies with well-fed nurses can replace workers lost through attrition. And, if food is abundant, the nurses can allow a colony to grow in population. *Source*: Photo courtesy of Randy Oliver.

It is by sharing jelly via trophallaxis that nurses "run" the show.

Jelly has three components: a lipid-rich secretion from mandibular glands, a protein-rich secretion from hypopharyngeal glands, plus nectar, or diluted honey from the crop. The two glandular components gel when mixed in a nurse bee's mouthparts. Differences in jelly mixture composition, or other added components, fed to worker or queen larvae ("royal jelly"), or to adults of various castes or age, are not yet fully understood (Figure 8.24).

Field Assessment of Colonies' Nutritional Status

Entrance Assessment

Nectar and pollen availability can be assessed by a brief observation of returning foragers at the hive entrance (Figure 8.25).

Figure 8.25 The abdomen of the returning forager at the lower left is distended with a crop full of nectar, indicated by a stretching of the first abdominal sclerites. The forager above it has a medium-sized load of yellow pollen packed in its corbicula (but does not appear to be returning with nectar). Another forager has a load of gray pollen, indicating that a mixed diet is coming in. If the majority of foragers are returning with similar loads, the colony is likely healthy, contains a functioning queen, and is experiencing good nutrition. *Source*: Photo courtesy of Randy Oliver.

In-Hive Assessment

Practical application: There are four important in-hive visual indicators of nutritional status:

- **Whether the colony is depleting, or adding to, its honey reserves;**
- **The amount of beebread in the pollen band around the brood;**
- **The presence of abundant larvae with good survivorship; and most importantly,**
- **The amount of jelly fed to larvae.**

Figure 8.24 A queen cell opened to show a second-instar larva in a queen cell floating in a surplus of jelly (which will eventually be completely consumed). Jelly is to honey bees as mother's milk is to mammals and serves as the "currency of protein" within the hive. Queen larvae, such as the one above, are fed to excess on jelly with more added nectar than that given to worker larvae. On this diet, a developing queen larva exponentially increases her weight an incredible 2000-fold in a period of only six days. *Source*: Photo courtesy of Randy Oliver.

Carbohydrate Income Assessment

Upon opening a hive, check for presence of "white wax" and freshly built comb (Figure 8.10), which suggests that the colony is enjoying a surplus of incoming nectar.

A colony that has unfilled drawn comb available generally cannot be stimulated to produce white wax. In such a case, the beekeeper can check to see if a nectar flow is occurring by holding a comb horizontally and giving it a quick vertical shake to see whether fresh nectar is dislodged. If so, the colony is said to be "shaking nectar" (Figure 8.26).

Practical application: the colony in the above photo is clearly on a strong nectar flow, but since there is no white wax apparent, it still has unfilled drawn comb available. On the other hand, the colony in Figure 8.10 has obviously filled all its combs and should be immediately given (preferentially) drawn comb, or foundation to be drawn.

Protein Status Assessment

A quick inspection of a center frame from the upper brood chamber (Figure 8.27) is generally sufficient to assess nutrients other than carbohydrates – notably protein.

The inspection of combs deserves more attention. Most everything that the vet needs to know about the colony's

Figure 8.26 This colony is said to be "shaking nectar," meaning it is bringing in nectar in excess of its immediate requirements. When "nectar shake" is observed, the beekeeper should ensure that there is enough empty comb space for its processing and storage. *Source*: Photo courtesy of Randy Oliver.

Figure 8.27 A typical central brood comb from the center of a brood chamber in early springtime. This colony will soon consume all the remaining aged honey, replacing it with brood. The presence of surplus beebread, a fairly solid brood pattern, well-fed young larvae (not visible), and honey reserves indicate good nutritional status. *Source*: Photo courtesy of Randy Oliver.

nutritional status can be gained from a quick glance of the honey-to-brood interface on a comb pulled from the center of the upper brood chamber (Figure 8.28).

The image below displays bright, dull-surfaced freshly stored pollen, and shiny and darker fermented beebread (Figure 8.29).

Practical application: Figures 8.29 and 8.30 illustrate where a vet should pay attention to at initial inspection, since a glance at the interface between honey reserves and the brood will inform them of

Figure 8.28 The interface above tells you most all you need to know about the nectar flow, pollen flow, and nutritional status of the colony. *Source*: Photo courtesy of Randy Oliver.

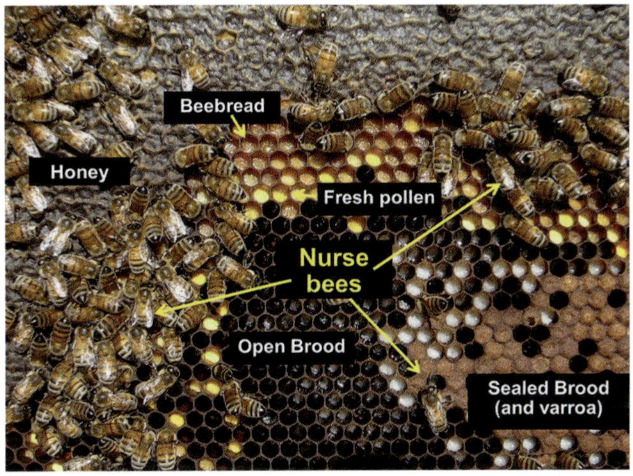

Figure 8.29 Pollen and nectar are initially placed throughout the brood area, but preferentially consumed by nurse bees from cells adjacent to brood. This dynamic removal results in both ready access to food for nurses (Schmickl and Crailsheim 2004), as well as the distinctive band of surplus beebread typically found at the interface between the brood (typically reared near the center of combs), and the honey (remaining at the periphery above and to the sides of the brood) (Camazine 1991).

most of what they need to know about colony condition and health.

Practical application: In addition to beebread, there are two additional protein reserves – the fat bodies of the nurse bees, and any unsealed brood, which is readily cannibalized and recycled in times of nutritional stress.

Ripened honey is capped with white beeswax above an air space. Over time, air diffuses through the capping, and the cap eventually sags and touches the honey – at which point it darkens. Thus, in the two figures above,

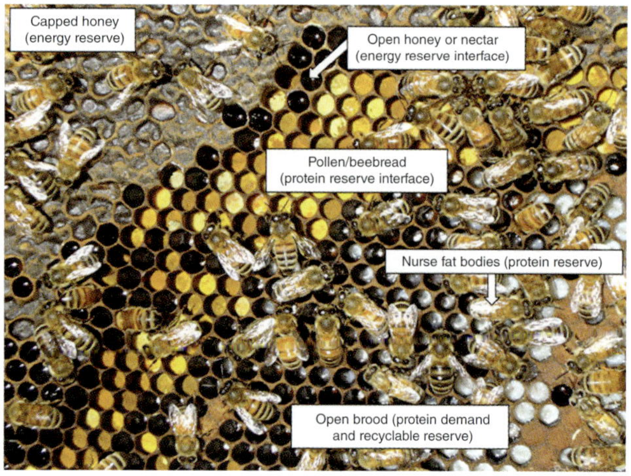

Figure 8.30 There are two dynamic interfaces with which one can assess the status of food consumption or storage as reserves. The "energy reserve interface" will be newly-uncapped honey if the colony is consuming its reserves, or fresh nectar if there is a surplus. The presence of a band of beebread at the "protein reserve interface" indicates that there is a reserve of pollen above the colony's immediate needs. *Source*: Photo courtesy of Randy Oliver.

Figure 8.31 Note the unripened shiny nectar surrounding the brood, the abundance of stored pollen, and the presence of drone cells at the bottom edge – all indications of a nectar flow and a colony in full-buildup mode. This rapidly growing (as evidenced by there being a far greater proportion of sealed than open brood), appears to have reached its limits for broodnest expansion, indicating a shortage of empty comb in the hive. Unless provided with more space in which to store nectar, this healthy and vigorous colony will likely initiate swarm preparation. *Source*: Photo courtesy of Randy Oliver.

one can tell the colonies are eating their way into honey reserves, rather than storing surplus nectar. This is normal during much of the year when there is not a surplus of incoming nectar, and indicates that the colony is losing, rather than gaining critical reserves. Compare this to Figure 8.31.

Figure 8.32 A colony living hand-to-mouth, as indicated by the small amount of beebread, and the honey being uncapped to provide energy. However, it has plenty of recently-stored honey reserves, as indicated by the light-colored cappings. *Source*: Photo courtesy of Randy Oliver.

Figure 8.33 A frame of healthy brood with a broad ring of pollen stored on the periphery. This would be typical of a rapidly expanding colony in spring, alarmingly with no honey reserves. The combination of a reserve of beebread, coupled with a solid brood pattern, indicates adequate nutrition. However, the only carbohydrates appear to be some fresh nectar immediately above the pollen band, the location indicated by the behavior of bees after being smoked to stick their heads into open cells of nectar to fill their crops.

A narrow band of pollen may indicate that a colony is living hand-to-mouth (i.e. barely meeting immediate protein needs); but this is not unusual for a growing colony (Figure 8.32). A rapidly-growing colony in springtime may consume a day's pollen harvest overnight – under such conditions, due to lack of reserves, the colony could cut back on broodrearing after only a day of rain.

Compare the frame in Figure 8.32 to the one in Figure 8.33.

One indicator of good nutritional status and freedom from brood disease is to confirm the survivorship of larvae (Figures 8.34 and 8.35).

Practical application: Keep in mind that the colony arranges its broodnest in three dimensions, as opposed to our frame-by-frame view. Thus, a central brood frame fully filled with brood may not present *any* pollen or honey reserves (as above). However, frames immediately adjacent to brood may be fully filled with beebread, and the next frame out, honey.

Another indicator of nutritional status is that a colony will rear drone brood only in times of good nutrition (Figure 8.36).

Figure 8.34 Decent larval survivorship is indicated by an even-age grouping of 4th-instar larvae, such as above – with few missed cells or replaced larvae. The proportion of empty cells – around 10% – is a bit higher than optimal and would be cause for further inspection. Compare to Figure 8.35. *Source*: Photo courtesy of Randy Oliver.

Figure 8.35 The above would generally be considered to be a "solid brood pattern," indicating a good queen and high larval survivorship. *Source*: Photo courtesy of Randy Oliver.

Figure 8.36 Colonies often build drone cells in the space between the two brood chambers. These cells are then unavoidably broken open when the brood chambers are separated for inspection. The presence of such drone brood indicates that the colony is enjoying good nutrition, since the nurses in a *hungry* colony will readily cannibalize drone brood, cease feeding drones, and push them out of the hive. *Source*: Photo courtesy of Randy Oliver.

Practical application: Seeing drone brood on the top bars is an indication that a colony is in good nutritional shape. The vet should immediately use the opportunity to inspect such exposed drone brood for the presence of varroa mites. If, early in the season, more than a small percentage of the pupae are infested, it would call for action against varroa at that time.

The honey bee colony treats any eggs and open brood as a protein reserve. During times of nutritional stress, the nurses will cannibalize that brood to convert its nutrients back into jelly and fat bodies to be used for long-term survival of the existing adult bees (Figure 8.37).

The colony in Figure 8.37 did not die, but the setback prevented it from later storing enough winter reserves (much less a surplus for harvest). Not all colonies are so lucky; below is the forensic sign for colony that died of starvation (Figure 8.38):

Practical application: Starvation is a common cause of colony loss, not so much during winter, but during spring buildup, when the colony is rapidly burning through its honey reserves. It is not unusual during springtime to find a colony with most of the workers not yet dead, but barely responsive. A dribble of sugar syrup over them will allow them to feed each other and rapidly recover, and then move to an inserted frame of honey or other sugar source.

Later in the season, when pollen is scarce, nurse bees will cannibalize eggs and larvae without hesitation, recycling their nutrients into jelly. At such times, nurses may follow

Figure 8.37 This rapidly growing colony was hit by a snowstorm during its spring buildup. This photograph was taken after four days of confinement, during which time the colony had consumed all its honey and pollen stores, and then cannibalized its unsealed brood (note that virtually all the cells, within and outside of the brood area, are devoid of contents). Similar hives which had been emergency fed did not engage in cannibalization and continued to build normally. *Source*: Photo courtesy of Randy Oliver.

Figure 8.38 In the final stages of starvation due to lack of honey, the workers pack themselves into empty cells headfirst. This sign presents a clear diagnosis of starvation, typically accompanied by a layer of dead workers on the bottom board. The entire colony population communally shares the last of their reserves, and typically starves as one. *Source*: Photo courtesy of Randy Oliver.

the queen, consuming a proportion (or all) of the eggs that she lays each day (rather than feeding the emerging larvae), resulting in a "spotty brood pattern" (Figure 8.39).

Practical application: During pollen and/or nectar dearth, the nurse bees will often cannibalize the eggs laid by the queen. The resulting spotty brood is an artifact of poor nutrition and should not be confused with disease. One can tell if the problem is dietary by examining for the *amount* of jelly being fed to the larvae.

Figure 8.39 Typical "spotty brood" during a pollen dearth after the main honey flow. Although the cappings of the honey at the top have already sunken, the lack of travel stain on the cappings indicates that the honey is fairly recent. Note that there is some fresh nectar just below the honey, but no signs of pollen. The brood pattern is "spotty" (i.e. has multiple empty cells) due to nurses cannibalizing eggs or larvae.

Monitoring the Jelly

Should there be any question on the nutritional status of a colony (such as when one questions the adequacy of incoming pollen), a quick and accurate assessment can be made by monitoring the amount of jelly being fed to 2nd instar larvae.

There are five larval instars, progressing at roughly a day each, with the 5th instar getting capped over. *The 1st instar larvae are typically fed only a small amount of jelly.* When they grow to second instar, nurses in a well-nourished colony will cover the bottoms of their cells completely with milky jelly (Figure 8.40).

However, if nurses are deficient in any essential nutrient, they will reduce the amount of jelly given to larvae, providing them with just enough to keep them growing. Larvae in a healthy colony are fed over a thousand times a day (Nelson et al. 1924), so lack of jelly indicates that the nurses are unable to feed them properly.

Practical application: The author has found that monitoring the *amount* of jelly being fed to the larvae is a quick and reliable method of determining colony nutritional status. Below are two examples each of good and poor nutritional status (Figures 8.41–8.44).

Practical application: Restriction of the amount of jelly being fed to the larvae indicates overall colony nutritional stress. This is also an indication that the bees are also likely immunocompromised and susceptible to stress. The feeding of a high-quality pollen substitute can turn the colony around in a matter of days – assuming that is the beekeeper's management goal at that time.

Figure 8.40 First- and second-instar larvae "swimmin' in jelly." Nurse bees in this colony sense an abundance of resources. Note that the jelly goes wall to wall. These larvae are being fed optimally, will exhibit high survivorship, have strong immunocompetence, and emerge to be longer-lived workers. *Source*: Photo courtesy of Randy Oliver.

Figure 8.42 The larvae above are being fed far less than optimally. Note the irregular size distribution – indicating that the nurses are likely cannibalizing some eggs, with the queen then returning to "fill in the holes." This is a strong indication of a nutritionally challenged colony. *Source*: Photo courtesy of Randy Oliver.

Figure 8.41 Adequately-fed larvae. Notice the shimmering reflection of jelly surrounding most of the larvae and the similarity in size of adjacent larvae, indicating high larval survivorship. However, some larvae are only being moderately fed, suggesting that the colony is at its nutritional edge. *Source*: Photo courtesy of Randy Oliver.

Figure 8.43 Another example of adequately-fed larvae. Although the cells are not awash in jelly, this amount of jelly indicates decent nutritional status. Do note the somewhat irregular larval development pattern, which calls for further inspection. *Source*: Photo courtesy of Randy Oliver.

Supplementary Feeding of Colonies

It is generally of benefit to a newly-established colony to feed sugar syrup in order to help the bees to draw out foundation so they can establish a broodnest and put away some reserves. Other than that, it is not necessary to feed a colony when adequate natural forage is being gathered by the foragers, nor necessarily at any other time of the year. However, in areas with poor forage, or under conditions of overstocking or unfavorable weather, supplemental feeding can be an important management tool.

As listed by Standifer et al. (1997), there are a number of management goals for which supplementary feeding can benefit (and be cost effective):

- To ensure continued colony development in places and times of pollen and/or nectar dearth.
- To develop colonies with optimum populations to take advantage of expected nectar flows.
- To develop colonies with optimum populations for crop pollination.
- To build colony populations for autumn or spring division.

Figure 8.44 Nutritional stress. The upper cells contain eggs, the lower two rows of cells contain larvae of mixed ages (indicating that the nurses are likely cannibalizing some larvae). The larvae shown are being fed the bare minimum amount of jelly, covering only a portion of the cell end. *Source*: Photo courtesy of Randy Oliver.

- To sustain brood rearing and colony development during inclement weather.
- To build colonies to high populations for queen and package-bee production.
- To maintain colonies and extend the season for drone production for queen matings.
- To maintain colonies in feedlot situations.
- To build colonies after pesticide losses.
- To provide adequate food reserves for overwintering colonies.

Practical application: *Always* understand the goal desired before making any management suggestions. Whether to feed, what to feed, and when to feed cannot be determined unless the vet knows the beekeeper's specific management goal(s).

Unless winter is extremely long, colonies provided with adequate food stores in autumn may not need supplemental foods in spring. However, before and after flowers bloom, especially if the weather is unusually cold and rainy, colonies may urgently need supplemental foods for subsistence and continued brood rearing until nectar and pollen collection is again adequate. A sudden curtailment of nectar and pollen income when brood rearing is in progress can set back colony buildup severely, or even result in colony loss.

Practical application – SPRINGTIME: Ideally, colonies enjoy an unbroken nectar and pollen flow.

Carbohydrates: Colony starvation typically occurs in early springtime, when colonies are burning through honey to maintain active broodrearing. Supplemental feeding of sugar in nearly any form may be indicated.

Protein: The momentum of population buildup in spring can be lost during just a few days of inclement weather, resulting in a colony suffering from protein shortage. Supplemental feeding (especially of pollen substitute) can benefit such a colony – especially if the beekeeper is willing to work in rain gear (Figure 8.45). Such feeding is often performed in preparation for almond pollination if there is no natural pollen coming in January and early February.

Practical application – AUTUMN: Ideally, colonies have stored ample honey reserves, and enter winter with a large amount young diutinus workers with well-developed fat bodies, and reserves of beebread in the combs.

Carbohydrates: Late-winter starvation can be averted by ensuring the hives go into winter with adequate honey reserves. Feed heavy syrup if indicated.

Protein: Of equal importance, is the rearing the last generations of brood, which will develop into workers exhibiting diutinus physiology to extend longevity (Mattila et al. 2001). This broodrearing under conditions of an autumn pollen flow takes place naturally in some areas, but not where autumn pollen is limiting. In that case, feeding pollen substitutes during the last two months before cessation of broodrearing in autumn can be of huge benefit (notably in preparation for almond pollination) (Figure 8.46). Note that, unlike as with natural pollen, bees will not store patty-fed pollen substitute in the combs.

Figure 8.45 Emergency feeding of colonies during the spring buildup can make the difference between colony losses and obtaining a good honey crop. *Source*: Photo courtesy of Randy Oliver.

Figure 8.46 The author's sons feeding chunks of pollen sub in an almond orchard upon placement of the hives shortly before bloom. The hives had been moved from a location with a natural pollen flow, to the orchard, which is a virtual desert as far as nutrition for the bees until the trees come into bloom. The feeding of pollen sub allows the colonies to continue their buildup until the trees commence flowering. *Source*: Photo courtesy of Randy Oliver.

Practical application: There are no "set rules" as to when colonies need feeding. An apiary consisting of two colonies, surrounded by a landscape with good forage, may never require feeding. On the other hand, commercial beekeepers are often completely dependent upon supplemental feeding to maintain their colonies.

Supplementary Feeding of Sugar

"Standard" suggestions for feeding sugar are:

- **For stimulative feeing:** Feed "light syrup" containing 50% solids (1: 1 weight: weight sugar: water). When using granulated sugar volume: volume gives the same results (note: that is not the same as topping off a bottle half-filled with sugar, which results in 38% sugar syrup).
- **For "putting on stores":** Feed "heavy syrup" containing 67% solids (2 : 1). Note that 2 : 1 sucrose syrup will rapidly granulate.

Results of studies and practical experience are mixed as to whether straight sucrose, or a blend of sucrose and inverted sugar (glucose and fructose) are best. Commercial suppliers currently sell a 77% solids sucrose/high fructose corn syrup (HFCS) blend that, due to the fructose, does not granulate during storage. Many beekeepers feed the concentrated syrup straight, but bees take it more readily if it is diluted with one-part water to seven parts concentrated syrup (thus resulting in a 67% sugar solution).

The most attractive sugar concentration for foragers appears to be in the 30–50% range (Waller, 1972), striking a balance between the bees' return on investment – calories gained versus the difficulty involved in licking it up. Once in the hive (likely more relevant as far as supplemental feeding is concerned), bees demonstrate a preference for 50% syrup (over 30% or 70%) for immediate consumption, tending to store the other concentrations (Eyer et al. 2016).

For winter feeding, concentrated syrup, or even granulated sugar or hard candy can be fed, although bees must have water access to dilute sugars to approximately 50% for transport in their crop (Simpson 1964). In the United States' East Coast, feeding baker's fondant is common practice, as are "candy boards," or even granulated sugar poured onto a sheet of newspaper resting on the top bars in contact with the cluster.

For spring feeding of nucleus hives, some California beekeepers use flaked bakery sugar Drivert®, which bees consume more readily than granulated sugar (which they may carry out of the hive if it isn't moistened). Drivert can also be made into a patty by mixing it with a bit of 77% commercial syrup blend.

Practical application: As a general rule, feed only "white" refined sugars, since the molasses in raw sugars may cause digestive issues. A good review of feeding methods can be found at (Somerville 2014).[13] Although some beekeepers add "feeding stimulants," acids, or other additives to sugar syrup, there is little if any evidence to support their use and, added scent may induce robbing.

13 https://www.dpi.nsw.gov.au/__data/assets/pdf_file/0018/532260/Feeding-sugar-to-honey-bees.pdf

Winter Stores

The warm-blooded honey bee superorganism requires a carbohydrate source to maintain its body temperature over winter. Local beekeepers can suggest the required amounts of winter honey stores for their region – typically 60 lbs in low latitude areas in which bees can forage during winter, and up to 90 lbs where winters are longer.

Practical application: Check hives for winter stores while weather is still warm; the simplest way is to heft them for weight (Figure 8.47). An experienced beekeeper judges with his or her fingers; the vet can use a pocket digital luggage scale. Any feeding should be with heavy syrup, early enough in autumn so the bees can concentrate it through evaporation to the approximately 83% sugar content necessary for storage as "honey."

Practical application: surprisingly, it is apparently impossible to overfeed a colony in preparation for winter. Canadian beekeepers routinely overwinter singles that have been fed heavy syrup until there is not an unfilled cell left in the combs.

Figure 8.47 Testing a hive for weight in order to assess the adequacy of winter stores. The tip weight (which is half the total weight) should be at least 65 lbs (130 lbs total hive weight) for a double-deep hive, or even more in higher latitudes. *Source*: Photo courtesy of Randy Oliver.

Feeding to Prevent Imminent Starvation

For emergency winter feeding when bees are unable to fly due to the cold, feeding dilute syrup can lead to dysentery. Unless colonies have access to water, or there is enough metabolic water condensation in the hive, they also may be unable to process dry sugar. As a result, in general, if feeding for winter rescue is necessary, it is generally best to feed heavy syrup or fondant.

Feeding for Stimulation

Feeding light syrup (Figure 8.48) will "stimulate" the colony to rear brood, but only if there are adequate pollen reserves. If not, feed pollen substitute to sustain broodrearing. Feeding light syrup along with pollen substitute increases colony response, as it simulates a nectar/pollen flow.

A problem with feeding light syrup is that during warm weather it can ferment, or "black mold" may grow in the feeders. Live yeasts may promote dysentery, and some molds release mycotoxins into syrup. To avoid spoilage at ambient temperatures above 20 °C, add unscented household bleach at a rate of ½-1 cup per 55 gal, or ½ tsp per gallon. Investigation by the author suggests that chlorine reacts with sugar in a reversible manner, yet retains its germicidal qualities, and causes no noticeable repellency or adverse effects (there may in fact be enhanced phagostimulation due to the chlorine or salt in bleach).

There are many ways to supply syrup or sugar to colonies; it is beyond the purview of this chapter to compare them (Figure 8.49).

Practical application: there are three main ways to feed sugar syrup: at the entrance, from the top, or

Figure 8.48 Colonies being fed sugar syrup during a nectar dearth. *Source*: Photo courtesy of Randy Oliver.

Figure 8.49 Close-up of supplemental sugar feeding jars on colonies. Note the small amount of "black mold" growing inside the nearest jar. *Source*: Photo courtesy of Randy Oliver.

with "insert feeders" that take the place of the frame. These are briefly reviewed at.[14]

Supplementary Feeding of Protein

Sugar income may "stimulate" a colony, but it is the supply of pollen that determines whether it rears brood. Thus, beekeepers may feed a pollen substitute ("pollen sub") – with or without natural pollen as a component – to promote or maintain a colony during dearth or unfavorable weather. An example is shown in Figure 8.50.

Practical application: a common question by recreational beekeepers is whether they "should" be feeding pollen sub. There are times when there is not only no benefit from encouraging colony growth, but actually potential harm, by increasing varroa reproduction. Any recommendation for feeding protein should be done only if a specific goal calls for it.

The author has operated several field trials comparing different pollen substitutes. One large trial indicated that current top-tier substitutes on the market performed nearly

14 http://scientificbeekeeping.com/fat-bees-part-3

Figure 8.50 Commercial pollen substitutes ("pollen subs") in patty form. High-quality formulations approach natural pollen in their ability to promote broodrearing. *Source*: Photo courtesy of Randy Oliver.

as well, on a weigh-to-weight basis, as did mixed natural pollen (Figure 8.51).

Composition of Pollen Subs

General Nutrient Content: Practical development of pollen substitutes suggests a 15–20% range for protein, around 5% lipid (typically canola oil), and around 50% sugar is likely to be a successful formulation.

Particle Size: Particle sizes of modern pollen substitutes are typically approximately 35 μm – most pollen grains collected by bees are under 50 μm. Since the bee gut is designed to quickly digest hollow pollen grains rather than granular animal feedstocks, better pollen substitutes are expensively ground to very small particle size, although scant research has been done along this line.

Lipids, Sterols, Vitamins, Minerals, Phagostimulants: Lipids and sterols are essential nutrients in a diet for any insect (Cohen 2015) although, in the short-term, nurse bees can transfer lipids and sterols from their own body reserves into jelly they produce (Robinson and Nation 1966). In practice, adding vegetable oil improves palatability (Manning et al. 2010) and texture of pollen sub. Acidifying the mix with citric acid, lemon juice, or vitamin C may be beneficial (Brighenti et al. 2017). Other phagostimulants are natural pollen, lemongrass oil, and some, but not necessarily all, vegetable oils. Although commercial pollen substitutes typically use canola oil, Manning (in addition to finding that rum was attractive) (Manning and Black 2013) found that:

> Honey bees showed a preference for linseed, coconut, grapeseed, almond, apricot, or evening primrose oils or mixtures of each when mixed in

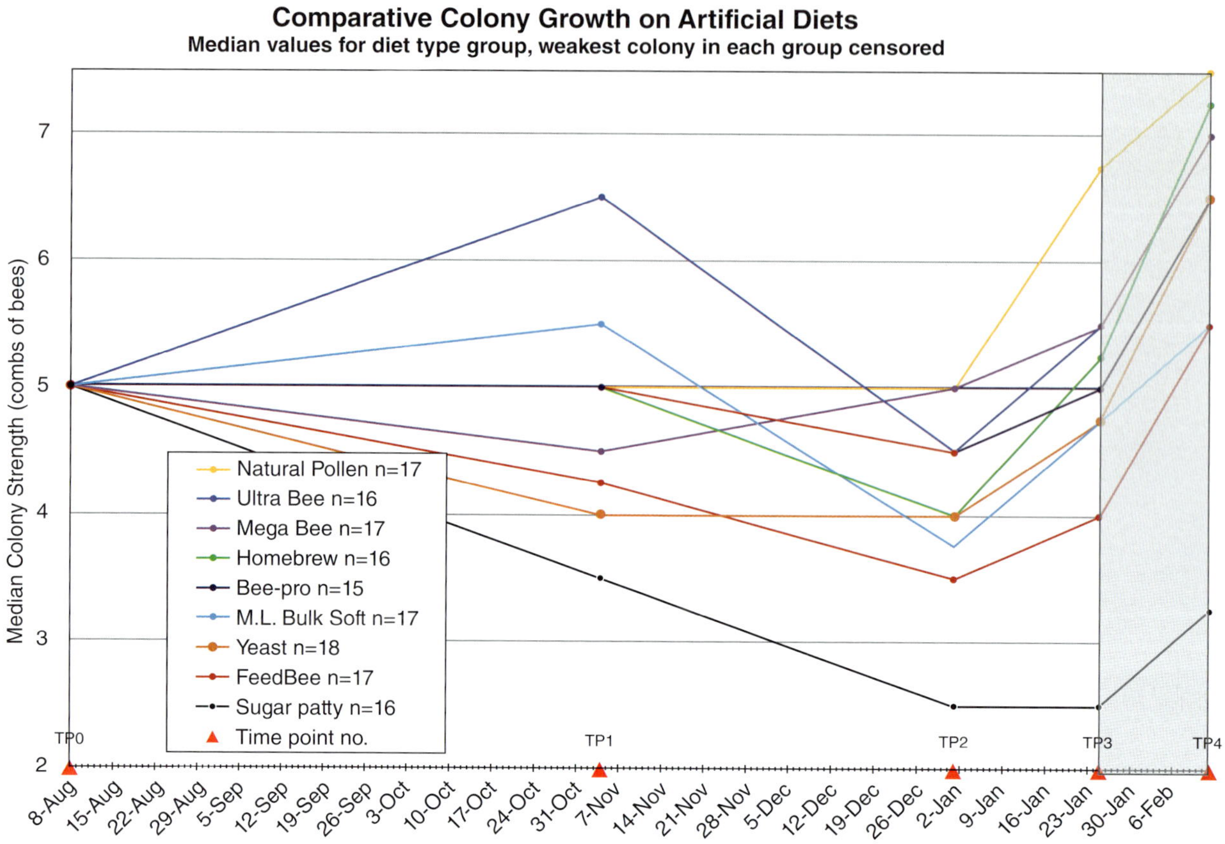

Figure 8.51 Comparison of colony buildup during an extended natural pollen dearth prior to almond pollination. The negative control was of sugar only; "Natural" was an equal weight of mixed corbicular pollen (but with less protein than some of the substitutes) (Oliver 2014a). Note that a natural pollen flow began in early January, at which point the addition of pollen sub no longer made a difference. *Source*: Illustration by Randy Oliver.

artificial diets at 5% or less. The bees showed a relatively low preference for the common vegetable oils from canola, sunflower, and soya bean seeds, which suggest these oils are not ideal for use in pollen substitutes.

Practical application: Despite the findings of Manning, the author, and manufacturers have found that bees eagerly consume, and colonies thrive on, substitutes containing canola or corn oil.

Pollen Sub Feeding Tips

- In general, feed stimulatory light syrup along with pollen supplement to encourage brood buildup. If there is no nectar flow and reserves are short, bees will sometimes consume the sugar out of the protein substitute patty and discard the protein. They may also do so if there is a natural pollen flow going on – the author has found no benefit from feeding pollen sub during a good natural pollen flow (Oliver 2014b).
- However, colonies with adequate honey reserves in either late summer or midwinter will typically respond strongly to feeding pollen sub, even in the absence of a nectar flow (whether natural or artificial).
- The best location to feed pollen substitute patties appears to be as close to the center of the brood cluster as possible, although bees will take the patty when fed in a rim at the top of the hive. Little research has been done on the actual utilization of protein with regard to sub placement; the issue being whether the patty is consumed as food by nurse bees, or simply removed as trash by housekeeping bees.
- Make sure a patty is in contact with the cluster, near brood – otherwise bees won't consume it (Figure 8.53). Uneaten substitute can harden or become food for wax moth or small hive beetle.
- Many commercial beekeepers prefer feeding pollen sub in chunks, rather than papered patties – allowing the patty to squeeze between the frames in the center of the cluster (Figures 8.52–8.56).

Figure 8.52 A strong, hungry colony can easily consume a 3-lb pollen sub patty in a few days. Note the nurse bees eagerly feeding on this patty, placed moments before. For late-summer production of brood in California, a 2½-lb patty fed every week to 10 days ensures a good population of well-fed "winter bees." Such late-summer feeding also indicates whether the colony is worth wintering. The failure of the odd colony to consume pollen sub indicates that the colony is queenless or has other issues. *Source*: Photo courtesy of Randy Oliver.

Figure 8.53 Commercial pollen sub fed in paper-sandwiched patty form. Note that the upper right-hand edge extends beyond the cluster (note the lack of bees between those frames), where the bees will generally not consume it, but scavengers may. *Source*: Photo courtesy of Randy Oliver.

Figure 8.54 As opposed to feeding paper-covered preformed patties, many commercial beekeepers prefer to feed cut chunks of soft bulk formulations. This photo is of a 1-kg chunk of pollen sub, showing how it got squeezed between the bottom bars of the upper brood chamber. This amount of patty, fed every 10 days, will support substantial brood rearing. *Source*: Photo courtesy of Randy Oliver.

Figure 8.55 Feeding chunks of pollen sub in the rain. The author's sons are here brushing the bees aside with a pine branchlet in order not to crush any bees, which could spread nosema spores. Working quickly in wet or cold weather does not appear to stress the colony to any great extent. Another method of feeding pollen sub is in powder form, which does not involve opening the hives, and can get as much protein into hives as by the feeding of patties (Figure 8.56). *Source*: Photo courtesy of Randy Oliver.

Bee "Health Products"

Beekeepers, like other livestock owners, always look for magical potions they can give to their colonies to "make their troubles go away." Companies oblige by offering a range of "feeding stimulants," "supplements," and bee

Figure 8.56 Pollen sub fed outside the hive in dry powder form is eagerly gathered by bees to be stored as beebread (as opposed to with patty form, which is not). However, this form of feeding has mixed results, with not all colonies in a yard benefitting equally, and may actually cause adverse effects during winter. *Source*: Maes et al. (2016).

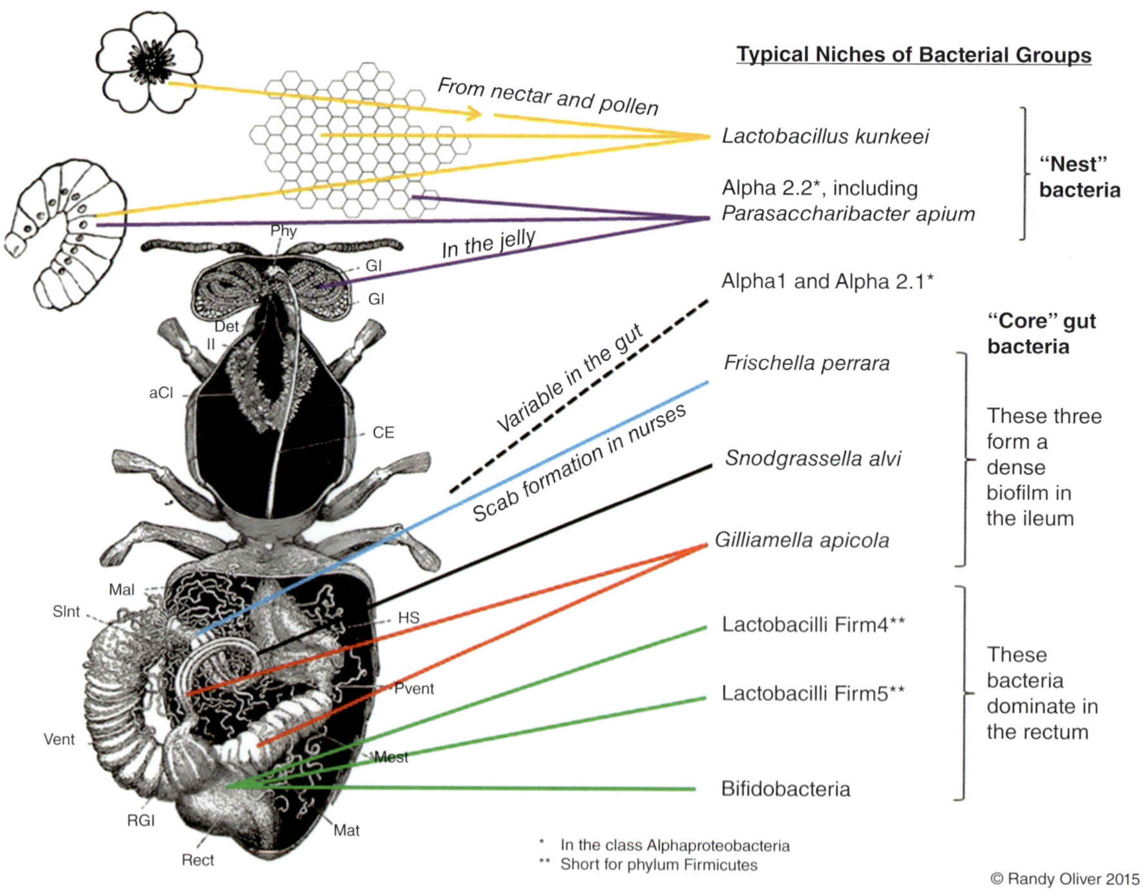

Figure 8.57 The "core" gut bacteria exhibit rather strict niche fidelity within the bee gut (Moran 2015). An adult honey bee may harbor a billion bacteria, with roughly 95% residing in the waste material in the hindgut; few survive in the microbe-hostile crop and midgut Drawing of bee from Snodgras, RE (1910) The Anatomy of the Honey Bee. USDA Technical Series, No. 18, Government Printing Office.

"health" feeds and additives, most of which have scant supportive data to back up their claims.

Probiotics

Several of popular bee "health products" involve probiotics, which are claimed by manufacturers to improve the honey bee microbiome. The combs, food stores, and bee mouthparts and gut are consistently colonized by a "core" microbiome of symbiotic bacteria (Figure 8.57),[15] established at first in the larvae, and then reestablished in the first days of a bee's adult life (since at pupation, the larval gut is shed during ecdysis).

A number of supply houses now sell "direct-fed microbial" (DFM) probiotic products for honey bees (some including fungi). There are, however, many currently unanswered questions concerning probiotics:

- Whether they are necessary, given the rapid natural establishment of the gut microbiome,
- Whether cultured bacteria are sufficiently strain-specific to establish in the gut, or to cause beneficial effects,
- Whether there is benefit in their feeding following treatment with antibiotics (Raymann et al. 2017), and
- The potential for harm from feeding inappropriate microorganisms.[16]

While research is ongoing, to date, there is little hard data to support claims of beneficial effect from feeding commercial probiotics.

15 Our knowledge on this subject is rapidly expanding. For current information I suggest following the publications from the following labs: Kirk Anderson, USDA; Nancy Moran, University of Texas; Irene Newton, Indiana University; and Quinn McFredrick, U.C. Riverside.

16 http://scientificbeekeeping.com/probiotics-and-colony-productivity

References

Anderson, K.E., Carroll, M.J., Sheehan, T. et al. (2014a). Hive-stored pollen of honey bees: many lines of evidence are consistent with pollen preservation, not nutrient conversion. *Molecular Ecology* 23: 5904–5917.

Anderson, K., Rodrigues, P.A.P., Mott, B.M. et al. (2015). Ecological succession in the honey bee gut: shift in lactobacillus strain dominance during early adult development. *Microbial Ecology* 71: 1008–1019.

Brighenti, D., Brighenti, C.R.G., and Carvalho, C.F. (2017). Life spans of Africanized honey bees fed sucrose diets enhanced with citric acid or lemon juice. *Journal of Apicultural Research* 56 (2): 91–99.

Camazine, S. (1991). Self-organizing pattern formation on the combs of honey bee colonies. *Behavioral Ecology and Sociobiology* 28: 61–76.

Carroll, M.J., Brown, N., Goodall, C. et al. (2017). Honey bees preferentially consume freshly-stored pollen. *PLoS One* 12 (4): e0175933. https://doi.org/10.1371/journal.pone.0175933.

Cohen, A. (2015). *Insect Diets*. CRC Press.

Corona, M., Velarde, R.A., Remolina, S. et al. (2007). Vitellogenin, juvenile hormone, insulin signaling, and queen honey bee longevity. *Proceedings of the National Academy of Sciences* 104 (17): 7128–7133.

Crailsheim, K. (1992). The flow of jelly within a honeybee colony. *Journal of Comparative Physiology B* 162: 681–689.

Detzel, A. and Wink, M. (1993). Attraction, deterrence or intoxication of bees (*Apis mellifera*) by plant allelochemicals. *Chemoecology* 4 (1): 8–18.

Eyer, M., Greco, M.K., Lang, J. et al. (2016). No spatial patterns for early nectar storage in honey bee colonies. *Insectes Sociaux* 63 (1): 51–59.

Ferioli, F., Armaforte, E., and Caboni, M.F. (2014). Comparison of the lipid content, fatty acid profile and sterol composition in local italian and commercial royal jelly samples. *Journal of the American Oil Chemists' Society* 91 (6): 875–884.

Frias, B., Barbosa, C.D., and Lourenço, A.P. (2016). Pollen nutrition in honey bees (*Apis mellifera*): impact on adult health. *Apidologie* 47 (1): 15–25.

de Groot, A.P. (1953). Protein and amino acid requirements of the honey bee (*Apis mellifica* L.). *Physiologia Comparata et Oecologia* 3: 1–83.

Haydak, M. (1970). Honey bee nutrition. *Annual Review of Entomology* 15: 143–156.

Johnson, B. (2009). Pattern formation on the combs of honeybees: increasing fitness by coupling self-organization with templates. *Proceedings of the Biological Sciences* 276 (1655): 255–261.

Keller, I., Fluri, P., and Imdorf, A. (2005). Pollen nutrition and colony development in honey bees: part 1. *Bee World* 86 (1): 3–10. https://doi.org/10.1080/0005772X.2005.11099641.

Kodai, T., Umebayashi, K., Nakatani, T. et al. (2007). Compositions of royal jelly II. Organic acid glycosides and sterols of the royal jelly of honeybees (*Apis mellifera*). *Chemical & Pharmaceutical Bulletin* 55 (10): 1528–1531.

Liao, L-H., Wu, W-Y., Berenbaum, M.R. (2017). Behavioral responses of honey bees (*Apis mellifera*) to natural and synthetic xenobiotics in food. *Scientific Reports* 7: 15924 DOI:10.1038/s41598-017-15066-5.

London-Shafir, I., Shafir, S., and Eisikowitch, D. (2003). Amygdalin in almond nectar and pollen – facts and possible roles. *Plant Systematics and Evolution* 238: 87–95.

Maes, P., Rodrigues, P.A.P., Oliver, R. et al. (2016). Diet-related gut bacterial dysbiosis correlates with impaired development, increased mortality and Nosema disease in the honeybee (*Apis mellifera*). *Molecular Ecology* 25: 5439–5450.

Manning, R & J Black (2013) Development of a pollen substitute meeting the nutritional needs of honey bees. RIRDC Pub. No. 13/004

Manning, R., Speijers, J., Harvey, M., and Black, J. (2010). Added vegetable and fish oils to low-fat pollen diets: effect on honey bee (*Apis mellifera* L.) consumption. *Australian Journal of Entomology* 49: 182–189.

Mao, W., Schuler, M.A., and Berenbaum, M.R. (2013). Honey constituents up-regulate detoxification and immunity genes in the western honeybee *Apis mellifera*. *Proceedings of the National Academy of Sciences of the United States of America* 110: 8842–8846.

Mattila, H., Harris, J.L., and Otis, G.W. (2001). Timing of production of winter bees in honey bee (*Apis mellifera*) colonies. *Insectes Sociaux* 48: 88–93.

Moran, N. (2015). Genomics of the honey bee microbiome. *Current Opinion in Insect Science* 10: 22–28.

Nelson, J, AP Sturtevant & B Lineburg (1924) Growth and feeding of honeybee larvae. US Department of Agriculture 1222.

Nelson, C., Ihle, K.E., Fondrk, M.K. et al. (2007). The gene vitellogenin has multiple coordinating effects on social organization. *PLoS Biology* 5 (3): e62.

Nicolson, S. and Human, H. (2013). Chemical composition of the 'low quality' pollen of sunflower (*Helianthus annuus*, Asteraceae). *Apidologie* 44 (2): 144–152.

Nolan, W.J. (1932). The development of package-bee colonies. *USDA Technical Bulletin* 309.

Oliver, R (2014a) http://scientificbeekeeping.com/a-comparative-test-of-the-pollen-sub

Oliver, R (2014b) http://scientificbeekeeping.com/when-to-feed-pollen-sub

Oliver, R (2020a), In prep.

Oliver, R. (2020b). An Experiment to Improve Pollen Sub, Part 2. *American Bee Journal* 160 (5): 555–560.

Pernal, S.F. and Currie, R.W. (2002). Discrimination and preferences for pollen-based cues by foraging honeybees, *Apis mellifera* L. *Animal Behaviour* 63 (2): 369–390.

Rashad, E. and Parker, R. (1958). Pollen as a limiting factor in brood rearing and honey production during three drought years, 1954, 1955, and 1956. *Transactions of the Kansas Academy of Science* 61 (3): 237–248.

Raymann, K., Shaffer, Z., and Moran, N.A. (2017). Antibiotic exposure perturbs the gut microbiota and elevates mortality in honeybees. *PLoS Biology* 15 (3): e2001861.

Rhoades, D.F. and Bergdahl, J.C. (1981). Adaptive significance of toxic nectar. *The American Naturalist* 117: 798–803.

Robinson, F. and Nation, J. (1966). Artificial diets for honey bees, *Apis mellifera*. *The Florida Entomologist* 49 (3): 176–184. Florida Agricultural Experiment Stations Journal Series No. 2284.

Salmela, H. (2015). Transfer of immunity from mother to offspring is mediated via egg-yolk protein vitellogenin. *PLoS Pathogens* 11 (7): e1005015.

Schmickl, T. and Crailsheim, K. (2004). Inner nest homeostasis in a changing environment with special emphasis on honey bee brood nursing and pollen supply. *Apidologie* 35: 249–263. https://doi.org/10.1051/apido:2004019.

Shapla, U.M., Solayman, M., Alam, N. et al. (2018). 5-Hydroxymethylfurfural (HMF) levels in honey and other food products: efects on bees and human health. *Chemistry Central Journal* 12: 35. https://doi.org/10.1186/s13065-018-0408-3.

Shaw, F (1956) Nectar and pollen plants of Massachusetts as observed in the central Connecticut Valley region. Special circular #27 University of Massachusetts.

Simpson, J. (1964). Dilution by honeybees of solid and liquid food containing sugar. *Journal of Apicultural Research* 3 (1): 37–40.

Smith, DB, Federico Solano, Laurel G. Woodruff, William F. Cannon, and Karl J. Ellefsen (2019) Geochemical and mineralogical maps, with interpretation, for soils of the conterminous United States: U.S. Geological Survey Scientific Investigations Report 2017-5118, doi:10.3133/sir20175118.

Somerville, Doug (2005) Fat Bees Skinny Bees, A manual on honey bee nutrition for beekeepers. Rural Industries Research and Development Corporation, Australia. Publication No 05/054.

Somerville, D (2014) Feeding sugar to honey bees. https://www.dpi.nsw.gov.au/__data/assets/pdf_file/0018/532260/Feeding-sugar-to-honey-bees.pdf.

Standifer, L. N., Moeller, F. E., Kauffeld, N. M., Herbert, E. W., Jr., and Shimanuki, H. 1977. Supplemental Feeding of Honey Bee Colonies. United States Department of Agriculture Agriculture Information Bulletin No. 413.

Svoboda, J., Herbert, E.W. Jr., Thompson, M.J., and Shimanuki, H. (1981). The fate of radiolabelled C28 and C29, phytosterols in the honey bee. *Journal of Insect Physiolcogy* 27 (3): 183–188.

Svoboda, J., Herbert, E.W. Jr., and Thompson, M.J. (1986). Sterols of organs involved in brood food production and of royal jelly in honey bees. *Insect Biochem* 16 (3): 479–482.

Tian, W., Li, M., Guo, H. et al. (2018). Architecture of the native major royal jelly protein 1 oligomer. *Nature Communications* 9 (1) https://doi.org/10.1038/s41467-018-05619-1.

Villette, C. (2015). Plant sterol diversity in pollen from angiosperms. *Lipids* 50 (8): 749–760.

Waller, G.D. (1972). Evaluating responses of honeybees to sugar solutions using an artificial-flower feeder. *Annals of the Entomological Society of America* 65: 857–862.

Wright, G., Nicolson, S.W., and Shafir, S. (2018). Nutritional physiology and ecology of honey bees. *Annual Review of Entomology* 63: 327–344.

Zhang, G., Zhang, W., Cui, X., and Xu, B. (2015). Zinc nutrition increases the antioxidant defenses of honey bees. *Entomologia Experimentalis et Applicata* 156: 201–210. https://doi.org/10.1111/eea.12342.

Further Reading

The following references have extensive bibliographies relevant to honey bee nutrition.

Herbert, E.W. Jr. (1992). *Honey Bee Nutrition*. Dadant and Sons: The Hive and the Honey Bee.

Schmickl, T. and Crailsheim, K. (2004). Inner nest homeostasis in a changing environment with special emphasis on honey bee brood nursing and pollen supply. *Apidologie* 35: 249–263.

Somerville, D. (2005). *Fat Bees Skinny Bees, a Manual on Honey Bee Nutrition for Beekeepers*. Australia: Rural Industries Research and Development Corporation. Available on the Web at www.rirdc.gov.au/reports/HBE/05-054.pdf.

Brodschneider, R. and Crailsheim, K. (2010). Nutrition and health in honey bees. *Apidologie* 41: 278–294.

Wright, G., Nicolson, S.W., and Shafir, S. (2018). Nutritional physiology and ecology of honey bees. *Annual Review of Entomology* 63: 327–344.

9

Honey Bee Microbiota and the Physiology of Antimicrobial Resistance

Kasie Raymann

Department of Biology, University of North Carolina at Greensboro, Greensboro, NC, USA

Overview

Honey bees (*Apis mellifera*) are subject to many infections and infestations caused by a broad range of organisms including mites, beetles, moths, fungi, bacteria, and viruses. These pests and diseases can severely impact colony health and, if untreated, can lead to colony loss. Because of the significance of honey bees to agricultural production, many treatments have been developed in an attempt to combat these maladies. Of all the existing honey bee medications, antibiotics are currently the only regulated substance. These regulations were introduced in 2017 with the FDAs implementation of the Veterinary Feed Directive (VFD) which included bees as livestock. This chapter provides an overview of the honey bee gut microbiome and its importance for bee health and the potential negative consequences of the use of antibiotics in beekeeping.

Antibiotics in Beekeeping

Antibiotics have been used in beekeeping for over 50 years to control or treat bacterial brood diseases. Oxytetracycline was introduced in the 1950s and is still the most commonly used antibiotic in apiculture. Two other antibiotics have been approved for use in beekeeping by the FDA – tylosin (approved in 2005) and lincomycin (approved in 2012), but they are rarely used in practice. All three of these antibiotics are broad spectrum bacteriostats that use similar mechanisms of action where they bind to the 50S ribosomal subunits of bacteria to inhibit protein synthesis and stop bacterial growth.

As mentioned above, oxytetracycline, which is marketed to beekeepers as Terramycin®, is the most widely used antibiotic by beekeepers both in the United States and other countries that allow the use of antibiotics in beekeeping. Oxytetracycline has historically been used to treat and control the spread of *Melissococcus plutonius* and *Paenibacillus larvae*, the causative agents of European Foulbrood (EFB) and American Foulbrood (AFB), respectively. AFB is particularly dangerous because *P. larvae* is a spore-forming bacterium. Antibiotics can kill the vegetative state of the bacterium (the symptom causing form) but not the spores, which can persist for up to 50 years or more in the hive (Genersch 2010).

Melissococcus plutonius (the causative agent of EFB) is not a spore-forming bacterium, so it can be more easily eradicated from infected hives by treatment with antibiotics. Both diseases can be spread from one hive to another when bees "drift" (go into the wrong colony) from an infected colony to an uninfected one, or if bees rob (steal resources) from an infected colony, or when beekeepers exchange or reuse infected equipment. In large commercial operations, where hundreds to thousands of hives are maintained and transported across the US for pollination services, the threat of disease spread is much more severe, which could ultimately result in the loss of hundreds of colonies. Therefore, in the case of large commercial operations, particularly when hives are highly concentrated in one location, it is not uncommon to medicate hives prophylactically. Please refer to Chapter 15 for more information regarding treatment of AFB and EFB.

Before the VFD was instituted in 2017, the purchase and administration of antibiotics in beekeeping was not regulated. Now, beekeepers must obtain a prescription or a VFD order from a veterinarian. Implementing a VFD for beekeepers is intended to help decrease the spread of antibiotic resistance, reduce the number of antibiotic residues in food and cosmetic products, and limit the negative impacts of antibiotic treatment on honey bees (see sections below). Please refer to Chapter 12 for a discussion of the VFD.

Honey Bee Medicine for the Veterinary Practitioner, First Edition. Edited by Terry Ryan Kane and Cynthia M. Faux.

The Honey Bee Gut Microbiome

The Worker Gut Microbiome

The microbes that live inside the gut of animals provide important functions to their hosts. For example, in many animals, including honey bees, the gut microbiota has been shown to aid in nutrient acquisition, metabolism, immune function, removal of toxins, impact behavior and development, and provide protection against pathogens (Raymann and Moran 2018). Most mammals acquire their microbiome vertically, or from mother-to-offspring, while invertebrates generally obtain their microbiome from the environment with each new generation. However, unlike most invertebrates, honey bee workers inherit their gut microbiota primarily through contact with their sisters within the hive via coprophagy and oral trophallaxis (Powell et al. 2014). Because of this bee-to-bee microbiota transmission, honey bee workers possess a highly specialized and conserved gut microbial community.

The worker microbiome has been the most extensively studied, since workers comprise over 95% of the colony's population. There is strong evidence that gut bacteria play a major role in worker health. For example, the bee gut microbiota has been shown to influence intestinal size, weight gain following emergence, insulin and vitellogenin signaling, and sucrose sensitivity (Zheng et al. 2017). In addition, it has been shown that the gut bacteria have major effects on the presence and abundance of short chain fatty acids in the guts of worker bees, which serve as the major energy source for intestinal epithelial cells (Zheng et al. 2017). Bee gut bacteria also aid in priming and stimulating the bee immune system (Emery et al. 2017; Kwong et al. 2017) and provide protection against pathogen infection (Raymann and Moran 2018). The gut microbiome is also involved in the metabolism of a variety of carbohydrates (Ellegaard et al. 2015; Ellegaard and Engel 2016; Kwong et al. 2014; Lee et al. 2015), including sugars that are generally toxic to bees (Zheng et al. 2016) and complex carbohydrates that are otherwise not digestible by bees (Engel et al. 2012). Taken together, the bee gut bacteria are crucial for bee development, digestion and metabolism, immune responsiveness, susceptibility to stress, and overall health (Raymann and Moran 2018).

The routes of acquisition of the worker bee gut microbiome are well known. During larvae development very few bacteria are present in the gut (Powell et al. 2014). Around day 10 pupation begins and the brood cells are capped with wax. During pupation the gut lining is shed, eliminating most, to all, microbes in the gut (Kwong and Moran 2016). A small number of microbes are acquired once the pupae begin to move their mouth parts and as they chew through the wax cap to emerge. However, most of the microbiome is acquired after emergence through three main routes: oral trophallaxis, contact with hive material and, most importantly, through fecal–oral exposure (Kwong and Moran 2016; Powell et al. 2014). The characteristic bee microbiome is not fully established and stabilized until approximately four days post-emergence (Powell et al. 2014).

The adult honey bee worker gut microbiome consists of five core bacterial species that can be found in all honey bee workers around the globe (*Snodgrassella alvi*, *Gilliamella apicola*, *Lactobacillus* sp. Firm 5, *Lactobacillus* sp. Firm 4, and *Bifidobacterium* spp.) (Kwong and Moran 2016, Figure 9.1). In addition, three other species (*Frischella perrara*, *Parasaccharibacter apium*, and *Bartonella apis*), are found in a large percent of honey bees (Kwong and Moran 2016). These eight bacteria (hereafter referred to as the dominant bacteria) make up ~99% of the bacteria present in the honey bee gut (Kwong and Moran 2016, Figure 9.1). Several related species are also found in *Bombus* (bumble bee) species (Kwong and Moran 2016), indicating that these bacteria have been maintained in their host for millions of years and have likely coevolved with one another and with their hosts (Kwong and Moran 2015). Moreover, most bacteria found in the honey bee gut are not found in the environment, meaning they are specialized species limited to living only in the digestive tracts of bees (Kwong and Moran 2016).

The gut of the honey bee is divided into four major sections, the honey crop, midgut, ileum, and rectum. Of these four regions, only the two hindgut sections, the ileum and the rectum, are colonized with large numbers of bacteria (Kwong and Moran 2016; Figure 9.1). Within the ileum and rectum, the bacteria are spatially organized. The two gram-negative bacterial species (*S. alvi* and *G. apicola*) dominate the ileum. *S. alvi* resides along the lining of the epithelium and *G. apicola* forms a layer directly on top of it (Martinson et al. 2012; Figure 9.1). The rectum is dominated by three gram-positive bacterial species: *Bifidobacterium* sp. and *Lactobacillus* sp. Firm 4 and Firm 5 (Martinson et al. 2012; Powell et al. 2014; Figure 9.1).

The structured organization of the different core species likely results from specialized interactions between the bacterial members and their honey bee host as well as interbacterial interactions. The five core species (*S. alvi*, *G. apicola*, *Lactobacillus* sp. Firm 5, *Lactobacillus* sp. Firm 4, and *Bifidobacterium* spp.) are stable across individuals and populations, but differences among individuals and colonies can be seen in the relative frequencies of the five-core species, the amount of strain (intraspecies) diversity within each core species, and the presence and abundance of non-core (*F. perrara*, *Pa. apium*, and *B. apis*) and atypical

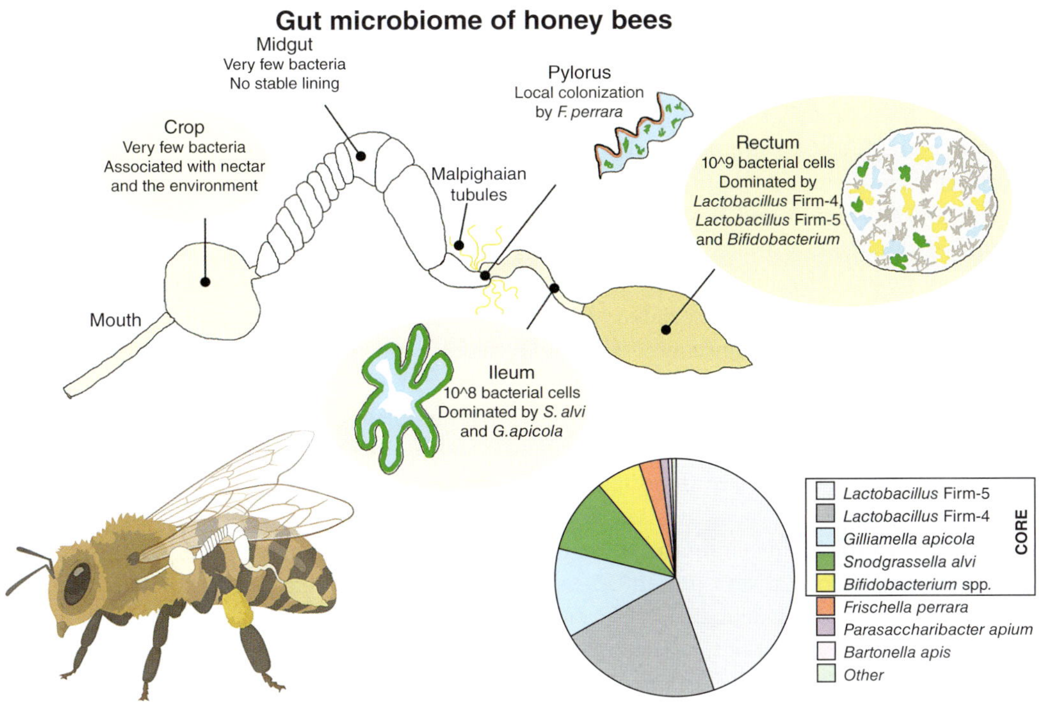

Figure 9.1 Structure and organization of the honey bee worker gut microbiome. The pie chart (bottom right) depicts the relative abundance of different bacterial species in a typical worker gut, with the five core species highlighted in the shaded box. *Source:* Adapted from Kwong and Moran (2016).

species (Raymann and Moran 2018). Atypical members found in the bee gut include members of the *Enterobacteriaceae* family such as *Serratia*, *Klebsiella*, and *Enterobacter* (Mattila et al. 2012; Moran et al. 2012), which are generally opportunistic pathogens and whose higher prevalence is thought to reflect disruption of the usual gut community (Raymann and Moran 2018); this potentially represents infection.

The Queen and Drone Gut Microbiome

In comparison to honey bee workers, the gut microbiome of queens and drones has been far less studied. From the few studies that have investigated the queen and drone gut microbiome it is clear that they do not possess the same conserved microbiome composition as workers, although they often contain some of the same bacteria in their hindguts as workers (Figure 9.2, Kapheim et al. 2015). Overall, queens and drones possess less diverse, unstable gut microbiomes, and do not appear to have a distinct microbiome community (Anderson et al. 2018; Elijah Powell et al. 2018; Tarpy et al. 2015), meaning the same bacterial composition has not been found across individuals.

The reason queens and drones have such different gut microbiome compositions than the workers who raise and care for them is not understood. It has been proposed that this could be due to a lower exchange of food between nurses and queens/drones when compared to the amount of food exchange between nurses and foragers (<6% versus ~52%) (Kapheim et al. 2015). The queen microbiome shifts with age and differences in queen rearing have also been shown to impact the overall composition of the microbiome (Anderson et al. 2018; Elijah Powell et al. 2018). No studies have investigated the drone microbiome over time or under different conditions.

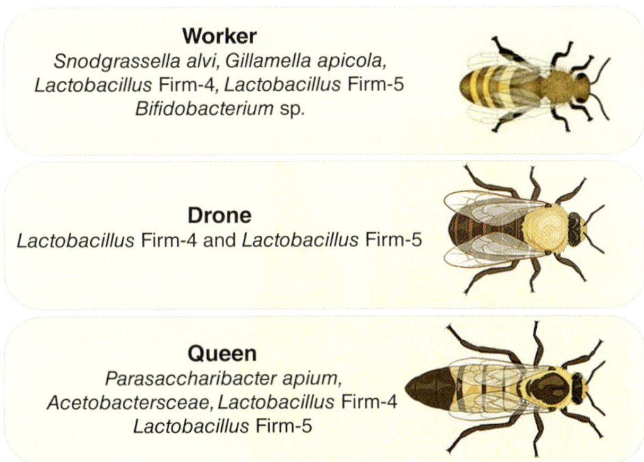

Figure 9.2 The dominant bacteria found in the gut of workers, drones, and queens.

Within the queen microbiome a few dominant bacteria have been somewhat consistently found such as *Pa. apium*, *Acetobacteraceae* (Alpha 2.1), and *Lactobacillus* sp. Firm 4 and Firm 5 (Figure 9.2, Anderson et al. 2018; Elijah Powell et al. 2018; Kapheim et al. 2015). *Pa. apium* is a non-core member of the worker microbiome that is only sporadically detected in worker guts (Raymann and Moran 2018). In queens, *Pa. apium* is significantly more abundant and consistently present in the gut microbiome. There is some evidence that *Pa. apium* could be associated with reproductive activity, which could explain why it is more prevalent in queens than workers. This evidence comes from a study that identified genes associated with membrane/envelope biogenesis (i.e. ferrodoxins and oxidoreductases) in *Pa. apium*, which are important for oogenesis (Engel et al. 2012). However, it has yet to be verified if *Pa. apium* plays a role in queen reproduction or heath. Although *Pa. apium* is not always present in the guts of workers, it has been detected in their crop and hypopharyngeal glands (secretes royal jelly) as well as in the royal jelly they feed to queens (Corby-Harris et al. 2014a,b). Therefore, queens might acquire *Pa. apium* when being fed by workers while the other bacteria in queens that are present in workers (e.g. *Lactobacillus* sp. Firm 4 and Firm 5) could be transferred through direct contact with fecal matter (Kapheim et al. 2015). To date, no studies have directly investigated how queens acquire their microbiome, thus these are merely hypotheses.

Natural Fluctuations in the Worker Microbiome

Although the core microbiome of workers is conserved across individuals, several studies have shown that the relative abundances of different gut bacteria change throughout a worker bee's lifetime (Anderson et al. 2018; Ellegaard and Engel 2019). However, disentangling the effects of age and lifestyle on changes in microbiome composition are difficult because division of labor is split across workers, with different tasks generally corresponding to the age of the bee (Johnson 2010). For example, the youngest workers are nurse bees. Nurse bees perform tasks within the hive, such as feeding brood and caring for the queen. The older workers are foragers and they take care of outside hive duties, e.g. collecting pollen and nectar. Not only do the tasks of nurses and foragers differ, they have different diets. The nurse diet is primarily composed of pollen, while the forager diet is mainly nectar and honey. Nurses and foragers are also exposed to different environmental conditions that could influence their microbiomes (e.g. nurses stay in the hive and foragers are exposed to the outside environment). The impact of division of labor was recently investigated in a study that compared the microbiome of age-controlled bees performing different hive tasks (Jones et al. 2018). In this study, the authors evaluated the relative abundance of gut bacteria in different castes of worker bees and found a correlation between the behavioral tasks they perform and their microbiome composition (Jones et al. 2018). This suggests that behavioral tasks do impact microbiome composition, but it is unclear if age alone has any impact. The honey bee worker gut microbiome composition also shifts with season, with bacterial loads being around 10 times higher in bees in the winter (Kešnerová et al. 2019). Differences have also been observed when comparing nurses and foragers. Nurses contain higher loads of bacteria in their gut than foragers, which was shown to result from the pollen diet of nurse bees (Kešnerová et al. 2019). Overall, current evidence suggests that season, diet, and behavioral tasks all impact the composition of the worker microbiome.

Strain-Level Diversity Within the Worker Microbiome

As mentioned above, all honey bee workers house the same five core bacterial species but, it has recently been shown that a high level of strain diversity (within species genetic variation) exists within each of these species, even within a single bee (Ellegaard and Engel 2019; Powell et al. 2016; Raymann et al. 2017a). As the number of available bee gut bacteria genomes has increased, the functional diversity of the strains within each core species has become more evident. This diversity is not insignificant, as different strains within the same bee gut species can possess various metabolic capabilities and gene repertoires (Ellegaard and Engel 2019; Engel et al. 2012; Kešnerová et al. 2017). In fact, extensive strain diversity of the honey bee core gut bacteria, corresponding to different functional capabilities, has been shown for all five of the core species. For example, some but not all strains of *G. apicola* have the ability to metabolize sugars that are toxic to honey bees, such as mannose (Zheng et al. 2016), different strains of the two *Lactobacillus* clades (Firm 4 and Firm 5) and *Bifidobacterium* possess different genes for carbohydrate utilization and transportation functions (Ellegaard et al. 2019; Ellegaard and Engel 2019; Kešnerová et al. 2017), and within *S. alvi*, differences in sensitivity to glyphosate have been shown across different strains (Motta et al. 2018), and some *S. alvi* strains – but not all – possess type VI secretion systems (T6SSs), which are frequently involved in strain and species competitions in natural communities (Steele et al. 2017). Moreover, bees of different ages possess different strains with different metabolic capabilities, corresponding to their age and diet (Ellegaard and Engel 2019). These functional differences suggest that strain diversity greatly contributes to host health. Loss of particular strains or species within the microbiome community could affect the ability of bees to metabolize and obtain nutrients from their diet as well as respond to environmental stressors or changes.

Impact of Antibiotics on the Honey Bee Microbiome

An alteration in microbial community composition within a host is referred to as dysbiosis. The term dysbiosis is commonly used to describe alterations in a host community that negatively impact the health of the organism, but the literal definition means any change in the microbial community. Therefore, there are many factors that could alter a community (Raymann and Moran 2018; Figure 9.3), and not all are necessarily bad.

As discussed above, the worker microbiome has been shown to change with division of labor, which is likely an important change to accommodate dietary and lifestyle changes. However, external factors such as exposure to pesticides and antibiotics have been shown to alter the microbiome in ways that have negative impacts on bee health (Motta et al. 2018; Raymann et al. 2017a,b). Whether negative or positive, dysbiosis of the gut undoubtedly impacts honey bee fitness (Figure 9.3), since the microbiome plays a major role in immune function, nutrition, metabolism, removal of toxins, growth and development, and protection against pathogens (Raymann and Moran 2018).

As explained above, antibiotics are frequently used in beekeeping to treat or prevent bacterial brood diseases. Unfortunately, antibiotics do not only target pathogenic bacteria, but also beneficial bacteria and the use of antibiotics in beekeeping can result in alterations of the natural gut microbiota. It has been shown that exposure to oxytetracycline severely disrupts the microbiome of honey bees. Following treatment, the overall gut community size was significantly diminished (Raymann et al. 2017b). Of the eight dominant species found in the worker microbiome, oxytetracycline exposure significantly reduced five species, *S. alvi*, *Lactobacillus* sp. Firm 4, *Lactobacillus* sp. Firm 5, *Bifidobacterium* sp., and *B. apis* (Figure 9.4; Raymann et al. 2017b). Moreover, oxytetracycline-exposed bees displayed a higher representation of atypical bacteria from the Enterobacteriaceae family (e.g. other bacteria; Figure 9.4), specifically the opportunistic pathogen *Serratia marcescens* (Raymann et al. 2017b).

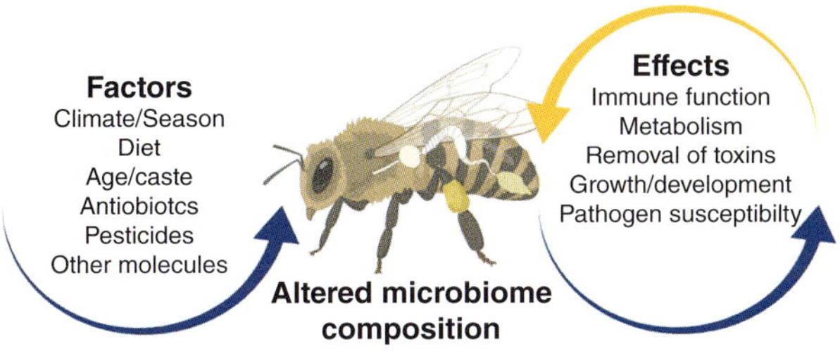

Figure 9.3 Factors that can alter the composition of the honey bee gut microbiome. *Source:* Adapted from Raymann and Moran (2018).

Alteration of the gut microbiome caused by antibiotics impacted honey bee survival both in the hive as well as in the laboratory (Raymann et al. 2017b). Microbiota-free (bees raised without their normal gut bacteria), did not exhibit increased mortality following oxytetracycline treatment in the laboratory, which suggests that the antibiotic treatment itself does not cause mortality but rather the disruption of the native gut microbial community (Figure 9.4; Raymann et al. 2017b). Furthermore, workers exposed to the opportunistic pathogen *Se. marcescens* following oxytetracycline treatment were significantly more susceptible to infection and death (Raymann et al. 2017b). In other animals, opportunistic infections are very common following antibiotic treatment. Therefore, increased susceptibility to *Se. marcescens* or other opportunistic pathogens could explain the increased mortality seen in the hive following treatment.

There is evidence that antibiotic treatment weakens the honey bee immune response, specifically by decreasing the expression levels of antimicrobial peptides (AMPs), key components of the immune response in honey bees (Li et al. 2019, 2017). Altered immune response due to antibiotic treatment has been correlated with increased infection by *Nosema*, a microsporidian parasite, (Li et al. 2017) and increased Deformed Wing Virus (DWV) titer (Li et al. 2019). The decrease in expression of AMPs is predicted to result from altered microbiome composition following antibiotic treatment, since the honey bee microbiome stimulates immune pathways (Kwong et al. 2017). Antibiotics have also been associated with nutritional deficiencies (Li et al. 2019), likely due to the disruption of the microbiome which is important for honey bee metabolism, growth, and development (Zheng et al. 2017).

Although there is evidence that antibiotic treatment can be harmful to honey bee health by disrupting the gut microbiome, the few studies that have investigated the impacts of antibiotic treatment have been limited to studying workers and colony level studies have not been

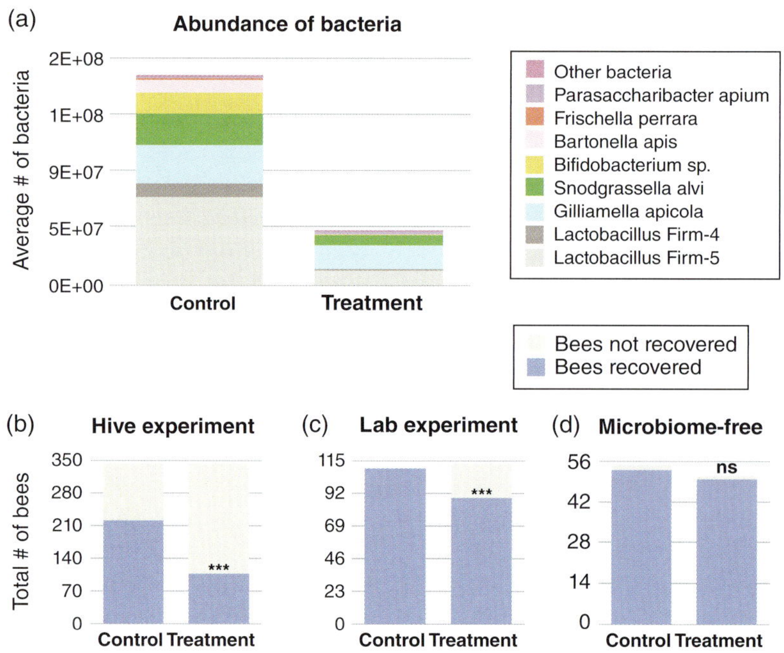

Figure 9.4 (a) Average bacterial abundance in the gut of honey bees that were not treated (control) or treated with oxytetracycline (treatment). Colors represent the abundance of each bacterial species. Number of bees alive or recovered three days following oxytetracycline treatment versus control bees. Bees were treated in the laboratory for five days and either (a) marked with paint, returned to the hive, and counted three days post-treatment or (b, c) kept in the laboratory and death monitored three days post-treatment. Stars (***) represent statistical significance (p value < 0.0001), ns means not significant. Bees used in experiments (a–c) were normal workers taken from the hive. Bees used in experiment (d) were raised in the laboratory without their gut microbiome in the lab and then treated. *Source:* Modified from Raymann et al. (2017b).

performed. Thus, it is unclear how antibiotics affect the gut microbiomes and overall health of an entire colony. The impacts of antibiotic treatment are likely less devastating than infection by brood diseases, but with knowledge of the negative consequences of antibiotic treatment it is important that research efforts be geared toward finding alternative treatments for AFB and EFB, such as management techniques, biological control methods, etc.

Probiotics for Honey Bees

Probiotic therapy, or the administration of live microbes in order to try to improve or restore the gut flora, has been proposed as a promising and cost-effective strategy to enhance bee health, particularly following antibiotic treatment. Several companies have created honey bee probiotics marketed and sold to beekeepers. In humans and other mammals, probiotics have not been proven to be effective in individuals with a healthy gut microbiome (Zmora et al. 2018), but there is some evidence that probiotics can be beneficial following antibiotic treatment (Korpela et al. 2018). Therefore, the use of probiotics could help mitigate the negative effects of antibiotic treatment on honey bees. However, all currently available probiotics for bees are composed of bacteria or yeast (from mammals or the environment) that are not native members of the honey bee gut and none have been scientifically proven to restore the gut microbiome. Moreover, even if the currently marketed probiotic microbes colonize the bee gut, it is unlikely they will provide the same functions as the native flora. The idea of using probiotics to restore the bee gut microbiome following antibiotic treatment is great, in theory, but probiotics must be created using the native gut microbes and properly tested for effectiveness. Several research groups are working on developing more natural and realistic probiotics for honey bees but, to date, no commercially sold probiotic has been proven effective.

Other Possible Impacts of Antibiotic Treatment on Honey Bee Health

Temperature Sensitivity

Honey bees maintain an optimal hive temperature of 33–36 °C year around. Workers thermoregulate the hive to maintain this temperature in the brood chamber. At low temperatures honey bees cluster and produce ectothermic heat to warm the hive and survive cold winters (Stabentheiner et al. 2010). Observations of honey bees in the laboratory following oxytetracycline treatment have revealed increased sensitivity to cold temperatures (Raymann, unpublished data). Honey bee workers (in groups of 20 more more) given oxytetracycline become immobile and die when exposed to cold (4 °C or 39 °F) much more quickly than groups of untreated bees. The reason for this observation has not been determined, but this could present a problem when treating colonies before winter if temperatures are suboptimal for bees (<50 °F). Further research needs to be conducted to determine if administration of antibiotics during cold periods impacts hive survival.

Potential Impacts on Reproductive Health

Although the impacts of antibiotics on honey bee reproductive health have yet to be studied, they have been investigated in other animals. In male rats, even low doses of oxytetracycline reduced spermatogenesis and reproductive fertility (Khaki et al. 2008) and in mice, oxytetracycline induced testicular damage (Elzeinová et al. 2013; Farombi et al. 2008). Oxytetracycline has also been shown to decrease human sperm motility and viability, *in vitro* (Hargreaves et al. 1998). Human male idiopathic azoospermia and other reproductive disorders have been associated with environmental exposures to endocrine disrupting chemicals during critical developmental periods (van den Driesche et al. 2015; Rahman et al. 2017). Like other antibiotics, oxytetracycline is classified as an endocrine disrupting chemical because of its mitochondrial inhibition in reproductive cells (Hou et al. 2019). The ability of antimicrobial substances to inhibit mitochondria may stem back to mitochondrial evolution from bacteria (Gray 2012). To date, no studies have been published on the impact of oxytetracycline on queen or drone reproductive health, but there is overwhelming evidence from other species that suggests that antibiotic treatment could lead to reproductive impairment in honey bees.

Antibiotic Resistance

The Bee Gut Microbiome and Bee Bacterial Pathogens

Laboratory studies have shown that bacteria can acquire resistance to antibiotics very quickly, i.e. within less than one week (Baym et al. 2016). As expected, the decades of oxytetracycline use in the United States has resulted in the bee gut bacteria acquiring resistance. A study investigating antibiotic resistance genes in honey bees from the US and countries that do not use antibiotics in beekeeping shows that oxytetracycline resistance genes are more abundant in US honey bees (Tian et al. 2012). In theory, resistance to oxytetracycline in the bee gut bacteria is a good thing, e.g. if all members of the gut microbiome are resistant then antibiotic treatment would not disrupt the microbiome. However, antibiotic resistance genes are costly (in terms of bacterial fitness) to maintain in the absence of antibiotics. Therefore, if hives are not treated frequently, the gut bacteria will start to lose resistance and, when treatments are applied, they will cause damage to sensitive members of the community.

Additionally, resistance genes can also be acquired by the pathogens that are being targeted by the antibiotic. This has been shown for several strains of *P. larvae* (Alippi et al. 2007; Evans 2003; Miyagi et al. 2000), which are resistant to oxytetracycline. In fact, tylosin was introduced in 2005 because *P. larvae* began exhibiting resistance to oxytetracycline. Opportunistic pathogens of honey bees, such as *Se. marcescens* are also resistant to oxytetracycline, allowing them to overgrow and become pathogenic following treatment (Raymann et al. 2018, 2017b).

Conclusion

The gut microbiome is important for honey bee health. The use of antibiotics in beekeeping can have negative impacts on bees by disrupting their intestinal flora. However, more research is needed to determine to what extent antibiotic treatment impacts the bee gut microbiome and overall colony health in field realistic situations. Overall, it is important to understand the implications of antibiotic usage on both honey bees and the environment and to work toward solutions. Additionally, it has become more evident that strain-level diversity is important for bee health and this has become a major avenue of bee microbiome research. It is possible that by understanding how individual strains and species impact bee health we will be able to create specific probiotic treatments that will restore/improve microbiome function and lead to healthy bees.

Acknowledgments

Kasie Raymann wrote this chapter while under support of the National Science Foundation: Grant No. DEB-1930776.

References

Alippi, A.M., López, A.C., Reynaldi, F.J. et al. (2007). Evidence for plasmid-mediated tetracycline resistance in Paenibacillus larvae, the causal agent of american foulbrood (AFB) disease in honeybees. *Veterinary Microbiology* 125: 290–303. https://doi.org/10.1016/j.vetmic.2007.05.018.

Anderson, K.E., Ricigliano, V.A., Mott, B.M. et al. (2018). The queen's gut refines with age: longevity phenotypes in a social insect model. *Microbiome* 6: 108. https://doi.org/10.1186/s40168-018-0489-1.

Baym, M., Lieberman, T.D., Kelsic, E.D. et al. (2016). Spatiotemporal microbial evolution on antibiotic landscapes. *Science* 353: 1147–1151. https://doi.org/10.1126/science.aag0822.

Corby-Harris, V., Maes, P., and Anderson, K.E. (2014a). The bacterial communities associated with honey bee (*Apis mellifera*) foragers. *PLoS One* 9: e95056.

Corby-Harris, V., Snyder, L.A., Schwan, M.R. et al. (2014b). Origin and effect of Alpha 2.2 Acetobacteraceae in honey bee larvae and description of *Parasaccharibacter apium* gen. nov., sp. nov. *Applied and Environmental Microbiology* 80: 7460–7472. https://doi.org/10.1128/AEM.02043-14.

van den Driesche, S., Macdonald, J., Anderson, R.A. et al. (2015). Prolonged exposure to acetaminophen reduces testosterone production by the human fetal testis in a xenograft model. *Science Translational Medicine* 7: 288ra80. https://doi.org/10.1126/scitranslmed.aaa4097.

Elijah Powell, J., Eiri, D., Moran, N.A., and Rangel, J. (2018). Modulation of the honey bee queen microbiota: effects of early social contact. *PLoS One* https://doi.org/10.1371/journal.pone.0200527.

Ellegaard, K.M. and Engel, P. (2016). Beyond 16S rRNA community profiling: intra-species diversity in the gut microbiota. *Frontiers in Microbiology* https://doi.org/10.3389/fmicb.2016.01475.

Ellegaard, K.M. and Engel, P. (2019). Genomic diversity landscape of the honey bee gut microbiota. *Nature Communications* https://doi.org/10.1038/s41467-019-08303-0.

Ellegaard, K.M., Tamarit, D., Javelind, E. et al. (2015). Extensive intra-phylotype diversity in lactobacilli and bifidobacteria from the honeybee gut. *BMC Genomics* 16: 284. https://doi.org/10.1186/s12864-015-1476-6.

Ellegaard, K.M., Brochet, S., Bonilla-Rosso, G. et al. (2019). Genomic changes underlying host specialization in the bee gut symbiont *Lactobacillus* Firm5. *Molecular Ecology* https://doi.org/10.1111/mec.15075.

Elzeinová, F., Pěknicová, J., Děd, L. et al. (2013). Adverse effect of tetracycline and doxycycline on testicular tissue and sperm parameters in CD1 outbred mice. *Experimental and Toxicologic Pathology* 65: 911–917. https://doi.org/10.1016/j.etp.2013.01.004.

Emery, O., Schmidt, K., and Engel, P. (2017). Immune system stimulation by the gut symbiont *Frischella perrara* in the honey bee (*Apis mellifera*). *Molecular Ecology* 26: 2576–2590. https://doi.org/10.1111/mec.14058.

Engel, P., Martinson, V.G., and Moran, N.A. (2012). Functional diversity within the simple gut microbiota of the honey bee. *Proceedings of the National Academy of Sciences of the United States of America* 109: 11002–11007.

Evans, J.D. (2003). Diverse origins of tetracycline resistance in the honey bee bacterial pathogen *Paenibacillus larvae*. *Journal of Invertebrate Pathology* 83: 46–50. https://doi.org/10.1016/s0022-2011(03)00039-9.

Farombi, E.O., Ugwuezunmba, M.C., Ezenwadu, T.T. et al. (2008). Tetracycline-induced reproductive toxicity in male rats: effects of vitamin C and N-acetylcysteine. *Experimental and Toxicologic Pathology* 60: 77–85. https://doi.org/10.1016/j.etp.2008.02.002.

Genersch, E. (2010). American foulbrood in honeybees and its causative agent, *Paenibacillus larvae*. *Journal of Invertebrate Pathology* 103 (Suppl): S10–S19.

Gray, M.W. (2012). Mitochondrial evolution. *Cold Spring Harbor Perspectives in Biology* 4 https://doi.org/10.1101/cshperspect.a011403.

Hargreaves, C.A., Rogers, S., Hills, F. et al. (1998). Effects of co-trimoxazole, erythromycin, amoxycillin, tetracycline and chloroquine on sperm function *in vitro*. *Human Reproduction* 13: 1878–1886. https://doi.org/10.1093/humrep/13.7.1878.

Hou, X., Zhu, L., Zhang, X. et al. (2019). Testosterone disruptor effect and gut microbiome perturbation in mice: early life exposure to doxycycline. *Chemosphere* 222: 722–731. https://doi.org/10.1016/j.chemosphere.2019.01.101.

Johnson, B.R. (2010). Division of labor in honeybees: form, function, and proximate mechanisms. *Behavioral Ecology and Sociobiology* https://doi.org/10.1007/s00265-009-0874-7.

Jones, J.C., Fruciano, C., Marchant, J. et al. (2018). The gut microbiome is associated with behavioural tasks in honey bees. *Insectes Sociaux* https://doi.org/10.1007/s00040-018-0624-9.

Kapheim, K.M., Rao, V.D., Yeoman, C.J. et al. (2015). Caste-specific differences in hindgut microbial communities of honey bees (*Apis mellifera*). *PLoS One* https://doi.org/10.1371/journal.pone.0123911.

Kešnerová, L., Mars, R.A.T., Ellegaard, K.M. et al. (2017). Disentangling metabolic functions of bacteria in the honey bee gut. *PLoS Biology* https://doi.org/10.1371/journal.pbio.2003467.

Kešnerová, L., Emery, O., Troilo, M. et al. (2019). Gut microbiota structure differs between honey bees in winter and summer. *The ISME Journal* https://doi.org/10.1038/s41396-019-0568-8.

Khaki, A., Novin, M.G., Khaki, A.A. et al. (2008). Comparative study of the effects of gentamicin, neomycin, streptomycin and ofloxacin antibiotics on sperm parameters and testis apoptosis in rats. *Pakistan Journal of Biological Sciences* 11: 1683–1689. https://www.ncbi.nlm.nih.gov/pubmed/18819619.

Korpela, K., Salonen, A., Vepsäläinen, O. et al. (2018). Probiotic supplementation restores normal microbiota composition and function in antibiotic-treated and in caesarean-born infants. *Microbiome* 6 https://doi.org/10.1186/s40168-018-0567-4.

Kwong, W.K. and Moran, N.A. (2015). Evolution of host specialization in gut microbes: the bee gut as a model. *Gut Microbes* 6: 214–220.

Kwong, W.K. and Moran, N.A. (2016). Gut microbial communities of social bees. *Nature Reviews. Microbiology* 14: 374–384.

Kwong, W.K., Engel, P., Koch, H., and Moran, N.A. (2014). Genomics and host specialization of honey bee and bumble bee gut symbionts. *Proceedings of the National Academy of Sciences of the United States of America* 111: 11509–11514.

Kwong, W.K., Mancenido, A.L., and Moran, N.A. (2017). Immune system stimulation by the native gut microbiota of honey bees. *Royal Society Open Science* 4: 170003.

Lee, F.J., Rusch, D.B., Stewart, F.J. et al. (2015). Saccharide breakdown and fermentation by the honey bee gut microbiome. *Environmental Microbiology* 17: 796–815. https://doi.org/10.1111/1462-2920.12526.

Li, J.H., Evans, J.D., Li, W.F. et al. (2017). New evidence showing that the destruction of gut bacteria by antibiotic treatment could increase the honey bee's vulnerability to Nosema infection. *PLoS One* 12: e0187505.

Li, J., Heerman, M.C., Evans, J.D. et al. (2019). Pollen reverses decreased lifespan, altered nutritional metabolism and suppressed immunity in honey bees (*Apis mellifera*) treated with antibiotics. *The Journal of Experimental Biology* 222 https://doi.org/10.1242/jeb.202077.

Martinson, V.G., Moy, J., and Moran, N.A. (2012). Establishment of characteristic gut bacteria during development of the honeybee worker. *Applied and Environmental Microbiology* 78: 2830–2840.

Mattila, H.R., Rios, D., Walker-Sperling, V.E. et al. (2012). Characterization of the active microbiotas associated with honey bees reveals healthier and broader communities when colonies are genetically diverse. *PLoS One* 7: e32962.

Miyagi, T., Peng, C.Y.S., Chuang, R.Y. et al. (2000). Verification of oxytetracycline-resistant American foulbrood pathogen *Paenibacillus larvae* in the United States. *Journal of Invertebrate Pathology* 75: 95–96.

Moran, N.A., Hansen, A.K., Powell, J.E., and Sabree, Z.L. (2012). Distinctive gut microbiota of honey bees assessed using deep sampling from individual worker bees. *PLoS One* 7: e36393.

Motta, E.V.S., Raymann, K., and Moran, N.A. (2018). Glyphosate perturbs the gut microbiota of honey bees. *Proceedings of the National Academy of Sciences of the United States of America* https://doi.org/10.1073/pnas.1803880115.

Powell, J.E., Martinson, V.G., Urban-Mead, K., and Moran, N.A. (2014). Routes of acquisition of the gut microbiota of the honey bee *Apis mellifera*. *Applied and Environmental Microbiology* 80: 7378–7387.

Powell, J.E., Ratnayeke, N., and Moran, N.A. (2016). Strain diversity and host specificity in a specialized gut symbiont of honey bees and bumble bees. *Molecular Ecology* 115 (41): 10305–10310.

Rahman, M.S., Kwon, W.-S., Karmakar, P.C. et al. (2017). Gestational exposure to bisphenol A affects the function and proteome profile of F1 spermatozoa in adult mice. *Environmental Health Perspectives* 125: 238–245. https://doi.org/10.1289/EHP378.

Raymann, K. and Moran, N.A. (2018). The role of the gut microbiome in health and disease of adult honey bee workers. *Current Opinion in Insect Science* https://doi.org/10.1016/j.cois.2018.02.012.

Raymann, K., Bobay, L.-M., and Moran, N.A. (2017a). Antibiotics reduce genetic diversity of core species in the honeybee gut microbiome. *Molecular Ecology* 10: 1589.

Raymann, K., Shaffer, Z., and Moran, N.A. (2017b). Antibiotic exposure perturbs the gut microbiota and elevates mortality in honeybees. *PLoS Biology* 15: e2001861.

Raymann, K., Coon, K.L., Shaffer, Z. et al. (2018). Pathogenicity of *Serratia marcescens* strains in honey bees. *MBio* 9 https://doi.org/10.1128/mBio.01649-18.

Stabentheiner, A., Kovac, H., and Brodschneider, R. (2010). Honeybee colony thermoregulation – regulatory mechanisms and contribution of individuals in dependence on age, location and thermal stress. *PLoS One* 5: e8967. https://doi.org/10.1371/journal.pone.0008967.

Steele, M.I., Kwong, W.K., Whiteley, M., and Moran, N.A. (2017). Diversification of Type VI secretion system toxins reveals ancient antagonism among bee gut microbes. *MBio* 8.

Tarpy, D.R., Mattila, H.R., and Newton, I.L.G. (2015). Development of the honey bee gut microbiome throughout the queen-rearing process. *Applied and Environmental Microbiology* 81: 3182–3191.

Tian, B., Fadhil, N.H., Powell, J.E. et al. (2012). Long-term exposure to antibiotics has caused accumulation of resistance determinants in the gut microbiota of honeybees. *MBio* 3.

Zheng, H., Nishida, A., Kwong, W.K. et al. (2016). Metabolism of toxic sugars by strains of the bee gut symbiont *Gilliamella apicola*. *MBio* 7 (6): e01326–e01316.

Zheng, H., Powell, J.E., Steele, M.I. et al. (2017). Honeybee gut microbiota promotes host weight gain via bacterial metabolism and hormonal signaling. *Proceedings of the National Academy of Sciences of the United States of America* 114: 4775–4780.

Zmora, N., Zilberman-Schapira, G., Suez, J. et al. (2018). Personalized gut mucosal colonization resistance to empiric probiotics is associated with unique host and microbiome features. *Cell* 174: 1388–1405.e21. https://doi.org/10.1016/j.cell.2018.08.041.

10

Honey Bee Pharmacology

Gigi Davidson

NC State College of Veterinary Medicine, Department of Clinical Pharmacy Services, Raleigh, NC, USA

Introduction

As a forager of pollen, nectar, tree resins, and water, *Apis mellifera* encounters a broad diversity of both natural and synthetic substances. In addition to the phytochemicals produced by forage, honey bees also encounter numerous agrochemicals either directly on crops or indirectly through drift of chemicals such as pesticides, herbicides, and fungicides. Compared to other pollinators, her practice of floral fidelity may intensify her exposure from treated forage. Honey bees may also experience exposure to agrochemicals and additional xenobiotics through runoff or direct application into water sources such as ponds, streams, lakes, creeks, fountains, and swimming pools. To further intensify exposures, the honey bee also encounters many natural and synthetic chemicals applied directly to the colony by humans for control of pests and pathogens such as mites and bacteria. With humans synthesizing and applying more and more new chemicals to all aspects of the landscape, *A. mellifera* often finds her somewhat limited metabolic capability to detoxify new chemicals challenged to the point of failure. For veterinarians to provide the best medical care for honey bees, a critical understanding of honey bee pharmacology and drug disposition in both the honey bee and hive products is imperative.

When embarking on a tutorial to explain drug disposition in a species, it is tempting to consider only the individual physical body of that organism while incorporating variations caused by life stage or pharmacogenetics. For example, pharmacology of the canine species is typically presented from the perspective of a stereotypical individual dog as a neonate, adult, or geriatric individual, and any relevant breed-specific differences are discussed comparatively to the average individual dog. When discussing social insects such as the honey bee, however, a different approach is required. While drug disposition in each of the 60 000 or more individual honey bees in a typical colony may be relatively similar between individuals, the impact of the caste system (different roles based on bee types), the hive body (the structure that the bees live in), and hive products (wax comb and food stores) on drug disposition and drug action in both brood and adult bees must also be considered. In this context, the honey bee colony, hive products, and hive body may be viewed collectively as an organism. The concept of a honey bee colony as a superorganism, and not just a large collection of individuals, dates to the nineteenth century with Herbert Spencer (1898), subsequently with Rudolf Steiner (1998), Bert Hölldobler and Edward O. Wilson (2009), and most recently, Thomas Seeley (2010). The analogy of the honey bee colony as a superorganism has been applied in several contexts such as group decision making and organ function and is appropriate for explaining honey bee pharmacology. It is from a perspective of the honey bee colony as an organism that this chapter attempts to describe drug disposition in *A. mellifera*.

Principles of Pharmacology

Before describing pharmacology specific to the *A. mellifera* organism, it is important to review the basic principles of pharmacology. Principles of pharmacology are largely divided into three areas of study: *pharmacokinetics* (*the study of the time course of drug absorption, distribution, metabolism, and excretion in an organism*), *pharmacodynamics* (*the relationship between the concentration of a substance in an organism and the effects that it causes*), and *pharmacogenetics* (*how changes in genes influence what a drug or substance does to an organism*). For this part of the chapter, the discussion will mostly be aimed at individual bees, however, the focus of our intent will still be the honey bee colony as an organism.

Honey Bee Medicine for the Veterinary Practitioner, First Edition. Edited by Terry Ryan Kane and Cynthia M. Faux.

Pharmacokinetics

The science of *pharmacokinetics* describes movement of drugs and other chemicals into, throughout, and out of a system. The pharmacokinetic time-course of the journey of a xenobiotic through a system is described in terms of absorption, distribution, metabolism, and excretion.

Drug and chemical *absorption* into a system is determined by the drug's chemical properties, formulation, and route of exposure. For example, the amount of chemical that enters a honey bee, honey, propolis, or wax comb depends on the chemical properties of the substance (e.g. lipophilic – attracted to lipid substances vs. hydrophilic – attracted to water, weak acid vs. weak base, etc.), the form of the substance (e.g. powder vs. solution vs. impregnated strip vs. volatile acid), and the route of exposure (e.g. direct ingestion vs. inhalation vs. cuticular contact).

Distribution is a description of where the xenobiotic travels in a system once absorption is accomplished. Distribution depends on the degree of perfusion to a tissue (e.g. blood or hemolymph flow), as well the degree of permeability between the vascular system and the tissue (e.g. passive diffusion vs. active transport or efflux pumps).

The extent of distribution throughout tissues is expressed as the substance's *volume of distribution* or *Vd*, an apparent or theoretical volume within an organism into which a substance disperses in order to produce an observed concentration within a sample from that organism. Substances with lower Vd are typically more localized to circulating plasma or hemolymph where they can be more easily metabolized and eliminated, while substances with high Vd distribute to many tissues beyond the vasculature and are more likely to persist longer in a given system. Knowing the Vd for a particular drug or chemical in the target species is a useful indicator of the likelihood of tissue penetration and persistence in a given organism, and therefore the potential for the substance to be able to perform its intended function within the organisms (e.g. for an antibiotic to attain enough of a concentration to kill a known microorganism, or for a toxic pesticide to damage a particular organ).

Metabolism, in our discussion for this chapter, is a process by which substances are made more suitable for elimination from a system (e.g. combined with existing substances to render the metabolite more electrically charged to enhance excretion in urine), or to alter the substance to change its function (e.g. by cleavage or addition of a new functional group through hydrolysis, oxidation, or reduction). Typically, metabolic processes result in an inactive product (detoxification); however, some substances may be metabolized to a metabolite that has more activity in a system (toxification). Toxification often results as a process of saturated metabolizing systems and multi-drug interactions where detoxification pathways are bypassed or overwhelmed.

Drugs and environmental chemicals are metabolized by four different kinds of reactions: oxidation, reduction, hydrolysis, and conjugation (Parkinson and Ogilvie 2008). The first three reactions (oxidation, reduction, and hydrolysis) are often grouped together and called "functionalization," or Phase I reactions, so that the conjugation reactions are called Phase II reactions. A third group of metabolic processes includes the Phase III systems that transport Phase II byproducts out of the cell, expediting the movement of toxins and metabolites toward excretion, and prevent infiltration of sensitive tissues by those compounds.

One of the major Phase I enzyme systems are the cytochrome P450 monooxygenases (CYP450) which are genetically coded for in an organism's DNA. Cytochrome P450 monooxygenase occurs in several different species-specific forms, or isozymes, which differ in how they act in an organism, including how they are regulated and distributed. Many CYP450 isozymes are ubiquitous, residing in multiple cells and tissues, while others must be spontaneously induced (upregulated) upon exposure to an insulting substance. Honey bees lack kidneys and liver; drug metabolizing enzymes are largely found in the fat body, Malpighian tubules, the midgut, mitochondria, and cuticle. CYP450 isoenzymes and transporters are important areas of investigation for understanding the toxicity of chemicals encountered by the honey bee. It is essential to note that drugs and chemical substances may enhance (i.e. inducers), or inhibit (i.e. inhibitors) the function of CYP450 isozymes and transporters. For example, azole fungicides are well known inhibitors of both the CYP450 and protein transporter systems in many species and disruption of these metabolic processes can cause normally non-toxic substances to exert significant toxicity.

Excretion describes the systems used to remove drug and chemical substances from an organism. Excretion is accomplished in several tissues including kidney, liver, spleen, lungs, and saliva and involves Phase III transporters. Since the honey bee lacks liver and kidneys (as well as spleen and lungs), the corresponding organs of excretion include the fat body, the Malpighian tubules, the hindgut, and the cuticle, with various Phase III substance-handling transporters located within these tissues (Figure 10.1).

Pharmacodynamics

Pharmacodynamics is the study of the biochemical, physiologic, and molecular effects of drugs on a system and involves receptor binding, post-receptor effects, and

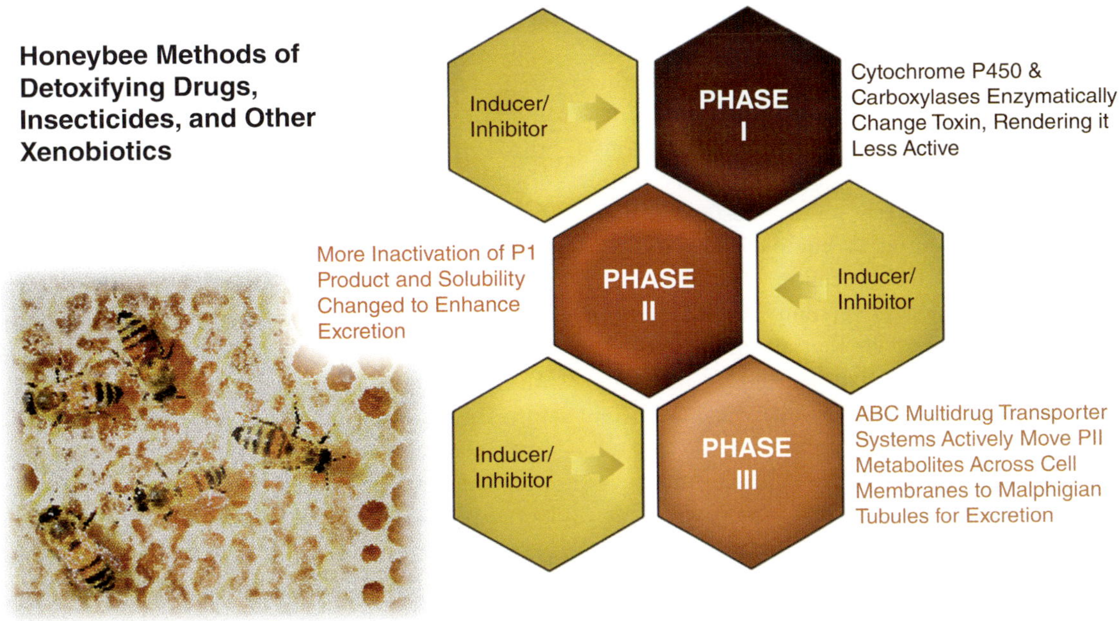

Figure 10.1 Honey Bee Methods of Detoxifying Drugs, Insecticides, and Other Xenobiotics. *Source:* Illustration by Patrick D. Wilson.

chemical interactions. For example, *A. mellifera* possesses receptors for the biological amines (e.g. most neurotransmitters and vasoactive compounds; dopamine, tyramine, octopamine, serotonin, and histamine) which may be acted upon by endogenous substances or, errantly, by substances with similar stereochemistry resulting in undesirable effects. Another important receptor in *A. mellifera* is the nicotinic acetylcholine receptor (nAChR) which plays a significant role in neurologic and cognitive processes. Neonicotinoid pesticides target this receptor and are highly toxic to honey bees, causing death at certain doses or severe disturbances in learning behaviors and foraging at sublethal doses.

The action of any substance can be accomplished at the cellular level, the organism level, or the population level. In the case of agricultural substances, pharmacodynamics is focused on mechanisms of toxicity either to the target (e.g. pests and pathogens) or to the host (e.g. the honey bee). For example, parasitic mites such as *Varroa destructor* are arthropods just as is the honey bee and figuring out how to kill one without harming the other can be quite a challenge. Specific pharmacodynamics and receptors for agrichemicals in the honey bee are described later in this chapter.

Pharmacogenomics

The science of *pharmacogenetics* describes how a single gene mutation in an organism can influence the response to an individual drug. The science of *pharmacogenomics* is broader and describes how the entire genome influences responses to drugs. The honey bee genome was successfully sequenced in 2006 (Honeybee Genome Sequencing Consortium 2006) and has offered significant insight into mechanisms of genes for the use of pollen and nectar, for immunity, detoxification, and for the development of the cuticle. Currently, the field of genomics regarding *A. mellifera* is primarily focused on sequencing and manipulation of genes that aid in the response to parasites and pathogens. For example, a transgenic honey bee resistant to the effects of neonicotinoid insecticides could potentially be developed. The ethical and long-term implications of such manipulations are highly controversial within and outside of the honey bee community; time will be the ultimate judge of the value or harm caused by such interventions.

Pharmacology of the Individual Honey Bee as it Impacts the Hive Organism

Pharmacokinetics

Absorption

Honey bees may be exposed to drugs and chemical substances by several mechanisms: trophallaxis (mouth-to-proboscis transfer of substances from one bee to another), direct ingestion of contaminated substances (pollen, nectar, honey stores, beebread, plant resins, etc.), inhalation (through the respiratory spiracles on the thorax and abdomen), and through direct contact with chemical substances (penetration of substances through the cuticular hydrocarbons in the exoskeleton). Figure 10.2. For any substance to penetrate a tissue membrane without the assistance of an active transporter pump, it must be in solution and not in an ionized form. Substances that facilitate dissolution of a drug or chemical will greatly enhance the likelihood of absorption. For example, substances with higher water content will facilitate dissolution of drugs or chemicals that are more polar or water soluble, and substances that are mostly made of lipids and are non-aqueous will facilitate dissolution of lipophilic drugs and chemicals. Absorption of drugs and chemicals also depends on the chemistry of the substance (e.g. acid-base chemistry, hydrophilicity or lipophilicity, particle size, and volatility) as well as the vehicle carrying the drug or chemical. For example, food jelly (royal or brood jelly) is strongly acidic and approximately 60–70% water. Honey is acidic with an average pH of 3.9 and typically contains up to 20% water. Beebread is also strongly acidic due to fermentation with lactic acid and is also somewhat polar due to water content. For this reason, glyphosate and other weak acids such as many herbicides, insecticides, and fungicides will not be ionized in this aqueous environment and will more easily diffuse across membranes. In contrast, sulfonylurea herbicides (e.g. chlorsulfuron, triflusulfuron) are weak bases and ionize in the acidic environment of these foodstuffs and are less absorbed. See Table 10.1 for Aqueous Solubility and Lipophilicity of Selected Substances for various agents that are used in or around the apiary.

Considering that every member of the colony receives jelly at some point in its life and that workers and drones consume honey and beebread most of their lives either through direct ingestion or through trophallactic contact, resultant exposure from foodstuffs contaminated with these agents is likely.

Apis mellifera secretes cuticular hydrocarbons (mono- and polyesters and free fatty acids) that coat the exoskeleton and provide a barrier against bacteria, fungi, and water loss. Substances that are lipophilic in nature (e.g. DDT, chlordane, pyrethroids, azole fungicides, macrolide antimicrobials, and avermectins) pass easily through the oily cuticle, where they may acutely cause toxicity or mortality, or eventually accumulate in the fat body where they may cause chronic morbidity or illness.

The fat body is built up over the summer and early fall and typically achieves the largest volume by late fall and early winter. Therefore, timing of application of lipophilic substances can impact the degree of distribution into the fat body. In addition to providing winter nutrients, the fat body also plays an important role in the secretion of wax scales, and it is likely that lipophilic contaminants may even appear in new wax comb from workers exposed to these agents. Likewise, partition of lipophilic contaminants may also occur into the wax comb and serve as a depot for continued exposure to brood and to young bees as they tend brood and comb. In 2010, a sampling of hive products including wax, pollen, and honey revealed 121 different pesticides and metabolites comprising 5519 total residues within 887 wax, pollen, bee, and associated hive samples (average of 6.2 detections per sample) from 23 states and 1 Canadian province with all wax foundation having significant accumulation of miticides. (Mullin CA, Frazier M, Frazier JL, Ashcraft S, Simonds R, et al. 2010).

The lipophilicity of a substance can somewhat be predicted by its octanol:water partition coefficient, expressed as logP. Substances with logP values less than 1 are hydrophilic, values around 0 are neutral, and values greater than 1 are considered lipophilic. Likelihood for partition into wax comb and across cuticular hydrocarbons might be predicted by those substances with logP values >0. Table 10.1 considers lipophilicity for various chemicals applied in or around the hive.

Other substances may also enter the honey bee through the respiratory system, via the spiracles located on the thorax and abdomen. Even though the spiracles are

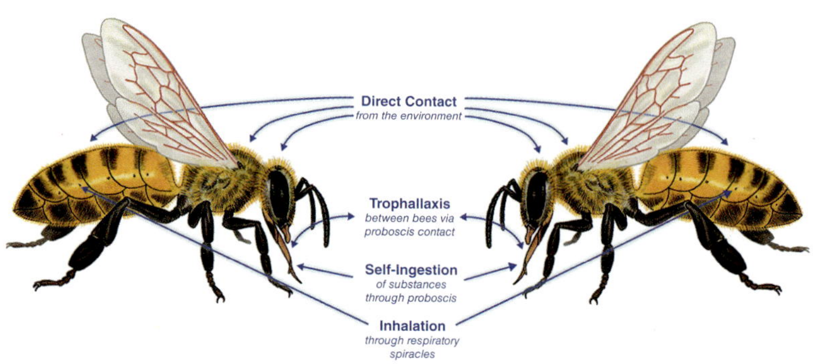

Figure 10.2 Routes of Exposure by Honey Bees to Drugs, Insecticides, and Other Xenobiotics. *Source:* Illustration by Patrick D. Wilson.

Table 10.1 Aqueous solubility and lipophilicity of selected substances.

Brand name	Active ingredients	Indication	Aqueous solubility[a]	Lipophilicity[b]
Agri-Flex	Abamectin	Pesticide	Insoluble	High
Apiguard	Thymol	Miticide	Very high	Moderate
Apilife VAR	Thymol (76%), eucalyptol (16.4%), menthol (3.8%), and camphor (3.8%)	Miticide	Very high	Moderate
Apistan	τ-fluvalinate	Miticide	Insoluble	High
Apivar	Amitraz	Miticide	Insoluble	High
Checkmite +	Coumaphos	Miticide	Very low	High
Fumadil-B	Fumagillin B	*Nosema*	Very low	High
Guardstar	Permethrin	Miticide	Very low	High
Kohinor, Admire, Advantage, Gaucho, Merit, Confidor, Hachikusan, Premise, Prothor, and Winner	Imidacloprid	Pesticide	Very high	Moderate
Lincomix	Lincomycin	Foulbrood	Very high	Low
Mite-a-thol	Menthol	Miticide	High	Moderate
Miteaway II	Formic acid	Miticide	Extremely high	(Hydrophilic)
Oxalic acid	Oxalic acid	Miticide	Extremely high	(Hydrophilic)
Paramoth	Paradichlorobenzene	Wax moths	Intermediate	Moderate
Regent 4 SC	Fipronil	Pesticide	Very low	High
RoundUp	Glyphosate	Herbicide	Extremely high	(Hydrophilic)
Stratego YLD	Prothioconazole	Azole fungicide	High	Low
Sucrocide	Sucrose octanoate	Miticide	Extremely high	(Hydrophilic)
Terramycin	Oxytetracycline	Foulbrood	High	Low
Tylan	Tylosin	Foulbrood	Very high	Low

[a] Insoluble $<1\,\text{mg}\,\text{l}^{-1}$, Very low $1–10\,\text{mg}\,\text{l}^{-1}$, Low $11–50\,\text{mg}\,\text{l}^{-1}$, Intermediate $51–150\,\text{mg}\,\text{l}^{-1}$, High $151–500\,\text{mg}\,\text{l}^{-1}$, Very high $500–5000\,\text{mg}\,\text{l}^{-1}$, Extremely high $>5000\,\text{mg}\,\text{l}^{-1}$ (Pereira et al. 2016).
[b] Based on logP <1 = hydrophilic, 0 = neutral, >1 = lipophilic.

guarded by fine hairs to prevent entry of dust particles, volatile chemicals easily penetrate open spiracles.

Vapor pressure is a common measure of the tendency of a solid or a liquid to transform into a gas (i.e. its volatility). The nearer a substance is to its boiling point, the higher is its vapor pressure. Chemicals with low boiling points, therefore, have higher volatility, which is maximized at higher temperatures.

The volatility of a substance, then, can be predicted based on its described boiling point and the ambient temperature in which it will be used. For example, menthol is used to treat tracheal mites by taking advantage of this route of exposure for positive therapeutic effects; however, other volatilizing substances that do not intend the honey bee as the primary target (e.g. thymol, formic acid, and oxalic acid used to target varroa mites) may also achieve systemic concentrations in the honey bee via spiracle inhalation.

The vapor pressure of a chemical and the ambient temperature should be carefully considered prior to application inside a colony. For example, volatile agrochemicals that are deposited in tree resins after spraying may also partition into propolis and be released in the hive body over time, especially during warmer temperatures.

As previously stated, the dosage form of the substance affects absorption. Placing dissolved drugs or chemical substances directly into the feed syrup of honey bees typically results in high levels of absorption while solid dosage forms sprinkled on top of frames are likely to have lower levels of absorption. Honey bees are repelled by natural phytotoxins and pesticides encountered in nectar and pollen (Reinhard et al. 2009; Ramirez-Romero et al. 2008) but are less repelled by toxins if they are encountered in food (Detzel and Wink 1993). For this reason, adding potentially toxic substances to sugar syrup may result in a higher degree of acceptance, and therefore, of absorption than if introduced by other routes of administration.

Distribution

Once absorption is accomplished, distribution of drugs and chemical substances to tissues is largely through hemolymph. The temporally (season and temperature) caused and dietary caused fluctuations in the hemolymph (protein content and osmotic pressure) may influence the speed and tissue target to which substances are distributed. For example, at 22–25 °C hemolymph circulates rapidly and concentrates in the abdomen and thorax, while at 5 °C the hemolymph circulates slowly and preferentially circulates to the head (Crailsheim 1985). Generally, substances that are more polar and water soluble will have a lower Vd and be more rapidly eliminated while those that are less polar must be biotransformed into excretable metabolites and take longer to eliminate. While the specific Vd for a substance in honey bees is rarely, if ever, available, comparative judgments may generally be made by considering the Vd in other species to potentially predict the likelihood of tissue distribution.

Metabolism

Compared to other insects, *A. mellifera* possesses only about half to one-third of the genes encoding for detoxification metabolism. While the honey bee is generally no more sensitive to individual toxins than other insects, the relative deficiency in detoxification enzymes likely makes the honey bee more susceptible to toxicity, especially when attempting to metabolize multiple substances simultaneously. Substance detoxification involves the conversion of less water-soluble substances to more water-soluble metabolites that can be excreted. Honey bee metabolism occurs in three different processes: Phase I, Phase II, and Phase III.

In Phase I detoxification, primarily through the actions of cytochrome P450 isoenzymes and carboxylesterases, the toxin structure is altered and rendered unsuitable to attach to lipophilic target receptors.

Phase II reactions typically involve conjugation of products of Phase I detoxification for (aqueous) solubilization and transport, and primarily consist of glutathione-*S*-transferases (GSTs) but also include glycosyltransferases, phosphotransferases, sulfotransferases, aminotransferases, and glycosidases (Xu et al. 2013).

Phase III transporters include ATP-binding cassette transporters (ABC transporters) (Heidel-Fischer and Vogel 2015) and are discussed in more detail later in this chapter.

Studies of both transporter systems and metabolizing enzymes for substances in the honey bee have become two of the most rapidly growing areas of scientific investigation. A comprehensive review of these investigations is too broad for the scope of this chapter; however, a general overview of metabolic and transport pathways for critical substances is presented.

Phase I Biotransformation Cytochrome P450 monooxygenase enzymes make up the majority of Phase I metabolism in the honey bee; however, compared to other insect species, *A. mellifera* is relatively lacking in a diversity of CYP families. While other insect species may have twice to three times as many CYP izoenzymes, *A. mellifera* only codes for 46 genes, most of which are in the CYP6 and CYP9 families (Claudianos et al. 2006). The CYP4 family is present but is significantly underrepresented compared to other insects. Considering the relatively pure and simple honey bee diet of pollen and nectar, it can be postulated that, teleologically, the honey bee did not require a large repertoire of isoenzymes to perform detoxification prior to the widespread presence of agrichemicals and that, when encountering novel phytochemicals, the necessary detoxification enzymes could be upregulated as required. Another possible theory for the paucity of CYP genes in the honey bee is that these bees form symbiotic partnerships with detoxifying organisms such as *Aspergillus* during beebread fermentation. This relationship between CYP genes and fungus farming has been demonstrated in the leafcutter ant which is also a known fungus farmer and has considerably fewer CYP genes than other ants (Nygaard et al. 2011; Suen et al. 2011).

Honey bees utilize cytochrome P450 enzymes for detoxification of honey and pollen flavonoids (Mao et al. 2009), pyrethroids (Johnson et al. 2006), organophosphate and phenoxypyrazole acaricides (Johnson et al. 2013), neonicotinoid insecticides (Iwasa et al. 2004), and mycotoxins (Niu et al. 2011). See Table 10.2 CYP450 Enzymes, Substrates, Inducers, and Inhibitors in *A. mellifera*.

Carboxylesterases, the other major components of Phase I Biotransformation, are responsible for the metabolism of both endogenous esters (e.g. neuro/developmental esters, pheromones) and exogenous esters and carbamates. Again, the honey bee is relatively deficient in these enzymes compared to other insects and only codes for 24 esterases compared to twice this number in some mosquito species (*Anopheles*) (Claudianos et al. 2006). The activity of carboxylesterases in *A. mellifera* varies with age and task of the various caste members with drones and newly emerged workers having the lowest concentrations and foragers having the highest (Liao 2016). Carboxylase activity is upregulated by

Table 10.2 Substrates, inducers, and inhibitors of CYP 450 families in *Apis mellifera*.

CYP isoenzyme family	Substrates	Inhibitors	Inducers
CYP4 family	Endogenous lipids, pheromones (Feyereisen 2005)	Quercetin (Mao et al. 2017)	Unknown for *A. mellifera* (resveratrol, rifampin, dexamethasone, progesterone, TNFα known in humans) (Edson and Rettie 2013)
CYP6 family	Quercetin (Mao et al. 2009)		Honey, pollen, propolis (Johnson et al. 2012) Thymol (Boncristianai et al. 2012) *p*-coumarate (Mao et al. 2013) Imidacloprid (Derecka et al. 2013) Coumaphos, τ-fluvalinate (Schmehl et al. 2014)
CYP9 family	τ-fluvalinate coumaphos (Mao et al. 2011)	Bifenthrin (Mao et al. 2011)	Honey, τ-fluvalinate, cypermethrin, bifenthrin (Mao et al. 2011)

exposure to organophosphates, neonicotinoids, pyrethroids, phenylpyrazoles, and spinosyns (Carvalho et al. 2013), as well as coumaphos (Schmehl et al. 2014).

Phase II Biotransformation Glutathione and GSTs are the primary components of Phase II metabolism in honey bees and are responsible for detoxifying reactive oxygen/nitrogen species and electrophiles (substances with positive or neutral valances that can accept electrons). GSTs are a superfamily of Phase II detoxification enzymes that detoxify primarily by catalyzing conjugation and redox reactions. Under normal conditions, neither GST enzyme activities nor glutathione levels operate at their maximum capacity, but instead are induced upon exposure to susceptible substrates. Pyrethroids (Nielsen et al. 2000) and coumaphos (Schmehl et al. 2014) are known inducers of GST.

Phase III Transporters ABC multidrug transporters utilize adenosine triphosphate (ATP) to actively move hydrophobic and conjugated metabolites across cell membranes. The honey bee encodes for 41 different ABC transporters in three subfamilies which can be found in multiple tissues (Dermauw and Van Leeuwen 2014). For example, the Malpighian tubules, responsible for elimination of waste and metabolites, contain five different ABC transporter isoforms (Chan et al. 2013). The exact substrates transported by different ABC transporters in the honey bee are relatively unknown due to the complexity of the process: ABC transporters operate on structurally unrelated substrates, so identifying one substrate often does not help to identify other candidate substrates. To determine which substances are potential substrates of ABC transporters, known ABC transporter pump blockers such as azole fungicides, verapamil, fumagillin, oxytetracycline, and quercetin (Wu et al. 2008) are used as pretreatments to determine if toxicity is enhanced for a substance in the face of transporter blockade. Using various ABC transporter pump inhibitors has revealed that neonicotinoids (acetamiprid, imidacloprid, and thiacloprid), avermectins (ivermectin), and acaricides (coumaphos and τ-fluvalinate) are substrates of the ABC transporter system and are, therefore, toxic in the presence of the above transporter blockading compounds (Guseman et al. 2016; Hawthorne and Dively 2011).

It is important to note that many ABC transporter pump inhibitors are encountered simultaneously in the landscape along with known transporter substrates, resulting in enhanced toxicity of those substrates. For example, oxytetracycline and τ-fluvalinate are commonly applied to honey bee colonies simultaneously in the early spring for control of foulbrood and mites. However, oxytetracycline disrupts the specific ABC transporter pump for which τ-fluvalinate is a substrate, resulting in toxicity of τ-fluvalinate, even at labeled doses.

Disruption of any of the Phase I, II, or III metabolism pathways can result in enhanced toxicity of chemical substances when used simultaneously. A comprehensive investigation into effects on the LD50 (The dose of a substance which will kill 50% of a sample population in a defined period of time. A high LD50 is associated with fairly innocuous substances, whereas a low LD50 is associated with more toxic substances) of combinations of fungicides, acaricides, and antimicrobials in the face of metabolizing enzyme inhibitors has revealed up to a 100-fold lowering of the LD50 for many chemicals tested (Johnson et al. 2013).

A future area of insecticidal research involves targeted blockade of Phase III transporter systems to combat insecticide resistance and increase kill rate in insect pests. Susceptibility of the honey bee to these mechanisms must be carefully considered prior to release of this class of insecticides for use in the landscape. See Table 10.3 for a list of known disruptors of ABC-transporters.

Nutritional factors also strongly affect metabolism in the honey bee. Adding honey, pollen, or propolis to sucrose

Table 10.3 Known disruptors of ABC-transporters.
Source: Modified from Buss and Callaghan (2008).

Arochlor 1254	Atrazine	Azinphos-ethyl
Chlorbenside	Chlorpyrifos	Coumaphos
Cypermethrin	DCPA	DDT
Diazinon	Endosulfan	Fenvalerate
Heptachlor	Itraconazole	Ketoconazole
Methylparathion		

feedings upregulates CYP isoenzyme genes, which encode enzymes that metabolize quercetin and other flavonoids. It has been demonstrated that diets consisting of only sucrose or high fructose corn syrup, however, do not induce upregulation of these important detoxification enzymes (Johnson et al. 2012). Naturally occurring substances in honey (*p*-coumaric acid) and in poplar bud resins (pinocembrin, pinobanksin, and pinobanksin 5-methyl-ester) upregulate all classes of detoxification enzymes in the honey bee, particularly genes which encode for enzymes that metabolize quercetin and acaricides (coumaphos and fluvalinate). Diets of sucrose and high fructose corn syrup do not contain these important enzyme inducers and, if fed, may contribute to increased toxicity.

Excretion

The main excretory organ of the honey bee is the Malpighian tubule which is responsible for maintaining homeostasis and excreting toxic and excess substances into the hindgut for elimination. Each Malpighian tubule is a blind-ending tube suspended in the hemolymph of the abdomen whose lumen is continuous with the lumen of the gut and is encircled with spiraling muscle cells that enable peristaltic emptying of the contents of the tubule into the hindgut. Nitrogenous waste and metabolites either passively diffuse into or are actively transported from the hemolymph into the Malpighian tubules by Phase III transporters and then expelled for elimination from the rectum.

Larvae of *A. mellifera* have 4 Malpighian tubules which are remodeled during the pupal stage and then completely replaced with 64 Malpighian tubules (Gonçalves et al. 2018) following emergence as adults. The lack of Malpighian tubules in pupating honey bees may indicate that the pupal honey bee is more vulnerable to exogenous toxins than are other life stages. Considering that many miticides are placed directly on brood frames, this potentially increased sensitivity to toxins could have a significant impact on developing pupae, without causing outward signs of problems in workers or young bees.

Pharmacodynamic Considerations in the Individual Honey Bee as they Impact the Hive Organism

Specific Receptors

Receptors for metabolic functions in *A. mellifera* are numerous and include receptors that are shared with other invertebrates and vertebrates such as 5HT2, nicotinic acetylcholine, muscarinic, dopaminergic, monoamine oxidase (MAO), NMDA, and GABA receptors. When targeting these receptors in pests and pathogens of the honey bee, unintended action on the honey bee receptors and resultant toxicity is likely. For example, the insect nAChR is a key factor in nerve system communication. It is shared by all insects, and competitive agonists for this receptor, the neonicotinoid pesticides are, therefore, neurotoxins in all insects. Hence, when farmers target destructive insect pests on their crops, honey bees are often unintended collateral casualties.

Additionally, use of multiple chemicals that target the same receptors simultaneously can result in additive agonistic toxicity (causing more toxicity when used together than either agent if used alone). Another type of agonistic toxicity is synergistic toxicity (where the toxic effects of different drugs or agrichemicals target different steps or receptors within a single metabolic pathway of an organism) and results when substances targeting different receptors are used simultaneously.

It is important for veterinarians to be familiar with the basic mechanisms of action of any chemical that is intended for use in or around the apiary, to both prevent toxicities so that if signs of toxicity arise, likely causative agents can be quickly identified and withdrawn. See Table 10.4: Chemical Combinations with Enhanced Toxicity to *A. mellifera* and Table 10.5: Mechanisms of Action of Selected Acaricides.

Specific Chemical Classes

Although the therapeutic uses of drugs and agrochemicals, and the effects on targeted pests and pathogens, have been thoroughly described in another chapter, it is also important to remember that these substances also have effects on the honey bee, either directly, or through the mechanisms previously discussed.

Antimicrobials

Antimicrobials are intended to kill bacterial pathogens in the honey bee colony, not the honey bee. The most commonly used antimicrobials in the US are lincomycin, oxytetracycline, tylosin, and fumagillin-B. Apparently, there are sources of fumagillin-B available again. Lincomycin and

Table 10.4 Chemical combinations with known toxicity in *Apis mellifera*.

	Aceptamiprid	Alpha-cypermethrin	Amitraz	Atrazine	Bosecalid	Carbaryl	Carbendazim	Chlorothalonil	Coumaphos	Cyprodinil	Deltamethrin	Dialios	Difenconazole	Difenoconazole + carbendazim	Dimethoate	Ebuconazole	Epoxiconazole	Ethyl parathion	Fenbuconazole	Fenpyroximate	Flumethrin	Flusilazole	Fumagillin-B	Imidacloprid	Lambda-cyhalothrin	Metconazole	Myclobutanil	Oxalic acid	Oxytetracycline	Prochloraz	Propiconazole	Pyraclostrobin	Tau-fluvalinate	Tebuconazole	Thiacloprid	Thiophanate-methyl	Tolyfluanid	Triadimefo	Triflumizole	Thymol	Uniconazole-P
Acetamiprid																	X						X								X							X	X		X
Alpha-cypermethrin								X					X									X								X	X			X							
Amitraz									X											X								X					X								
Atrazine						X																																			
Boscalid																																	X								
Carbaryl				X																																					
Carbendazim									X																								X								
Chlorothalonil		X							X																X								X								
Coumaphos			X				X	X				X	X		X			X		X	X	X							X	X	X		X	X		X				X	
Cyprodinil																																			X						
Deltamethrin													X	X																X											
Dialios									X																																
Difenconazole		X							X																X								X								
Difenoconazole + carbendazim											X																														
Dimethoate									X																																
Ebuconazole																																			X						
Epoxiconazole	X																																								
Ethyl parathion									X																																
Fenbuconazole																																	X								
Fenpyroximate			X						X																			X		X			X								
Flumethrin									X																																
Flusilazole		X							X																X								X								
Fumagillin-B	X																							X																	
Imidacloprid																							X								X								X		
Lambda-cyhalothrin								X					X									X	X							X	X			X		X					
Metconazole																																	X								
Myclobutanil																																	X								
Oxalic acid			X																	X		X											X							X	
Oxytetracycline									X																								X								
Prochloraz		X							X		x									X					X								X		X						
Propiconazole	X	X							X															X	X								X		X						
Pyraclostrobin																																	X								
Tau-fluvalinate			X		X		X	X	X				X						X	X		X				X	X	X	X	X	X	X		X		X					
Tebuconazole		X																							X								X								
Thiacloprid										X						X															X						X		X		
Thiophanate-methyl									X																X								X								
Thymol									X																			X					X								
Tolyfluanid																																			X						
Triadimefo	X																																								
Triflumizole	X																							X											X						
Uniconazole-P	X																																								
	Aceptamiprid	Alpha-cypermethrin	Amitraz	Atrazine	Bosecalid	Carbaryl	Carbendazim	Chlorothalonil	Coumaphos	Cyprodinil	Deltamethrin	Dialios	Difenconazole	Diffenoconazole + carbendazim	Dimethoate	Ebuconazole	Epoxiconazole	Ethyl parathion	Fenbuconazole	Fenpyroximate	Flumethrin	Flusilazole	Fumagillin-B	Imidacloprid	Lambda-cyhalothrin	Metconazole	Myclobutanil	Oxalic acid	Oxytetracycline	Prochloraz	Propiconazole	Pyraclostrobin	Tau-fluvalinate	Tebuconazole	Thiacloprid	Thiophanate-methyl	Tolyfluanid	Triadimefo	Triflumizole	Thymol	Uniconazole-P

tylosin are macrolide antimicrobials that target protein synthesis (by binding reversibly to the P site on the 50S subunit of the bacterial ribosome). Oxytetracycline inhibits bacterial cell growth (by binding to the 30S ribosomal subunit and preventing amino-acyl tRNA from binding to the A site of the ribosome). Oxytetracycline, as described previously, is also a known substrate of ABC-transporter systems and has been shown to decrease efflux of other ACB-transporter substrates such as coumaphos, and tau-fluvalinate, and oxalic acid (Zhu et al. 2014). Nonetheless, toxicity studies for all three of these antimicrobials administered at up to five times the labeled dose in honey bee colonies demonstrated no toxicity to the honey bee. (FDA Freedom of Information Summaries 2019) On the other hand, Fumagillin-B shows no direct toxicity against honey bees at labeled doses for *Nosema*, however subtherapeutic doses of

Table 10.5 Mechanisms of action of selected acaricides.

Chemical	Brand[a]	Pesticide class	Mechanism
Amitraz	Apivar Strips	Monoamine oxidase inhibitor	MAO inhibition
Fluvalinate	Apistan® Strips	Pyrethrin	Sodium channel blockade
Coumaphos	CheckMite+® Strips	Organophosphate	Acetylcholinesterase inhibition
Thymol	Apiguard®, Thymovar®	Essential organic oil	Unknown[b]
Thymol, eucalyptol, menthol, camphor	Api Life Var®	Essential organic oil	Unknown[b]
Formic acid	Mite Away Quick Strips®	Formate ester organic acid	Unknown[b]
Oxalic acid	Oxalic acid (wood bleach)	Organic acid	Unknown[b]
Hops beta acids	HopGuard II®	Biochemical pesticide	Unknown[b]

[a] Inclusion of a brand name should not be taken as a recommendation of that brand.
[b] Possible destruction of cuticle based on disruption of cell walls as in antimicrobial effect (Lopez-Romero et al. 2015).

fumagillin-B demonstrate an alteration in honey bee gut protein expression (Huang et al. 2013).

Acaricides

Formic acid reduces survival of all life stages of the honey bee and decreases hive productivity (Fries 1991; Giovenazzo and Dubreuil 2011). Oxalic acid affects honey bee behaviors and productivity if ingested (Schneider et al. 2012). The essential oil thymol is a GABA receptor agonist in both mites and the honey bee (Tong and Coats 2012), which may lead to drug interactions with other substances acting at the GABA receptor such as avermectins and fipronil. Bees ingesting a diet with thymol at a level of 50 ppm demonstrate a reduced production of the hormone vitellogenin which is critically important in queen egg production, and in worker resilience, behavior, and foraging activities (Charpentier et al. 2014).

Fungicides

Fungicides, while not intentionally used to kill insects, do show direct toxicity on honey bee brood (Zhu et al. 2014). The azole class fungicides are ABC-transporter pump inhibitors and lead to increased toxicity of simultaneously exposed ABC-transporter substrates. Fungicides, in an indirect toxic effect, may also disrupt the beneficial effects of *Aspergillus* spp. in fermenting beebread, as well as inhibiting the beneficial chemical detoxification provided by other non-pathogenic fungi living within the hive.

Herbicides

Herbicides do not intend insects as the targets of their action; however, some herbicides are toxic to honey bees, often through their effect on the bees' nervous system. Paraquat is known to kill or significantly reduce the lifespan of all life stages of *A. mellifera* (Moffett et al. 1972; Cousin et al. 2013). Glyphosate significantly disturbs the gut microbiome in honey bees (Motta et al. 2018); however, the impact of this disruption on honey bee colonies remains to be determined. Glyphosate also reduces short term memory and disrupts foraging behaviors in workers (Dai et al. 2018) (Figure 10.3).

Figure 10.3 Honey bees foraging for water on a leaf. *Source:* Photo courtesy of Zachary Y. Huang.

Insecticides

Insecticides are clearly intended to kill insects, and many are lethal to honey bees at labeled doses or, when used simultaneously with other agents, may increase toxicity. Miticidal insecticides are intended to be non-toxic to honey bees, and include coumaphos (an organophosphate), pyrethrins (voltage gated sodium channel agonists), amitraz (octopamine receptor antagonist), organic acids (inhibit mitochondrial respiration), and essential oils (GABA receptor agonists); however, as previously discussed, accumulation of these agents in wax comb or simultaneous use with other agents can increase mortality and morbidity within the colony. Other insecticides are considerably more toxic to honey bees and include active ingredients commonly used by veterinarians to treat ectoparasites on mammals such as neonicotinoids and spinosyns (nAChR), fipronil (GABA gated chloride channel agonist), and lufenuron (insect growth regulators).

Pharmacologic Considerations for Honey Bees as Food-Producing Animals

Honey bees are food-producing animals; they produce honey, which has, historically, been the most widely sought commodity from honey bees by humans. Because honey is a human food, in most countries, drug or other chemical treatments intended for use in the colony have designated time periods, called withdrawal times, during which honey cannot be collected for human consumption. The following discussion will focus on the laws and regulations in the United States, as an example of such regulations.

In the USA, these withdrawal periods are usually expressed on the packaging of Food and Drug Administration (FDA)-approved or Environmental Protection Agency (EPA)-registered products, typically in terms of days following application during which honey collection supers may not be placed on the hive. The reasoning behind such regulations is due to the lack of "metabolism" in products stored within wax comb. Once honey is deposited and capped, depletion of any contained substance is unlikely. Unlike other animal-sourced foods, such as eggs and milk, honey stores in a hive cannot be practically culled on a daily basis, therefore withdrawal periods are mechanisms used to prevent residues of these products occurring in honey. For example, the Apivar® (amitraz 3.33%) package instructions state to remove honey supers from the hive, place Apivar strips in the hive for a minimum of six weeks, and then to not place honey supers for another 14 days after the Apivar strips are removed (De Toledo 2017, FDA 2019). Although it is true that the regulations state the withdrawal time for this product is 14 days after the last treatment, in an operational-sense the actual time that honey is quarantined from treated hives is at least 56 days.

EPA has established maximum tolerances for EPA-registered products for use in honey bee hives. Tolerances for these EPA pesticides in honey and wax comb are presented in Table 10.6. Updates on EPA tolerances for chemicals in honey and wax can be found in the US Code of Federal Regulations Title 40, subchapter E, Part 180.

The FDA approves antimicrobials for control of disease in honey bee colonies . Honey tolerances for these approved medications are 500 ppb for tylosin and 750 ppb for oxytetracycline and lincomycin. FDA lists no maximum tolerances for any of these antimicrobials in any honey bee hive products. See US Code of Federal Regulations Title 21, subchapter E, part 556.

Table 10.6 Maximum tolerances of US EPA registered miticides in honey and wax.

Miticide	EPA honey tolerance (ppm)	EPA wax tolerance (ppm)	Treatment period without supers + labeled WDT
Amitraz	0.2	9	42 d + 14 d
Coumaphos	0.15	45	42 d + 14 d
τ-fluvalinate	0.02	—	42–56 d + 14 d
Fenpyroximate	0.10	Revoked[a]	NA
Bacillus thuringiensis	Exempt[b]	Exempt	NA
Menthol	Exempt	Exempt	21–30 d + 30 d
Formic acid	Exempt	Exempt	NA
Oxalic acid	Exempt	Exempt	NA
Thymol	Exempt	Exempt	30 d + 0 d
Eucalyptus oil	Exempt at <2 g/hive @ concentrations >80%	Exempt at <2 g/hive @ concentrations >80%	NA
Potassium salts of hops β-acids	Exempt	Exempt	NA

[a] Revoked means that the tolerance level was formerly in place but has been withdrawn by EPA.
[b] Exempt means that EPA does not list a minimum tolerance for this substance in this hive product.

In comparison, Canada, China, the European Union, India, Japan, and New Zealand also list recommended concentrations (RCs) or maximum residue limits (MRLs) for drugs used in honey bee hives. Other hive products such as pollen, propolis, and royal jelly are also consumed by humans. Guidance for MRLs in royal jelly is available in Japan, but guidance on how to avoid drug or chemical residues in these products is largely lacking in other countries.

It is very important to note that safety for the honey bee, and the withdrawal times for honey and wax, are determined for a product under conditions assuming a use as a single agent, and exactly as labeled (for the labeled target species, indication, dose, route, and duration of treatment); manufacturers of hive treatments are not currently required in the USA to study the impact of simultaneous use of chemical combinations. Due to the paucity of metabolizing enzymes in the honey bee, simultaneous use of drug or chemicals in the hive, even if used exactly as labeled, would likely alter pharmacokinetic and pharmacodynamic profiles of these agents, likely leading to unexpected outcomes. Veterinarians caring for honey bee colonies must instill a keen sense of vigilance in beekeepers that they be mindful of any overlap in agents applied to the beehive as well as in the surrounding landscape where workers will forage. The care plan should involve a delicate balance of optimizing honey bee health, while minimizing hive and honey contamination.

A common approach to achieve this dual goal, due to the relatively limited times of the year when nectar is not flowing, and honey is not being stored is that beekeepers typically choose to apply multiple treatments simultaneously to honey bee colonies (e.g. antimicrobial and miticide preventatives) during the short early spring and late fall periods when no honey supers are placed on colonies. However, given the previously discussed synergistic and additive toxicities occurring between and across the different potential treatments for colonies, the practices of preventative treatment should be examined carefully and, if still a reasonable therapeutic goal, should consider the benefits of using only one agent at a time with no overlap.

Summary

Although *A. mellifera* is a member of the large taxonomic class *Arthropoda-Insecta*, its pharmacology is significantly different compared to other insects. Indeed, for the worker bee, her pharmacology and ability to detoxify drugs and other substances encountered in her landscape are also significantly different even from other members of the Apidae family such as bumble bees and stingless bees. While some natural and synthetic chemicals are beneficial to the honey bee, many substances used in the landscape by humans exert serious toxicities on honey bee colonies. As a result, veterinarians caring for honey bees must acquire a firm foundational understanding of the absorption, distribution, metabolism, and excretion of these substances in the honey bee, as well as an understanding of their mechanisms of action and pharmacokinetics as they interact with the hive and with hive products. There is little doubt that humans will continue to develop insecticidal, fungicidal, and herbicidal drugs and other chemicals for use on agricultural crops and landscaping; however, research must be conducted to ensure that these products do not negatively impact this vitally important pollinator of more than one-third of the world's human food supply.

References

Boncristianai, H., Underwood, R., Schwarz, R. et al. (2012). Direct effect of acaricides on pathogen loads and gene expression levels in honeybees *Apis mellifera*. *Journal of Insect Physiology* 58: 613–620.

Buss, D.S. and Callaghan, A. (2008). Interaction of pesticides with *p*-glycoprotein and other ABC proteins: a survey of the possible importance to insecticide, herbicide and fungicide resistance. *Pesticide Biochemistry and Physiology* 90 (3): 141–153.

Carvalho, S.M., Belzunces, L.P., Carvalho, G.A. et al. (2013). Enzymatic biomarkers as tools to assess environmental quality: a case study of exposure of the honeybee *Apis mellifera* to insecticides. *Environmental Toxicology and Chemistry* 32 (9): 2117–2124.

Chan, Q.W., Chan, M.Y., Logan, M. et al. (2013). Honey bee protein atlas at organ-level resolution. *Genome Research* 23 (11): 1951–1960.

Charpentier, G., Vidau, C., Ferdy, J.-B. et al. (2014). Lethal and sub-lethal effects of thymol on honeybee (*Apis mellifera*) larvae reared *in vitro*. *Pest Management Science* 70 (1): 140–147.

Claudianos, C., Ranson, H., Johnson, R.M. et al. (2006). A deficit of detoxification enzymes: pesticide sensitivity and environmental response in the honeybee. *Insect Molecular Biology* 15 (5): 615–636.

Cousin, M., Silva-Zacarin, E., Kretzschmar, A. et al. (2013). Size changes in honey bee larvae oenocytes induced by exposure to paraquat at very low concentrations. *PLoS One* 8 (5): e65693.

Crailsheim, K. (1985). Distribution of haemolymph in the honeybee (*Apis mellifica*) in relation to season, age and temperature. *Journal of Insect Physiology* 31 (9): 707–713.

Dai, P., Yan, Z., Ma, S. et al. (2018). The herbicide glyphosate negatively affects midgut bacterial communities and survival of honey bee during larvae reared *in vitro*. *Journal of Agricultural and Food Chemistry* 66 (29): 7786–7793.

Derecka, K., Blythe, M.J., Malla, S. et al. (2013). Transient exposure to low levels of insecticide affects metabolic networks of honeybee larvae. *PLoS One* 8: e68191.

Dermauw, W. and Van Leeuwen, T. (2014). The ABC gene family in arthropods: comparative genomics and role in insecticide transport and resistance. *Insect Biochemistry and Molecular Biology* 45: 89–110.

Detzel, A. and Wink, M. (1993). Attraction, deterrence or intoxication of bees (*Apis mellifera*) by plant allelochemicals. *Chemoecology* 4 (1): 8–18.

Edson, K.Z. and Rettie, A.E. (2013). CYP4 enzymes as potential drug targets: focus on enzyme multiplicity, inducers and inhibitors, and therapeutic modulation of 20-hydroxyeicosatetraenoic acid (20-HETE) synthase and fatty acid ω-hydroxylase activities. *Current Topics in Medicinal Chemistry* 13 (12): 1429–1440.

FDA. (2019). Freedom of Information Summaries for Tylosin Tartrate (https://animaldrugsatfda.fda.gov/adafda/app/search/public/document/downloadFoi/292.), Lincomycin (https://animaldrugsatfda.fda.gov/adafda/app/search/public/document/downloadFoi/372), and Oxytetracycline Hydrochloride (https://animaldrugsatfda.fda.gov/adafda/app/search/public/document/downloadFoi/959). accessible at www.animaldrugsatfda.gov. Viewed June 1, 2019.

Feyereisen, R. (2005). Insect cytochrome P450. In: *Comprehensive Molecular Insect Science – Biochemistry and Molecular Biology*, vol. 4 (eds. L.I. Gilbert, K. Iatrou and S.S. Gill), 1–77. Oxford: Elsevier.

Fries, I. (1991). Treatment of sealed honey bee brood with formic acid for control of *Varroa jacobsoni*. *American Bee Journal* 131: 313–314.

Giovenazzo, P. and Dubreuil, P. (2011). Evaluation of spring organic treatments against *Varroa destructor* (Acari: Varroidae) in honey bee *Apis mellifera* (Hymenoptera: Apidae) colonies in eastern Canada. *Experimental and Applied Acarology* 55 (1): 65–76.

Gonçalves, W.G., Fernandes, K.M., Santana, W.C. et al. (2018). Post-embryonic development of the Malpighian tubules in *Apis mellifera* (Hymenoptera) workers: morphology, remodeling, apoptosis, and cell proliferation. *Protoplasma* 255 (2): 585–599.

Guseman, A.J., Miller, K., Kunkle, G. et al. (2016). Multi-drug resistance transporters and a mechanism-based strategy for assessing risks of pesticide combinations to honey bees. *PLoS One* 11 (2): e0148242.

Hawthorne, D.J. and Dively, G.P. (2011). Killing them with kindness? In-hive medications may inhibit xenobiotic efflux transporters and endanger honey bees. *PLoS One* 6 (11): e26796.

Heidel-Fischer, H.M. and Vogel, H. (2015). Molecular mechanisms of insect adaptation to plant secondary compounds. *Current Opinion in Insect Science* 8: 8–14.

Hölldobler, B. and Wilson, E.O. (2009). *The Superorganism: The Beauty, Elegance, and Strangeness of Insect Societies*. WW Norton & Company.

Honeybee Genome Sequencing Consortium (2006). Insights into social insects from the genome of the honeybee *Apis mellifera*. *Nature* 443 (7114): 931.

Huang, W.-F., Solter, L.F., Yau, P.M., and Imai, B.S. (2013). *Nosema ceranae* escapes fumagillin control in honey bees. *PLoS Pathogens* 9 (3): e1003185.

Iwasa, T., Motoyama, N., Ambrose, J.T., and Roe, R.M. (2004). Mechanism for the differential toxicity of neonicotinoid insecticides in the honey bee, *Apis mellifera*. *Crop Protection* 23 (5): 371–378.

Johnson, R.M., Wen, Z., Schuler, M.A., and Berenbaum, M.R. (2006). Mediation of pyrethroid insecticide toxicity to honey bees (Hymenoptera: Apidae) by cytochrome P450 monooxygenases. *Journal of Economic Entomology* 99 (4): 1046–1050.

Johnson, R.M., Mao, W., Pollock, H.S. et al. (2012). Ecologically appropriate xenobiotics induce cytochrome P450s in *Apis mellifera*. *PLoS One* 7 (2): e31051.

Johnson, R.M., Dahlgren, L., Siegfried, B.D., and Ellis, M.D. (2013). Acaricide, fungicide and drug interactions in honey bees (*Apis mellifera*). *PLoS One* 8 (1): e54092.

Liao, L.H. (2016). Behavioral and biochemical processing of natural and synthetic xenobiotics in the western honey bee *Apis mellifera*. Doctoral dissertation, University of Illinois at Urbana-Champaign.

Lopez-Romero, J.C., González-Ríos, H., Borges, A., and Simões, M. (2015). Antibacterial effects and mode of action of selected essential oils components against *Escherichia coli* and *Staphylococcus aureus*. *Evidence-based Complementary and Alternative Medicine* 2015: 1–9.

Mao, W., Rupasinghe, S.G., Johnson, R.M. et al. (2009). Quercetin-metabolizing CYP6AS enzymes of the pollinator *Apis mellifera* (Hymenoptera: Apidae). *Comparative Biochemistry and Physiology Part B: Biochemistry and Molecular Biology* 154 (4): 427–434.

Mao, W., Schuler, M.A., and Berenbaum, M.R. (2011). CYP9Q-mediated detoxification of acaricides in the honey bee (*Apis mellifera*). *Proceedings of the National Academy of Sciences of the United States of America* 108 (31): 12657–12662.

Mao, W., Schuler, M.A., and Berenbaum, M.R. (2013). Honey constituents upregulate detoxification and immunity genes in the western honeybee *Apis mellifera*. *Proceedings of the National Academy of Sciences of the United States of America* 110: 8842–8846.

Mao, W., Schuler, M.A., and Berenbaum, M.R. (2017). Disruption of quercetin metabolism by fungicide affects energy production in honey bees (*Apis mellifera*). *Proceedings of the National Academy of Sciences of the United States of America* 114 (10): 2538–2543.

Moffett, J.O., Morton, H.L., and MacDonald, R.H. (1972). Toxicity of some herbicidal sprays to honey bees. *Journal of Economic Entomology* 65 (1): 32–36.

Motta, E.V., Raymann, K., and Moran, N.A. (2018). Glyphosate perturbs the gut microbiota of honey bees. *Proceedings of the National Academy of Sciences of the United States of America* 115 (41): 10305–10310.

Mullin, C.A., Frazier, M., Frazier, J.L. et al. (2010). High Levels of Miticides and Agrochemicals in North American Apiaries: Implications for Honey Bee Health. PLoS ONE 5 (3): e9754.

Nielsen, S.A., Brødsgaard, C.J., and Hansen, H. (2000). Effects on detoxification enzymes in different life stages of honey bees (*Apis mellifera* L., Hymenoptera: Apidae) treated with a synthetic pyrethroid (flumethrin). *Alternatives to Laboratory Animals* 28 (3): 437–443.

Niu, G., Johnson, R.M., and Berenbaum, M.R. (2011). Toxicity of mycotoxins to honeybees and its amelioration by propolis. *Apidologie* 42 (1): 79.

Nygaard, S., Zhang, G., Schiøtt, M. et al. (2011). The genome of the leaf-cutting ant *Acromyrmex echinatior* suggests key adaptations to advanced social life and fungus farming. *Genome Research* 21 (8): 1339–1348.

Parkinson, A. and Ogilvie, B. (2008). Biotransformation of xenobiotics. In: *Casarett & Doull's Toxicology: The Basic Science of Poisons*, 7e (ed. C.D. Klaassen), 161–304. New York, NY: McGraw-Hill.

Pereira, V.J., da Cunha, J.P.A.R., de Morais, T.P. et al. (2016). Physical-chemical properties of pesticides: concepts, applications, and interactions with the environment. *Bioscience Journal* 32 (3).

Ramirez-Romero, R., Desneux, N., Decourtye, A. et al. (2008). Does Cry1ab protein affect learning performances of the honey bee *Apis mellifera* L. (Hymenoptera, Apidae)? *Ecotoxicology and Environmental Safety* 70 (2): 327–333.

Reinhard, A., Janke, M., von der Ohe, W. et al. (2009). Feeding deterrence and detrimental effects of pyrrolizidine alkaloids fed to honey bees (*Apis mellifera*). *Journal of Chemical Ecology* 35 (9): 1086–1095.

Schmehl, D.R., Teal, P.E.A., Frazier, J.L., and Grozinger, C.M. (2014). Genomic analysis of the interaction between pesticide exposure and nutrition in honeybees (*Apis mellifera*). *Journal of Insect Physiology* 71: 177–190.

Schneider, S., Eisenhardt, D., and Rademacher, E. (2012). Sublethal effects of oxalic acid on *Apis mellifera* (Hymenoptera: Apidae): changes in behaviour and longevity. *Apidologie* 43 (2): 218–225.

Seeley, T.D. (2010). *Honeybee Democracy*. Princeton University Press.

Spencer, H. (1898). *Principles of Sociology*, vol. 6. D. Appleton and Company.

Steiner, R. (1998). *Bees*. SteinerBooks.

Suen, G., Teiling, C., Li, L. et al. (2011). The genome sequence of the leaf-cutter ant *Atta cephalotes* reveals insights into its obligate symbiotic lifestyle. *PLoS Genetics* 7 (2): e1002007.

Tong, F. and Coats, J.R. (2012). Quantitative structure–activity relationships of monoterpenoid binding activities to the housefly GABA receptor. *Pest Management Science* 68 (8): 1122–1129.

De Toledo, V.A.A. (2017). *Honey Analysis*. BoD–Books on Demand.

Wu, C.P., Calcagno, A.M., and Ambudkar, S.V. (2008). Reversal of ABC drug transporter-mediated multidrug resistance in cancer cells: evaluation of current strategies. *Current Molecular Pharmacology* 1 (2): 93–105.

Xu, J., Strange, J.P., Welker, D.L., and James, R.R. (2013). Detoxification and stress response genes expressed in a western North American bumble bee, *Bombus huntii* (Hymenoptera: Apidae). *BMC Genomics* 14 (1): 874.

Zhu, W., Schmehl, D.R., Mullin, C.A., and Frazier, J.L. (2014). Four common pesticides, their mixtures and a formulation solvent in the hive environment have high oral toxicity to honey bee larvae. *PLoS One* 9 (1): e77547.

Section II

Beekeeping Principles for Veterinarians

11

Equipment and Safety

Adam J. Ingrao[1,2]

[1] *Extension Specialist, Michigan State University Extension, MI, USA*
[2] *Michigan Co-Coordinator, USDA North Central Region Sustainable Agriculture Research and Education Program (NCR-SARE), MI, USA*

Introduction

The proper care of honey bees requires a beekeeper to have knowledge of the equipment used in the trade and the potential safety concerns that can arise from beekeeping. An understanding of the equipment used in beekeeping as well as the risks associated with working honey bees allows individuals to make informed decisions in a dynamic apiary and to take the proper precautions for the care of honey bees and personal safety of the beekeeper. As with any animal husbandry, understanding the organism and how to care for them properly is aligned with the knowledge and proper use of beekeeping tools and an awareness of potential safety concerns.

Langstroth Hive

The most common bee hive used in the United States by commercial, sideliner and hobbyist beekeepers is the Langstroth hive (Figure 11.1).[1] Patented in 1852 by Lorenzo Langstroth of Philadelphia, PA, this hive has been adopted widely by beekeepers because of its incorporation of standardized spacing, modular design, and removable frames that allows a beekeeper to easily work the hive one frame at a time and control space for brood and honey production (USPTO 1852).

Bee Space

Proper space within a hive is a key component to ensuring the hive is easy to manage and that bees produce comb on individual frames that are removable without being destructive to other combs in the hive. **Bee space**, approximately 1/4–3/8 in., is defined as the minimum distance between two surfaces necessary for bees to move unimpeded and/or the distance between two surfaces within the hive that discourages comb building or propolis hoarding between those surfaces (Figure 11.2). The incorporation of bee space into the design of the Langstroth hive was revolutionary and resulted in a hive that limits the formation of non-uniform combs, such brace and bridge comb, built between two opposing frames or surfaces within the hive.

Components of a Hive

Over time, Langstroth's original hive design has been improved upon as new technologies and materials have become available. The modern Langstroth hive consists of many modular components that create a hive that allows a beekeeper to easily handle the hive in sections and gives bees the proper space necessary to create uniform combs on removable frames. These design improvements continue to this day as beekeepers seek novel solutions to modern challenges; however, because the measurements for Langstroth hive components have been standardized in the United States, supplies can be purchased from multiple manufacturers without concern of hive component compatibility issues.

Bottom Board

The base of the modern Langstroth hive is the bottom board. The **bottom board** is the floor of the hive that includes the hive entrance and bottom board landing,

1 Photos by Adam Ingrao unless otherwise noted.

Honey Bee Medicine for the Veterinary Practitioner, First Edition. Edited by Terry Ryan Kane and Cynthia M. Faux.

Figure 11.1 A modern Langstroth bee hive equipped with medium hive bodies and a telescoping top cover.

Figure 11.2 Bee space incorporated into a hive design allows for bees to move easily through the hive and discourages formation of non-uniform combs.

which allows bees to alight and take off from the hive (Figure 11.3). Two common designs for bottom boards are solid and screened (Figure 11.4). Solid bottom boards have no openings on the floor of the hive and are the most commonly used in the industry.

Many commercial beekeepers use special pallets that incorporate four solid bottom boards on one pallet (referred to as a four-way pallet) which allows for easier securing of hives and maximizes the use of space during transport (Figure 11.5).

Screened bottom boards utilize wire mesh, usually #8 hardware cloth or smaller, as the floor of the interior portion of the bottom board, and were developed for the monitoring of hive pests, such as varroa mites. Although screened bottom boards offer opportunities for collection of pest management data, like mite drop after a miticide treatment, they have some disadvantages when it comes to some mite treatments such as formic and oxalic acid. Organic acids are heavier than air and can exit readily when applied to hives with unsealed screened bottom boards during treatments. Conversely, screened bottom boards do offer increased ventilation, which can be beneficial during winters in northern climates to prevent condensation build up in the colony as winter bees respire.

Entrance Reducer

During the season, a beekeeper will manage the size of the hive entrance on the bottom board with an entrance reducer. The **entrance reducer** is a piece of material, usually wood, inserted in the bottom board entrance to reduce the size of the entrance according to the needs of the colony (Figure 11.6). Its use limits the space guard bees must protect to prevent predation or robbing of hives. The beekeeper may increase the size of the opening used by the bees over the course of the season as the colony population increases. At the peak of the season, a reducer may not be used at all, but as the population begins to decrease into the fall, the reducer is again applied to prevent robbing behavior. Generally, the smallest opening is used during the winter to discourage pests, such as mice, while still allowing for a lower entrance for bees and air flow. However, some motivated pests may be able to penetrate this defensive measure and additional equipment, such as a metal mouse guard, may be used by some beekeepers.

Hive Body

Each bee colony is housed in one to many hive bodies of varying size on a single bottom board (Table 11.1). A **hive body** is a wooden box that houses frames and serves as shelter for the colony (newer hive body materials such as polystyrene and polypropylene are available). Hive bodies are stacked on top of one another over the season to allow colonies to increase in population by offering space for

Figure 11.3 The bottom board landing provides a space for bees to alight and take off as they enter and exit the hive.

Figure 11.4 Solid (left) and screened (right) bottom boards commonly used by beekeepers.

expansion in the brood chamber and storage space for honey. Hive bodies commonly come in three depths (deep, medium, and shallow) and two widths (8- or 10-frame) (Figure 11.7a,b).

The classic arrangement, used by many beekeepers, is two deep hive bodies for the brood chamber and medium or shallow hive bodies for honey storage (referred to as honey supers) (Figure 11.8). Ultimately, the hive body type used by a beekeeper is a personal preference and many beekeepers, particularly those who are older and/or have physical limitations, may choose to use medium or shallow hive bodies for the entire hive to limit weight of the hive body and individual frames (Figure 11.9). In these situations, a brood chamber may occupy three to four medium hive bodies or shallows.

Frames

Frames are removable components of the hive that reside in a hive body upon which beeswax honeycomb is built (Figure 11.10). Each hive body contains a set of wooden or plastic frames (usually 8 or 10) that are matched with the appropriate hive body depth (Table 11.1) and width. The correct frames must be matched with the corresponding hive body to maintain proper bee space in the hive. Excess space inside the hive, due to improper equipment use or pairing of frames and hive bodies, will result in the construction of non-uniform comb by the bees (Figure 11.11). The beekeeper will be forced to remove this comb regularly and will struggle with working such hives in a non-destructive manner.

Figure 11.5 Four-way pallets used by commercial beekeepers incorporate the bottom board into the pallet design to make transporting bees easier and more secure.

Figure 11.6 Entrance reducer installed on a Langstroth hive.

Table 11.1 Standardized hive body and frame dimensions found in the U.S.

Hive body type	Depth of hive body (in.)	Depth of frames (in.)
Deep	9 5/8	9 1/8
Medium	6 5/8	6 1/4
Shallow	5 11/16	5 3/8

Each frame in the hive body contains a center insert called foundation (note that some beekeepers may use foundationless frames). **Foundation** is a thin sheet of beeswax or plastic, stamped with the size and shape of a normal worker cell, which serves as the base upon which bees will build wax combs (Figure 11.12). Use of either beeswax or plastic foundation is the beekeeper's choice but there are some advantages and disadvantages to each.

Beeswax foundation has a long history of use in the beekeeping industry. It is a familiar medium and is readily accepted by bees upon which to build comb. However, beeswax foundation requires special wiring equipment as well as more time to install into frames. In addition, beeswax is more susceptible to damage from poor handling by a beekeeper than plastic foundation.

Plastic foundation requires no special equipment to install and is more durable than beeswax foundation. Most plastic foundation is coated with a thin layer of beeswax to encourage bees to draw comb on the foreign surface; however, some bees may not readily accept plastic foundation.

(a) (b)

Figure 11.7 Hive bodies in 8 and 10 frame designs are widely used by beekeepers.

Figure 11.8 Bee hives using two deep hive bodies for the brood chamber (bottom boxes) and medium hive bodies as honey supers.

Specialty Frames

Specialty frames can serve specific purposes for managing a hive and producing desirable bee products. Examples of such frames include drone comb and "cut comb" frames. **Drone comb frames** are fitted with plastic foundation where the stamped cell pattern is of the larger size and shape of drone cells (Figure 11.13). These frames promote drone rearing by the colony and are used as part of an integrated pest management program. Drone frame use aids the management of varroa mite populations as mites prefer the larger drone cells for reproduction (Calderone 2005). Once the frame is predominantly full of capped drone brood, a beekeeper will remove the frame and destroy (usually by freezing) the drone larvae and associated mites.

Cut comb frames are, generally, wooden frames of varying (usually shallow) depths (note special hive bodies are utilized in cut comb systems) that use thin beeswax foundation and/or sections to produce uniform beeswax comb containing honey. This comb is cut from the foundation and packaged in containers. Cut comb honey is a highly desirable product that fetches premium prices in many honey markets.

Figure 11.9 Some apiaries may incorporate several approaches to hive body use. This apiary has hives with all medium hive bodies (far left and right hives) and combinations of deep and medium hive bodies (center hives).

Figure 11.10 Removable frames allow for easy management of the hive by dividing the hive into many frames which can be individually inspected.

Queen Excluder

Queen excluders are flat metal or plastic devices with slotted openings that are large enough for worker bees to pass through but small enough to prevent queens from passing through and are placed between hive bodies (Figure 11.14). Although the brood chamber tends to occupy two deep hive bodies (or three to four medium or shallow hive bodies), some beekeepers seek to ensure the queen does not lay eggs above a brood chamber if they are managing brood chamber size and/or are ensuring honey supers only contain honey. This barrier ensures that hive bodies above the queen excluder only contain honey when the queen is sequestered below the excluder. They are also used for a variety of queen-rearing processes.

Figure 11.11 This hive had a top deep hive body (removed here) that had medium frames incorrectly installed. This resulted in non-uniform combs constructed to bridge the gap between the frames of the two hive bodies.

Inner Cover

The **inner cover** is an optional piece of equipment most often used with telescoping outer covers to provide proper bee space between the frames of the top hive body and the outer cover, and to improve ventilation in the hive (Figure 11.15). Beekeepers will often notch a top entrance into the inner cover to increase ventilation within the hive and to provide a second entrance for bees at the top of the hive, which is particularly important for proper ventilation during summer and winter.

Figure 11.13 Drone frame brood *Source:* Photo by Terry Ryan Kane.

Outer Cover

To ensure the colony is protected, an outer cover must be added as the final piece of equipment. The **outer cover** is a solid board placed on the top of the hive to prevent exposure from the elements and potential predators/robbers. Two widely used styles are telescoping and migratory covers (Figure 11.16). Telescoping covers are generally used by hobbyists or sideliners. They are usually wooden with edges that overhang the hive body on all four sides (hence, the hive body "telescopes" into the cover) and is covered by a thin metal sheet. Migratory covers are flat sheets of material (typically wood) with no overhangs or two overhangs on the front and back of the hive body. This elimination (or limiting) of overhangs allows commercial beekeepers to

(a) (b)

Figure 11.12 A frame with plastic foundation installed (left) and the formation of beeswax upon the foundation surface after installation into a hive (right). *Source:* Photo credits: Adam Ingrao and Lacey Ingrao.

Figure 11.14 Queen excluder. *Source:* Photo by Cynthia Faux.

Figure 11.15 The inner cover allows for proper bee space and ventilation at the top of the hive and is most frequently used with a telescoping outer cover.

stack hives closely together on pallets for transport to pollination sites. Migratory covers are also used by hobbyists and sideliners because they are inexpensive to purchase and/or easy to build with limited tools.

Hive Stand

Once a beekeeper has sourced the necessary equipment to care for and house bees, they must then place the hive on a suitable **hive stand**. A hive stand is a base that elevates the hive to prevent direct contact with the ground.

Hives should never be installed directly on the ground for several reasons. Hives on the ground are more vulnerable to pests, such mice and ants, and predators, such as skunks and raccoons. Direct contact with the moist ground will also increase moisture in the woodenware, decreasing the life of the equipment and increasing moisture content within the hive. Beekeepers have developed many solutions for keeping bee hives elevated in the field which include placing hives on pallets, cinder blocks, wooden stands, or hive stands produced by beekeeping supply companies (Figure 11.17). Most often, commercial beekeepers will keep bees on pallets so that they can be loaded easily with a fork lift for transportation to pollination events.

Alternative Hives

For commercial beekeepers and sideliners in the United States, the Langstroth hive is ubiquitous throughout the industry due to its standardized nature, modular design, and early adoption by commercial beekeepers. However, there are many alternatives to the traditional Langstroth hive that are primarily used by hobbyists. Common alternatives, such as the Warre hive, Kenyan top bar hive, the Slovenian AZ hive, and the classic skep hive, offer some attractive features and/or cultural relevance to many small-scale beekeepers. For example, the AZ hive is utilized by many beekeepers with mobility limitations because it eliminates the use of hive bodies and makes Langstroth style frames accessible from the rear of the hive rather than the top, unlike the Langstroth hive. Additionally, it is the classic hive used by beekeepers throughout Slovenian countries and therefore has cultural significance to many beekeepers. However, there are drawbacks to some of these alternative hives.

Warre hives, top bar hives, and skep hives do not use Langstroth style frames. Instead, they offer a top bar from which large combs are drawn and suspended (Warre and top bar hives) or just allow combs to be built on the inner surfaces of the hive (skep hives). These conditions prevent comprehensive hive inspections, limits the beekeeper's ability to properly monitor pests and pathogens, increases disturbance to the colony during honey harvest, and eliminates the ability of honey frames to be extracted using a radial or tangential extractor.

Figure 11.16 Top covers protect hives from the elements and potential robbers. This apiary uses both telescoping outer covers (two left hives) and migratory covers (3rd and 4th hives from the left). *Source:* Photo credit: Lacey Ingrao.

Figure 11.17 Hive stands keep hives off the ground to reduce moisture accumulation and predator/pest intrusion. This simple hive stand consists of two cinder blocks.

Figure 11.18 The Flow Hive uses a honey super and proprietary plastic frames which allow the beekeeper to harvest honey directly from the hive. Problems with this design can arise due to propolis and wax deposits which can make the mechanical components of the frames difficult to move over time.

More recently, alternative hives have come on the market that offer novel solutions to honey harvesting for beekeepers. Perhaps the most well-known is the Flow® Hive, a Langstroth style hive that uses a honey super that incorporates proprietary plastic combs for honey storage that can be manipulated through a cranking mechanism to harvest honey from the hive without removing frames (Figure 11.18). Although an attractive feature to a beekeeper without access to a honey extractor, Flow Hives have several drawbacks such as; the cost is two to three times that of a standard Langstroth hive, frames in the honey super are made of clear plastic with movable parts that become difficult to use over time because of wax and propolis deposits, and bees may not readily store honey in the cells of frames because the surface is foreign to bees and is not coated in wax.

Tools of the Trade

As with any form of animal husbandry, beekeeping has specific tools that serve to increase efficiency, safety, and minimize stress on the animal. Safety should always be a priority when working bees. Proper use of the tools developed for safe beekeeping is the best way to ensure that safety. Some tools have a long history of use in beekeeping, such as the use of smoke, while others are relatively new, like pollen traps, but all serve important roles in the modern beekeeping practice.

Personal Protective Equipment

Minimizing bee stings is key to the safety and protection of the beekeeper and the bees. As a beekeeper works a hive, their attention should be focused on the task at hand and not on the concern of being stung. Therefore, being outfitted with proper personal protective equipment (PPE) will directly aid in improving hive handling skills and the comfort level of the beekeeper, especially those new to beekeeping.

Veils

The beekeeper's **veil** is PPE worn over the head to prevent bee stings and is made of a mesh material (typically nylon) that can be seen through. Veils come in a variety of styles from simple face veils to veils that incorporate full coveralls into their design (Figure 11.19). The type of veil a beekeeper will use is a personal preference and many beekeepers will change styles as their comfort level around bees increases over time. Common styles are the round veil and the fencer style veil. The face is a particularly vulnerable place for a beekeeper to be stung and is a focal point for honey bees when attacking a predator. Honey bees have long shared the environment with predators with thick fur, like bears and skunks, making them relatively immune to attacks by bees. However, areas of the face are exposed and vulnerable to bee stings, such as the nose and eye regions. Honey bees can recognize carbon dioxide (CO_2) and will track CO_2 plumes to the source (the breath of a predator) in response to a disturbance that warrants an attack (Crane 1990). Beekeepers often note that they can tell a hive is becoming agitated when bees begin to hit their veils.

Figure 11.19 Veils that incorporate coveralls provide maximum protection to beekeepers and are one of many styles available.

Gloves

Beekeepers can be stung in the hands while working the frames of a hive or lifting hive bodies. As a result, many beekeepers wear gloves as part of their regular PPE. Gloves used by beekeepers vary in style from thick leather gloves to latex or nitrile gloves and, like most beekeeping PPE, comes down to personal preference (Figure 11.20a,b).

Leather beekeeping gloves offer maximum protection from stings by offering a thick barrier to stingers, but they limit dexterity, which can lead to less meticulous hive handling. Conversely, latex or nitrile gloves offer excellent dexterity when working hives and offer some protection from stings; however, the material will not prevent stings as well as leather gloves. Latex and nitrile gloves can offer a balance between protection and dexterity when working hives.

Use of disposable latex and nitrile gloves can also be incorporated into a beekeeper's integrated pest management program as a component of their biosecurity practices. Changing gloves after examining a diseased hive can help prevent pathogen transmission between apiaries and/or individual hives. This practice is especially important for veterinarians examining hives.

Beekeepers typically wear one pair of leather gloves ubiquitously. Over a period of time, the gloves may become potential harbors for serious pathogens, such as American foulbrood (AFB) and European foulbrood (EFB), and act as fomites that pass pathogens between hives and/or apiaries. Leather gloves are difficult to disinfect so those that are contaminated with serious pathogens should be discarded or destroyed.

(a) (b)

Figure 11.20 Gloves provide protection from stings. The most common styles used by beekeepers are leather gloves (left) and nitrile/latex gloves (right).

Hive Tool

The multipurpose tools used by beekeepers to separate hive bodies and frames, pry apart propolis seals, and scrape hive surfaces are called **hive tools**. Honey bees utilize propolis to seal openings and cracks throughout a bee hive as part of their regular activities and thus they glue the hive body surfaces and frames together, making working the hive difficult without tools. A hive tool will usually offer one end to pry apart hive bodies and scrape surfaces (e.g. burr comb removal) while another end aids in the removal of frames. Many hive tool designs are available to meet the needs and preferences of beekeepers; however, a few have become industry standards and are frequently used by commercial, sideliner, and hobbyist beekeepers alike. These include the standard hive tool and the J-hook hive tool.

The standard hive tool offers a straight edge end for prying apart hive bodies and scraping surfaces and a curved edge end for splitting apart frames and removing frames by leveraging the tool between the edge of the frame and the hive body. The J-hook hive tool also offers a straight edge end used to pry apart hive bodies and scrape surfaces; however, this end also incorporates a small ledge to the straight edge end that allows the beekeeper to split apart frames. The opposite end of the hive tool bears the J-hook used to remove frames. One slides the J portion of the tool under the edge of a frame and rests the small ledge of the tool on the opposing frame to create leverage to lift the frame edge (Figure 11.21a,b).

Hive tools are also a source of pathogen transmission between hives and apiaries. Proper sanitization, such as heating hive tools or cleaning with a disinfecting solution, after examination of a diseased hive is crucial to prevent transmission and should be a component of a regular biosecurity practice for all veterinarians.

Smoker

The use of smoke to collect honey and manage bees has been used by cultures across the world and may be the most recognizable tool in the beekeeper's tool box (Crane 1985). The modern **smoker** is a ventilated device in which smoldering material is maintained to direct smoke at bees through use of a bellows (Figure 11.22). The smoker is frequently used by beekeepers to minimize the defensive response of the bees during hive inspections.

Honey bees that perceive smoke in their environment will gorge themselves on honey in preparation for evacuation of the hive, thus smoking a hive distracts bees from the beekeeper (Newton 1968, 1969). Additionally, smoke causes the chemical receptors of the antennae to become clogged with smoke particulates which masks the alarm pheromones being emitted by other bees (Visscher et al. 1995).

One common misperception of the honey bee response to smoke is that it calms the bees. Although beekeepers perceive the bees "calming" when smoke is introduced, the behavior is not calming but rather a stress response. The misunderstanding of the response and of the term "calming" may result in overusing smoke during inspections and smoking hives before they are opened for inspections.

Figure 11.21 The J Hook hive tool incorporates a ledge on the prying end of the tool to separate frames (left) while the J Hook allows for easy removal of frames from a hive. *Source:* Photo credits: Adam Ingrao and Lacey Ingrao.

Beekeepers should prioritize minimizing stress on bees during inspections and avoid preemptively using, or overusing, smoke with European honey bees.

Feeders

Supplemental feeding of sugar, in syrup or solid form, may be necessary when establishing new colonies, strengthening weak colonies, medicating colonies, preparing colonies for winter, or supplemental feeding of colonies during the winter. Syrup feeding is usually used during the beekeeping season and is accomplished in a variety of ways using several feeder designs. Typical syrup feeders found in hobbyist beekeeping operations include entrance feeders, top feeders, bucket feeders, and frame feeders (Figure 11.23). Sideliners may use bucket or frame feeders because of their large volume. Commercial beekeepers will most often use frame feeders as they are incorporated into the interior of the hive, which avoids any problems maximizing space for transport of hives. It should be noted that external feeders, such as entrance feeders or bucket feeders (used for group feeding), may induce robbing of nearby hives and should be monitored during their use to prevent this behavior.

Solid feeding is an activity typically used during the winter. Feeding of fondant, candy boards, or dry sugar (e.g. the Mountain Camp method) allows the beekeeper to augment food supplies when honey reserves are exhausted. However, solid feeding requires an additional spacer to be added to the top of the hive to allow for space to incorporate the supplemental food. In most cases this spacer also serves as a top entrance during the winter (Figure 11.24).

Pollen Traps

Many beekeepers will collect pollen from bees. Locally sourced pollen has become a highly sought-after product for many health-conscious consumers because of its diverse protein, amino acid, mineral, vitamin and lipid profile (T'ai and Cane 2000). Beekeepers may also collect pollen to limit the amount entering the hive when pollen availability is high but space for storing pollen in the hive is limited. In this situation, if the volume of pollen entering the hive is not reduced using a pollen trap, a hive may

Figure 11.22 A smoker is a tool that uses smoke directed at bees through a bellows to minimize defensive responses of bees during hive inspections.

Figure 11.23 A top feeder is a common way to feed bees sugar syrup during the season and is incorporated into the interior of the hive by placing an empty hive body around the feeder and putting the top cover over the feeder. This style of feeder prevents robbing because access is limited to bees on the inside of the hive.

become "pollen bound," which can trigger swarming due to limited space for the queen to lay eggs.

Pollen traps either mount to the front of the hive or are incorporated into the hive. A front mounted pollen trap removes pollen from the pollen baskets as bees enter the hive (Figure 11.25). Pollen traps that are incorporated into the hive, like the Sundance pollen trap, require returning forager bees to navigate a series of screens which knocks the pollen from the pollen baskets. Both pollen trap designs result in reduced pollen volume entering the hive and allows the beekeeper to more effectively manage space in the brood chamber to avoid swarming.

Safety

Beekeeping can be a hazardous activity if proper safety precautions are not followed by the beekeeper. Understanding the potential safety risks and how to minimize them should be a priority for beekeepers of all experience levels.

Bee Stings

Honey bees are venomous insects and stings may result in serious systemic medical conditions. When a bee sting occurs, the stinger should be removed immediately by scraping the stinger out of the skin with a sharp-edged object, like a hive tool, finger nail, or credit card.

Approximately 32% of beekeepers will develop an allergy to stings over their career (Antolín-Amérigo et al. 2014; Ludman and Boyle 2015). These allergies can vary greatly in severity, but the most serious cases may result in anaphylaxis or even death. It is important that beekeepers consult their physicians if an allergy is suspected or begins to develop over time. If a severe allergy is detected by a physician often a prescription for epinephrine will be issued to the patient. In such cases, epinephrine should be carried with the beekeeper in the apiary at all times. Apiaries are

Figure 11.24 Supplemental feeding of dry sugar, candy boards, and fondant during the winter require an additional spacer be added to the top of the hive to allow enough space to add feed. In this photo, the Mountain Camp method is being used to feed dry sugar to bees.

Figure 11.25 Pollen traps allow a beekeeper to manage the amount of pollen entering the hive to prevent a colony from becoming pollen bound, which may cause a colony to swarm.

often in remote locations without immediate access to medical facilities, so having the appropriate medication in close proximity to a beekeeper with potentially serious allergic reactions should be a top priority.

Apiary Site Selection

Apiary site selection and development should be part of a beekeeper's safety considerations, as remote sites used for apiaries may have safety hazards that can cause severe injury. Important safety aspects of good apiary site selection are to ensure an apiary is located on level ground without tripping hazards and is accessible by vehicle.

An apiary site that is relatively level will reduce stress on the body during lifting and moving of hives and will limit encounters with uneven terrain (Figure 11.26). Tripping hazards like rocks, tree stumps, brush and thick grass all present great risks to a beekeeper when walking through an apiary, especially when carrying equipment. The area should be cleared to remove any potential tripping hazards prior to hive installation.

Risk of fires and flooding should also be considered when selecting apiary sites. Brush and thick grass should be regularly trimmed to minimize risk of fire during the warm seasons and apiaries should not be placed in lowlands to prevent flooding during the wet seasons.

Beekeeping equipment is heavy and often cumbersome, and this should be considered when selecting apiary sites. A good rule of thumb is to ensure that you can drive up to the apiary in a vehicle, whether a passenger vehicle or off-road vehicle (Figure 11.27). By ensuring equipment can be delivered directly to the site by a vehicle, the beekeeper will limit the distance and time a load must be carried and will thereby reduce chances of injury.

Pesticide Use

Maintaining healthy colonies is the key to honey bee survival and requires a beekeeper to have a firm understanding of how to monitor and treat for pests and pathogens. Beekeepers should always purchase pesticides from a reputable dealer and follow the labeled instructions on proper use and PPE exactly. Improper use of pesticides can harm bees, beekeepers, and the environment.

Summary

Beekeeping is an activity enjoyed by many, from the backyard hobbyist to the commercial beekeeper. Regardless of the level of experience or scale of operation, every beekeeper should have a firm understanding of the equipment and tools used in beekeeping and the associated risks of keeping bees. With experience and knowledge of these factors, comes the ability to think critically and make informed decisions in an apiary which will increase success and enjoyment of the beekeeper and the health and productivity of their bees.

Figure 11.26 An apiary site should be on flat ground with all tripping hazards removed. Some beekeepers may incorporate a weed barrier to create a hazard free walking surface.

Figure 11.27 An apiary site that is accessible by car or off-road vehicle will eliminate heavy lifting and limit potential injuries associated with moving beekeeping equipment. This DR Power Wagon allows for transporting of beekeeping equipment to and from the apiary.

References

Antolín-Amérigo, D., Moreno Aguilar, C., Vega, A., and Alvarez-Mon, M. (2014). Venom immunotherapy: an updated review. *Current Allergy and Asthma Reports* 14 (7): 449.

Calderone, N.W. (2005). Evaluation of drone brood removal for management of *Varroa destructor* (Acari: Varroidae) in colonies of *Apis mellifera* (Hymenoptera: Apidae) in the northeastern United States. *Journal of Economic Entomology* 98 (3): 645–650.

Crane, E. (1985). *The Archeology of Beekeeping*. Ithaca, NY: Cornell University Press.

Crane, E. (1990). *Bees and Beekeeping: Science, Practice, and World Resources*. Ithaca, NY: Cornell University Press.

Ludman, S.W. and Boyle, R.J. (2015). Stinging insect allergy: current perspectives on venom immunotherapy. *Journal of Asthma and Allergy* 8: 75–86. https://doi.org/10.2147/JAA.S62288.

Newton, D.C. (1968). Behavioural response of honeybees to colony disturbance by smoke. I. Engorging behaviour. *Journal of Apicultural Research* 7 (1): 3–9.

Newton, D.C. (1969). Behavioural response of honeybees to colony disturbance by smoke. II. Guards and foragers. *Journal of Apicultural Research* 8 (2): 79–82.

T'ai, H.R. and Cane, J.H. (2000). Pollen nutritional content and digestibility for animals. In: *Pollen and Pollination*, 187–209. Vienna: Springer.

United States Patent and Trademark Office. (1852) Patent Number 9300 Bee Hive. https://patentimages.storage.googleapis.com/71/b8/4e/0122ed14d4ec59/US9300.pdf.

Visscher, P.K., Vetter, R.S., and Robinson, G.E. (1995). Alarm pheromone perception in honey bees is decreased by smoke (Hymenoptera: Apidae). *Journal of Insect Behavior* 8 (1): 11–18.

12

The Apiarist

Katie Lee and Gary S. Reuter

Department of Entomology, University of Minnesota, Bee Research Facility, St. Paul, MN, USA

The Apiarist

The apiarist, or beekeeper, is the person who owns and cares for the honey bees. In most cases, as the veterinarian, you will be in contact with the apiarist or, in the case of a large operation, a designated employee. Just like any other animal husbandry, the owner is ultimately responsible for the bees' care. This section will describe the different sizes and types of operations to help you understand where the bees, and beekeepers, may be at any time during a diagnosis and treatment.

Apiarists are often classified by the number of colonies they own or operate. A hobby beekeeper is someone with fewer than 50 colonies (Figure 12.1). A commercial beekeeper manages more than 500 colonies. Beekeepers in between are known as "sideliners." The numbers are somewhat arbitrary and can vary depending on the purpose of the classification. These distinctions, based on number of colonies, will be the main classification used in this chapter. The bees are the same, but the management of the bees may differ and your interactions with these different groups may vary.

In addition to operation size, the geography and climate of a region results in differences in management. Where a northern beekeeper may move their colonies to the south for winter, a southern beekeeper being in a more temperate climate may not move their colonies. The timing of the application of a treatment product may also vary due to differences of factors linked to climate. For example, certain treatments like antibiotics cannot be applied when the bees are collecting nectar that will be made into honey for human consumption.

The type of beekeeping operation, or the way the bees are used for income, is another important difference between beekeepers. Honey producers operate their bees to produce honey for sale. Some beekeepers, called pollinators, concentrate on leasing their bees for agricultural pollination contracts. All honey bees pollinate flowers as they collect nectar and pollen, but these beekeepers lease their colonies to growers or farmers to pollinate a crop. Beekeepers that concentrate on rearing queens for sale are called queen producers. Of course, some beekeepers use their bees for a combination of these tasks. Many hobby beekeepers do not seek to generate an income from their colonies and primarily keep honey bees for enjoyment, or to produce honey for personal consumption or gifts (Table 12.1).

Classification by Operation Size

Commercial Beekeepers

Demographics

Commercial beekeepers are the fewest in number, however the majority (>75%) of the honey bee colonies in the United States are managed by these beekeepers. The largest commercial operation manages more than 70 000 colonies, although most commercial operations have fewer than 10 000 colonies. Commercial beekeeper operations tend to be migratory, following blooming plants and using their colonies for honey production, pollination, and/or queen production. Commercial beekeepers tend to apply intensive management practices, as their livelihood depends on healthy colonies to generate an income. Due to the large scale and migratory nature, commercial beekeepers spend much of their time on logistics and they rely more on equipment and trucks compared to the other beekeeper groups.

Honey Bee Medicine for the Veterinary Practitioner, First Edition. Edited by Terry Ryan Kane and Cynthia M. Faux.

Figure 12.1 Hobby beekeeper apiary. *Source:* Photo by Dr. Marla Spivak.

Table 12.1 Examples of different types of beekeeping operations.

Class	# Colonies	Type of operation
Commercial	2000	Pollination – honey production
Commercial	20000	Pollination – honey production
Commercial	5000	Pollination – honey production – queens – nucs
Commercial	20000	Pollination – honey production – nucs
Sideliner	400	Pollination – honey production
Sideliner	400	Honey production – queens – nucs
Sideliner	200	Honey production
Sideliner	100	Honey production – queens – nucs
Hobby	1	Honey production
Hobby	10	Honey production
Hobby	25	Honey production
Hobby	15	Honey production

Management

Commercial beekeepers can vary widely in their management practices. Compared to the other beekeeper groups, they tend to apply more treatments and feed their bees to improve the health of their colonies. Ease of application of a practice or treatment, effectiveness, and cost are main considerations in their management decisions. Commercial beekeepers often feed their colonies with supplemental sugar and protein (pollen patties) to increase colony populations or in times of food dearth. In addition to feeding, commercial beekeepers are more likely to apply treatments to sick colonies, prevent symptoms of a common disease, or prevent the parasitic mite *Varroa destructor* from reaching detrimental levels.

Colony Configuration

Commercial beekeepers in the United States overwhelmingly use Langstroth-style equipment. Colony configurations are made up of one or a few different sized boxes. Typical combinations include one deep box (9 5/8″ in height), two deep boxes, or one deep box and one medium box (6 5/8″ in height). Boxes sizes can vary in width, holding either 8 or 10 frames. Colonies are often kept on pallets, with four or six colonies per pallet. The pallets allow a beekeeper to load and unload colonies on a truck using a forklift. Commercial beekeepers that use individual bottom boards instead of pallets often have a specialized boom truck with a mechanized lift that helps them pick up and move the colonies on and off the truck. Boxes used to collect honey for harvest are put on top of the main box(es) and can be various sizes in depth. These boxes are generally removed prior to the colonies being moved to a new location. Colonies tend to have the same pallet and box configuration across an operation to decrease complications in management, easily swap equipment, and to prevent an uneven load on a truck when moving colonies.

Apiary Size and Transportation

For relative ease of management and logistics, commercial beekeepers keep their colonies in apiaries, sometimes called yards. An apiary is often treated as a unit by the beekeeper as colonies within an apiary are frequently managed the same. There are generally 24–60 colonies in each apiary although some apiaries may contain >100 colonies (Figure 12.2). A flatbed truck is used to move the colonies in and out of a location. Apiary sizes will add up to the size of the beekeeper's truck: if the truck holds 64 pallets (256 colonies if four colonies per pallet), then the beekeeper will place colonies in about seven apiaries each with approximately 7–10 pallets. A commercial beekeeper with 2000 colonies will have about 50 apiaries if they average 10 pallets each with four colonies at each location.

The size of an apiary can depend on the estimated amount of bee friendly forage surrounding a location (Figure 12.3). Being able to estimate the holding capacity of a location is a skill that is honed with time and experience. If colonies at a location have historically collected large amounts of nectar and pollen, then a beekeeper may increase the number of colonies at that location. If colonies do poorly each year at a location, then the beekeeper may reduce the number of colonies. The quality of a location may change each year depending on the weather, the availability of nectar and pollen producing flowers, and the number of competing honey bee colonies within foraging range. Apiaries are usually spaced at least 2 mi apart with the ideal being 5 mi apart. This distance is necessary to prevent overlap of the apiaries'

Figure 12.2 Commercial beekeeper apiary with 32 colonies.

(a)

(b)

Figure 12.3 Blooming monoculture fields of planted sunflowers (a) and canola (b), both of which are bee forage.

foraging range as honey bees can fly up to 5 mi, but tend to stay within a 2–3 mi range. Too many colonies foraging on the same area reduces the amount of harvestable honey from all colonies.

Another factor that can influence apiary size is the ease of management. Smaller apiary sizes mean there are more locations to visit, which can increase the amount of time spent driving. Smaller apiaries may be ideal if there is little forage around for the bees. However, if there is not enough forage to sustain even a small apiary, beekeepers may keep their colonies in very large apiaries to reduce the amount of time spent driving in order to feed the bees with supplemental nectar and protein. For example, in California colonies are sometimes kept in apiaries containing hundreds of colonies from approximately November to just prior to the almond bloom in February. With little forage at that time of year, colonies are kept in large apiaries that allow for frequent feedings and easy management. In contrast, some plants produce so much nectar that beekeepers will also keep very large apiaries to take advantage of the bloom. For example, apiaries often have >100 colonies in north eastern North Dakota during canola bloom.

Colonies are transported long distances on semi-trucks holding about 400–700 colonies, depending on truck size, the number of boxes comprising a colony, and the weight per colony. Colonies are covered by a large net during transport across long distances (Figures 12.4 and 12.5).

Before and after transport, colonies are kept in large apiaries called holding yards that can contain >1000 colonies in one location (Figure 12.6). Sometimes colonies need to be held for a few weeks or more in a holding yard until orchards are ready to receive them or until locations dry out and are accessible after rain. The colonies are then loaded on smaller trucks and taken to the individual apiaries or orchards. The concentration of colonies in holding yards cause concern in terms of disease spread and difficulty treating colonies because the colonies may be stacked four high. However, holding yards can be a management opportunity for a beekeeper to perform management tasks to all colonies in a short amount of time. Beekeepers can inspect colonies to cull weak ones, feed, or treat all colonies for a pest or pathogen, reducing the disease burden and potential of spread within an operation.

Health Monitoring

Commercial beekeepers monitor their colonies for health, including signs of queen failure, population size, food stores, and diseases and pests. In general, beekeepers monitor colonies in order to determine if management is needed based on colony status or to fix an issue. Commercial beekeepers look for clinical signs of diseases and pests. The success of detecting if symptoms are present is dependent on how well the person is trained and commonness of the diseases. If a crew member is not trained on identifying signs of certain diseases, then the individual may not know the causal pathogen, but may see a poor brood pattern (lots of missing capped pupae) or brood that looks abnormal. Inquiring about the symptoms present can help identify a pathogen.

Figure 12.4 Commercial beekeeper loading honey bee colonies onto a truck.

Figure 12.5 Commercial beekeeper colonies covered with a net to prevent excess flying during the drive.

Figure 12.6 Commercial beekeeper holding yard used to store colonies awaiting transport to another location.

Commercial beekeepers may monitor for diseases and pests that require a field test to determine the colony burden. Some commercial beekeepers use their state apiary inspector program or hire other trained individuals to inspect colonies for diseases and pests, especially if the detection requires a field (*V. destructor*) or lab test (*Nosema* spp.). However, an inspector may not be able to visit an operation when the beekeeper requests one, so training beekeepers and their crews to monitor can help fill in the gaps and identify problems at an earlier stage.

Pathogen and Pest Transmission

Colonies in commercial operations are often kept and transported in high densities. This proximity can increase the potential of horizontal transmission of pathogens and pests through the co-mingling of bees within an apiary, holding yard, and operation. For example, if a colony has a pathogen, then other colonies in that apiary or holding yard are at higher risk of exposure to that pathogen. Transmission of pests and pathogens can also occur among beekeeper operations when apiaries are close to each other.

When possible, a larger distance separating apiaries, especially apiaries of different beekeepers, can reduce horizontal transmission.

Moving equipment and colonies can increase the transmission of pathogens and pests. After a colony dies, a beekeeper may place the old equipment on a healthy colony. If that equipment contains a pathogen, then that pathogen can spread to the healthy colony. In addition, transporting colonies can be important for both the transmission and susceptibility of a colony to pathogens and pests. Transported colonies can have elevated levels of stress, resulting in decreased defense against pathogens and pests. For example, moving bees from a warm location to a cold location can increase the occurrence of the symptoms of chalkbrood and European foulbrood. Transportation can also result in higher exposure to pesticides or poor nutrition resulting in more susceptible colonies. For example, pollination of crops like blueberries and cranberries have been associated with increased colony mortality and morbidity. In particular beekeepers have identified higher levels of European foulbrood after pollinating blueberries, potentially a result of increased exposure to fungicides. Transport on the back of trucks can increase transmission risk through the mixing of bees from different colonies, especially during warmer weather when the bees move around more. To decrease bee movement during transportation and reduce the number of bees left behind at the starting location, beekeepers can move colonies during times bees are less likely to fly (not during daylight hours, in the cold and/or rain).

The current and prior location and the number of colonies infected can inform management recommendations provided to a beekeeper to reduce the spread of a pathogen. The beekeeper can increase the monitoring of populations at higher risk, and they can remove a sick colony through or quarantine or euthanization. Commercial beekeepers may have an option of setting up an isolation yard for diseased colonies until they recover. However, placing colonies in quarantine is not a common practice. A sick colony is often treated in the location it was found or it is culled. In addition, moving a colony to an isolation yard during the day can leave behind foragers and these infected bees can then drift to other colonies thus spreading the pathogen.

Despite the increased risk of horizontal transmission due to high colony density and stress of transport, commercial beekeepers in the United States tend to have lower colony mortality than the other two beekeeper groups according to annual colony loss survey performed by the non-profit group the Bee Informed Partnership. The lower losses are likely due to the higher use of more intense management practices that increase colony health, especially treating for *V. destructor* mites.

Visiting Apiaries

Visiting the colonies for a commercial beekeeper may be challenging. The colonies will likely not be at the beekeeper's home and all the affected colonies may not be together. A beekeeper may need to move colonies during or after a treatment, so following up on a colony's health can be an issue. Communication with the beekeeper is key to understanding how many colonies to inspect, the time it will take to inspect, the treatment window, and how to follow-up if necessary.

Typically, apiaries are on other landowners' properties. If you are given directions or coordinates to an apiary to examine colonies without the beekeeper present, then ask how to enter the location and if there are any special considerations with the property or landowner. For example, there could be an electric fence to deactivate if the location is in an area with bears (Figure 12.7), some locations may be behind a gate that requires a key or special knot, needs to be kept closed at all times due to the presence of livestock (Figure 12.8), or there may be a chronically wet route to avoid (Figure 12.9). Beekeepers may assume you are in a vehicle with high clearance and

Figure 12.7 In locations where bears are known to be present, apiaries may have an electric fence to prevent bear damage. Before entering the apiary, be sure to turn off the fence and turn on when you leave.

(a)

(b)

Figure 12.8 Entrance to an apiary that requires undoing a knot to open the gate (a). If necessary, take a picture of the knot to replicate it upon exiting the location. If livestock are present or if you are not certain they are not present, close the gate while in the apiary. It is not uncommon for colonies to be kept with cattle (b).

(a)

(b)

Figure 12.9 If it has recently rained, take caution when entering a location since the roads can be softer than they look. Here a truck is stuck in a muddy road *Source:* a; photo credit: Ellen Topitzhofer and a location where ruts are evident, indicating a soft spot to avoid driving on (b).

Figure 12.10 Transporting strapped colonies into a yard with difficult access using an all-terrain vehicle.

four-wheel drive, so ask if your specific vehicle can make the route.

Other locations can be at an individual's house and that person may need notification. Whenever possible, drive a vehicle that identifies your profession. Always leave the apiary in the condition that you found it by eliminating any trace that you were present, including removing any trash or equipment, checking that all hive boxes are flush, and leaving any gate how it was found, unless instructed otherwise.

Hobby Beekeepers

Demographics

The largest number of beekeepers are in the hobby beekeeper category (also called small-scale or backyard beekeepers). Hobby beekeepers are a diverse group of people who keep honey bees for a variety of reasons that include honey production, an interest in the biology, pollinating gardens or fruit trees, feeling more in touch with nature, and just for fun to learn something new. There is a wide range of experience levels in the hobby beekeeper group, with some keeping bees for multiple decades and others new to the activity.

Management

As many hobby beekeepers do not derive an income from their bees, they tend to have a less intensive management style than commercial beekeepers in terms of the application of treatments and moving colonies. However, there is an enormous range of beekeeping management philosophies. Some beekeepers are unwilling to add anything to their colonies, including antibiotics, *V. destructor* treatments, sugar syrup, or protein patties. Other beekeepers apply more intensive treatments and feed their bees more like a commercial beekeeper would do. Hobby beekeepers may be more willing to apply time-intensive or costly management practices if it helps them reduce the levels of a pest or pathogen. For example, a beekeeper may keep a smaller colony by creating a break in the brood cycle to reduce the population growth of *V. destructor* mites, even if it takes time and reduces the amount of honey the colony can produce.

Colony Configuration

Many hobby beekeepers use Langstroth hive bodies, but this group is more likely to use a variety of different types of equipment. The other two most common hive types are top-bar and Warré, both consisting of horizontal bars where the bees build comb. Skeps and other hive designs lacking moveable frames are infrequently used since they do not allow for colony inspections.

Apiary Size and Transportation

Hobby beekeepers tend not to be migratory, and rather keep their bees in one location all year. Most hobby beekeepers have 10 or fewer colonies. Any movement of colonies is often done by hand and sometimes in the back of pick-up trucks or enclosed vehicles where colonies are screened to prevent bee movement during transport and overheating. Straps can be used to keep the hive components together and a one-person hive mover or two-person hive lifter can decrease back strain. Hives can be moved by wagon or an all-terrain vehicle if the equipment is adequately secured (Figure 12.10).

Health Monitoring

Colony monitoring for health can range from an inspection every week to once or twice a year. Beekeepers may inspect colonies to inform management decisions, to learn, or just to view the bees. However, colony health inspections are an important part of colony management and should be done throughout a year when possible to identify problems and inform practices. Issues in colony health can be regional and based on time of year. A typical yearly cycle in the United States usually starts in March-May when colonies are growing and when new colonies are most commonly available for sale. To start new colonies or to replace dead colonies in spring, hobby beekeepers generally purchase bees from a commercial beekeeper. If their colonies survived winter,

Figure 12.11 Colonies in the floodplain of a nearby river. Access to the location required a truck with high clearance and tall mud boots to walk the remaining 100 yards to the colonies. Five of the eight colonies survived, with two of the dead colonies swept around the corner of the building.

they may start another colony by removing about half the bees and brood from the surviving colony to a new box and adding a queen. Colonies grow over the season when blooming flowers are abundant, typically in summer. The honey gathered may be harvested if excess is produced. Far north beekeepers make sure the colonies have enough honey for winter, up to 100 pounds depending on the climate and usually put an insulating wrap on their colonies in October. Those colonies are not inspected again until March. Southern beekeepers do not need as much honey for their bees to get through winter. They can feed, if necessary, in late winter early spring if flowers do not bloom.

Experienced beekeepers often monitor for health and can determine when an issue is present. Newer beekeepers need more guidance on correctly identifying a problem and how to correct it. First year beekeepers may know very little about how to inspect a colony for health. Second year beekeepers may have a false sense of confidence if a colony survives. It takes years to understand just how much there is to learn about bees and beekeeping. Asking for the symptoms in a colony with issues can inform you of the state of the colony and help gauge how familiar the beekeeper is with performing inspections.

Pathogen and Pest Transmission

As colonies tend to be kept in lower density apiaries, there can be less risk of transmission within an apiary. However, in cities or towns where hobby beekeeping is common, colony density can be high. The mismanagement of a pest or pathogen in a hobby colony may result in an outbreak that leads to the infection of multiple neighboring beekeepers' colonies as well as their own colonies.

Visiting Apiaries

Hobby beekeepers usually keep their bees at their home or close by. Visiting these colonies will be easier to arrange and they will most likely stay in the same location throughout a treatment. Again, there may be special considerations with the location (Figure 12.11).

Sideliner Beekeepers

Sideliner beekeepers, being between the two other operation groups in colony numbers, tend to be a combination of commercial and hobby beekeepers in management style, colony configuration, apiary size, transportation, and pest and pathogen transmission. Sideliner beekeepers tend to be adept at monitoring colony health. The name "sideliner" can be misleading as some of these beekeepers make their living off their colonies. Sideliners with more colonies will do many of the things commercial beekeepers may do such as move their bees for pollination and honey production. Those with fewer colonies tend to not move their colonies because it is not cost effective. Sideliners may not keep all their colonies at their home but the bees are usually kept fairly close by. Arrangements to see the colonies should be straightforward.

Classification by Operation Type

Honey Producers

Honey production is a common use of honey bee colonies by all operational sizes. Hobby beekeepers may produce honey for personal use or for secondary income. Sideliner and commercial beekeepers may generate a more substantial income from honey production. To take advantage of the growing season, many colonies may be moved to locations to follow the bloom of different nectar producing plants. The top honey producing states in the United States for the past 10 years are North Dakota, South Dakota, California, Florida, Montana, Texas, and Minnesota. The ranking may shift from year-to-year, although North Dakota has reliably been the leader. Beekeepers who are in the Midwest for the summer to produce a honey crop will often move their colonies south for the winter for another income source and/or more successful wintering. Important floral sources for these beekeepers while in the Midwest include clovers, alfalfa, canola, basswood, and dandelions. Beekeepers based in more temperate climate areas of the country do not migrate their colonies for over-wintering but many move their colonies multiple times per year to follow the bloom of different plants or different crops for pollination services. Some of the blooms, in Florida for example, are orange, galberry, palmetto, tupelo, palm, sea grape, mangrove, Brazilian pepper, and cotton. There are many other important sources for honey production information on the mainland United States and Hawaii. Local beekeepers, as well as the state apiary inspector (if applicable), personnel at state or regional bee labs, local or state beekeeping clubs, master beekeeping programs, and master gardeners, are good resources to learn the local floral resources.

Pollinators

Commercial and sideliner beekeepers are the primary beekeeper groups that bring colonies to crops for pollination services. Some beekeepers focus on a single crop; others may move their colonies multiple times to follow the bloom of various crops. In particular, many beekeepers from all across the mainland United States transport their bees to

Figure 12.12 Colonies placed in almond orchards just before bloom (a), during almond bloom (b), and a honey bee pollinating almond blossoms (c).

orchards in California starting in mid-January for almond pollination – the largest commercial pollination event in the world (Figure 12.12).

According to the California Department of Food and Agriculture, in 2018 there were more than 1 100 000 acres of almond trees in production and 1 800 000 honey bee colonies were brought into California for almond pollination. Beekeepers in the western United States may travel from farm-to-farm to use their colonies to pollinate various seed crops like onion and carrot, and/or fruit crops like cherries and apples. Beekeepers in the eastern United States may use their colonies to pollinate crops like blueberries, cranberries, and cucurbits. Colonies are moved to a predetermined location in a crop to meet a quota of colonies per acre needed for the highest fruit or nut set, then moved out when the bloom ends. Colonies used for pollination may have increased exposure to insecticides and fungicides that can result in bee or brood death or increased susceptibility to pathogens.

Queen and Package Bee Producers

Queen and package production for sale is predominantly done by commercial and sideliner beekeepers. Queen producers are usually located in locations with a temperate climate: Texas, Mississippi, Louisiana, Georgia, Florida, California, or Hawaii. Queen producers generally start raising queens in late February. Strong colonies are needed to raise queens. Briefly, the beekeepers move larvae from selected breeder colonies to artificial queen cups (grafting) and then put them in special colonies to encourage the bees to raise queens called a mating nuc. Once capped, a ready to emerge queen cell is installed into a special queenless small colony. The queen emerges, goes out on a mating flight then returns and starts to lay eggs. The queen producer removes the laying queen and can sell her or use her in a different colony. A new queen cell is put into the queenless mating nuc and the process begins again. A large producer may raise thousands of queens.

Package producers are usually also queen producers. Packages are a way to start a new colony that mimics a swarm of bees. A package of bees is a screened box containing typically two or three pounds of bees (7000 or 10 000 bees), with a mated queen in a separate cage and a can of sugar syrup for food. The completed package is sent to customers that can include all operation types. Beekeepers may also sell nucleus colonies or nucs, which are different from mating nucs. A nuc is a small colony, containing a laying queen and adult bees on usually four or five frames that contain developing bees and food. The bees and frames from a nuc can be introduced into a larger box where the colony can grow.

Example of a Year in the Life of a Commercial Beekeeper

To better understand how commercial beekeepers operate, here is an example of a typical year in the life of commercial beekeepers from the upper Midwest. As with most beekeepers, the year for the upper Midwest commercial beekeepers begins in spring (March and April). The colonies are typically in a southern location, like California or Texas during the winter. The beekeeper raises or purchases new queens, then divides the bees and brood from strong colonies into new boxes and adds a new queen to each colony to make new colonies. These new colonies help replace a beekeeper's winter losses, increase their operation size, or can be sold to other beekeepers.

Beginning in May, these beekeepers bring their colonies to the upper Midwest for honey production. Colonies are examined for health and food supplies. A beekeeper may need to feed the bees if there is no nectar flow at the destination. When flowers are blooming and producing nectar, the beekeeper adds boxes, called honey supers, for the bees to fill with honey for harvest. Midwest beekeepers usually leave their colonies at one location for the summer.

At the end of summer (August through September), the honey is harvested. During this process the beekeeper visits the apiary and removes the boxes of honey and brings them to their extracting facility to remove the honey (Figure 12.13). After the honey is harvested, the bees are inspected for health and food stores. The bees are prepared for winter: they are fed sugar syrup if they do not have enough honey for winter and are either shipped to their winter location or wrapped with an insulation material to remain in place. The vast majority of commercial beekeepers do not overwinter their colonies in the upper Midwest.

In fall, some Midwest beekeepers put their colonies in "cold storage buildings" usually from October through January. These are climate-controlled buildings that can hold thousands of colonies. These buildings are kept at about 45 °F and have circulating air, with temperature, humidity, oxygen, and CO_2 levels controlled to ensure the health of the colonies. The purpose is to allow the bees to winter with the lowest stress while eating the least amount of food (honey). Other beekeepers transport their colonies south or to the west and place them in apiaries or large holding yards (hundreds of colonies in one apiary) to inspect and feed as needed.

Many commercial beekeepers move their colonies to California for the almond bloom in mid-January. Almonds generally bloom from mid-February to the beginning of March. Colonies are often large once the bloom ends since almond pollen and nectar stimulate colonies to grow. Smaller colonies or brood death at this time indicates a

(a) (b)

Figure 12.13 Smaller scale beekeepers may use hand-powered tools to extract honey from the honey boxes, including a hand-crank centrifuge extractor and an electric hot knife to manually remove the wax cappings to expose the honey (a). Commercial beekeepers use a large-scale, mechanized extraction system to extract the honey from the honey boxes (b).

problem like poor weather or pesticide exposure. After the bloom is over, beekeepers move their colonies out of the orchards to the southern states or to other locations in the western states to start the process of dividing their colonies. At this time, beekeepers check the status of all their colonies, feeding pollen substitute and/or sugar syrup if there are not enough stores in the colony and no resources yet for the bees to collect. Beekeepers who do not transport their colonies to California for pollination may do so to escape the risks that accompany pollination and transportation or to produce a different income, like raising queen bees.

Summary

Beekeepers can be classified by operation size, type, and geography. These differences can lead to differences among beekeeper groups in the logistics of applying treatments, the tolerance of beekeepers to disease in their colonies, and risk of exposure to pathogens and pests. While these classifications are commonly used to talk about beekeepers, the descriptions are generalizations and may not fit all the beekeepers you may encounter.

13

Basics of Apiary Design

Brandon K. Hopkins

Department of Entomology, Washington State University, Pullman, WA, USA

The placement of hives and the layout of the apiary can have major impacts on the health and survival of the colonies. Beginner beekeepers have a plethora of resources to help in deciding on the placement of their hives when they start their foray into beekeeping such as The Hive and the Honey Bee (2015).

The first decision in the designation of an apiary (the location where hives are located) is often dictated by logistical and safety concerns such as vehicle access or consideration for the unaware passerby. It is often best to place hives where they are not readily seen, as any interaction with bees will be blamed on the visible colonies (even if it is actually wasps that are causing the trouble). Additionally, placing colonies out of sight reduces the risk of vandalism.

General Rules

- Keep the apiary clean – Remove, wrap, or tightly cover unused frames/equipment old comb and woodware, as they can harbor disease, and attract pests. The old comb or equipment can incite robbing in the apiary, increasing opportunity for disease transmission
- Hives facing south or southeast (northern hemisphere) – Colonies benefit from the sun's warmth in the morning to help initiate foraging as soon as possible
- Sun or partial sun (warmer places need more shade or installation of shade cloth) – Beekeeping practices are highly regional/local. In general colonies perform better when they are placed in sunny locations compared to shady spots. However, in some places of Southern California or Arizona direct sun all day can be too much.
- Access to water – If there is no obvious source of water available near the apiary, it can be important to provide a water source. Providing water at the time the apiary is established can prevent issues with neighbors' pools, birdbaths, etc.
- Level or mild sloping ground – You want a slight forward slope to the individual hives so that water doesn't accumulate on the bottom board, but not so much that as additional hive components are added they slide forward or fall forward.
- Avoid low-lying or flood prone areas
- Consideration for wind protection – For areas with strong winds colonies are best placed in an area that will protect the colonies from sustained high winds and this is especially true for winter months
- Clear flight path – Placing the hives where the bees don't have to struggle with major obstacles as they come and go is valuable. Also consider any human traffic as to avoid bees running into people passing by.
- Hives elevated from the ground – Bees generally prefer to nest in cavities off the ground. Elevating your hives prevents rotting of the woodenware, saves your back, and helps prevent mice and other pests from accessing the hives. They should not be elevated higher than what is comfortable for the beekeeper to work (Figure 13.1).

Common Rules and Regulation (Vary by Country, State, and City)

- Registration of colonies (number of colonies and location)
- Registration of apiary site
- Limits to the number of colonies in a given apiary
- Fencing or signage to protect both bees and unaware community members

Honey Bee Medicine for the Veterinary Practitioner, First Edition. Edited by Terry Ryan Kane and Cynthia M. Faux.

Figure 13.1 Hives placed on stands can keep the colonies out of potentially wet condition, can help protect bees from potential pests, and makes work a little easier for the beekeeper.

Effect of Apiary Location and Design on Honey Bee Health

The fungal brood pathogen chalkbrood (*Ascosphaera apis*) is an example of a honey bee disease affected by environmental factors of the apiary location. Fungal pathogens can thrive in cool damp locations; therefore, colonies placed in cold, damp, and low-lying areas are more prone to incidence of chalkbrood (Sanford et al. 2019).

Hive placement/arrangement in an apiary has the potential to impact disease transmission as the arrangement of colonies can dramatically impact the amount of drifting that occurs. (Drifting is when foragers return to the wrong hive.) Varroa mites and American foulbrood are two common examples of diseases that can spread rapidly via drifting bees.

In an apiary where bees are drifting between colonies, there is an increased risk that diseases and parasites will be spread through the yard rapidly before they can be isolated and managed. To minimize drifting, hives should be arranged so they are not in neat rows or repetitive patterns. When hives are in rows, bees from hives in the center of the row tend to drift to colonies at the end of the rows. Placing colonies in a circular pattern can help reduce drifting. If colonies are arranged in a row, the hives can be faced in slightly different directions or intentionally misaligned. When space is available colonies should ideally be spaced a meter or more apart (Figures 13.2 and 13.3).

Hive design can also aid in the reduction of drifting. Honey bees are known to recognize geometric shapes and color (not red) and will use those to help them orient to their hive. Painting hives different colors or with unique designs can help reduce drifting (Free 1958) (Figure 13.4).

A major factor in colony health is the availability of forage and ideally a diversity of plant species. The location of an apiary site in proximity to high quality/diverse forage should be considered. When bees are nutritionally stressed it exacerbates many disease and parasite issues.

Windbreaks in an apiary are particularly important if colonies are going to be overwintered in that location. Natural windbreaks are preferred such as shrubs or a tree line, however fencing or buildings can also be used as effective windbreaks. Windbreaks should be evaluated for their potential as shade for the colonies. The arrangement along a windbreak can increase the colonies time in the shade in hot climates and reduce time in the shade in cooler climates.

Figure 13.2 When space allows hives can be placed to maximize distance between groups of hive on pallets. Offsetting pallet alignment can aid in bee orientation and reduce drifting.

Figure 13.3 Sometimes the beekeeper is forced to lay out hives in a linear fashion, which can increase drifting. After hives are set and bees orient to their new location bees will often drift in large numbers to the ends of the linear rows. Maximizing the distance between the hives can help minimize drift.

Figure 13.4 Painting hives different colors, large geometric shapes on the hives, and structure in the apiary can reduce drift and help bees orient to their home hive.

Bears can be a major problem for beekeepers and their colonies in certain locations. The most common and effective strategy to prevent bear damage where bears and bees commingle is to use high-voltage electric fencing (Figure 13.5).

Figure 13.5 Hives enclosed within electric fencing.

Considerations for Size of Beekeeping Operation

Apiary layout/design and location considerations are very different for a beekeeper with five hives compared to a beekeeper with 5000 hives. Most beekeepers with 20 or more hives do not usually have enough land/forage on their own property to support those colonies. Sideliner and commercial beekeepers often rely heavily on relationships with private landowners and government leased land to supply enough apiary locations to support large numbers of colonies. Larger beekeepers who are involved in pollination events might also have apiary location and design dictated at some level by the needs/restrictions of the grower. The beekeeper will often work closely with the manager of pollination dependent crops to ensure proper hive layout but it is often a compromise between what is best for the bees/beekeeper and farm needs (tractor access, irrigation, etc.). The guidelines above are ideal situations that cannot be applied to beekeeping at a commercial scale. For example, placing colonies a meter or more from each other, orienting the entrances at different angles, and place hives facing sun are impossible when colonies are placed on pallets. However, beekeepers can still space pallets as much as possible and offset pallet placements so the entrances are not directly opposite each other.

References

Anon (2015). *The Hive and the Honey Bee.* Hamilton, IL: Dadant & Sons Inc.

Free, J. (1958). *The drifting of honey-bees. The Journal of Agricultural Science* 51 (3): 294–306. https://doi.org/10.1017/S0021859600035103.

Sanford, M.T., Cameron J. Jack C.J., Jamie Ellis, J. (2019) Chalkbrood recommendation. University of Florida IFAS Extension Publication #ENY116

14

Clinical Examination of a Honey Bee Hive

Jerry Hayes

Bee Culture Magazine, Medina, OH, USA

You spent years in school learning about animal health and how that is impacted by disease, then years in your business practice learning the finer points of how to identify, respond, and treat animal pests, parasites, and diseases. Now you have merged these "learnings" into a suite of knowledge, skills, and abilities to treat the injuries and illnesses of pets and food animals successfully.

You are now reading this either because you are a beekeeper and are already fascinated, you have been approached by a beekeeper for help because of VFD requirements, or you want to start your own hives and learn all you can.

The Veterinary Feed Directive (VFD) requires that certain antibiotics for honey bees and other food producing livestock be available **only** through a veterinarian prescription or through a VFD order from a veterinarian. This is a change, after decades of beekeepers diagnosing bee diseases themselves, and purchasing an antibiotic either from a beekeeping supply company or over the counter at the local farm store.

Beekeeping and honey bee management is a visual sport. To understand or have an inkling of the sounds and odors associated with honey bee vocabulary that relates to colony health, you must tie the visuals – what you can see – to the colonies' production of sounds, odors, and movement that can overwhelm our human senses.

Visuals come first. To perform a visual examination well, confidently, and comfortably, you must be able to see the hives' structural moving parts and pieces along with individual and colony activity holistically and from a distance. Then, focus on certain areas so that over time you can build a visual picture that will allow you to compare and contrast what you see in this ordinarily closed dark world, called a bee hive. When you, the vet, or your beekeeper client, open a honey bee colony to inspect and diagnose, you are doing something the majority of the population would never agree or, have the opportunity, to do.

As generally entomophobic (fear of insects) beings, we humans have been taught to respond to insects quickly and physically by swatting, running, and screaming. No matter how you have explained the vast array of honey bee pollinated fruits, nuts, and vegetables found every day, the average person does not connect the honey bee to their health and well-being. Once you are exposed to the world of the honey bee, or 50 000 honey bees in a colony, your view of them changes.

Let's Open a Bee Hive

In a perfect world pick a warm, sunny, calm (no wind) day. On a day like this, about a third of the colony, the worker foragers, will have left the colony. They will be off searching for flower nectar, pollen food resources, water, or propolis (tree gums saps and resins) they can collect and bring back to the colony to feed their developing siblings and use to make the hive a livable sustainable environment (Figure 14.1).

1) At a moderate distance from the apiary location, light the most important tool a beekeeper has. . .the smoker. Honey bees communicate with each other using odors called pheromones. A cool white smoke puffed gently into a colony disrupts and masks this honey bee pheromone language. If the bees in the colony cannot "talk" to each other they cannot organize a group defensive action toward you and the beekeeper (Figure 14.2).
2) Wear personal protective clothing (PPE) such as a bee suit, including veil and disposable gloves. The color white is considered a neutral color, not attracting honey bee attention. Honey bees use their sting organs for defense. Having a protective barrier between you and a grumpy

Honey Bee Medicine for the Veterinary Practitioner, First Edition. Edited by Terry Ryan Kane and Cynthia M. Faux.

Figure 14.1 The bee yard on a nice day. *Source:* Photo courtesy of Cynthia M. Faux.

Figure 14.2 Lighting a smoker. A variety of fuels can be used as long as the fumes are not toxic to bees or beekeepers and produces a cool smoke. *Source:* Photo courtesy of Cynthia M. Faux.

worker bee armed with a defensive stinger is a stress reducing tool for both you and the colony (Figure 14.3).

3) All your movements around an individual bee hive, or in an apiary with many bee hives, should always be measured, slow, calm, and deliberate. With this in mind, approach the hive from a direction that is not in line with the colony's main entrance. You do not want to disrupt their access to their "home" or alert them that there is somebody outside disturbing the returning and leaving foragers' flight. Look around you, be aware, and keep out of the honey bees' flight path in and out of the hive (Figure 14.4).
4) For a few moments, stand closely to one side or the other of the hive you are interested in and look down at the entrance. Is there a lot of flight activity in and out of the entrance, indicating population strength and vitality? Or, is the flight activity relatively modest or inactive, perhaps indicating a colony health issue? (Figure 14.5).
5) With the smoker, apply a few puffs of cool white thick smoke at the entrance. This is where those worker "guard" bees are located. The calming influence of the

Figure 14.3 Full bee suit with gloves. *Source:* Photo courtesy of Terry Ryan Kane.

Figure 14.4 (a) Standing in front of the hive entrance (not to do). Returning foragers will cause a "traffic back-up" if the entrance you block their flight path. They may "bump" you in annoyance, encouraging you to move aside. (b) Stand to the side of the entrance so bees can be observed flying into the entrance. *Source:* Photo courtesy of Cynthia M. Faux.

Figure 14.5 Observe activity at the entrance, including the pollen loads being brought back to the hive and defensive behavior by the guard bees. Bees are returning to this active hive. Loads of pollen can be readily observed. *Source:* Photo courtesy of Cynthia M. Faux.

Figure 14.6 A few puffs of smoke at the entrance will help calm the defensive guard bees. *Source:* Photo courtesy of Cynthia M. Faux.

smoke will allow you entrance into their world safely. Just a few puffs of smoke are better at first, rather than many (Figure 14.6).

6) The next most important beekeeper tool is the "hive tool." Insert it under the lid gently and, if an inner cover is present, loosen this as well. Apply a few puffs of smoke inside the colony and then slowly close the cover. This allows the smoke to permeate the inside of the hive (Figures 14.7 and 14.8).
7) Wait 30–60 seconds for the smoke to have its communication calming effects.
8) Remove the outer and inner cover and place them upside down, flat on the ground, or on an empty box or stand, behind the colony. A puff of smoke may be appropriate over the open exposed frames (Figures 14.9 and 14.10).
9) You will now be looking at the top bars of the frames that contain beeswax comb. The first box could either be a "honey super," or a "brood box," where baby bees are developing.
10) Using the hive tool, loosen the box as you did with the cover and tilt it up a couple of inches and apply several puffs of smoke into the hive. Bees from inside may fill this area. Move slowly and avoid crushing bees when

Figure 14.7 Lift the lid and apply a few puffs of smoke. *Source:* Photo courtesy of Cynthia M. Faux.

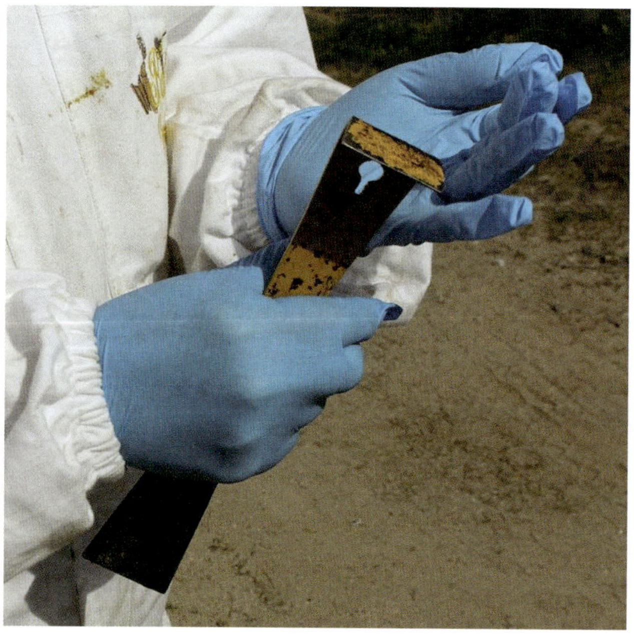

Figure 14.8 A hive tool. *Source:* Photo courtesy of Cynthia M. Faux.

you lower the box. The damaged or killed bees may emit an "alarm" pheromone that can alert other members of the colony, or adjacent colonies, into taking more defensive actions. Use smoke freely but judiciously. If this box is a super, remove it. Note its weight and set it down on top of the upturned lid you placed on the ground behind the hive. If there are multiple supers, do the same thing with each (Figures 14.11–14.13).

11) If you find a "queen excluder" at some point in box removal, take care to not damage it when removing, in case it has been propolized by the bees. Examine the queen excluder to ensure the queen is not on it. She

Figure 14.9 The top of the hive placed upside down (inside is up) on an empty hive body. This provides a surface for resting hive boxes. Avoid placing hive boxes in the grass, as the queen can fall into the grass and become lost. *Source:* Photo courtesy of Cynthia M. Faux.

Figure 14.10 Applying a few puffs of smoke over the top of the exposed frames. *Source:* Photo courtesy of Cynthia M. Faux.

Figure 14.11 Tipping the box to apply smoke. *Source:* Photo courtesy of Cynthia M. Faux.

Figure 14.12 If the box is a honey super, remove and place on the upturned lid. *Source:* Photo courtesy of Cynthia M. Faux.

will usually be in the boxes below the excluder. You are now in the brood boxes.

12) Using the hive tool, smoothly and slowly loosen the frame closest to the outside wall of the brood box.

Figure 14.13 If there is one, remove the queen excluder. *Source:* Photo courtesy of Cynthia M. Faux.

Usually the frame in this location is full of stored honey and probably does not have the queen on it. Keeping the queen safe in this inspection process is paramount. Remove the frame and set in down next to the hive, standing it up on end (Figure 14.4).

13) With removal of the outer frame, there is now more maneuvering room available in the box. Each

Figure 14.14 (a–c) Remove the outer frame first and set it down on end. This will provide more working space within the box. *Source:* Photo courtesy of Cynthia M. Faux.

additional frame can now be loosened, then slowly and gently pulled out from the box. Examine each frame for normal brood pattern, evidence of a healthy queen, capped brood, pollen stores, capped honey, etc. Look for the queen or evidence of eggs on every frame you examine in the brood box. When examining the frame/comb, hold the frame up in front of you with the sun at your back so each cell can be examined with as much light reaching the bottom of the cell as possible (Figures 14.15 and 14.16).

14) Examine each frame for abnormalities such as abnormal brood patterns such as the shotgun brood pattern, dead/dying larvae and pupae, etc. After examination, the frame can be returned to the box in the position and orientation it was when you removed it (Figure 14.17).

Figure 14.15 Sequentially remove and examine each frame. *Source:* Photo courtesy of Cynthia M. Faux.

Figure 14.16 Examining a frame with the sun at your back to best visualize eggs and larvae within the cells. *Source:* Photo courtesy of Cynthia M. Faux.

Figure 14.17 Pollen and nectar. *Source:* Photo courtesy of Cynthia M. Faux.

15) When **ALL** the frames/comb have been examined, reassemble the colony in reverse order that you disassembled it and go to the next colony and repeat. Now you have an idea of how to do it. After 100 times you will be an expert.

You may see other types of movable frame colonies such as top bar hives. The procedure is similar for opening and inspection.

If you encounter a hive that does not have movable frames for easy comb management, examination to ascertain pest, parasite and disease levels will be more challenging. Please explain to the beekeeper that movable frame colonies were designed so that the beekeeper manager could manage the hive efficiently for the health and safety of the honey bee colony.

When you experience opportunities to be present when a honey bee hive is opened and you see those 50 000 workers, take a moment to look at them. How many different subtle or dramatic color differences do you see? Some workers may be very dark almost black or brown or lighter browns, dark yellow or very light yellows, and more. These color combinations are indications of different drone fathers. Look in the same colony three weeks later and you may see different percentages of colors as other drones' sperm previously stored was used. All this to say honey bees are not going for the genetic "home run" as you may see in selective breeding of other animals. The selective breeding process in honey bees has not been as successful as in cows, chickens, dogs, cats, sheep, etc. There is tremendous genetic diversity in a honey bee colony.

Beekeepers use a variety of techniques to manage their honey bee colony(ies). Many of these techniques are in this book. This is a journey, in which you may not feel immediately comfortable. Each time you open a hive for inspection, you will increase your understanding and confidence in this procedure.

15

Veterinary Regulations

Christopher J. Cripps

Betterbee, Greenwich, NY, USA

Introduction

Traditionally, veterinarians in the US and Canada had not worked professionally with honey bees until the new antibiotic usage rules went into effect in the US on 1 January 2017 and in Canada on 1 December 2018. In many other countries, veterinarians study honey bees as part of their formal training. The broad comparative medicine training of veterinarians makes them ideal medical providers for honey bees. As more veterinarians in the US and Canada work with honey bees and beekeepers, several regulatory issues should be understood, most related to Federal antibiotic regulations.

Health Inspections

For centuries, beekeepers have recognized disease in their bee colonies. Foulbrood was described as early as the 1700s. In 1877, San Bernardino County in California passed the first US law for apiary disease inspection. By the early 1900s, many states had passed laws to control foulbrood in honey bee colonies. These laws required regular inspections and burning of affected combs; they also prohibited keeping or moving infected bees. In New York State, this type of law was passed in 1928 when an estimated 7% of hives were infected with American foulbrood (AFB). By the 1950s, New York reported AFB infected 2% or less of colonies inspected.

Each state has its own regulations regarding honey bee disease; there are no federal laws or regulations. The inspectors and apiarists who oversee enforcement of the state laws have been successful in decreasing the incidence of AFB nationally. The state bee inspectors may have experience with bees, but little formal education or they might have advanced degrees in entomology. However, there are few, if any, veterinarians in these positions. Because veterinarians in the US have traditionally not been involved with honey bees, veterinary services for honey bees have not been addressed in state laws. One regulatory issue veterinarians face is they do not have the authority to issue interstate health papers. If a veterinarian has a beekeeper client planning to move honey bee colonies to another state, the state apiculturist, or one of their bee inspectors, must do the inspection and supply the required interstate health certificate.

Antibiotic Control

In 2015, the US Food and Drug Administration (FDA) offered guidelines to change the approval status of many medically important antibiotics from Over the Counter (OTC) to Prescription (RX) or Veterinary Feed Directive (VFD). By changing the status to RX or VFD, the FDA planned to bring under veterinary supervision any antibiotic deemed medically important to human medicine and used in animal feed. The effective date for these changes was 1 January 2017. After that date, a veterinarian needed to issue an antibiotic order (Either an RX or VFD) for any food-producing animal to receive these antibiotics in its food or water. Honey bees are food-producing animals.

Veterinary Feed Directive Background

Prior to 1996, all animal drugs were approved as either Over the Counter or Prescription. Drugs given to animals in their feed were all approved for use as OTC. Changing those approvals to Prescription to ensure veterinary

Honey Bee Medicine for the Veterinary Practitioner, First Edition. Edited by Terry Ryan Kane and Cynthia M. Faux.

oversight would have required many feed mills to follow the rules of their state Pharmacy Boards. Very few, if any, feed mills would have been able to comply with those rules. The Animal Drug Availability Act of 1996 created the VFD. The FDA could now require veterinary supervision for the use of certain drugs while still ensuring the availability of the drugs from feed mills and not require feed mills to operate as pharmacies.

In 2015, the FDA sought cooperation from drug manufacturers to apply for updated labels for medically important antibiotics used in animal feed prior to 1 January 2017. All label claims for production enhancement uses were eliminated. All OTC approvals were changed to either VFD or RX Approvals. VFD drugs are categorized as Type I or Type II. Type II drugs had previously been defined as any drug with a withholding period. Type II drugs could only be mixed into medicated feed in a licensed feed mill.

In December 2016, the FDA changed VFD categorization so Type II drugs are only those with a withholding period in major species at their lowest concentration. The major species are defined as dogs, cats, horses, cattle, pigs, and chickens. Honey bees are classified as a minor species. This change allowed the manufacturer of oxytetracycline to register the drug as Type I, so it would be available from all feed mills. This change also allowed the manufacturer to maintain the approval for honey bees, which requires removing any application of oxytetracycline six weeks prior to collecting honey to prevent oxytetracycline residues in honey. Without that concession from the FDA, the drug sponsor would have been likely to remove the honey bee approvals from its oxytetracycline label in order to maintain Type I classification and the ability to prepare oxytetracycline medicated feed in any feed mill, not just the few mills licensed to prepare Type II antibiotics.

Many veterinarians were not familiar with VFDs because previously VFDs were used only in swine and chicken production. The FDA guidelines state that the FDA would approve antibiotics provided in food as VFD status and antibiotics used in water as RX status. All antibiotics approved for use in honey bees, whether RX or VFD, can be administered by mixing with confectioners' sugar and sprinkling in the hive where the bees eat the sugar and antibiotic mixture (Figure 15.1). Thus, many veterinarians concluded that they would need to write a VFD for any antibiotic use in honey bees. This conclusion is incorrect as the approved label of each antibiotic includes a status. That status dictates what type of antibiotic order to issue, either a VFD or RX. The status on those labels is based on the use in major species. The veterinarian does not decide if the antibiotic is VFD or RX; the FDA determined the status in the approval process.

Figure 15.1 All three antibiotics approved for honey bees can be fed by mixing the antibiotic in powdered sugar and applying to the top of the frames inside the hive. The sugar and antibiotic mixture is applied to the outside edges as it can be toxic to uncapped larvae and the uncapped larvae are typically found in the center of the box. This application method is the same for Prescription and Veterinary Feed Directive drugs. *Source:* Courtesy of Christopher Cripps.

The approved labels for animal drugs are available on the FDA website https://animaldrugsatfda.fda.gov. A search box is provided where the user can type "bees" to get a list of all antibiotics approved for use in honey bees. As this chapter is being written in early 2020, there are 11 approvals of various forms of three antibiotics – oxytetracycline, tylosin, and lincomycin. Oxytetracycline has both RX and VFD status approvals. Tylosin and lincomycin have only RX status approvals.

Non-medically Important Antibiotics

The new VFD rules are specifically for "medically important" antibiotics and the FDA maintains a list of antibiotics deemed important for human medicine. An example of a non-medically important antibiotic used by beekeepers is fumagillin, used for treatment of nosema. This antibiotic was not on the list of medically important antibiotics, and thus was not required to have a label change.

Antibiotic Approvals

Oxytetracycline, tylosin, and lincomycin are the three antibiotics approved for use in honey bees as of 2020. The FDA approves indications for antibiotic uses that are therapeutic in nature, which are for treatment, control, or prevention of disease. No antibiotic approvals for growth promotion in livestock exist as of January 2017. Treatment is when the animals being given antibiotics are infected with the target bacteria. Control is when there are infected individuals within the population, and antibiotics are given to decrease or stop the spread of the bacteria to non-infected animals. Prevention is when there is no disease present but giving antibiotics is expected to stop an outbreak from happening.

Only the pioneer oxytetracycline brand Terramycin® has an antibiotic approval which includes "treatment" of a honey bee disease (NADA 008-622). The generic approval of Pennox 343® brand of oxytetracycline was based on that pioneer drug approval and does not include "For Treatment" on its label. None of the approvals for antibiotics in honey bees include a "For Prevention of. . ." indication.

All antibiotic approvals for honey bees are for the control of American foulbrood (AFB) (and possibly European foulbrood (EFB)). Control means that the disease has been diagnosed in the target population and use of the antibiotic will decrease its spread.

To work within the spirit of the label, there are two scenarios to consider. One is that the bees may travel between colonies. If a hive is found with an infection in one apiary, the label allows giving antibiotics to other hives in the apiary to control the spread of disease from infected to non-infected hives. Honey bees often fly miles away from their hive and may enter other hives, especially to rob honey. It is thus possible that a veterinarian could know about a diagnosis in one apiary and then issue an antibiotic order for control of a disease in a second apiary, even if the disease has not been diagnosed in the second apiary because they are trying to control spread between colonies. In a second scenario, uninfected bees exist in a colony that has signs of disease. The application of antibiotics to the hive with signs of infection is to control the spread of disease from infected larvae to non-infected larvae in the same hive.

Beekeepers' Definition of How They Use Antibiotics

While the FDA may list Prevention, Control, and Treatment as the allowed indications, the beekeeper tends to think of using antibiotics for prophylaxis/prevention or treatment. Many beekeepers understand that antibiotics for treatment of AFB is ultimately ineffective because AFB forms a highly resistant spore that can remain viable for decades in the environment. So, many beekeepers use antibiotics only for prevention of AFB; prevention is an off-label use. Prevention use is driven by beekeepers who have never had the disease in their colonies, and hope to keep it that way, or by those beekeepers who have experienced AFB, in order to suppress the clinical signs and colony losses.

Antibiotic Resistance

The laws about antibiotic use in food-producing animals are being driven in order to decrease the development of resistant strains of bacteria and the spread of the genes that give bacteria resistance to antibiotics to other bacteria species. The laws are being applied to all food-producing animals, including the minor species, such as honey bees. The Minor Use and Minor Species Animal Health Act of 2004 was passed to help make more medications legally available for these species. Beekeepers have argued that because, as an industry, they are such small users of antibiotics, compared to cattle and swine, that beekeepers should be exempted from needing an RX or VFD to obtain antibiotics. The FDA, however, did not exempt any species from these laws. They did create an Office of Minor Use and Minor Species to help ensure antibiotics and other drugs are available for animals such as honey bees, but only under proper supervision and review.

Bacterial diseases of honey bees have developed resistance to oxytetracycline (Miyagi et al. 2000; Evans 2003; Alippi et al. 2007). For decades, beekeepers have applied oxytetracycline to beehives. Some beekeepers have applied antibiotics spring and fall to "prevent" AFB. In the 1990s, beekeepers reported oxytetracycline not working as expected to prevent clinical signs of AFB. In 2000, Miyagi et al. confirmed resistance to oxytetracycline in *Paenibacillus larvae*, the causative agent of AFB.

At the time resistance was discovered, only oxytetracycline was approved for honey bees. Testing of other antibiotics began, and in 2005, Elanco received FDA approval for Tylan® soluble brand of tylosin to be used for the control of AFB in honey bees. In 2012, Pfizer received FDA approval of the Lincomix® brand of lincomycin hydrochloride for the control of AFB in honey bees.

Withholding

Honey bees are food-producing animals. They collect nectar and convert it into honey. Veterinarians must ensure there are no violative antibiotic residues in food produced by animals. All the approvals for antibiotics for honey bees include withholding time information. The withholding statements specify a time period prior to the honey flow when you must not give antibiotics; for example, "at least six weeks prior to main honey flow." Veterinarians must be aware of when the honey flows occur for their particular area. The honey flow happens when plants are blooming, and the bees collect nectar from their flowers. The dates of the honey flows vary based on weather and climate. For some areas there may be few plants that bloom, so the honey flow is limited. Other areas may have a variety of plants that bloom at staggered times so the honey flow is extended.

For example, if you must treat for two weeks, and then stop treatment six weeks before the honey flow begins, treatment must start at least eight weeks before the honey flow begins. This might require treating bees when the weather is quite cold in the northern states. For this reason, fall time treatments may be preferred because they can be done after any fall honey flow and well before the spring honey flow. To be able to effectively time treatments, the veterinarian must know about the blooming plants in their area. To learn about the blooming plants, the veterinarian can search for local publications by Cooperative Extension or beekeeping clubs, or they can ask their clients.

Veterinary Client Patient Relationships Required for a VFD

To issue a lawful VFD, several requirements must be met. First, the veterinarian must be licensed to practice veterinary medicine and must be operating within the course of the veterinarian's professional practice. If the veterinarian can legally practice in their state, they are operating within the course of their professional practice and can issue a VFD for bees. A veterinarian with a professional license is subject to supervision by their state's Veterinary Board. Governmental supervision of the veterinary overseers of antibiotic use is therefore available.

Next, the VFD must be issued from within the context of a Veterinary Client Patient Relationship (VCPR). An acceptable VCPR to the FDA includes: the veterinarian has assumed the responsibility for making medical judgments regarding the health of the animals and the need for medical treatment; the client has agreed to follow the instructions of the veterinarian; the veterinarian has sufficient knowledge of the animals to make a diagnosis; and the veterinarian is available for follow up in case of adverse reactions or treatment failures. These requirements are listed in Code of Federal Regulations Title 21 Section 530.3(i).

Some states include a definition of a VCPR including these points in their practice act, other states do not. If the state does not define a VCPR to at least the level of the FDA, then the regulations require the veterinarian to use the federal definition of a VCPR.

To form a VCPR for honey bees, the FDA has specifically stated the veterinarian must physically visit the apiary for which they issue the antibiotic order. The veterinarian should be personally familiar with the husbandry and medical issues of the bees. However, once the veterinarian and beekeeper have established the VCPR, they are free to maintain that relationship by whatever means they want. Telephone conversations, pictures, lab reports, or inspection reports can all be used as a basis for issuing an antibiotic order when a VCPR already exists.

The federal regulations do not require any documentation of the VCPR, but a medical record may be proof of the relationship. Although some organizations, such as milk processors, require producers to supply this kind of documentation if they use antibiotics, most beekeepers will not be subjected to this requirement from their market.

The VFD Form

A blank VFD form is shown (Figure 15.2).

The information required on the VFD form is not extensive, but there are specific statements required by the statute.

The upper sections are name, address, and phone number of the veterinarian and of the client. The client can be the owner of the hives or the beekeepers that are working these hives.

The drug name, level in the feed, and duration of use are listed next. For honey bees, only oxytetracycline has a VFD approval. The label says to use 200 mg per ounce of powdered sugar when mixing, so the drug level is listed as 6400 g per ton. It is unlikely that anyone is going to mix a ton of this mixture at once, but that is the standard unit required by the FDA.

The species and production class is simply "honey bees." The approvals do not separate different production classes of honey bees in the same way as, for example, chickens where a farm could be feeding laying hens or broilers.

Number of Refills Authorized differs from a prescription. The FDA expects this information to come from the label. Oxytetracycline for honey bees has no refills allowed on the label. As of June 2019, no VFD drugs allowed refills. A VFD is good for one feeding. If a beekeeper needs to feed the bees again after doing it once, then a new VFD is needed.

The indicated use for oxytetracycline on the label is for the "control of AFB caused by *Bacillus larvae*, and European foulbrood caused by *Streptococcus pluton* susceptible to oxytetracycline." These two bacteria have had name changes in recent years. The causative agent of AFB is now called *P. larvae* and the causative agent of European foulbrood (EFB)is now called *Melissococcus pluton*.

When the FDA considers "For Control of," there are a few things for the veterinarian to consider. Control of a disease intimates that the disease has been diagnosed in the population being medicated. Because bees travel so far and may

Veterinary Feed Directive for Honey Bees

All parties must retain a copy of this VFD for 2 or more years after the date of issuance in original form

Veterinarian: ______________________ **Client:** ______________________

Address: ______________________ **Address:** ______________________

______________________ ______________________

Phone: ______________________ **Phone:** ______________________

Fax or Email: ______________________ **Fax or Email:** ______________________

Drug Name: Oxytetracycline **Drug Level:** 6400 grams/ton **Duration of use:** about 15 days

Species and Production Class: Honey Bees **Number of Refills Authorized:** NONE

Indication for use: For the control of American Foulbrood and European Foulbrood susceptible to oxytetracycline

Caution: Do not apply to open, uncapped brood to avoid larval death.

Use of feed containing this veterinary feed directive (VFD) drug in a manner other than as directed on the labeling (Extra Label Use) is not permitted.

Number of Honey Bee Colonies: ______________________

Address or Premises where Honey Bees will be treated: ______________________

Special Instructions: There are 3 recommended feeding directions:

1. **Dusting – Apply one ounce of this feed (200mg active ingredient) by applying to the outer ends of the frames repeat every 4 to 5 days for 3 times. Do not apply to open brood.**
2. **Syrup – Mix one ounce of this feed in a 5lb honey jar (57 fluid ounces) full of 1:1 sugar syrup (by weight) and feed to the bees. Repeat treatment every 4 to 5 days for 3 times using freshly prepared syrup each treatment.**
3. **Extender Patty: Mix 4 ounces of this feed (800mg active ingredient) with 185 g vegetable shortening (such as Crisco™) and 330g sugar. Place this extender patty on top of the top bars in brood chamber.**

Affirmation of intent for combination VFD Drugs:

This VFD only authorizes the use of the VFD drug cited in this order and is not intended to authorize the use of such drug in combination with any other animal drugs.

Withdrawal Time:

This VFD Feed must be withdrawn (stop feeding to the bees) 6 weeks prior to the main honey flow. Honey stored during medicated periods in combs for surplus honey must not be used for human consumption.

VFD Date of Issuance: ______________________ **(Month/Day/Year)**

VFD Expiration Date: ______________________ **(Month/Day/Year) (Cannot exceed 6 months after issuance)**

Veterinarian's Signature: ______________________

Figure 15.2 Example VFD form designed by the author and available on the internet from the Honey Bee Veterinary Consortium (https://www.hbvc.org). *Source:* Courtesy of Christopher Cripps.

drift into other hives, the diagnosis does not have to be made in the particular apiary but could be made in an apiary a few miles away. Finding information about surrounding apiaries and their diagnoses can be difficult for a veterinarian. The state apiarist or the discussions among neighboring beekeepers might be resources. If considering writing a VFD for a hive that has been diagnosed with EFB, the veterinarian should consider that not all the bees and brood in the hive are infected. Consequently, the antibiotic will be used to control the spread of bacteria from the infected individuals to the uninfected individuals within the hive. The use of oxytetracycline in hives affected by EFB is thus according to the label indications and is not for treatment, which would be a use not included on the label.

Unlike EFB, because of the spore forming nature, using antibiotics in hives infected with AFB will only mask clinical signs temporarily and is illegal in many states. The combs and honey in the infected colony treated with antibiotics will still harbor AFB spores and clinical disease can recrudesce after antibiotics are stopped.

The Caution line on the VFD is for any special cautions the veterinarian wants the beekeeper to know about. For oxytetracycline, the beekeeper should know to not apply the dust to the open larvae as oxytetracycline can kill larvae when applied directly. Normally, the nurse bees are expected to pick up the antibiotic-treated sugar and process it into the food they are supplying the larvae. The nurse bees do not move it directly into the larvae.

The "number of animals" is not the number of individual bees but the number of colonies.

The address where the honey bees are being treated can be a street address or it can be Global Positioning System (GPS) coordinates. The address must enable someone following up on the VFD order to find the location where the treatments occurred.

The required statements, which are different from a prescription, are about Extra Label Drug Use (ELDU) and the Affirmation of Intent. The ELDU statement is "Use of feed containing this veterinary feed directive (VFD) drug in a manner other than as directed on the labeling (extra-label use), is not permitted."

The affirmation of intent is for how the medication is to be used with other drugs that might be fed. The label does not allow this in oxytetracycline for bees, so should read "This VFD only authorizes the use of the VFD drug cited in this order and is not intended to authorize the use of such drug in combination with any other animal drug."

Because honey bees produce food, we must also note a withdrawal time. The drug's label reads to stop giving this feed six weeks prior to the main honey flow. The manufacturer-provided example "Blue Bird" labels expand this even more.

Federal law requires that when a VFD drug is approved, the drug's sponsor must provide representative labeling for the medicated feeds to be produced that the actual feed mills producing the medicated feed can use as a guide for developing their own labels. The FDA Center for Veterinary Medicine refers to these labels as Blue Bird labels. Blue Bird Feed Mill is the fictitious mill name used on these labels.

The Blue Bird labels address feeding early in the spring or in the fall to avoid contaminating honey. "Early in the spring" means applying the medicated feed early enough so that the bees can consume it at least six weeks prior to the flowers blooming and the bees collecting nectar to make honey. "In the fall" means after all the flowers have bloomed, the honey supers have been removed, and the bees are no longer collecting nectar that will be made into honey for human consumption. The Blue Bird label directions go on to specify the removal of any honey that the bees actually do produce while being medicated and making sure it is not used for human food. The prescription label is also different in that it does not give a specific time for the treatment, but rather says that the drug should be fed early in the spring or in fall to avoid contamination of production honey.

The special instructions section includes directions for feeding. In the accompanying sample VFD, the three feeding methods are listed.

The date of issuance and the date of expiration of the VFD must both be written in. The expiration date cannot be any more than six months after the date of issuance.

Finally, the veterinarian must sign the form. The veterinarian then supplies copies to the VFD distributor or feed mill and, by federal statute, must keep the original for at least two years. Some states may require veterinary records be kept longer.

Extra Label Drug Use

Prior to passage of the Animal Medicine Drug Use Clarification Act (AMDUCA) of 1994, issuing prescriptions for uses not listed on the label was a crime; though it was seldom enforced. The items on the label that had to be followed included specific species, disease indication, dose, duration, frequency, and route of administration. AMDUCA established an algorithm for veterinarians to use to prescribe drugs for reasons that were not on the label. The main points of that algorithm ensure the veterinarian has made a careful diagnosis and evaluation of the situation, looks for other drugs that are labeled for the species or that contain the same active ingredient, maintains the identity of the animals treated, and establishes substantially extended withdrawal times so as to not market adulterated food products from the animal.

The AMDUCA criteria also specifically prohibit ELDU in animal feed. One ELDU that has created issues is feeding tylosin mixed in sugar syrup. Mixing in sugar syrup is on label for oxytetracycline but is not for tylosin. Medicating bees with tylosin mixed in sugar syrup instead of dry confectioner's sugar can result in extended withholding times that when they are not followed lead to violative residues in honey.

When using a VFD, ELDU is prohibited by the law. One FDA advice says manufacturers can make available pre-filled VFD forms as there is no leeway in following the directions. The label indications and antibiotic name and level can be filled in ahead of time.

The FDA does acknowledge that there are minor species or minor uses that might not be included on the label. Compliance Policy Guide Section 615.115, issued in December 2016, is a non-binding advisory that says they might not enforce prohibitions against ELDU in VFD drugs when used in minor species, such as bees, if certain precautions are taken. These precautions follow a similar algorithm as AMDUCA. The main points include there is a valid VCPR, there is no other labeled drug, the treated animals are identified, there will be no adulteration of human food products, and any problems are reported to the FDA. When using a drug in this manner, the FDA also recommends adding a statement to the VFD that "this use is prescribed under Compliance Policy Guide Section 615-115" so the feed mill knows that the veterinarian knows it is an ELDU.

In looking at what drugs might be used in an extra label manner, the veterinarian must remember there are drugs that are prohibited from any use in food-producing animals. That list is published by the FDA and readily available on the internet and includes chloramphenicol and fluoroquinolones among others.

How to Obtain Antibiotics

When the veterinarian issues an antibiotic order, there are several sources of antibiotics to consider. The source most well-known to veterinarians will be to use a prescription and to buy the antibiotics from their distributor and then sell it to the beekeeper. The veterinarian can also issue a paper or electronic prescription that the beekeeper could fill at a local or internet pharmacy.

When a VFD is issued, the beekeeper usually takes the lead in obtaining antibiotics. There are beekeeping supply companies that have registered with the FDA as feed mills to make antibiotic feed. The feed is supplied in units as small as one pound that could treat five hives.

For most veterinarians, those are the common choices available, but there are other choices for larger operations. Type A medicated article is the term used to describe the concentrated drug that will be mixed in the feed. In this case, the drug is too concentrated to feed directly. Type B medicated feeds can be thought of as "premixes." To make a Type B medicated feed, the Type A medicated article is mixed with some feed to start to dilute it, but the drug is still too concentrated to feed directly. Type C medicated feeds include the Type A medicated article diluted to the level where they are ready to feed. These types of medicated articles are regulated differently. The Types B and C medicated feeds are available from a feed mill only with a VFD from a veterinarian. The Type A medicated article is available for purchase by feed mills. There is no legal requirement for any paperwork for a drug distributor to sell Type A medicated article to a feed mill.

A veterinarian or a beekeeper could establish a feed mill or VFD distributorship and carry VFD feeds. They simply need to send a letter to the FDA declaring they are making and/or distributing VFD feeds. With this letter, they become subject to FDA inspections and audits of the VFD paperwork. The company selling the oxytetracycline Type A medicated article can sell to any feed mill because oxytetracycline is a category I drug. Though not a legal requirement, they may require the feed mill to supply a copy of a VFD because they have a legal requirement to report to the FDA which animal species each bag of Type A medicated article was used to feed. If the beekeeper makes their own feed mill to be able to obtain the Type A medicated article, they will need to have a VFD in their files before feeding to their bees any Type C medicated feed they prepare.

The original drug sponsor, in their application to the FDA for drug approval, must provide example labels of feeds that may be made with their medication. These labels are produced as if a fictitious company, the Blue Bird Feed Mill, made the feed. These so-called Blue Bird labels are available on the FDA website. Honey bee labels can be found in the miscellaneous species area.

Identity of Hives

Many of the drug use rules require that the animals are identified. For other species, individual ear tags or radio frequency chips are used, or animals can be identified by groups and not broken up or dispersed until the withdrawal period is exhausted. For honey bees, individual identification of bees is nonsensical, and individual identification of hives has not been a standard practice. One hive could be made up of several boxes and contain thousands of individual bees. In certain states, beekeepers do have to mark the beekeeper's name or registration number on some hives or parts of hives. For antibiotic treatments, beekeepers could consider writing numbers on hives or components or attaching numbered ear tags, like those used for cattle, to the hives to aid in record keeping (Figure 15.3). New York State's Apiary Inspection Service applies yellow stickers to hives to identify ones they have sampled for disease testing. The beekeeper should also agree not to disperse the parts of the treated hives during the treatment or withdrawal period.

Size of the Beekeeping Operation and Duties of the Veterinarian

Beekeepers may keep from one to several thousand hives. A beekeeper with few hives will expect typical companion animal style veterinary services. This means the beekeeper will expect the veterinarian to do a thorough physical exam, come to a diagnosis, work with the beekeeper to complete treatment, and explain all of this in easy to understand terms. This beekeeper is paying for services that might cost more than the value of the bees. The bees are pets however, and value of the pet is not their main consideration.

The beekeeper with dozens to thousands of hives will be looking at their operation differently. They have a lot of money invested in bees and equipment trying to make money through bee production, pollination services, and/or honey production. If they have more than 100 hives, they likely have employees. The owner of the hives will want to do things in a systematic way that ensures the bees stay healthy and therefore exposes the beekeeper to less financial risk. Costs of treatments make a difference on the bottom line. A treatment's effectiveness and ease of use matter as well. The veterinarian should be prepared to help institute protocols for antibiotic use in such a system. The veterinarian will not be called to spend an hour on one or two hives. They might be called to spend several hours developing a protocol that the employees can follow to detect and treat disease appropriately. Key areas of a food production system protocol are; who is responsible for carrying it out, what diagnostic signs are they looking for, what treatment are they allowed to administer, what are the problems to avoid during treatment, how do they identify the hives treated, what do they do to avoid creating residues in food, and when do they call for follow up by the veterinarian. During visits by the veterinarian, the records created by the beekeepers should be reviewed and any modifications needed can be considered. The state and federal rules and laws don't give explicit time frames for veterinary visits, but it is reasonable to expect a physical visit to the apiary should be made by the veterinarian at least once per year.

Figure 15.3 Identification of animals treated with antibiotics to avoid residues in food is important. Beehives traditionally are not identified individually. There are marker systems available that could be used. *Source:* Courtesy of Christopher Cripps.

There are several regulatory issues the honey bee veterinarian must keep in mind. The lack of authority to write a health inspection report for interstate travel is one that is different from most other species. Most of the others are routine for food-production veterinarians. These issues revolve around using antibiotics in a safe manner so as not to create human food contamination nor resistance to antibiotics while alleviating disease in the animals being treated. Veterinarians in the USA and Canada have only recently been pulled into supplying services to honey bees. As more is discovered about honey bees, antibiotics, and drug resistance, these issues are sure to evolve. The reader is encouraged to check for changes before issuing antibiotic orders.

Honey Bee Veterinary Regulations in Canada

While much of what has been stated for the US is the same in Canada – veterinarians are now required to provide oversight of medically important antibiotics used in honey bees and ensure labels, and in particular, withdrawal times are properly followed. Also, the same three antibiotics are registered for honey bees – oxytetracycline, tylosin, and lincomycin as in the US. There are, however, a few notable differences that may impact the veterinarian.

Veterinarians in Canada do not have to complete the VFD. All medically important antibiotics for use in food animals, which includes honey bees, are to be sold by prescription only. The label indications are different, as some of the drugs are labeled for prevention and/or treatment, while others are for control. Veterinarians should ensure they are familiar with the individual drug labels prior to prescribing. The Provincial Apiary Programs, in many instances, have recommendations on antibiotic use within their provinces. There is variability between the provinces in these recommendations, and also in the protocols for disease inspection and management, particularly in regard to AFB. Another major difference of importance to veterinarians is with respect to the VCPR. In Canada, in order for a veterinarian to prescribe antibiotics to a beekeeper they must establish a VCPR, as required by the Provincial Veterinary Regulatory Body. Some Provincial Veterinary Regulatory bodies have put in an exception for honey bees. This means that the requirements for a valid VCPR with respect to honey bees may be different than the standard VCPR requirements for all other species of veterinary concern. It is important for veterinarians working with honey bees in Canada to understand the specific VCPR requirements within their province (Figure 15.4).

Thank you to Dr. Britteny Kyle for the Information on Canadian regulations.

Managing AFB: A Beekeeper-Veterinarian Discussion Guide

While consulting with your veterinarian, they may ask you to describe the methods you use to prevent American foulbrood (AFB) in your colonies, and ensure you know what to do should a suspected case be found. This Discussion Guide is intended to help facilitate that conversation, as well as assist in identifying areas where you are proactive and potentially areas where you can improve.

Biosecurity for AFB Prevention

Check all that you do in your hive management / beekeeping operation:

- ☐ Regularly disinfect hive tool(s) and smoker(s). Describe the method(s) you use: ______
- ☐ Inspect tools and equipment when moving between bee yards. Describe the method(s) you use: ______
- ☐ When using gloves, they are disposable and a new pair is worn at each location
- ☐ Have a comb replacement strategy in place. Describe: ______
- ☐ Purchase bees, queens, or used equipment only from sellers with a permit issued by OMAFRA.

Have you taken the Biosecurity workshop available through the Ontario Beekeepers' Association (OBA) Tech Transfer Program (TTP)? If so, what year? ______

Monitoring for AFB

I am, and/or individuals conducting colony inspections on my hives are, familiar with the signs and symptoms of clinical AFB: Yes No I'm not sure

I and/or individuals conducting colony inspections on my hives have taken Integrated Pest Management (IPM) workshops available through the Ontario Beekeepers' Association (OBA) Tech Transfer Program (TTP)? If so, what year?: ______

Reporting and Disposal of Clinical AFB

If you discovered a suspected case of AFB in your colony(ies), describe the steps you would take[1]:

______

---

[1] Requirements for reporting suspected cases of AFB can be found here: http://www.omafra.gov.on.ca/english/food/inspection/bees/afb-mgmt.htm

Figure 15.4 Example of a Canadian provincial (Ontario) VCPR form. *Source:* Courtesy of Britteny Kyle.

Establishing a Veterinarian-Client-Patient Relationship (VCPR)
A Guide for Ontario Beekeepers

The Veterinarian-Client-Patient Relationship (VCPR) describes the professional working relationship between a veterinarian, a client, and the patient. This must be established before a veterinarian can extend professional services to a client, except in emergency circumstances. In order to establish the VCPR with a beekeeper the veterinarian must, at a minimum:

- confirm the provincial registration of the beekeeper
- confirm the number of colonies held by the beekeeper
- confirm the production management practices of the beekeeper
- confirm the standard operating procedure / protocol for use in a disease requiring an antimicrobial drug[1]

Requesting additional information is at the discretion of the veterinarian. This Guide has been developed to assist beekeepers in Ontario prepare for their initial discussion with a veterinarian. More resources for beekeepers looking to access antibiotics can be found here: [OBA link to "Resources for Antibiotic Access"].

Name: ____________________ Company Name (if applicable): ____________________
Address: ____________________ Phone 1: ____________________
____________________ Phone 2: ____________________
____________________ Email: ____________________

Beekeeper Registration Number[2]: ____________________
Number of colonies to be preventatively treated in spring: __________
Number of colonies to be preventatively treated in the fall: __________
Total[3]: __________

Production management practices (check all that apply):

- ☐ Honey producer
- ☐ Queen / nuc producer
- ☐ Pollination services provider
- ☐ Other (specify): ____________________

In your beekeeping management, do you incorporate[4]

- ☐ Biosecurity for AFB prevention
- ☐ Monitoring for AFB
- ☐ Reporting and disposal of clinical AFB

---

[1] This criteria for SOPs and protocol for use will be determined during discussions with your veterinarian.

[2] By law, all beekeepers in Ontario must register their colonies on an annual basis with the Ontario Ministry of Agriculture, Food and Rural Affairs (OMAFRA). For more information, and to register your colonies, follow this link: http://www.omafra.gov.on.ca/english/food/inspection/bees/info_registration.htm

[3] At the discretion of the veterinarian, prescriptions may be obtained for one treatment period (spring **or** fall), or for two treatment periods (spring **and** fall).

[4] At the discretion of the veterinarian, prescriptions may be obtained for one treatment period (spring **or** fall), or for two treatment periods (spring **and** fall).

Figure 15.4 (Continued)

References

Alippi, A.M., López, A.C., Reynaldi, F.J. et al. (2007). Evidence for plasmid-mediated tetracycline resistance in *Paenibacillus larvae*, the causal agent of American Foulbrood (AFB) disease in honeybees. *Veterinary Microbiology* 125: 290–303. https://doi.org/10.1016/J.VETMIC.2007.05.018.

Evans, J.D. (2003). Diverse origins of tetracycline resistance in the honey bee bacterial pathogen *Paenibacillus larvae*. *Journal of Invertebrate Pathology* 83: 46–50. https://doi.org/10.1016/S0022-2011(03)00039-9.

Miyagi, T., Peng, C.Y.S., Chuang, R.Y. et al. (2000). Verification of oxytetracycline-resistant American foulbrood pathogen *Paenibacillus larvae* in the United States. *Journal of Invertebrate Pathology*. http://dx.doi.org/10.1006/jipa.1999.4888.

16

Medical Records

Marcie Logsdon[1] *and Terry Ryan Kane*[2]

[1] *Department of Veterinary Clinical Sciences, College of Veterinary Medicine, Pullman, WA, USA*
[2] *A2 Bee Vet, Ann Arbor, MI, USA*

Figure 16.1 A beekeeper performing a routine hive inspection using an electronic tablet to record findings.

With the advent of the Veterinary Feed Directive (VFD), responsibility for antibiotic treatment decisions regarding honeybee colonies shifted from apiarists to veterinarians. The requirements for establishment of a valid Veterinary Client Patient Relationship (VCPR) and associated record keeping are now the same as for any other veterinary species. Although the legal burden for record keeping falls to the veterinarian, adequate colony health tracking requires participation from both the veterinarian and the apiarist (Figure 16.1).

Veterinary Client Patient Relationship (VCPR)

As with any other food producing animal, a veterinarian's job is not just overseeing antibiotic use and writing VFDs/Rxs. Bee doctors must know biology, behavior, diseases, and management issues for this species. Veterinarians bring significant value to honeybee health; many herd health management tools apply. Veterinary services include a number of important aspects of bee husbandry: education, recordkeeping, integrated pest management, pesticide review/use, nutrition, food safety of hive products, disease management, and biosecurity.

A VCPR denotes a mutual agreement established between an owner and a veterinarian. In it, the veterinarian agrees to treat and provide care for their animal (or herd or colony) to the best of their ability and the owner agrees to follow the veterinarian's guidance on the administration of medications and treatments. The American Veterinary Medical Association (AVMA) Board of Veterinary Governors describes the basics of a VCPR in their Principles of Veterinary Ethics guiding document. All practicing veterinarians in the United States are subject to these ethical guidelines. Additionally, most states (and provinces) have established a legal working definition of a VCPR. Recommending veterinary treatments or prescribing medications without a valid VCPR can result in sanctions against a veterinarian's license and may carry legal ramifications (https://www.fda.gov/animal-veterinary/development-approval-process/does-state-or-federal-vcpr-definition-apply-lawful-vfd-my-state).

Many owners are unaware of the reciprocal nature of a VCPR and, for years, most honey bee producers have self-medicated their hives. Because of this, many honeybee veterinarians require clients to sign a VCPR document or medication use agreement before dispensing treatments. A sample document is provided in Chapter 15.

The field of telehealth is rapidly developing and affecting the landscape of veterinary medicine as new technologies are developed and incorporated. Telemedicine uses digital tools

Honey Bee Medicine for the Veterinary Practitioner, First Edition. Edited by Terry Ryan Kane and Cynthia M. Faux.

to exchange medical information electronically using modalities such as Skype, phone apps, tablets, digital cameras, and videos, to communicate with clients and visualize patient issues. Telemedicine has the potential to enhance our ability to care for patient health, particularly for follow up "virtual exams" of the bee yard, hives, frames, and bees. Telemedicine can only be practiced when a legal VCPR exists; a VCPR cannot be established by telephone or electronic means. Without a VCPR a veterinarian may provide general health advice but must specifically avoid diagnosing, prognosing, or treating; thus, digital communications should be in the non-clinical realm of general advice. Always consult your state (and provincial) regulations regarding the rules, constraints, and ethical considerations of telemedicine. The AVMA provides a Telehealth Resource Center to help veterinarians implement telehealth options and stay abreast of new developments (http://www.avma.org/telehealth).

It is advisable to know the names and contact information for the state apiarists, assuming your state has one designated (https://apiaryinspectors.org/us-inspection-services and for Canada, https://apiaryinspectors.org/provincial-inspection-services).

Be aware of legislative issues that impact managed bees and producer's interests. With the increasing popularity of beekeeping, more cities, counties, and states have ordinances affecting beekeepers including registration requirements, inspection requirements, limitations on the maximum number of hives allowed, and location restrictions. Be familiar with new farm bill programs benefiting beekeepers, the honey market, labeling requirements, pesticide and labor issues, and any state pollinator protection plans.

Hive Records

Veterinarians should encourage beekeepers (commercial, sideliners, as well as hobbyists) to maintain good records of their hives. Bricks stacked in various configurations (up, down, sideways) while useful as hive markers and providing dual function as weights to keep the hive cover on, are not records. Colored buttons, markers, or tags are not records, even if the beekeeper understands what they mean. Hive records should be able to be examined, and understood, by veterinarians and/or inspectors. Recording the data for beekeeping practices will only enhance colony management and may help to determine the causes of health issues. Also, hive data can be pooled for the benefit of all beekeepers by submitting information to Bee Informed Partnership's annual Mitecheck and Colony Loss surveys.

Hive records are written or digital notes on each hive. These records can be incorporated, as needed, in the veterinary medical record and can provide important history and background to the veterinarian, especially when examining an operation for the first time. Many resources are available to new beekeepers to help them establish a habit of regular record keeping. Smartphone and web-based apps are now available and many offer "starter" and "pro" versions to accommodate both small- and large-scale operations. Alternatively, many organizations, bee clubs, and beehive equipment suppliers offer free downloadable colony heath forms or hive recording forms. Large scale producers may prefer to establish their own record format to better suit their needs. Whatever method is selected, ideally a record entry should be made every time the hive is opened.

For the beekeeper, records should not be so complex that they are disincentivized from using them. Records are supposed to improve beekeeping practices and to also provide useful information supporting the goals of the apiary. A hobbyist with few hives may only need a few note pages. Apiarists with multiple bee yards, hundreds of hives, mating nucs, queen rearing/breeding programs, or are transporting hives for pollination will require a more manageable system.

A record should be kept for each hive and the following criteria should be included:

- Date of inspection and time of day
- Weather/ambient temperature
- Hive Identifier
- Queen status
- Brood status
- Amount of food stores
- Space available for colony
- Presence/absence external pests/parasites
- Mite count data
- Temperament (particularly in some areas of the country)
- Removal/addition of any supers
- Any food added
- Any treatments performed
- General notes – important to take into account seasonality of bee population. Relative to food supply, i.e. honey flow? dearth? What is plant biodiversity within 1–3 km. Water source?

We advise exploration of the numerous examples of hive inspection forms, written and/or digital. Recommendations can be tailored to the apiary size and needs. Hive technology and colony management tools/apps are in constant development.

Medical Records

"If it isn't documented, it didn't happen." Generation of a complete problem oriented medical record is not only required of the veterinarian but will also streamline consultations with other veterinarians if needed. Although there are species-specific considerations, honeybee medical records should first and foremost conform to the basics of a sound veterinary medical record. Also, all veterinary records should conform to the applicable state and/or country administrative codes. A sound medical record, regardless of species, should contain:

- **Date**. All record entries should contain the day, month, and year.
- **Owner/operation information**. This should include current address, including E911 address, and contact information.
- **Type of Operation and Goal of Operation**
- **Macro and microenvironment (1–3 km)**
- **Biosecurity**
- **Patient information and signalment**. This should include any appropriate identifiers. While for companion animals and pets this usually comprises name/breed/age/reproductive status and microchip or tattoo information, a hive descriptor or identification number is usually sufficient in honeybee operations. Some commercial hives are now equipped with GPS tracking. Additional information for a particular hive can include information such as subspecies/strain of bee, queen age and number, if she is marked.
- **Presenting complaint**. What is the reason they are consulting with a veterinarian?
  - What is the length of duration of the problem?
  - Have they encountered this issue in past seasons?
  - How many hives are affected?
  - Has the owner tried any previous treatments to address the issue, and if so, have they seen any improvement?
- **History**. A thorough operation history should be obtained when establishing a VCPR and should be updated annually, or when addressing a new complaint. The following honeybee-specific questions should be obtained when gathering a history:
  - Hive type (Langstroth, Top Bar, Warre, display, other).
  - Bee use (backyard/hobby – no products sold at all, commercial pollination, honey production, nuc breeder, education/display, other). The end product of the hives will have important treatment implications regarding the use of medications and withdrawal times. Hives used for commercial pollination often travel large distances and may have a higher risk of exposure to varroa mites and other parasites.
  - Age of the hive.
  - Length of ownership.
  - How many hives are on the premises? If hives are housed at more than one location, what is the total number of hives managed?
  - How were the bees acquired (wild caught/swarm, breeder, local apiary, etc.)? If they were purchased from a breeder, what variety/species? Were they marketed as a hygienic strain? As varroa-resistant? If a recent acquisition, did they purchase a "package" of bees (worker bees with an as-yet unintroduced queen), or a "nuc" (nucleus colony containing worker bees, established queen, and a store of honey and brood).
  - Has there been any recent acquisition of bees?
  - What is the travel history of the hives?
  - When was the last addition of a queen and what was the source?
  - Has there been any pesticide application in the surrounding area?
  - How long has the owner raised bees?
- **Exam findings**. This portion of the record will closely resemble a beekeeper's hive inspection report. Comments should be made on the location of the hive, presence of the queen, brood status, presence of any parasites, indications of current treatments such as grease patties or mite treatment strips, and any signs of disease or poor health.
- **Treatments performed or prescribed**. This includes recommendations for "at home" remedies as well. Be sure to include instructions given to the owner including treatment amount, administration route, duration of treatment and withdrawal times.
- **Samples collected**. If bees were collected for testing this should be recorded along with number and source (hive identifier).
- **Laboratory or other test results**. These will usually comprise a later entry made into the record once results are available. If a send-out test, be sure to include the laboratory.
- **Client communications**. These can be encompassed chronologically within the relevant problem write ups or occur in a separate document. Electronic or written communications such as emails are easily added in their entirety while additional effort must be made to record in-person or phone conversations.
- **Written discharge/instructions/summary**. Providing the client with a written copy of the diagnosis and treatment recommendations will improve client compliance. Keeping a copy of information given to the client is just good record-keeping.

Veterinary Feed Directives/Prescriptions

Administration of medications to hives consists predominantly of combining medications with powdered or cane sugar, sugar water, or supplemental feed. In the United States, use of medicated feeds containing medically important antibiotics now requires issuance of a Veterinary Feed

Directive or a prescription. We recommend you refer to the Food Animal Residue Databank (FARAD) for the most up to date drug information for honey bees (http://www.farad.org/vetgram/honeybees.asp). The Federal Drug Administration (FDA) site is available at https://www.fda.gov/animal-veterinary/development-approval-process/using-medically-important-antimicrobials-bees-questions-and-answers. Honeybees are classified as a minor species. The labels were written for the major livestock species. Check the label for honeybee application and withdrawal times.

Although there are many similarities, VFDs differ from veterinary prescriptions. As with prescriptions, the FDA requires the presence of a valid Veterinary-Client-Patient-Relationship for issuance of a VFD. In states with no superseding VCPR requirements the federal definition applies. Only licensed veterinarians are allowed to issue VFDs. The FDA provides clear guidelines on the information that must be provided on a VFD and the AVMA offers a template for veterinary use. The forms and the FDA requirements for information that must be found on a VFD are reviewed in Chapter 15.

Some VFD suppliers are now offering digital VFD record keeping.

Bee veterinarians can also use a digital spreadsheet for apiary drug records.

Resources

AVMA State Summary Report Records Retention, https://www.avma.org/advocacy/state-local-issues/records-retention

Bee Culture, https://www.beeculture.com/electronic-record-keeping-the-path-to-better-beekeeping, https://www.beeculture.com/better-beekeeping-with-better-records

Bee Informed Partnership (BIP), https://beeinformed.org/2013/07/12/keeping-records, https://beeinformed.org/citizen-science/mitecheck, https://beeinformed.org/citizen-science/loss-and-management-survey

Bees Need Vets, https://pollinators.msu.edu/programs/bees-need-vets

Hive Tracks, https://go.hivetracks.com/commercial

Keep Bees Alive, https://pollinators.msu.edu/keep-bees-alive

National Veterinary Accreditation Program, https://nvap.aphis.usda.gov/VFD/index.htm

Sweet Pines Apiary, https://www.tianca.com/record-keeping

Veterinary Feed Directive Requirements for Veterinarians, https://www.fda.gov/animal-veterinary/development-approval-process/veterinary-feed-directive-requirements-veterinarians

VPCR Relationship, AVMA, https://www.avma.org/policies/principles-veterinary-medical-ethics-avma#III (accessed 25 January 2020)

Appendix 16.1A

Veterinary – Client Management Agreement

I. **Producer**

Owner's Name ______________________________

Apiary Name ______________________________

Address ______________________________

____

E 911 address or GPS location ______________________________

Contact Information: Phone ______________________________

email ______________________________

website ______________________________

Type of Operation ______________________________

Number of Yards ______________________________

Yard under Veterinary Care ______________________________

Number of Hives in Yard ______________________________

Number of Hives under Veterinary Care ______________________________

I have agreed to implement the following management practices as part of bee herd health and to comply with drug usage directions.

- ☐ Maintain Communication
- ☐ Follow Treatment Protocols (IPM and/or drug usage)
- ☐ Keep and Maintain Hive Records

Owner's Signature ______________________ Date ____________________

II. **Veterinarian**

Name __________________________ Practice/Clinic Name__________________

Address ______________________________

Phone (s)______________________________

I and/or other personnel from my veterinary practice will work with the above apiary as needed.

Veterinarian Signature_________________________________Date_________

Additional Clarifications/Comments:

____

____

____

____

____

____

____

____

____

____

____

____

Appendix 16.1B

Sample Hive Record

Hive Record Form

Name of Operation ______________________________

Date	Ambient Temp.	Hive #	Queen Status	Brood Status	Food Stores	Food Added	External Parasites	Disposition	Supers +/-	Treatments/ General Notes

17

Epidemiology and Biosecurity

Kristen K. Obbink and James A. Roth

Center for Food Security and Public Health, Iowa State University College of Veterinary Medicine, Ames, IA, USA

Honey bees serve a vital role in agriculture and food security both nationally and globally. However, their health can be negatively impacted by a variety of pathogens including viruses, bacteria, fungi, protozoa, and parasites. Like other social organisms, honey bees are at risk for pathogen transmission through numerous routes, and co-infections with multiple pathogens are not uncommon (Figure 17.1).

Disease management in honey bees requires a combination of best management practices (BMPs) that keep bees healthy, enhance disease resistance, and provides biosecurity to reduce disease exposure. Since honey bees are free ranging, it is not possible to completely exclude exposure to endemic diseases through biosecurity; however, a combination of good management practices and improved biosecurity can reduce the likelihood, and consequences, of infection in a colony. Knowledge and implementation of BMPs with an emphasis on good biosecurity is a beekeeper's best defense in protecting the health and vitality of their colonies.

This presents veterinarians with an opportunity to provide education and professional services to beekeepers, particularly hobbyists and sideline apiarists who often have less experience and could benefit from professional assistance. As food-producing animals, bees are considered livestock by the US Food and Drug Administration (FDA) as well as by the Food and Agriculture Organization (FAO) of the United Nations and by the World Organization for Animal Health (OIE). Indeed, FAO and OIE state that livestock owners should establish a working relationship with a veterinarian to ensure animal health and welfare as well as prompt identification and reporting of disease issues (OIE and FAO 2009; OIE 2014).

Apis mellifera Disease Epidemiology

Although not all inclusive, this chapter will focus on those pathogens of interest to *Apis mellifera* that had the highest potential or actual impact on the US honey bee industry at the time of writing. Specific details on these and other honey bee diseases can be found in Section 17.1.3 of this text. The US Department of Agriculture (USDA) Agricultural Research Service (ARS) Bee Research Laboratory in Beltsville, MD, offers bee disease diagnostic services for the US and its territories.

Following large-scale, unexplained losses of managed US honey bee colonies during the winters of 2006–2007 and 2007–2008 (vanEngelsdorp et al. 2007, 2008), investigators identified the cause as a multifactorial, common set of clinical signs that were termed colony collapse disorder (CCD) (vanEngelsdorp et al. 2009). During 2009, in the aftermath of CCD, the USDA Animal and Plant Health Inspection Service (APHIS) began conducting the National Honey Bee Disease Survey (National Survey) to establish an epidemiologic honey bee disease baseline in the US and its territories and verify the absence of exotic threats to the honey bee. This annual survey, funded by USDA-APHIS and conducted in collaboration with the University of Maryland, USDA ARS, and State Apiary Specialists, has expanded from three participating states in 2009 to 38 states and territories in 2019 and represents the most comprehensive data set of US honey bee pest and health information to date (USDA-APHIS Honey Bee Pests and Diseases Survey Project Plan for 2019 (2019)). All survey data, including historic and ongoing research, is incorporated into the nationwide Bee Informed Partnership (BIP) database. Results of the

Honey Bee Medicine for the Veterinary Practitioner, First Edition. Edited by Terry Ryan Kane and Cynthia M. Faux.

Figure 17.1 Warning sign. *Source:* Photo by Cynthia Faux.

National Survey have and will continue to provide valuable insight into the epidemiology of honey bee diseases.

Exotic Diseases of Honey Bees

Longitudinal monitoring through the National Survey has continued to provide strong evidence for the absence of important exotic diseases of honey bees (*Tropilaelaps* spp., *Apis cerana*, and Slow Bee Paralysis Virus [SBPV]) in managed US honey bee colonies (USDA-APHIS Honey Bee Pests and Diseases Survey Project Plan for 2019 (2019)). *Tropilaelaps* mites are an ectoparasite of immature stages of honey bees. Infestations are regulated worldwide for trade purposes and are reportable in uninfested regions, including the US (Vidal-Naquet 2015). *A. cerana*, or the Asian honey bee, has extended beyond its native range of southern Asia. It is considered an invasive species in many parts of the world because it serves as a natural host to both *Varroa* and *Nosema* spp. so could increase the spread of these parasites and also has the potential to compete with *A. mellifera* for nectar, pollen, nesting sites, and other vital resources (Plant Health Australia, Hort Innovation, AgriFutures, and Wheen Bee Foundation 2016; Egelie et al. 2015). SBPV causes paralysis of the forelegs of bees and is vectored by the *Varroa* mite which transmits the virus to both pupae and adult bees (Santillàn-Galicia et al. 2010). This virus has been linked to high mortality of colonies infested with the *Varroa* mite (Carreck et al. 2010; Martin et al. 1998).

Absence of these potentially devastating exotic pathogens suggests that current methods to prevent their introduction into the US have been successful; however, continued vigilance is paramount to the protection of the honey bee industry.

Varroa destructor

The parasitic mite, *Varroa destructor*, is the greatest single threat to managed colonies of *A. mellifera* on a global scale (Rosenkranz et al. 2010). Originally, *A. cerana* served as the natural host of the *Varroa* mite; however, during the middle of the twentieth century, *V. destructor* transferred to a new host, the honey bee *A. mellifera* (Vidal-Naquet 2015). *Varroa* are now ubiquitously present in US honey bee colonies and have a dramatic impact on honey bee health, not only due to the direct effects of their feeding on brood and adult bees but also their ability to serve as a vector of numerous viruses. According to samples collected by Fahey et al. through the 2016–2017 National Survey, the presence of *V. destructor* mites has remained relatively constant in survey samples since 2010, with an average of 90% of samples testing positive each year. The average load of *Varroa* peaked during the 2012–2013 survey year at 5.5 mites per 100 bees and has gradually decreased since that time, now averaging 3.3 mites per 100 bees. While *Varroa* load has decreased over time, there has been little to no change in prevalence. Possible explanations include the success of nationwide outreach and extension efforts targeting beekeepers regarding appropriate monitoring and treatment of *Varroa*. Alternatively, viruses vectored by *V. destructor* may have become more virulent, resulting in higher colony loss and thus a drop in mite populations. (Fahey et al. 2016–2017).

Nosemosis

Nosemosis, or nosema disease, can affect all three castes of honey bees and is caused by two species of spore-forming microsporidian parasites, *Nosema apis* and *Nosema ceranae* (Vidal-Naquet 2015). According to the 2016–2017 National Survey, *Nosema* spp. spore prevalence has remained historically consistent, being detected on average in 50% of samples. Similar to *V. destructor*, the average load of *Nosema* spp. has decreased over time in obtained samples while prevalence has remained about the same. It should be noted that in 2013, speciation of *N. apis* and *N. ceranae* was discontinued after no detections of *N. apis* from 2009 to 2010, detections of 1.3% in 2011, and 0.7% in 2012. Therefore, all *Nosema* counts since 2013 are attributed to *N. ceranae* (Fahey et al. 2016–2017). The average *Nosema* spore load during the 2016–2017 survey revealed

0.54 million spores per bee which is slightly lower than the previous five years of the survey where *Nosema* spore load averaged 0.66 million spores per bee. This trend will continue to be monitored through the survey in coming years to determine its significance to overall honey bee health.

Multi-year Disease Baseline

In addition to annual reports, endemic disease results from the 2009–2014 National Survey (including *V. destructor*, *Nosema* spp., and eight honey bee viruses) were assessed and quantified by Traynor et al. (2015) providing a multi-year disease baseline to assist in identifying drivers of poor bee health. Several observations were made including the identification of significant differences in disease prevalence between migratory hives (those transported to different locations for pollination services) and stationary hives. Migratory beekeepers had significantly lower *Varroa* prevalence and loads than stationary colonies, while the opposite was seen with *Nosema* spp. which were more prevalent in migratory colonies. One hypothesis offered by the authors was that migratory beekeepers may be treating for mites more frequently or that the physical movement of transporting bees for pollination purposes may somehow interfere with mite reproduction. Seasonal differences were also noted in *V. destructor* and *Nosema* spp. loads. *Nosema* spore counts peaked annually in April, when colonies are often nutritionally stressed after the winter months and bees may be confined within the hive due to spring rains. Conversely, *Varroa* infestations peaked during the months of August through November, a critical time of year in temperate climates when colonies must rear bees in preparation of winter. Not only does this result in poor overwinter colony survival rates, but without intervention from the beekeeper, the collapse of highly infested colonies has been demonstrated to cause neighboring colonies to experience a surge in mite populations. Peck and Seeley (2019) found that the collapsing hives likely serve as "robber lures" for neighboring hives, thus spreading mites to robber bees which then transport them back to their original hive. In addition, mite-infested workers from collapsing colonies that drift into neighboring colonies also likely play a role in *Varroa* spread. An important biosecurity practice, therefore, is to identify and remove collapsing colonies early.

Honey Bee Viruses

Viral diseases are a major threat to honey bee health and the ability of the *Varroa* mite to vector a multitude of viruses has been a major driver of honey bee decline (Brosi et al. 2018; Traynor et al. 2015). Honey bee viruses can infect all developmental stages of honey bees including the egg, brood, and adults and colonies can be co-infected with multiple viral pathogens simultaneously (Chen et al. 2005; Chen et al. 2006; Shen et al. 2005). For example, Traynor et al. (2015) found that *Varroa* levels were significantly elevated in sampled bees that concurrently tested positive for both acute bee paralysis virus (ABPV) and deformed wing virus (DWV). Both ABPV and DWV increased in viral load linearly with mite levels, indicating that mite presence is directly linked to viral replication for these two viruses (Traynor et al. 2015). These complex host–parasite interactions are a source of increasing concern and intensified research in the beekeeping community.

Viral screens conducted through the National Survey suggested a concerning trend of escalation in the prevalence of several viruses during 2009–2014 with black queen cell virus (BQCV), chronic bee paralysis virus (CBPV), Kashmir bee virus (KBV), and Lake Sinai Virus 2 all increasing in prevalence during this time (Traynor et al. 2015). In addition, CBPV was undetected in 2009 but doubled annually between 2009 and 2014 (Traynor et al. 2015). This increasing trend in the prevalence of numerous viruses may suggest that the honey bee's immune system is compromised resulting in the inability to protect itself against a number of stressors including but not limited to decreasing habitat, increased pressure from pesticides, and poor nutrition (Archer et al. 2014; Bryden et al. 2013; Higes et al. 2009; Pettis et al. 2013; Sanchez-Bayo and Goka 2014; Simon-Delso et al. 2014; van der Sluijs et al. 2013), ultimately leading to increased colony mortality (Johnson et al. 2010; Spleen et al. 2013; Steinhauer et al. 2014; vanEngelsdorp et al. 2008, 2012).

Viral Transmission

According to Chen et al. (2006), transmission processes are a crucial aspect of the dynamics of viral infections and determine the spread and persistence of disease in a population. Viral transmission generally occurs through either horizontal or vertical routes and honey bee viruses utilize both of these strategies (Amiri et al. 2017; Chen et al. 2005; Chen et al. 2006; Shen et al. 2005). Horizontal transmission can be either direct or indirect. Direct horizontal transmission includes airborne, foodborne, or venereal routes whereas indirect horizontal transmission involves an intermediate biological host, such as a mite vector, which acquires the virus and then transmits it from one host to another. This type of transmission occurs among individuals of the same generation. Vertical transmission takes place when viruses are passed from mother to offspring via the egg, either within the egg (transovarian) or on the surface of the egg (transovum). Horizontal transmission typically favors apparent disease expression and infection prevalence often increases in conditions of high host population density and high pathogen replication rates that result in increased opportunity for viruses to spread from organism to organism. Conversely, vertical transmission allows for long-term persistence of the virus within the population and favors a more benign disease process which allows offspring to survive and continue to propagate spread of the virus on to the next generation. Viral infections within a population often reflect a complex balance in evolutionary trade-offs between the two transmission processes. (Chen et al. 2006)

Knowledge and Management Practices as Drivers of Honey Bee Health

Despite increased knowledge of honey bee disease epidemiology as well as numerous local, regional, and national initiatives to promote pollinator health, a comprehensive understanding of the drivers of poor honey bee health and colony loss remains minimal. Specifically, the impact of beekeeper knowledge and management practices on colony health is poorly understood. Jacques et al. (2016) conducted a descriptive epidemiological study across 17 European countries to identify key risk factors surrounding honey bee colony mortality. Results demonstrated that the winter mortality rate of hobbyist beekeepers with small apiaries and little beekeeping experience was double that of professional beekeepers. In addition, observed disease prevalence, including *Varroa* infestation, was significantly higher for hobbyist beekeepers than professional beekeepers. Overall, the researchers concluded that while climactic conditions and other biological variables were drivers of honey bee health, the main factors in determining the health of honey bee colonies were beekeeper background and management practices. Thus, they recommended a stronger focus on the need for beekeeper training to promote good management practices and enable beekeepers to detect clinical signs of disease earlier and intervene appropriately.

Biosecurity

As demonstrated by Jacques et al. (2016), the successful management of honey bee colonies requires knowledge and implementation of good husbandry practices, including a focus on biosecurity. The field of biosecurity as it relates to honey bee veterinary medicine serves as an opportunity for veterinarians to provide biosecurity education and audits to beekeeping clients as part of a comprehensive honey bee health program. To assist veterinarians and beekeepers in conducting such an audit, a *Beekeeping Biosecurity and Best Practices Checklist* is included at the conclusion of this chapter as Appendix 17.A and can also be downloaded as a fillable PDF from the Center for Food Security and Public Health at the following web address: www.cfsph.iastate.edu.

Biosecurity is defined as a series of management practices used to prevent the introduction and spread of pathogenic agents (Center for Food Security and Public Health 2015). Just as with other animal and livestock premises, biosecurity principles can be implemented in apiaries to protect honey bee colony health. Failure to apply good biosecurity techniques can result in the introduction of exotic pathogens or the establishment of endemic pathogens which may spread to surrounding apiaries and could have detrimental effects on the honey bee industry at the regional, national, and even international level. As such, the responsibility of practicing good biosecurity begins at the individual level and is in the best interest of every beekeeper; however, the protective effects are cumulative, widespread, and serve the greater good of the beekeeping and agricultural industries.

General Management Practices for Biosecurity

Training

All beekeepers and their employees should obtain training on BMPs and should make every effort to remain current on evolving developments in the beekeeping industry. Over the years, a number of organizations have developed resources on BMPs for honey bees. One such resource compiled by the Honey Bee Health Coalition (2019) offers valuable BMPs on several topics including beekeeper safety, honey bee nutrition, hive and apiary set up and maintenance, pesticide exposure, treatment of parasites and diseases, and queen health, breeding, and stock selection. This guide, entitled "Best Management Practices for Hive Health: A Guide for Beekeepers" can be accessed free of charge at the following website: https://honeybeehealthcoalition.org/hivehealthbmps. It is also recommended that beginning beekeepers complete a beekeeping course and join local and state beekeeping associations in their area as this can serve as an excellent networking opportunity and allow beginners to learn from more experienced and knowledgeable beekeepers.

According to OIE and FAO (2009), beekeepers should actively seek and complete relevant training opportunities and keep records of all training completed by them and their employees. As part of this training, beekeepers should be aware of potential exotic and endemic threats to honey bee health in their area, and areas where their hives may be transported, as well as regulations, reporting requirements, and legal obligations for detection of certain pathogens (Plant Health Australia, Hort Innovation, AgriFutures, and Wheen Bee Foundation 2016; Vidal-Naquet 2015).

Record Keeping and Traceability

Record keeping is essential for good colony management and allows the beekeeper to better understand and identify the root cause of health or sanitary issues. Records should be kept on the following:

- Number of hives and apiaries (and identification of individual hives),
- Introduction of new stock whether originating through purchased packages or nucleus colonies, splits, or swarms,
- Migratory beekeeping movements, including routes taken and dates,
- Date, origin, and type of feeding supplements administered,
- Date, origin, dose, and use of any natural or medicinal treatments administered,
- Date, origin, and use of any chemicals or cleaning products used on hive equipment, and
- Notes and findings of all hive inspections conducted, including but not limited to normal and abnormal findings, suspected/diagnosed diseases, mortalities, colonies affected, and methods to correct/remove problems (Vidal-Naquet 2015).

In addition, all colonies and apiaries should be marked with unique identification to allow for traceability through record keeping. Queens should also be marked according to the international color code to allow for traceability (if she leaves with a swarm) and easier identification during hive inspections (Vidal-Naquet 2015). There are five internationally recognized marking colors for queens with the color sequence depending on the year that the queen was born. Because queens rarely live more than five years, the color code starts over again in year six. Marking queens according to the international color code not only allows for them to be more easily identified within a hive but also allows for a consistent method within the beekeeping industry to indicate which year the queen was introduced into the colony (Piedmont Beekeepers' Association 2017).

Apiary Placement and Signage

The location of the apiary and hives within it will play an important role in colony health. To promote healthy colonies, the beekeeper should:

- Ensure easy, but controlled, access to the apiary,
- Ensure the apiary and surrounding area is well maintained,
- Ensure that hives can be placed in a manner in which they will be protected from inclement weather and other hazards to the extent possible,
- Ensure the presence of diverse, natural food sources,
- Ensure the presence of a good quality water supply,
- Know, if possible, of the farming practices used nearby, including pesticide use, and
- Know, if possible, the presence of nearby colonies and the management practices utilized in those apiaries (Vidal-Naquet 2015).

Signage is also important and should be well designed and clearly posted. Biosecurity signs providing the apiary owner's name and contact information should be placed at apiary or property entrances to alert people to contact the apiary owner and obtain permission before entering the property. The apiary owner's name and contact information will also be essential in the event of a biosecurity incident (such as an exotic pest detection) or emergency. In the case of migratory colonies, signs should accompany hives to their new location. (Plant Health Australia, Hort Innovation, AgriFutures, and Wheen Bee Foundation 2016)

Sourcing

Proper sourcing of stock, equipment, and supplementary feed products is an important part of a good biosecurity program and will play a vital role in the foundational health of the colony. All purchased queens and bees should come from trusted sources and should be pest free to the extent possible (Plant Health Australia, Hort Innovation, AgriFutures, and Wheen Bee Foundation 2016). Thought should also be given to the honey bee strain chosen to ensure it is appropriate for the environment in which it will be placed and the pathogens to which it will potentially be exposed (Vidal-Naquet 2015). Strains vary in a number of characteristics, including but not limited to their temperament, ability to build up honey stores, hygienic behavior, disease resistance, and ability to overwinter. Honey bees in the US and Canada are a genetic blend of several strains/subspecies introduced from Europe and many new hybrids have been introduced in an attempt to optimize the available genetics (Mid-Atlantic Apiculture Research and Extension Consortium (MAAREC) 2020). It is important to holistically consider honey bee genetic traits when determining which strain is best for the environment in which the hives will be placed. Just as with purchased stock, wild swarms can serve as a potential hazard as they pose the risk of disease introduction. Recovered swarms should be isolated from the apiary until a thorough inspection can be conducted and any necessary treatments, such as miticides, can be applied (Vidal-Naquet 2015).

Hive equipment should also be purchased from a reputable source and the choice of materials used should be appropriate for the local environment. Hives must remain dry inside and any paints or waxes applied to the materials must be safe and compliant with regulations for food-producing animals (Vidal-Naquet 2015). Borrowed and secondhand equipment should also be cleaned and disinfected thoroughly before use (Plant Health Australia, Hort Innovation, AgriFutures, and Wheen Bee Foundation 2016). Equipment must also be maintained appropriately as the hive boxes, frames, and foundations will deteriorate over time, providing conditions favorable for disease and pests.

Frame management is especially important as pathogenic agents and residues of treatments and pesticides can accumulate over time. Dark wax combs should be removed and hive frames should be renewed every two–three years (Vidal-Naquet 2015). To the extent possible, all frames and supers should be associated with one hive and exchange of hive boxes between colonies and apiaries should be minimized. All empty hives should be immediately removed from the apiary, cleaned to remove visible propolis and other debris, disinfected with a mild bleach solution, and rinsed thoroughly before storing. Equipment should be stored without chemical treatment in a well-ventilated area. Hive boxes with frames should be stored in a crisscross pattern to allow for ventilation and opening in the top and bottom of the box to protect against wax moth infestation (Vidal-Naquet 2015).

Supplementary feed products should be obtained from trusted sources and should be stored and handled appropriately.

Colony Management Practices

Animal management is a key part of any livestock biosecurity program and honey bees are no exception. The biology of honey bees and our inability to protect them from the environment and predation presents unique challenges in this arena. All-in/all-out approaches cannot be adapted to colony management as bees are the one species of livestock where the animal, rather than the farmer, controls the food supply of the colony (Vidal-Naquet 2015). As such, there is an even greater reliance on husbandry practices to ensure the health of the colony.

Hive Inspections

Routine hive inspections are of utmost importance and apiarists must manage hives year-round. Overall activity of the bees both outside and within the hive, brood patterns, pollen and honey storage, and signs of disease and pests should be assessed. Due to their ubiquitous nature in the US and detrimental impacts on colony health, *Varroa* mite infestations should be monitored and managed appropriately throughout the year as part of an integrated pest management program. This component of beekeeping has become increasingly vital to ensure honey bee health and the success of the beekeeping industry.

During hive inspections, any abnormal findings or signs of disease should be recorded and carefully monitored. All regulations and legal obligations to report disease suspicions should be followed as necessary. In some cases, diseased colonies that have the potential to recover may be worth saving. If feasible and allowable by law, these colonies should be isolated from the apiary as this will limit the risk of disease transmission to other healthy colonies within the apiary. Weak hives, or those that may pose a danger to other colonies within the apiary, should be humanely euthanized (Vidal-Naquet 2015).

Sanitation and Hygiene

Practicing good sanitation and hygiene is a key component of any biosecurity program. All small tools and clothing (e.g. gloves, beekeeping suits, hive tools, brushes) should be disinfected with a mild bleach solution or alcohol between inspections of different apiaries and after inspection of any colony that appears to be diseased (Honey Bee Health Coalition 2019; Vidal-Naquet 2015). For the veterinarian, disposable gloves are a preferred option. In addition, workers, visitors, and vehicles can be contaminated and serve as routes of pathogen introduction and transmission to other hives and should be cleaned before entering and leaving the apiary (Plant Health Australia, Hort Innovation, AgriFutures, and Wheen Bee Foundation 2016).

Special Considerations for Migratory Colonies

The movement of hives for pollination services increases the risk of disease and pest spread to other regions, so special precautions should be taken. To the extent possible, movements should be minimized as constant transportation results in increased stress for the colony and makes them more susceptible to disease or infestation. For risk reduction purposes, the following management practices should be implemented whenever hives are transported:

- Before moving any hives or products, contact the destination State Department of Agriculture/Apiary Inspector to determine any requirements for health certification as these vary by state/territory (a list of US Honey Bee Laws by state is available at https://apiaryinspectors.org/state-laws; however, you should always verify their accuracy with the Department of Agriculture/Apiary Inspector prior to transport),
- Assess any potential disease threat that might be posed by poorly managed hives near the new location,

- To minimize stress on the colony, hives should be moved at night or in the early morning when bees are inside,
- Transported colonies should be fed a carbohydrate supplement prior to shipment to prepare them for the journey,
- Be knowledgeable about all endemic and exotic pathogens, as well as disease reporting regulations, in the region to which the hives are being moved,
- Ensure that hives and equipment are covered and secured to prevent robbing by other bees,
- Keep accurate records of all hive movements for traceability purposes, and
- Follow all transportation regulations (Plant Health Australia, Hort Innovation, AgriFutures, and Wheen Bee Foundation 2016; Vidal-Naquet 2015).

Summary

Beekeepers and honey bee veterinarians have an important role to play in ensuring the health of bees and protecting the honey bee industry. As described by Plant Health Australia in their *Biosecurity Manual for Beekeepers* (2016), there are six easy steps that beekeepers and veterinarians can take to reduce threats from exotic and endemic pathogens:

1) Be aware of biosecurity threats in your region including all exotic and endemic pathogens.
2) Use healthy honey bee stock and clean equipment from trusted sources.
3) Practice good sanitation and hygiene and ensure that all workers, visitors, vehicles, and equipment are clean before entering and leaving the apiary.
4) Monitor hives carefully and report any unusual findings. Keep written and photographic records of all abnormal observations as constant vigilance is necessary for early detection of disease.
5) Be knowledgeable of and abide by all laws and reporting regulations in your region.
6) If you suspect an exotic pest or pathogen or other reportable disease, contact your State Apiary Inspector or State Department of Agriculture immediately.

The Role of the Veterinarian in Biosecurity

As noted above, veterinarians working in honey bee medicine can provide valuable services to clients related to biosecurity and management for disease control. Using the biosecurity threats and principles detailed herein, veterinarians should consider developing biosecurity and management checklists with which they could conduct audits of their clients' apiaries (see Appendix 17.A as an example). The goal of the audits would be to provide clients with effective strategies to prevent the introduction of exotic and endemic pathogens and to mitigate the impact of endemic pathogens that enter the hive. This approach would be particularly helpful to inexperienced hobbyist and sideline beekeepers who would likely benefit most from professional assistance with their beekeeping education and management practices.

References

Amiri, E., Kryger, P., Meixner, M. et al. (2017). Quantitative patterns of vertical transmission of deformed wing virus in honey bees. PLoS One 13 (3): e0195283.

Archer, C.R., Pirk, C.W.W., Wright, G.A., and Nicolson, S.W. (2014). Nutrition affects survival in African honeybees exposed to interacting stressors. Functional Ecology 28: 913–923.

Brosi, B., Delaplane, K., Boots, M., and de Roode, J. (2018). Ecological and evolutionary approaches to managing honey bee disease. Nature Ecology and Evolution 1 (9): 1250–1262.

Bryden, J., Gill, R.J., Mitton, R.A.A. et al. (2013). Chronic sublethal stress causes bee colony failure. Ecology Letters 16: 1463–1469.

Carreck, N., Ball, B., and Martin, S. (2010). Honey bee colony collapse and changes in viral prevalence associated with *Varroa destructor*. Journal of Apicultural Research 49: 93–94.

Center for Food Security and Public Health. (2015). Biosecurity: Overview. Just In Time Training. http://www.cfsph.iastate.edu/Emergency-Response/Just-in-Time/03-Biosecurity-Overview_HANDOUT.pdf

Chen, Y., Pettis, J., Collins, A., and Feldlaufer, M. (2005). Prevalence and transmission of honeybee viruses. Applied and Environmental Microbiology 72 (1): 606–611.

Chen, Y., Evans, J., and Feldlaufer, M. (2006). Horizontal and vertical transmission of viruses in the honey bee, *Apir mellifera*. Journal of Invertebrate Pathology 92: 152–159.

Egelie, A., Mortensen, A., Gillett-Kaufman, G., and Ellis, J. (2015). Asian Honey Bee (Apis cerana). Entomology and Nematology Department, University of Florida. Publication Number: EENY-616.

Fahey, R., Rennich, K., Nessa, A.et al. (2016–2017). National Honey Bee Disease Survey Report. USDA-APHIS. https://www.aphis.usda.gov/plant_health/plant_pest_info/

honey_bees/downloads/2016-2017-National-Survey-Report.pdf

Higes, M., Martin-Hernandez, R., Garrido-Bailo, E. et al. (2009). Honeybee colony collapse due to *Nosema ceranae* in professional apiaries. Environmental Microbiology Reports 1: 110–113.

Honey Bee Health Coalition (2019). *Best Management Practices for Hive Health: A Guide For Beekeepers*, 1e. https://honeybeehealthcoalition.org/hivehealthbmps.

Jacques, A., Laurent, M., EPILOBEE Consortium et al. (2016). A pan-European epidemiological study reveals honey bee colony survival depends on beekeeper education and disease control. PLoS One 12 (3): e0172591.

Johnson, R.M., Ellis, M.D., Mullin, C.A., and Frazier, M. (2010). Pesticides and honey bee toxicity – USA. Apidologie 41: 312–331.

Martin, S., Hogarth, A., van Breda, J., and Perrett, J. (1998). A scientific note on *Varroa jacobsoni* Oudemans and the collapse of *Apis mellifera* colonies in the United Kingdom. Apidologie 39: 369–370.

Mid-Atlantic Apiculture Research and Extension Consortium (MAAREC). (2020). Selecting the Right Type of Bee. http://agdev.anr.udel.edu/maarec/beginning-beekeeping-2/selecting-the-right-type-of-bee/

OIE (2014). Bee Health and Veterinarians. Rome, Italy: OIE.

OIE and FAO (2009). Guide to Good Farming Practices for Animal Production Food Safety. Rome, Italy: OIE and FAO.

Peck, D.T. and Seeley, T.D. (2019). Mite bombs or robber lures? The roles of drifting and robbing in *Varroa destructor* transmission from collapsing honey bee colonies to their neighbors. PLoS One 14 (6): e0218392.

Pettis, J.S., Lichtenberg, E.M., Andree, M. et al. (2013). Crop pollination exposes honey bees to pesticides which alters their susceptibility to the gut pathogen *Nosema ceranae*. PLoS One 8 https://doi.org/10.1371/journal.pone.0070182.

Piedmont Beekeepers' Association. (2017). International Queen Bee Marking Colors. https://www.piedmontbeekeepers.com/queen-bee-marking-colors

Plant Health Australia, Hort Innovation, AgriFutures, and Wheen Bee Foundation. (2016). Biosecurity Manual for Beekeepers – Reducing the risk of exotic and established pests affected honey bees. Version 1.1

Rosenkranz, P., Aumeier, P., and Ziegelmann, B. (2010). Biology and control of *Varroa destructor*. Journal of Invertebrate Pathology 103 (Suppl 1): S96–S119.

Sanchez-Bayo, F. and Goka, K. (2014). Pesticide residues and bees – a risk assessment. PLoS One 9: e94482. https://doi.org/10.1371/journal.pone.0094482.

Santillàn-Galicia, M., Ball, B., Clark, S., and Alderson, P. (2010). Transmission of deformed wing virus and slow paralysis virus to adult bees (*Apis mellifera* L.) by *Varroa destructor*. Journal of Apicultural Research 49: 141–148.

Shen, M., Cui, L., Ostiguy, N., and Cox-Foster, D. (2005). Intricate transmission routes and interactions between picorna-like viruses (Kashmire bee virus and sacbrood virus) with the honeybee host and the parasitic varroa mite. The Journal of General Virology 86: 2281–2289.

Simon-Delso, N., San Martin, G., Bruneau, E. et al. (2014). Honeybee colony disorder in crop areas: the role of pesticides and viruses. PLoS One 9 https://doi.org/10.1371/journal.pone.0103073.

van der Sluijs, J.P., Simon-Delso, N., Goulson, D. et al. (2013). Neonicotinoids, bee disorders and the sustainability of pollinator services. Current Opinion in Environment Sustainability 5: 293–305.

Spleen, A.M., Lengerich, E.J., Rennich, K. et al. (2013). A national survey of managed honey bee 2011–12 winter colony losses in the United States: results from the Bee Informed Partnership. Journal of Apicultural Research 52 https://doi.org/10.3896/ibra.1.52.2.07.

Steinhauer, N.A., Rennich, K., Wilson, M.E. et al. (2014). A national survey of managed honey bee 2012–2013 annual colony losses in the USA: results from the Bee Informed Partnership. Journal of Apicultural Research 53: 1–18. https://doi.org/10.3896/ibra.1.53.1.01.

Traynor, K., Rennich, K., Forsgren, E. et al. (2015). Multiyear survey targeting disease incidence in US honey bees. Apidologie 47: 325–347.

USDA-APHIS Honey Bee Pests and Diseases Survey Project Plan for 2019 (2019). https://www.aphis.usda.gov/plant_health/plant_pest_info/honey_bees/downloads/SurveyProjectPlan.pdf

vanEngelsdorp, D., Underwood, R., Caron, D., and Hayes, J. Jr. (2007). An estimate of managed colony losses in the winter of 2006–2007: a report commissioned by the Apiary Inspectors of America. American Bee Journal 147: 599–603.

vanEngelsdorp, D., Hayes, J. Jr., Underwood, R.M., and Pettis, J. (2008). A survey of honey bee colony losses in the U.S., fall 2007 to spring 2008. PLoS One 3: e4071.

vanEngelsdorp, D., Evans, J.D., Saegerman, C. et al. (2009). Colony collapse disorder: a descriptive study. PLoS One 4 (8): e6481. https://doi.org/10.1371/journal.pone.000648.

vanEngelsdorp, D., Caron, D., Hayes, J. et al. (2012). A national survey of managed honey bee 2010–2011 winter colony losses in the USA: results from the Bee Informed Partnership. Journal of Apicultural Research 51: 115–124.

Vidal-Naquet, N. (2015). Honeybee Veterinary Medicine: Apis mellifera L, 1e. Sheffield, United Kingdom: 5m Publishing.

Appendix 17.A Beekeeping Biosecurity and Best Practices Checklist

Audit Conducted by____ **Date**____________

Apiary Owner and Contact Information____

____

Apiary Address/GPS Coordinates____

Recommended best practices	In place	In progress	Not in place	Comments
Training				
Complete training/maintain current knowledge of beekeeping through CE (beekeeper and employees)				
Possess knowledge of and be able to recognize all exotic and endemic threats to honey bee health in all hive locations				
Know current regulations and disease reporting requirements for all hive locations				
Record keeping/traceability				
Document completion of all training programs for beekeeper and employees				
Maintain current contact information for State Apiary Inspector/State Department of Agriculture for all hive locations				
Record number of apiaries and number of hives within each apiary				
Record any introduction of new stock, including source				
Record migratory hive movements, including routes taken and dates				
Record date, origin, and type of feeding supplements administered				
Record date, origin, dose, and use of any natural or medicinal treatments administered				
Record date, origin, and use of any chemicals or cleaning products used on hive equipment				
Record findings of all hive inspections conducted throughout the season including normal/abnormal findings, suspected/diagnosed diseases, mortalities, colonies affected				
Mark all colonies and apiaries with unique identification				
Mark all queens according to international color code				
Apiary placement				
Ensure easy, yet controlled, access to apiary				
Ensure apiary/surrounding area is well maintained				
Ensure hives are protected from inclement weather/other hazards to the extent possible				
Ensure presence of diverse, natural food sources				
Ensure presence of good quality water supply				
Possess knowledge of nearby farming practices, including pesticide use				
Possess knowledge of nearby colonies and management practices utilized				
Sourcing				
Purchase queens/bees from trusted sources and ensure they are pest-free to the extent possible				
Choose appropriate honey bee strain for regional environment of hive location/potential pathogen exposures				
Isolate recovered wild swarms from apiary until thoroughly inspected/miticides applied (if needed)				

(*Continued*)

Recommended best practices	In place	In progress	Not in place	Comments
Purchase hive equipment from reputable source made with appropriate materials				
Clean and disinfect any borrowed/secondhand equipment before use				
Maintain hive boxes/frames/foundations over time including removal of dark wax combs and replacement of hive frames every three years				
Minimize exchange of frames and supers between colonies and apiaries to the extent possible				
Remove all empty hives from apiary immediately				
Store equipment in well-ventilated/chemical free area and store hive boxes in crisscross pattern				
Obtain supplementary feed products from trusted source and store/handle appropriately				
Ensure supplemental feed is appropriate for needs of colony and time of year (winter/early spring = candy; late spring/fall = syrup; honey sourced only from colony/apiary in which hive is located)				
Hive inspections				
Conduct routine hive inspections				
Observe and document the following during inspection: queen status, activity of bees outside and within hive, brood patterns, pollen/honey storage, signs of disease/pests				
Implement and document presence of an integrated pest management program				
Isolate any diseased colonies from apiary that can be saved (as allowable by law)				
Humanely euthanize any weak hives or those that pose danger to other colonies				
Sanitation/hygiene				
Disinfect all small tools/clothing between inspection of different apiaries and after inspection of any apparently diseased colony				
Ensure that all workers/visitors/vehicles are clean before entering and leaving the apiary				
Display signage at apiary/property entrances with apiary owner's name/contact information and a request to obtain permission before entering				
Migratory colonies				
Minimize movement of hives to the extent possible				
Contact destination State Department of Agriculture/Apiary Inspector prior to any hive movement to determine health certification requirements				
Move hives only at night/early morning				
Feed colonies carbohydrate supplement prior to moving				
Possess knowledge of all established/exotic pathogens and disease reporting regulations in region to which hives are being moved				
Cover and secure hives/equipment prior to moving				
Assess any potential disease threat due to poorly managed hives near new location				
Keep accurate records of all hive movements				
Follow all transportation regulations				

Section III

Honey Bee Diseases, Disorders, and Special Topics

18

Parasite Transmission Between Hives and Spillover to Non-*Apis* Pollinators

Scott McArt

Department of Entomology, College of Agriculture and Life Sciences, Cornell University, Ithaca, NY, USA

Introduction

Unsustainable honey bee colony losses have been documented over the past several years in the US, Europe, and other parts of the world. While many factors are contributing to losses, a key challenge is parasites (vanEngelsdorp et al. 2009; Evans and Spivak 2010; Pettis et al. 2015). The varroa mite (*Varroa destructor*), Deformed Wing Virus (DWV), European Foulbrood (*Melissococcus plutonius*), American Foulbrood (*Paenibacillus larvae*), and chalkbrood (*Ascosphaera apis*) are major threats to honey bees and beekeepers, to name just a few.

When a colony of honey bees gets sick, can bees from the infected colony transmit parasites to other colonies within an apiary? What about transmission to nearby apiaries? Are all apiaries at risk, or is there a maximum distance at which transmission can occur? What about risk to wildlife? Can parasite spillover occur between managed colonies of *Apis mellifera* to wild, non-*Apis* bees or other insects? Such spillover between domestic animals and wildlife occurs in other systems, with classic examples such as the transmission of brucellosis (caused by *Brucella* bacteria) between cattle and wild elk (Kauffman et al. 2016; Rayl et al. 2019), and rinderpest (caused by *Rinderpest morbillivirus*) transmission, which devastated domestic cattle, wildebeest, and other ungulate wildlife in Africa (Kock et al. 1999). Does similar cross-species risk of transmission occur for parasites of honey bees and wild pollinators?

The past several years have seen a dramatic increase in research addressing how parasites of honey bees are transmitted. This is an important development, since knowledge of transmission is critical for any attempts to predict, and ultimately control, the spread of disease. While important parasites such as varroa or American Foulbrood may seem like impossible problems for beekeepers to surmount, remember that world-wide eradication of rinderpest occurred in 2011 (OIE 2011). Notably, the eradication of rinderpest occurred because of participatory epidemiological methods that allowed veterinarians to interact closely with cattle herders and more effectively implement control measures (Mariner et al. 2012). Indeed, a major goal of this book is to provide information that will hopefully facilitate similar interactions between veterinarians and beekeepers.

Parasite Transmission Within an Apiary

Before they were domesticated, honey bee colonies occurred naturally at a relatively low density in the landscape. Today, feral colonies of honey bees are still found at this low density, primarily in tree cavities. For example, colonies inhabiting trees in forests around the city of Nizhny Novgorod, Russia were found to be present at a density of 1–2 colonies per km^2 (Galton 1971). Similarly, Tom Seeley has conducted numerous surveys of feral honey bee colonies in trees surrounding Ithaca, NY, USA, finding they exist at a density of about 1 colony per km^2 (Seeley 2007; Seeley et al. 2015).

In contrast, modern beekeeping practices often encourage colonies to be placed 1 m apart or less (Crane 1990). Commercial beekeepers typically stack multiple colonies side-by-side on a single pallet so they can be moved efficiently from location to location (e.g. orchards for crop pollination, followed by fields for honey production). This crowding of multiple colonies at a given location greatly improves the ease of beekeeping, of course. In fact, there are common beekeeping terms used to describe a location with multiple colonies: an *apiary* or a *bee yard*. However, crowding of colonies in an apiary comes with some costs.

Honey Bee Medicine for the Veterinary Practitioner, First Edition. Edited by Terry Ryan Kane and Cynthia M. Faux.

There are two main reasons why crowding colonies in an apiary can increase risk of disease transmission. First, foraging bees from neighboring colonies can have difficulty distinguishing between their own colony and a neighboring colony, resulting in "drifting" between hives (Free 1958). While intuitively this may seem like a rare occurrence for anyone who's watched highly intelligent honey bee foragers leaving and entering a hive, it is not. In fact, when colonies of the same color are placed next to each other and face the same direction, it is common for 40% of the forager bees to be from a neighboring colony (Jay 1965, 1966a,b). Drifting by foragers can be greatly reduced by increasing the spacing between colonies, painting them different colors, and having them face different directions (Jay 1965, 1966b). However, some drifting between colonies that share a typical apiary will occur regardless of spacing between colonies.

Second, colonies within an apiary are at increased risk of having their honey stolen by foragers from other colonies when food is scarce (Free 1954; Downs and Ratnieks 2000). This phenomenon is called "robbing" and it is most common for weak hives to be robbed by foragers from strong hives (Figure 18.1). Because a leading reason why colonies become weak is due to disease, colonies that have high parasite levels are often at increased risk of being robbed (Sakofski et al. 1990; Greatti et al. 1992; Frey et al. 2011; Peck and Seeley 2019).

Both drifting and robbing increase contact rates between bees from different colonies in an apiary, which has the potential to increase parasite transmission. Indeed, one recent study found that as varroa mite abundance in a colony increases, the mites are increasingly likely to climb onto foragers from another colony that are robbing (Cervo et al. 2014). So, what can a beekeeper do to minimize risk of parasite transmission in their apiary? Perhaps the best study to date that has investigated how parasite transmission is impacted by spatial positioning between colonies within an apiary was conducted by Seeley and Smith (2015).

Figure 18.1 Bees robbing a hive. *Source:* Photo courtesy of Emma Walters.

For their study, Seeley and Smith (2015) arranged 12 colonies as would be typical in a modern "crowded" apiary (spaced ~1 m apart and facing the same direction), then placed another 12 colonies much further apart in a "dispersed" apiary (~33 m apart, on average). Swarming occurred in many of the colonies early in the season, which reduced mite counts due to the brood interruption. However, in late summer, colonies that had swarmed in the crowded apiary, but not those that had swarmed in the dispersed apiary, developed high mite counts and died over winter. At the same time, the authors documented that drifting between colonies during the summer occurred at a rate of 35–48% in the crowded apiary, while drifting was reduced to 0–6% in the dispersed apiary colonies. These results suggest that early-season swarming reduced a colony's mite load, but when colonies were crowded in apiaries, this reduction was erased over the summer as mites were spread easily via drifting and/or robbing (Figure 18.1).

In addition to spatial positioning, anything that reduces contact rates between bees from different colonies in an apiary has the potential to reduce parasite transmission, including painting hives different colors and facing them different directions, both of which have been shown to reduce drift and decrease parasite prevalence (Dynes et al. 2019). Lower hive densities in apiaries have been linked to lower parasite levels (Mõtus et al. 2016; Dynes et al. 2019), which could be due to reduced transmission or other factors, such as nutritional stress from competition for food (Dolezal and Toth 2018). But hive density has been found to be a poor predictor of parasite levels in other studies (e.g. Giacobino et al. 2014). A recent epidemiological model by Bartlett et al. (2019) suggests that changes in density or spatial positioning of colonies within apiaries are unlikely to impact prevalence of parasites that are able to infect bees at all life stages, such as *Nosema*. Their model suggests only parasites with a base R_0 (the basic reproduction number of a parasite) around 3 are likely to be impacted in a meaningful way by the number or arrangement of colonies in an apiary. Unfortunately, several major honey bee parasites (*Nosema*, American Foulbrood, DWV, the Acute Paralysis Virus complex) appear to have base R_0 values much higher than 3. For example, *Nosema ceranae* is estimated to have a base R_0 around 23 (Higes et al. 2008, 2009; Paxton 2010). While this new modeling work by Bartlett et al. is excellent, further epidemiological models are needed to understand within-apiary dynamics for parasites that are more likely (or only likely) to infect bees at certain life stages (e.g. varroa, American Foulbrood, European Foulbrood). Indeed, empirical results show that apiary density and configuration can impact prevalence of some of these parasites, as outlined above.

Figure 18.2 Monitoring a hive for varroa mites. *Source:* Photo courtesy of Emma Walters.

Finally, it is important to note that beekeepers often transfer frames (brood combs, frames of honey, etc.) or adult bees between colonies for various reasons, such as building up the strength of a weak colony, or creating a nucleus colony for mating a new queen. This practice greatly increases the probability of transmission for all parasites that can be transmitted via contaminated wax, such as *Nosema* (Bailey 1953), American Foulbrood (Ratnieks 1992), and chalkbrood (Koenig 1987). Furthermore, brood parasites such as varroa, American Foulbrood, European Foulbrood, and chalkbrood are particularly prone to transmission when brood frames are shared between colonies. It should go without saying that good hive inspection and parasite identification skills are critical for any beekeeper who wants to minimize disease in their operation (Figure 18.2).

Parasite Transmission Between Apiaries

In all countries with domesticated honey bees, detections of American Foulbrood must be reported (De Graaf et al. 2006). In some areas of the world, the detection of American Foulbrood leads to quarantine of that apiary, euthanizing all bees in the affected colony or colonies, and destruction of the hives (Pettis et al. 2015). This is not true everywhere, but the reason for the practice is simple and important: there is risk of transmission between apiaries.

While less is known about factors governing transmission between apiaries compared to within an apiary, numerous studies have found that distance between apiaries, robbing by foragers from other hives, and beekeeper identity can be important. For example, Lindström et al. (2008) found that transmission of American Foulbrood occurred at a distance of 1 km from clinically diagnosed colonies that were allowed to be robbed, but was significantly lower at a distance of 2 km or further. This study found *P. larvae* spores in colonies that were 3 km from the affected apiary, but no clinical symptoms developed in those colonies. The importance of distance was further confirmed in a study that modeled spread of American Foulbrood on the island of Jersey, which is off the northwest coast of France. In this study, both distance between apiaries and ownership of the apiaries by particular beekeepers were predictors of disease spread (Datta et al. 2013). The significant effect of ownership (i.e. beekeeper operation) as a predictor of American Foulbrood spread is noteworthy, as a beekeeper's tools and clothes can become contaminated when inspecting an infected hive, then transmitted to colonies in the next apiary the beekeeper visits. If such contamination occurs, it is important to wash hive tools and smokers with bleach water and scrub them with a steel wool pad. This washing will not destroy *P. larvae* spores, but it will dilute them by removing materials such as wax, honey, and propolis that may contain larger concentrations of spores. Bee jackets and leather gloves can be washed in a similar manner, but if the contamination cannot be eliminated, it is safer to destroy them.

The importance of distance and robbing have also been shown for between-apiary transmission of European Foulbrood and varroa. For example, a study by Belloy et al. (2007) found that adult forager bees carrying *M. plutonius* were detected in 30% of colonies, on average, in apiaries without European Foulbrood symptoms but located near apiaries with clinical symptoms. When apiaries were located 500–1000 m from the apiary with clinical symptoms, forager bees carrying *M. plutonius* were detected in 40% of colonies, while this decreased to 11% when apiaries were located >1000 m from the apiary with clinical symptoms. Similarly, distance between apiaries was found to reduce varroa loads in one recent study (Nolan and Delaplane 2017). In this study, "receiver" colonies placed further away from experimental colonies artificially supplemented with varroa were more likely to remain at low varroa levels compared to colonies that were closer to the varroa-supplemented colonies.

While the above examples likely highlight direct transmission of parasites via robbing of colonies in different apiaries or the occasional long-distance mistake by a drifting forager, another route of transmission is also possible. Specifically, the shared use of flowers (Figure 18.3). While this indirect route of transmission is typically thought to be less important than robbing (Fries and Camazine 2001), new data suggest floral transmission may in fact be an efficient way for parasites to be transmitted between apiaries. For example, one recent study found the varroa mite can

Figure 18.3 Honey bee foraging alongside a sweat bee on Echinacea flower. Parasites such as DWV can be transmitted at flowers. *Source:* Photo courtesy of Emma Walters.

quickly attach itself to honey bees as they forage at flowers (Peck et al. 2016). Furthermore, honey bees can deposit and acquire numerous parasites from flowers as they forage (Graystock et al. 2015), and DWV is often found on flowers near apiaries when colonies in those apiaries are positive for DWV (Alger et al. 2019).

Just how prevalent are parasites of honey bees on flowers? A postdoc in my lab, Pete Graystock, became interested in this question and decided to screen 2624 flowers representing 89 plant species near Ithaca, NY, USA. He found that 9% of individual flowers and 70% of flower species were positive for at least one bee parasite, and there was significant variation in the prevalence of parasites among different plant genera and species (Graystock et al. 2020; Figure 18.4). In other words, somewhere around 1 out of every 11 individual flowers is contaminated with a bee parasite. With the average adult forager honey bee often visiting over 1000 flowers d^{-1} (Winston 1987), this means that flowers may be an important transmission venue even if only a small fraction of visits to contaminated flowers result in successful transmission. Indeed, flowers may act as a stopover for parasites as they're transmitted from one colony (or apiary) to another, or from one bee species to another, as discussed in the next section.

Evidence of Parasite Spillover From *Apis mellifera* to Non-*Apis* Insects

There is a growing body of literature showing that many parasites of honey bees are also found in wild, non-*Apis* pollinators, and other insects. For example, *N. ceranae* was recently reported in association with bees in the genera *Osmia* and *Andrena* (Ravoet et al. 2014), and is also commonly found in bumble bees (Plischuk et al. 2009; Graystock et al. 2013; Gamboa et al. 2015). Chalkbrood (*A. apis*) has been found in association with bees in the genera *Andrena* and *Halictus* (Evison et al. 2012). New data from Graystock et al. (2020) indicate that honey bees near Ithaca, NY, USA share parasites with at least 46 species of bees representing 14 genera (Figure 18.4a). Perhaps the broadest screening that has occurred in both honey bees and other insects is for the RNA viruses. DWV and numerous other RNA viruses have been found in bumble bees, mason bees, sweat bees, and mining bees (Evison et al. 2012; Gamboa et al. 2015; McMahon et al. 2015; Dolezal et al. 2016). DWV in particular appears to be highly cosmopolitan, with detections in 65 arthropod species spanning eight insect orders and three orders of Arachnida (Martin and Brettell 2019) (Figure 18.3).

As any parasitologist will tell you, simply detecting parasite DNA or RNA in an organism does not mean it is causing an infection. In fact, virulence in unknown hosts is often difficult to predict (Dybdahl and Storfer 2003); therefore data on parasite replication and/or effects on putative hosts are needed. Numerous studies have begun to investigate parasite replication and effects on non-honey bee hosts. For example, Graystock et al. (2013) found that *N. ceranae* can replicate in bumble bees, reduce survival by 48%, and also had detrimental sub-lethal impacts on behavior. Müller et al. (2019) also found evidence that *N. ceranae* could potentially replicate in adults of the mason bee, *Osmia bicornis*. While this study did not find impacts on survival, a similar study that assessed impacts of *N. ceranae* on larval *O. bicornis* did observe reduced survival (Bramke et al. 2019). Additional studies have investigated impacts of viruses on non-honey bee pollinators. Fürst et al. (2014) found that DWV reduces bumble bee worker survival by six days, causing deformed wings and non-viable offspring. Conversely, experimental inoculations with a suite of RNA viruses known to be lethal to honey bees did not impact short-term survival of the cellophane bee, *Colletes inaequalis*, or the leafcutter bee, *Megachile rotundata* (Dolezal et al. 2016).

Is there evidence that parasite spillover occurs between honey bees and wild pollinators? In an excellent study by Fürst et al. (2014), the authors genotyped DWV strains in both honey bees and bumble bees across the UK, finding that bumble bees and honey bees were positive for very

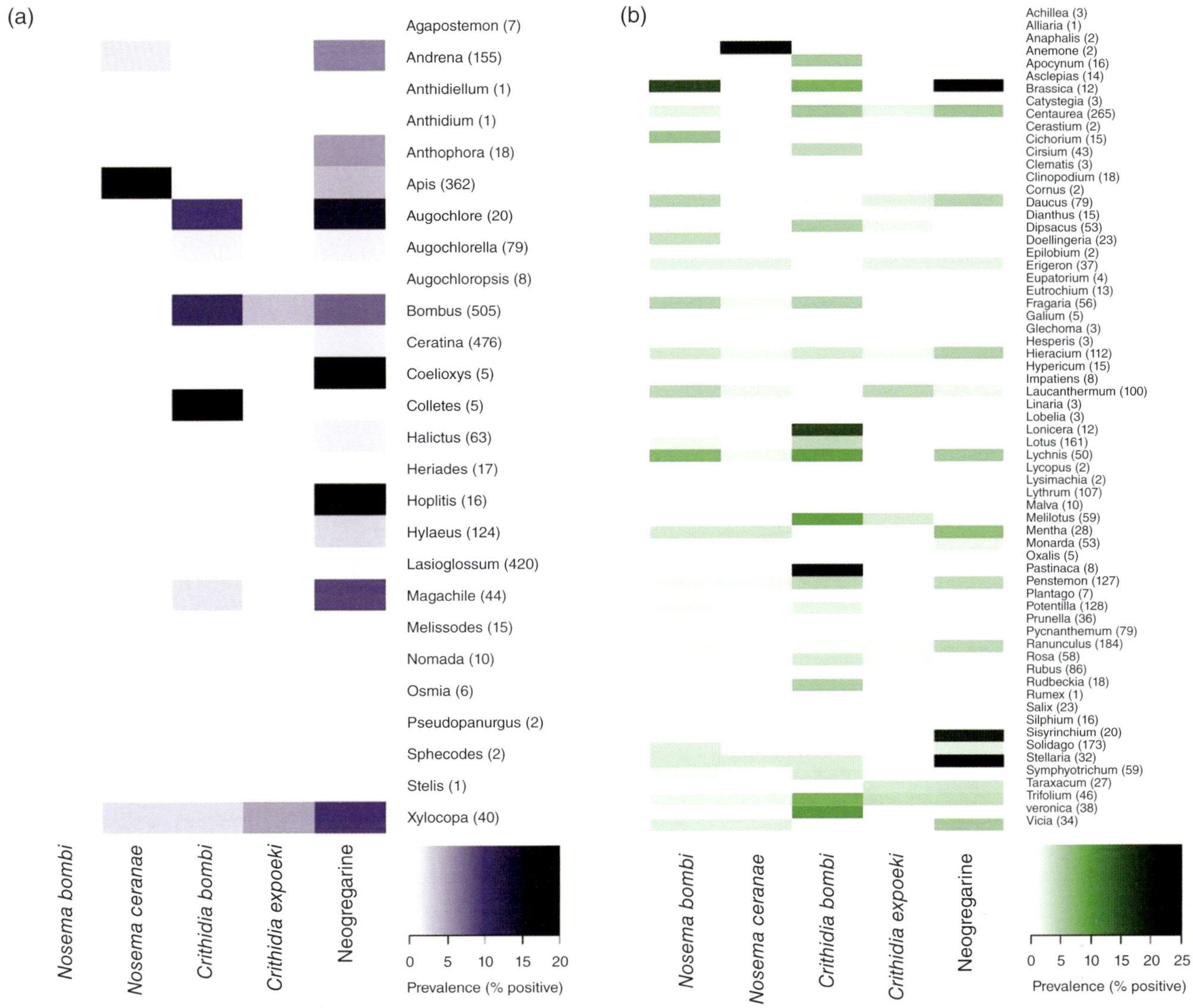

Figure 18.4 Parasite prevalence in bee (a) and on flower (b) genera across three old-field communities near Ithaca, NY, USA. Screenings of 2672 bees representing 26 genera (at least 110 species), and 2624 flowers representing 65 genera (89 species). Rows represent bee (a) or plant (b) genus with sample number in parenthesis. Overall, 12% of individual bees and 42% of bee species were positive for at least one parasite, while 9% of individual flowers and 70% of flower species were positive for at least one parasite. All screenings were conducted using molecular methods (multiplex Polymerase Chain Reaction) as described in Graystock et al. (2020). Reproduced with permission of Springer Nature.

similar viral strains when they were located in similar geographic regions. A study by Alger et al. (2019) took this one step further, finding evidence for viral replication in over a quarter of two-spotted bumble bees (*Bombus bimaculatus*) and nearly 10% of the half-black bumble bees (*Bombus vagans*) that were sampled in Vermont. The negative viral strand was found for both Black Queen Cell Virus and DWV, indicating both viruses could cause active infections in the bumble bees. Furthermore, replicating DWV was found more often in bumble bees near apiaries compared to bumble bees that weren't near apiaries. Thus, bumble bees were more likely to have active DWV infections when near apiaries. Finally, a link between honey bees, varroa, DWV, and prevalence of this virus in bumble bees was made recently by Manley et al. (2019). This study found that bumble bees on islands located between the UK and France where varroa was present in honey bee colonies were more likely to have DWV, and higher viral titers, compared to bumble bees on islands where varroa was absent (i.e. all honey bee colonies were varroa-free on this subset of islands). The results for bumble bees mirrored viral prevalence and titers in the honey bees on each island. Thus, this study provides strong evidence that the recent world-wide varroa epidemic is an important factor shaping spillover of DWV from honey bees to bumble bees. Given that DWV has been found in 65 arthropod species, the studies above may represent the tip of the iceberg in terms of varroa's indirect impact on other insects.

Because many wild pollinator species have experienced well-documented range contractions and extinctions over recent decades (Kosior et al. 2007; Goulson et al. 2008; Williams and Osborne 2009; Williams et al. 2014), and some of these declines have been linked to multi-host parasites (Cameron et al. 2011; Meeus et al. 2011; Schmid-Hempel et al. 2014), there is concern that inadequate beekeeper management practices and parasite spillover from honey bees could be contributing to declines in the health of wild pollinators. As mentioned at the outset in this chapter, a major goal of this book is to provide information that will hopefully facilitate positive interactions between veterinarians and beekeepers, thus paving the way to improved health of managed honey bees. These efforts may indeed have knock-on effects for the health of wild pollinators and other insects.

References

Alger, S.A., Burnham, P.A., Boncristiani, H.F., and Brody, A.K. (2019). RNA virus spillover from managed honeybees (*Apis mellifera*) to wild bumblebees (*Bombus* spp.). *PLoS One* 14: e0217822.

Bailey, L. (1953). The transmission of *Nosema* disease. *Bee World* 34: 171–172.

Bartlett, L.J., Rozins, C., Brosi, B.J. et al. (2019). Industrial bees: the impact of apicultural intensification on local disease prevalence. *Journal of Applied Ecology* 56: 2195–2205.

Belloy, L., Imdorf, A., Fries, I. et al. (2007). Spatial distribution of *Melissococcus plutonius* in adult honey bees collected from apiaries and colonies with and without symptoms of European foulbrood. *Apidologie* 38: 136–140.

Bramke, K., Müller, U., McMahon, P.D., and Rolff, J. (2019). Exposure of larvae of the solitary bee *Osmia bicornis* to the honey bee pathogen *Nosema ceranae* affects life history. *Insects* 10: 380.

Cameron, S.A., Lozier, J.D., Strange, J.P. et al. (2011). Patterns of widespread decline in North American bumble bees. *Proceedings of the National Academy of Sciences of the United States of America* 108: 662–667.

Cervo, R., Bruschini, C., Cappa, F. et al. (2014). High *Varroa* mite abundance influences chemical profiles of worker bees and mite–host preferences. *The Journal of Experimental Biology* 217: 2998.

Crane, E. (1990). *Bees and Beekeeping: Science, Practice and Worldwide Resources*. Ithaca, NY: Cornell University Press.

Datta, S., Bull, J.C., Budge, G.E., and Keeling, M.J. (2013). Modelling the spread of American foulbrood in honeybees. *Journal of the Royal Society Interface* 10: 20130650.

De Graaf, D.C., Alippi, A.M., Brown, M. et al. (2006). Diagnosis of American foulbrood in honey bees: a synthesis and proposed analytical protocols. *Letters in Applied Microbiology* 43: 583–590.

Dolezal, A.G. and Toth, A.L. (2018). Feedbacks between nutrition and disease in honey bee health. *Current Opinion in Insect Science* 26: 114–119.

Dolezal, A.G., Hendrix, S.D., Scavo, N.A. et al. (2016). Honey bee viruses in wild bees: viral prevalence, loads, and experimental inoculation. *PLoS One* 11: e0166190.

Downs, S.G. and Ratnieks, F.L.W. (2000). Adaptive shifts in honey bee (*Apis mellifera* L.) guarding behavior support predictions of the acceptance threshold model. *Behavioral Ecology* 11: 326–333.

Dybdahl, M.F. and Storfer, A. (2003). Parasite local adaptation: Red Queen versus Suicide King. *Trends in Ecology and Evolution* 18: 523–530.

Dynes, T.L., Berry, J.A., Delaplane, K.S. et al. (2019). Reduced density and visually complex apiaries reduce parasite load and promote honey production and overwintering survival in honey bees. *PLoS One* 14: e0216286.

Evans, J.D. and Spivak, M. (2010). Socialized medicine: individual and communal disease barriers in honey bees. *Journal of Invertebrate Pathology* 103: S62–S72.

Evison, S.E.F., Roberts, K.E., Laurenson, L. et al. (2012). Pervasiveness of parasites in pollinators. *PLoS One* 7: e30641.

Free, J.B. (1954). The behaviour of robber honeybees. *Behaviour* 7: 233–240.

Free, J.B. (1958). The drifting of honey-bees. *The Journal of Agricultural Science* 51: 294–306.

Frey, E., Schnell, H., and Rosenkranz, P. (2011). Invasion of *Varroa destructor* mites into mite-free honey bee colonies under the controlled conditions of a military training area. *Journal of Apicultural Research* 50: 138–144.

Fries, I. and Camazine, S. (2001). Implications of horizontal and vertical pathogen transmission for honey bee epidemiology. *Apidologie* 32: 199–214.

Fürst, M.A., McMahon, D.P., Osborne, J.L. et al. (2014). Disease associations between honeybees and bumblebees as a threat to wild pollinators. *Nature* 506: 364.

Galton, D. (1971). *Survey of a Thousand Years of Beekeeping in Russia*. London: Bee Research Association.

Gamboa, V., Ravoet, J., Brunain, M. et al. (2015). Bee pathogens found in *Bombus atratus* from Colombia: a case study. *Journal of Invertebrate Pathology* 129: 36–39.

Giacobino, A., Cagnolo, N.B., Merke, J. et al. (2014). Risk factors associated with the presence of *Varroa destructor* in honey bee colonies from east-central Argentina. *Preventive Veterinary Medicine* 115: 280–287.

Goulson, D., Lye, G.C., and Darvill, B. (2008). Decline and conservation of bumble bees. In: *Annual Review of Entomology*, 191–208. Palo Alto, CA: Annual Reviews.

Graystock, P., Yates, K., Darvill, B. et al. (2013). Emerging dangers: deadly effects of an emergent parasite in a new pollinator host. *Journal of Invertebrate Pathology* 114: 114–119.

Graystock, P., Goulson, D., and Hughes, W.O.H. (2015). Parasites in bloom: flowers aid dispersal and transmission of pollinator parasites within and between bee species. *Proceedings of the Royal Society Series B* 282: 7.

Graystock, P., Ng, W.H., Parks, K. et al. (2020). Dominant bee species and floral abundance drive parasite temporal dynamics in plant–pollinator communities. *Nature Ecology and Evolution*. https://doi.org/10.1038/s41559-020-1247-x.

Greatti, M., Milani, N., and Nazzi, F. (1992). Reinfestation of an acaricide-treated apiary by *Varroa jacobsoni* Oud. *Experimental and Applied Acarology* 16: 279–286.

Higes, M., Martín-Hernández, R., Botías, C. et al. (2008). How natural infection by *Nosema ceranae* causes honeybee colony collapse. *Environmental Microbiology* 10: 2659–2669.

Higes, M., Martín-Hernández, R., Garrido-Bailón, E. et al. (2009). Honeybee colony collapse due to *Nosema ceranae* in professional apiaries. *Environmental Microbiology Reports* 1: 110–113.

Jay, S.C. (1965). Drifting of honeybees in commercial apiaries 1. Effect of various environmental factors. *Journal of Apicultural Research* 4: 167–175.

Jay, S.C. (1966a). Drifting of honeybees in commercial apiaries II. Effect of various factors when hives are arranged in rows. *Journal of Apicultural Research* 5: 103–112.

Jay, S.C. (1966b). Drifting of honeybees in commercial apiaries. III. Effect of apiary layout. *Journal of Apicultural Research* 5: 137–148.

Kauffman, M., Peck, D., Scurlock, B. et al. (2016). Risk assessment and management of brucellosis in the southern greater Yellowstone area (I): a citizen-science based risk model for bovine brucellosis transmission from elk to cattle. *Preventive Veterinary Medicine* 132: 88–97.

Kock, R.A., Wambua, J.M., Mwanzia, J. et al. (1999). Rinderpest epidemic in wild ruminants in Kenya 1993–1997. *Veterinary Record* 145: 275.

Koenig, J.P. (1987). *Factors Contributing to the Pathogenesis of Chalk Brood Disease in Honey Bee Colonies*. Madison, WI, USA: University of Wisconsin, Madison.

Kosior, A., Celary, W., Olejniczak, P. et al. (2007). The decline of the bumble bees and cuckoo bees (Hymenoptera: Apidae: Bombini) of Western and Central Europe. *Oryx* 41: 79–88.

Lindström, A., Korpela, S., and Fries, I. (2008). Horizontal transmission of *Paenibacillus larvae* spores between honey bee (*Apis mellifera*) colonies through robbing. *Apidologie* 39: 515–522.

Manley, R., Temperton, B., Doyle, T. et al. (2019). Knock-on community impacts of a novel vector: spillover of emerging DWV-B from *Varroa*-infested honeybees to wild bumblebees. *Ecology Letters* 22: 1306–1315.

Mariner, J.C., House, J.A., Mebus, C.A. et al. (2012). Rinderpest eradication: appropriate technology and social innovations. *Science* 337: 1309.

Martin, S.J. and Brettell, L.E. (2019). Deformed wing virus in honeybees and other insects. *Annual Review of Virology* 6: 49–69.

McMahon, D.P., Furst, M.A., Caspar, J. et al. (2015). A sting in the spit: widespread cross-infection of multiple RNA viruses across wild and managed bees. *Journal of Animal Ecology* 84: 615–624.

Meeus, I., Brown, M.J.F., De Graaf, D.C., and Smagghe, G. (2011). Effects of invasive parasites on bumble bee declines. *Conservation Biology* 25: 662–671.

Mõtus, K., Raie, A., Orro, T. et al. (2016). Epidemiology, risk factors and varroa mite control in the Estonian honey bee population. *Journal of Apicultural Research* 55: 396–412.

Müller, U., McMahon, D.P., and Rolff, J. (2019). Exposure of the wild bee *Osmia bicornis* to the honey bee pathogen *Nosema ceranae*. *Agricultural and Forest Entomology* 21: 363–371.

Nolan, M.P. and Delaplane, K.S. (2017). Distance between honey bee *Apis mellifera* colonies regulates populations of *Varroa destructor* at a landscape scale. *Apidologie* 48: 8–16.

OIE. (2011). No more deaths from rinderpest. World Organization for Animal Health.

Paxton, R.J. (2010). Does infection by *Nosema ceranae* cause "Colony Collapse Disorder" in honey bees (*Apis mellifera*)? *Journal of Apicultural Research* 49: 80–84.

Peck, D.T. and Seeley, T.D. (2019). Mite bombs or robber lures? The roles of drifting and robbing in *Varroa destructor* transmission from collapsing honey bee colonies to their neighbors. *PLoS One* 14: e0218392.

Peck, D.T., Smith, M.L., and Seeley, T.D. (2016). *Varroa destructor* mites can nimbly climb from flowers onto foraging honey bees. *PLoS One* 11: e0167798.

Pettis, J.S., Chen, Y., Ellis, J. et al. (2015). Diseases and pests of honey bees. In: *The Hive and the Honey Bee* (ed. J.M. Graham). Hamilton, IL: Dadant & Sons.

Plischuk, S., Martin-Hernandez, R., Prieto, L. et al. (2009). South American native bumblebees (Hymenoptera: Apidae) infected by *Nosema ceranae* (Microsporidia), an emerging pathogen of honeybees (*Apis mellifera*). *Environmental Microbiology Reports* 1: 131–135.

Ratnieks, F.L.W. (1992). American foulbrood: the spread and control of an important disease of the honey bee. *Bee World* 73: 177–191.

Ravoet, J., De Smet, L., Meeus, I. et al. (2014). Widespread occurrence of honey bee pathogens in solitary bees. *Journal of Invertebrate Pathology* 122: 55–58.

Rayl, N.D., Proffitt, K.M., Almberg, E.S. et al. (2019). Modeling elk-to-livestock transmission risk to predict hotspots of brucellosis spillover. *The Journal of Wildlife Management* 83: 817–829.

Sakofski, F., Koeniger, N., and Fuchs, S. (1990). Seasonality of honey bee colony invasion by *Varroa jacobsoni* Oud. *Apidologie* 21: 547–550.

Schmid-Hempel, R., Eckhardt, M., Goulson, D. et al. (2014). The invasion of southern South America by imported bumblebees and associated parasites. *Journal of Animal Ecology* 83: 823–837.

Seeley, T.D. (2007). Honey bees of the Arnot Forest: a population of feral colonies persisting with *Varroa destructor* in the northeastern United States. *Apidologie* 38: 19–29.

Seeley, T.D. and Smith, M.L. (2015). Crowding honeybee colonies in apiaries can increase their vulnerability to the deadly ectoparasite Varroa destructor. *Apidologie* 46: 716–727.

Seeley, T.D., Tarpy, D.R., Griffin, S.R. et al. (2015). A survivor population of wild colonies of European honeybees in the northeastern United States: investigating its genetic structure. *Apidologie* 46: 654–666.

vanEngelsdorp, D., Evans, J.D., Saegerman, C. et al. (2009). Colony collapse disorder: a descriptive study. *PLoS One* 4.

Williams, P.H. and Osborne, J.L. (2009). Bumblebee vulnerability and conservation world-wide. *Apidologie* 40: 367–387.

Williams, P.H., Thorp, R.W., Richardson, L.L., and Colla, S.R. (2014). *Bumble Bees of North America: An Identification Guide*. Princeton, NJ: Princeton University Press.

Winston, M. (1987). *The Biology of the Honey Bee*. Cambridge: Harvard University Press.

19

Colony Collapse Disorder and Honey Bee Health

Jay D. Evans and Yanping (Judy) Chen

USDA-ARS, Bee Research Laboratory, Beltsville, MD, USA

Introduction

Honey bees (genus *Apis*) form highly structured colonies with a single queen and thousands of largely sterile workers. The western honey bee, *Apis mellifera*, has a longstanding partnership with humans, thanks to the harvesting of wax and honey and, more recently, an appreciation for the economic impacts of these bees on crops that benefit from pollination. *A. mellifera* (hereafter "honey bee") colonies hold 30 000–60 000 worker bees when mature. Honey bees thrive in artificial nest boxes. Thanks in large part to the invention of the "Langstroth" removable-frame hive in the 1800s, beekeepers were actively managing millions of honey bee colonies worldwide by the mid-1900s. In the United States, approximately 2.5 million honey bee colonies are maintained across the season, most of which are transported for the pollination of almonds and other crops. This number has remained steady for 20 years (Figure 19.1).

In the latter part of 2006, several prominent US commercial beekeepers reported an odd decline in the worker populations of their honey bee colonies. These losses were positively associated at the level of apiaries (defined as aggregations of from one to several hundred honey bee colonies). The defining trait of this event, soon named colony collapse disorder (CCD, vanEngelsdorp et al. 2009), was the rapid loss of female workers in honey bee colonies, from tens of thousands of individuals to several hundred or fewer over a time period from one to several weeks. These numbers were unsustainable and colonies impacted by CCD eventually died completely. Unlike many colony loss events, those facing CCD had an active queen present, healthy developing (larval and pupal) bees, and no overt signs of disease caused by parasitic mites (*Varroa destructor* or *Acarapis woodii*) or microbes. The enigmatic nature of CCD and the genuine fear that an essential agricultural species was at risk led to intense media and industry attention. This attention was followed by forensic and experimental efforts to understand the stresses on individual bees that shorten their lives and place colonies at risk, efforts that continue today.

Was CCD a Unique and Significant Phenomenon?

Sudden declines of domesticated honey bee colonies have been described since at least the 1800's. Most prominent among these events is "Isle of Wight" disease, a phenomenon that presented itself early in the twentieth century on this island, near the southern England town of Portsmouth. In records of the time, bee losses were noticeable in 1904 and widespread two years later (F 1916). Strikingly, after 95% of colonies were lost on the Isle of Wight, mysterious die-offs followed on the English mainland. Debates soon formed over the cause, with three plausible biotic agents receiving the most attention: the newly identified microsporidian parasite *Nosema apis*, the parasitic tracheal mite *A. woodii*, and as-yet unidentified "paralytic" viruses. Some 60 years after Isle of Wight and the ensuing colony losses throughout England, there was still great disagreement over the relative importance of viruses, gut parasites, and mites, as relayed by Leslie Bailey (1964) and Brother Adam (1968), a leading bee pathologist and breeder, respectively. As will be seen for CCD, the best inferences came from ruling out causes that did not fit the epidemiology. For Isle of Wight, these candidates included seasonal climate and the as-yet scarce agrochemicals. Isle of Wight had an impact on beekeepers worldwide, driving regulatory actions such as the "Honey Bee Act of 1922" in the US, placed into law to effectively prohibit legal importation of honey bees for decades. This Act likely

Honey Bee Medicine for the Veterinary Practitioner, First Edition. Edited by Terry Ryan Kane and Cynthia M. Faux.

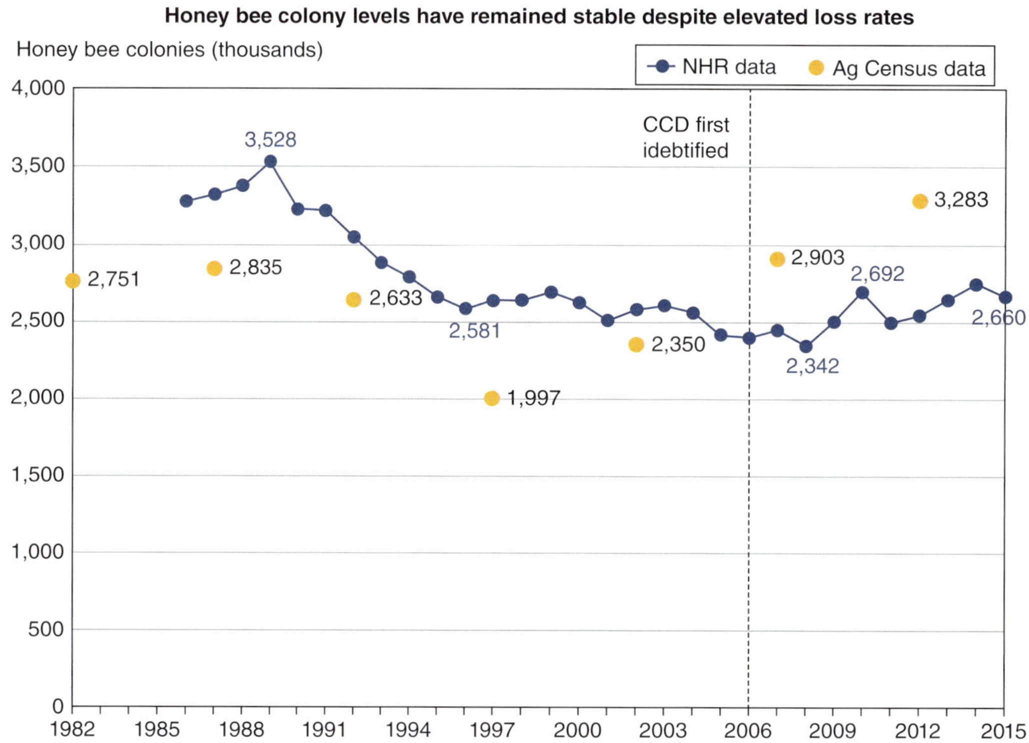

Figure 19.1 Steady honey bee colony numbers in the United States since 1996. *Source:* USDA, Economic Research Service using data from USDA, National Agriculture Statistics Service Agriculture Census data and USDA, Agriculture Marketing Service National Honey Report Data.
Notes: CCD = Colony Collapse Disorder. NHR = National Honey Report.

played a role in delaying the introduction of the tracheal mite into the US by 60 more years.

In the US, widespread cases of "disappearing disease" were described in the 1970's. Wilson and Menapace (1979) describe the phenomenon and its aftermath for US beekeepers, while also summarizing historical severe losses across the globe. In many ways these events all matched CCD in terms of impact, range, colony traits, and human responses.

After reaching significant levels in 2006, CCD reports were widespread by the Spring of 2007. In the absence of true historical data, it was hard to compare these reports with normal loss rates, but there was a real sense among experienced beekeepers that both the nature and severity of CCD were unusual. As in past events, initial labels for CCD included vague phenomenological themes such as "disappearing disease" and "fall dwindle." The name CCD was coined to emphasize the rapidity of the decline, to make clear that disease symptoms were not present, and as reports continued, to emphasize that these events were not limited to a particular place or season.

Even during the time of peak awareness for CCD, this was not a frequent event. In the US, the most accurate, or at least consistent, estimates for the frequency of CCD come from annual beekeeper surveys initiated by the Bee Informed Partnership in the wake of 2007 (www.beeinformed.org). Survey results from 2008 indicate that 7.5% of beekeepers felt their colonies had declined in 2007 as a result of CCD (based on a survey of 228 commercial and "sideline" beekeepers with 20 or more hives, Figure 19.2). Numbers were the same for hobbyist beekeepers who generally managed from one to several bee hives. By contrast, even during the "peak" years in terms of awareness of CCD (reports from 2007 to 2009), 30% of beekeepers traced their colony losses to queen failures, and "starvation" and mites were identified as causes nearly three times as often as CCD.

CCD reports have remained more or less steady for the past 10 years, as have the reported losses due to pesticides, queen failure, starvation, and mites (Figure 19.2). This does not diminish the great impact of CCD on beekeepers, especially those making their livings with thousands of bee

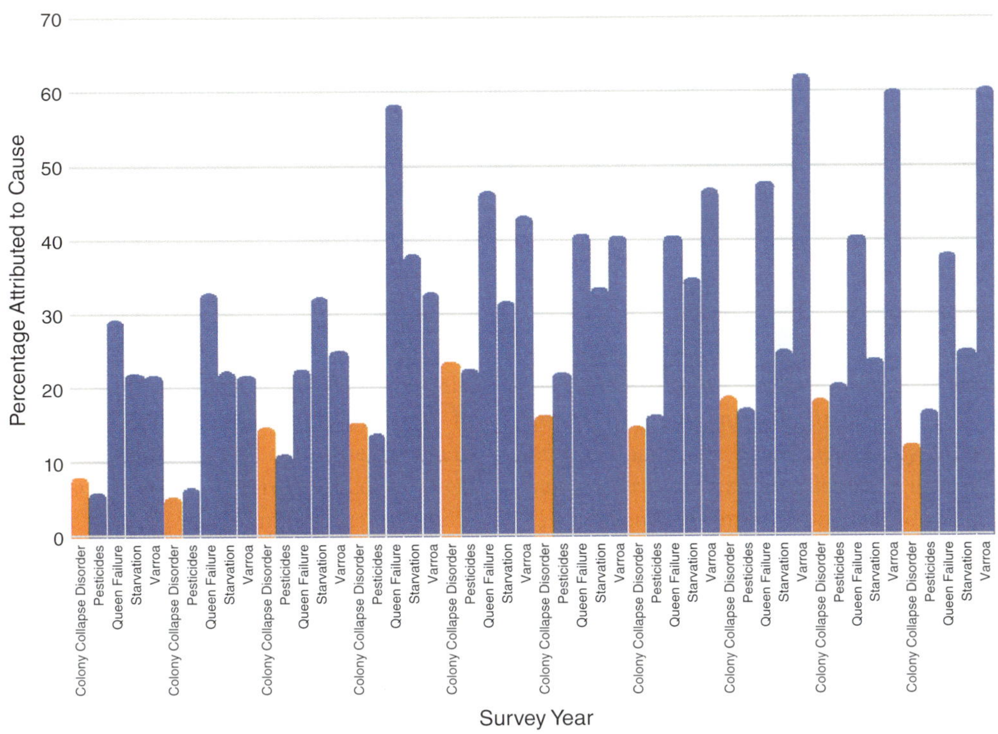

Figure 19.2 Self-reported causes of bee colony losses, by year (www.beeinformed.org).

colonies. Beekeepers are ingenious at splitting colonies in order to compensate for lost colonies. These splits are costly in terms of labor and opportunity costs for both incipient and source colonies, but this method is standard for most loss recoveries. CCD is far more extreme when it occurs, often impacting 90% or more of the colonies in an apiary. This variance, more than average loss rates, is arguably hardest on the industry, since individual beekeepers have been driven out of the trade or have been unable to fulfill valuable pollination contracts. For that reason alone, CCD and other "disappearing" events deserve the attention they have received.

Figure 19.3 Colonies marked in a commercial apiary for having CCD symptoms. *Source:* Photo courtesy of Jay Evans.

Searching for a Cause: Biotic Actors

By definition, CCD events were not linked to parasitism by *Varroa* mites, a main driver themselves of colony losses then and presently. Other likely causes of bee mortality and colony declines include fungal diseases of developing bees, bacterial diseases of larvae and adults, protozoa including trypanosomatids, amoeboids, and gregarines, microsporidian gut parasites and a diverse array of RNA viruses. As soon as CCD was apparent, samples were collected from impacted apiaries (Figure 19.3), leading to laboratory analyses of microbes tied to healthy or declining colonies (Figure 19.4).

In a landmark paper, Cox-Foster and colleagues (2007) surveyed samples of worker bees from canonical CCD colonies in Pennsylvania, California, and Florida using the emergent "454" light-based high-throughput sequencing platform. This survey identified known suspects in terms of bee parasites alongside several novel discoveries.

Chief among "known" parasites to earn some validation in this survey were the microsporidian species *N. apis* and *N. ceranae*, which together were significantly over-represented in bees sampled from collapsed colonies.

Figure 19.4 Field collection of honey bee samples for laboratory analysis. *Source:* Photo courtesy of Jay Evans.

N. ceranae was especially compelling, since this gut parasite is a relatively recent arrival in North America (or at any rate was not detected until recently, e.g. Chen et al. 2008). *N. ceranae* is sure to have an impact on worker bee longevity and the contributions of these bees to their colonies.

Among the novel identifications, the RNA virus Israeli acute paralysis virus (IAPV, Dicistroviridae) was also overrepresented in collapsing colonies (Cox-Foster et al. 2007).

While this virus has long been present in the US (Chen and Evans 2007) it is possible that novel strains are indeed involved with colony losses (Palacios et al. 2008). Cox-Foster et al. set the stage for numerous additional discovery-based surveys of field honey samples and arguably describes a pipeline that is generally useful for forensic studies of disease in inadequately studied taxa. Using a partly overlapping but expanded dataset, Cornman et al. (2012) interrogated numerous healthy and collapsed colonies using ILLUMINA next-generation sequencing of extracted RNAs. While copy numbers of IAPV and other earlier candidates were not supported in this sequencing effort and subsequent quantitative-PCR screens, the results highlighted an important difference between worker bees in collapsed and healthy colonies. Bees from collapsed colonies carried, on average, severalfold higher levels of parasites and pathogens. The identified species varied by geography, and presumable beekeeper, for migratory colonies. Rather than finding a single link with CCD, this survey instead indicated distinct "chords" of parasites or pathogens that arise in the face of CCD. Decoupling causal agents from opportunists is ongoing, with tedious experimental efforts to test some candidates and broad surveys of their dynamics across space and time. As indicated by the relative rarity of CCD as opposed to other loss events, these surveys have been expanded to include longitudinal surveys of healthy colonies in order to tag those species that appear *prior* to negative fates.

Abiotic Causes of Honey Bee Declines

Coincident with the great effort made to find a biological cause of CCD have been many efforts to identify abiotic causes of bee declines. These efforts have focused on long-present agrochemicals to which honey bees are exposed as well as products whose use has expanded greatly in the past decade. Mullin et al. (2010) identified a diverse set of agrochemicals and other anthropogenic products in bees, honey, and wax. Specifically, 121 distinct chemicals were found above testing thresholds in hundreds of samples. This study, and the advertisement of methodologies capable of quantifying agrochemicals to the level of parts-per-billion, have had a profound effect on honey bee science. While no single chemical or chemical class has been linked to CCD on the whole, it is clear that honey bees and other beneficial insects are at risk from chemicals used to control pest insects and microbes in agriculture. Importantly, honey bees show lower redundancy in gene families known to be involved in detoxification of chemical stress actors. Many enzymes in the cytochrome p450 family, for example, help to disarm synthetic chemicals faced by both pest and beneficial insects. Honey bees have at most half of the members in this broad family when compared to other insect taxa, and especially insects that feed on plants (i.e. many crop pests (Claudianos et al. 2006)). These differences have a likely evolutionary root in that honey bees are "rewarded" with pollen and nectar plants which depend on bees for pollination. It has been shown that these rewards contain lower levels of secondary chemical defenses than do the vegetative parts of plants, and especially lower than damaged plants that can mount a defensive response to herbivores.

Arguably the best-studied agrochemical compounds that impact honey bees are the neonicotinoids. Synthetic derivatives of natural plant defenses (i.e., the chemicals produced by tobacco and related plants against caterpillars and other insects), neonicotinoids are profoundly effective as crop defense products and their use in agriculture has soared in the past decade (Krupke et al. 2012). Honey bees are extremely sensitive to some members of this class, with LD50's at the level of several parts per billion (Fairbrother et al. 2014; Lundin et al. 2015; Williamson et al. 2013). Neonicotinoids are applied either as a broadcast spray commonly used in agricultural row crops or systemically either by treating seeds prior to planting (common for corn and soybean, the row crops with the highest acreage in the US) or by drenching soil surrounding crops. Along with acute effects on bees, neonicotinoids are known to disrupt colonies by affecting honey bee learning and foraging (Henry et al. 2012). Current research on the impacts of pesticides on honey bees and other pollinators is focus on field-level assays of colonies positioned near agricultural fields or colonies fed prescribed chemical doses (Alburaki et al. 2018;

Lu et al. 2014). These are challenging experiments and there remains significant work to resolve how to protect honey bees and other pollinators from abiotic stress.

Interactive Effects

As in medicine and veterinary science, co-acting factors are likely to be involved in colony losses, and CCD in particular. Due to the prevalence of pathogens and parasites in collapsing colonies it was natural to search for interactive stresses that compromised honey bee defenses. Indeed, there is experimental and mechanistic evidence that pesticide exposure can impair the abilities of honey bees to fend off disease agents (Sánchez-Bayo et al. 2016). Interactive effects can also occur between two abiotic stresses. Certain crop protection products have synergistic impacts on bee health and identifying such "toxic couples" can lead to regulatory changes that protect honey bees. Reed Johnson (Ohio State University) and colleagues identified an important synergism between the fungicide propiconazole and two insecticides, chlorantraniliprole and diflubenzuron. When applied together, the fungicide and chlorantraniliprole were sevenfold more toxic for bees than the fungicide alone (Wade et al. 2019). This information was used by the almond industry, a mainstay for commercial beekeeping, to develop guidelines for growers against using both synergists while bees were foraging. More widely, climatic stress has been found to interact with pesticide stress in bees, putting colonies at risk (Monchanin et al. 2019).

Conclusions

CCD has been an almost mystical phenomenon for beekeepers and those who seek to protect bees. While CCD was, and is, infrequent, the impacts on effected beekeepers were severe. Beekeepers who suffered CCD often had insufficient survivors to rebuild and were forced to buy replacements from competitors or distant bee suppliers. In the long term, by posing a threat to the very existence of honey bees, CCD galvanized the public and the research community. CCD also exposed strong divisions between beekeepers, researchers, and crop producers, while also forcing increased dialogue across these groups. In one provocative book, Dr. Sainath Suryanarayanan and Daniel Kleinman argue that the controversy of CCD reflects a general difference between the more "experiential" or holistic beekeepers and scientists seeking to explain phenomena in discrete, and at times biased, terms (Suryanarayanan and Lee Kleinman 2016). In terms of veterinary science, CCD greatly expanded the list of biotic causes for bee declines. While the list of registered medicines for bees is still small, targets for diagnostics and management decisions have tripled since 2006 and, for the first time, national surveys are funded to document parasites and pathogens across multiple years (Traynor et al. 2016). Honey bee colonies are arguably more resilient to stress and disease than are pollinators that do not live in large colonies. If the causes of honey bee CCD are general to other pollinators it is likely that significant losses are occurring and are under-reported in these less studied species.

References

Adam, B. (1968). "Isle of Wight" or acariñe disease: its historical and practical aspects. *Bee World* 49: 6–18. https://doi.org/10.1080/0005772X.1968.11097180.

Alburaki, M., Chen, D., Skinner, J.A. et al. (2018). Honey bee survival and pathogen prevalence: from the perspective of landscape and exposure to pesticides. *Insects* 9: 65. https://doi.org/10.3390/insects9020065.

Bailey, L. (1964). The 'Isle of Wight disease': the origin and significance of the myth. *Bee World* 45: 32–37. https://doi.org/10.1080/0005772X.1964.11097032.

Chen, Y. and Evans, J.D. (2007). Historical presence of Israeli acute paralysis virus in the United States. *American Bee Journal* 147: 1027–1028.

Chen, Y., Evans, J.D., Smith, I.B., and Pettis, J.S. (2008). *Nosema ceranae* is a long-present and wide-spread microsporidian infection of the European honey bee (*Apis mellifera*) in the United States. *Journal of Invertebrate Pathology* 97: 186–188.

Claudianos, C., Ranson, H., Johnson, R.M. et al. (2006). A deficit of detoxification enzymes: pesticide sensitivity and environmental response in the honeybee. *Insect Molecular Biology* 15: 615–636.

Cornman, R.S., Tarpy, D.R., Chen, Y. et al. (2012). Pathogen webs in collapsing honey bee colonies. *PLoS One* 7: e43562.

Cox-Foster, D.L., Conlan, S., Holmes, E.C. et al. (2007). A metagenomic survey of microbes in honey bee colony collapse disorder. *Science* 318: 283–287.

F (1916). Isle of Wight disease in bees. *Nature* 97: 161–161. https://doi.org/10.1038/097161a0.

Fairbrother, A., Purdy, J., Anderson, T., and Fell, R. (2014). Risks of neonicotinoid insecticides to honeybees. *Environmental Toxicology and Chemistry* 33: 719–731.

Henry, M., Béguin, M., Requier, F. et al. (2012). A common pesticide decreases foraging success and survival in honey bees. *Science* 336: 348–350.

Krupke, C.H., Hunt, G.J., Eitzer, B.D. et al. (2012). Multiple routes of pesticide exposure for honey bees living near agricultural fields. *PLoS One* 7: e29268.

Lu, C., Warchol, K.M., and Callahan, R.A. (2014). Sub-lethal exposure to neonicotinoids impaired honey bees winterization before proceeding to colony collapse disorder. *Bulletin of Insectology* 67: 125–130.

Lundin, O., Rundlöf, M., Smith, H.G. et al. (2015). Neonicotinoid insecticides and their impacts on bees: a systematic review of research approaches and identification of knowledge gaps. *PLoS One* 10: e0136928. https://doi.org/10.1371/journal.pone.0136928.

Monchanin, C., Henry, M., Decourtye, A. et al. (2019). Hazard of a neonicotinoid insecticide on the homing flight of the honeybee depends on climatic conditions and Varroa infestation. *Chemosphere* 224: 360–368. https://doi.org/10.1016/j.chemosphere.2019.02.129.

Mullin, C.A., Frazier, M., Frazier, J.L. et al. (2010). High levels of miticides and agrochemicals in North American apiaries: implications for honey bee health. *PLoS One* 5: e9754.

Palacios, G., Hui, J., Quan, P.L. et al. (2008). Genetic analysis of Israel acute paralysis virus: distinct clusters are circulating in the United States. *Journal of Virology* 82: 6209–6217. https://doi.org/10.1128/JVI.00251-08.

Sánchez-Bayo, F., Goulson, D., Pennacchio, F. et al. (2016). Are bee diseases linked to pesticides? – a brief review. *Environment International* 89–90: 7–11. https://doi.org/10.1016/j.envint.2016.01.009.

Suryanarayanan, S. and Lee Kleinman, D. (2016). *Vanishing Bees: Science, Politics, and Honeybee Health*. Rutgers University Press.

Traynor, K.S., Rennich, K., Forsgren, E. et al. (2016). Multiyear survey targeting disease incidence in US honey bees. *Apidologie* 47: 325–347. https://doi.org/10.1007/s13592-016-0431-0.

vanEngelsdorp, D., Evans, J.D., Saegerman, C. et al. (2009). Colony collapse disorder: a descriptive study. *PLoS One* 4: e6481. https://doi.org/10.1371/journal.pone.0006481.

Wade, A., Lin, C.H., Kurkul, C. et al. (2019). Combined toxicity of insecticides and fungicides applied to California almond orchards to honey bee larvae and adults. *Insects* 10: 20. https://doi.org/10.3390/insects10010020.

Williamson, S.M., Baker, D.D., and Wright, G.A. (2013). Acute exposure to a sublethal dose of imidacloprid and coumaphos enhances olfactory learning and memory in the honeybee *Apis mellifera*. *Invertebrate Neuroscience* 13: 63–70.

Wilson, W.T. and Menapace, D.M. (1979). Disappearing disease of honeybees: a survey of the United States. *American Bee Journal* 119: 118–119, 184–186, 217.

20

The Parasitic Mite *Varroa destructor*: History, Biology, Monitoring, and Management

David T. Peck

Department of Entomology, Cornell University, Ithaca, NY, USA

Figure 20.1 An adult female *Varroa destructor* perched on an experimentally removed drone pupa. *Source:* Photo courtesy of David Peck.

Introduction

Varroa mites are consistently cited by beekeepers as among the greatest health threats faced by their bees. Most researchers agree that recent increases in honey bee colony loss are attributable to multiple health stressors that harm bees, but when metanalyses are conducted to try to identify primary causes of honey bee colony loss, varroa mite population levels and levels of varroa-associated bee viruses are the best supported causes of colony death (Guzmán-Novoa et al. 2009; Le Conte et al. 2010; Staveley et al. 2014). Colonies of honey bees that have not been treated to control mites typically succumb to their infestation and die before the end of their second winter – with many untreated colonies succumbing and dying at the end of their first autumn season

While reading this chapter, it is important to remember that *Varroa destructor* is not yet a vanquished foe. Much remains unknown about the mite, and best practices for mite management are not unanimously agreed upon and may change dramatically from year to year. Beyond that, the mites are actively evolving resistance to multiple miticidal chemicals (Sammataro et al. 2005), meaning that current best practices may not be effective in the future. Because we do not have definitive solutions for dealing with varroa mites, some of the information in this chapter may become outdated, and some of the described management practices will evolve as research continues into the *Varroa–Apis* parasite–host system. Our knowledge of *V. destructor* (Figure 20.2) is changing far more quickly than our knowledge of longer-established bee diseases. Though this chapter is meant as a guide to understand and manage mite infestation, the diligent bee doctor should make particular efforts to update their knowledge of mite management regularly.

History

To understand *Varroa destructor* consider its recent evolutionary history, which begins with the Eastern (or "Asian") honey bee species, *Apis cerana*. Visually, behaviorally, and physiologically similar to the Western (or "European") honey bee *Apis mellifera*, the Eastern honey bee harbors a number of parasites which live in, on, and around the bees' bodies and nests. Among these parasites is a genus of mites first documented in 1904 and named *Varroa* after ancient Roman scholar and beekeeper Marcus Terentius Varro (Oudemans 1904). The first mite species described, *Varroa jacobsoni*, was eventually detected in *Apis cerana* colonies all across the Eastern honey bees' range, but is not known to cause any major harm to infested colonies, as the mite populations within a colony tend to remain quite small.

Honey Bee Medicine for the Veterinary Practitioner, First Edition. Edited by Terry Ryan Kane and Cynthia M. Faux.

Figure 20.2 An adult mite perched on a toothpick. *Source:* Photo courtesy of David Peck.

In the 1950s and 1960s, beekeepers in the USSR and the Philippines detected unknown red mites infesting their Western honey bee colonies. After investigation, these mites were identified as *Varroa jacobsoni*. Exactly how and when the mites made their transition from infesting *A. cerana* to infesting *A. mellifera* is unknown, but the process was likely driven by the shipment of non-native *A. mellifera* into regions occupied by native *A. cerana* (in the case of the Soviet Union, the movement of bees and beekeeping equipment along the Trans-Siberian Railway between Western and Eastern Russia). Genetic analyses suggest that the mites probably shifted from *A. cerana* to *A. mellifera* more than once (Oldroyd, 1999). Whatever the original circumstances, some interaction between the two bee species allowed *V. jacobsoni* to shift hosts, undergo a period of rapid evolutionary change, and speciate into the harmful parasite *V. destructor*. This apt species name was formalized in 2000 (Anderson and Trueman 2000), after the harmfulness of the mite had manifested itself. However, because this speciation event was recognized relatively recently, decades of invaluable research on the biology of *V. destructor* is published using the name *V. jacobsoni*.

The mite species that typically infests the Western honey bee (*Apis mellifera*) is *Varroa destructor*. The mite species that typically infests the Eastern honey bee (*Apis cerana*) is *Varroa jacobsoni*. However, *V. destructor* can also infest the Eastern honey bee (and does so in East Asia) while *V. jacobsoni* can infest *Apis mellifera* (indeed, the species now known as *V. destructor* arose when *V. jacobsoni* encountered and infested Western honey bees.) The common name for both mite species is variously written as "*Varroa* mite," "Varroa mite," or "varroa mite." Throughout this chapter, the terms "varroa mites" or more often simply "mites" are used to specifically describe *Varroa destructor*, and discussion of the progenitor species *Varroa jacobsoni* is clearly distinguished using the term "*V. jacobsoni*."

The international spread of the mites over the last seven decades has been driven by natural transmissions in some cases (such as honey bee swarms flying over international borders or settling on ships in transit) and by accidental importation by humans in many other cases (such as legally or illegally shipping mite-infested bees).

"Varroa Action Plans" were developed and deployed in a number of countries, but most collapsed soon after the mites invaded, replaced with management guidelines that have been regularly updated and adjusted as we have learned more about these parasites (Table 20.1).

Despite the mites' impact on beekeeping, low levels of infestation are not obvious to inexperienced beekeepers. In addition, many previously mite-free regions did not invest heavily in rigorous and sensitive mite surveillance. For these reasons, in many cases (notably the United States, Ireland, New Zealand, and Nova Scotia) it is estimated that the mites may not have been detected until three to five years after they initially arrived in the region.

Table 20.1 The detection of mites in key countries around the world.

The Global Spread of Varroa destructor
Notable dates of arrival/detection
1950s – Detected in Soviet Union, in what is now Russia
1957 – Philippines
1960s – Japan and eastern Europe
1970s – South America
1977 – Germany
1987 – USA (Wisconsin and Florida, in either liked or independent events)
1989 – Canada
1992 – U.K. (Britain)
2000 – New Zealand
2001 – Ireland
2007 – Hawaii (Oahu)
2010 – Madagascar
2014 – Mauritius
2017 – La Réunion (Indian Ocean)
2018 – Fiji

At this time, the only mite-free regions with a population of *A. mellifera* are Australia, the Isle of Man in the United Kingdom, the island and isolated far-northern mainland region of Newfoundland and Labrador in Canada, New Zealand's Chatham Islands, and a few other isolated regions. Of these, Australia and the Isle of Man have successfully detected and eliminated mite incursions before they developed into catastrophic biosecurity breaches.

Biology

V. destructor is a mesostigmatid mite in the family Varroidae, which only contains the genus *Varroa*. As an arachnid, the mite has eight legs, though like many mites and ticks the first pair of legs are used for both locomotion (grasping and climbing onto bees) as well as chemosensation and mechanosensation (Rosenkranz et al. 2010) making these legs functionally similar to antennae in insects. The mite lives its life in the darkness of the hive and lacks eyes. Due to the reproductive biology of *Varroa* (described below) essentially all mites likely to be encountered by a beekeeper will be females. The adult mites appear to be red to reddish-brown ovals from above –1.5–2 mm wide, and only 1–1.8 mm long. The shape, size, and dorsoventral flattening of the mites allow them to evade grooming by bees, especially when the mites insert themselves between and underneath the overlapping exoskeletal plates (sclerites) on the ventral surface of a bee (Rosenkranz et al. 2010) (Figure 20.3).

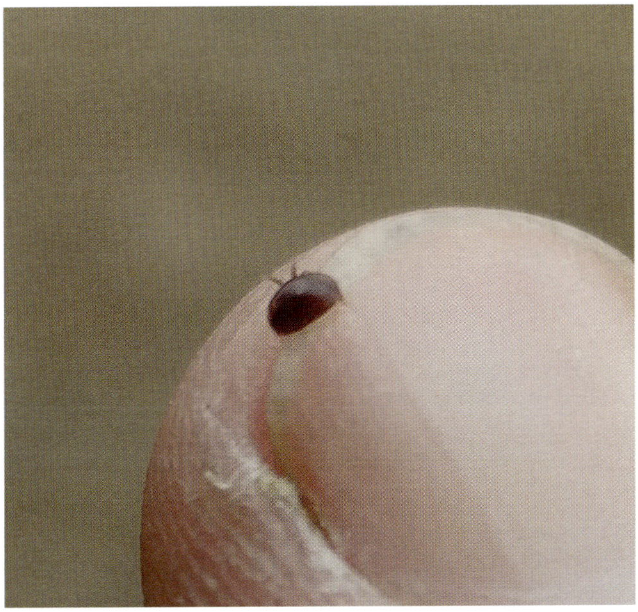

Figure 20.3 A mite on the author's finger, extending her chemosensory forelegs into the air. *Source:* Photo courtesy of David Peck.

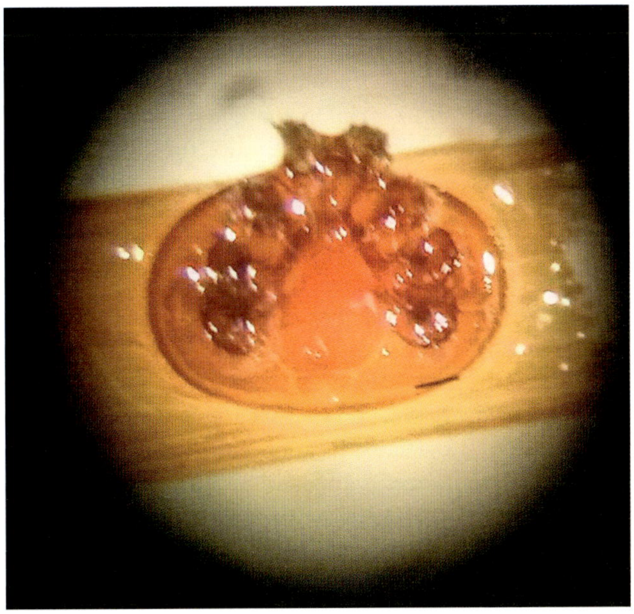

Figure 20.4 The ventral surface of a mite suspended in oil. The large dorsal idiosoma covers and protects the six locomotory legs and the two anterior sensory and locomotory legs of the mite. *Source:* Photo courtesy of David Peck.

The mites have a wealth of adaptations that make them excellent parasites of honey bees. The hydrocarbon cues of the mites incorporate odors from their host colony, which serves to chemically "cloak" the mites (Kather et al. 2015). Even when they are not wedged between the overlapping sclerite armor of a bee's exoskeleton, mites will quickly and nimbly position themselves on the regions of a bee's body that she is least able to groom (Peck et al. 2016). Even their reddish coloration is likely beneficial, since honey bees lack the ability to detect light reflected in the red wavelengths, rendering the mites partially camouflaged to the bees (Figure 20.4).

Mite Life History

Female mites live alternately between two life history phases – the "reproductive" and "phoretic" phases. In the reproductive phase, the mites enter cells containing developing immature bees and produce the next generation of mites, while in the phoretic phase the mites ride on the bodies of bees in the hive, feeding on their fat bodies (not their hemolymph, as had been reported for decades [Ramsey et al. 2018]).

Reproductive Phase

Before the reproductive phase, a mature female mite will position herself on a nurse bee or other young hive bee, since such bees will spend their time in the vicinity of developing larval bees. When the mite detects chemical cues from a large, prepupal larva developing in a nearby cell, the mite will climb from the bee into the cell and will position herself in the larval food at the bottom of the cell. After the workers cap the cell with wax, the last of the larval food is eaten, and the young bee pupates, the mite begins feeding and laying eggs. The mother mite lays the first egg about 70 hours after the cell is capped, and it will be an unfertilized haploid egg that will develop into a haploid male mite. Subsequent eggs, laid at intervals of about 30 hours, will be fertilized, diploid, and will develop into females (Martin 1994). Between laying each egg the mother mite will feed on the pupating bee and will repeatedly excrete waste in the same location on the wall of the cell. This collection of feces is added to by the young male and female mites as they feed and grow through a series of nymphal molts, until they are finally ready to mate (which also takes place at the fecal aggregation site) (Martin 1994). In cells infested by a single mother mite (which is most common) the lone male will impregnate his sisters. (Multiple mother, or "foundress," mites in the same cell allows for genetic exchange between mite lineages.)

When the weakened bee emerges from the cell, the mother mite and the mature daughter mites will cling to its body or crawl out of the cell on their own. The male mite and any immature female mites generally remain behind in the cell and are cleaned out with the rest of the post-pupation refuse. The clearest indication that a cell has been used for mite reproduction are the white remnants of the fecal pile on one of the cell walls, usually at least halfway down the cell's depth and typically on the top surface, since gravity keeps the pupating bee toward the bottom of the cell (Martin 2001). Since mite reproduction is characterized by inbreeding, their population carries few harmful recessive alleles. This no doubt assisted their global spread, since a single mite could arrive in a new country and give rise to hundreds, then thousands, and then millions ofdescendant mites without noteworthy inbreeding depression or other consequences arising from the genetic bottleneck.

Because drone bees in the genus *Apis* generally take about three days longer to develop than workers (Winston 1987), the mites can produce significantly more offspring if they infest a drone cell than a worker cell. This time difference may seem minor, but Martin and Cook (1996) elucidate the benefit to the mites of reproducing on males. It is therefore no surprise that the mites have evolved a strong preference to infest developing drone brood when available (Martin 2001). On the Eastern honey bee, *V. jacobsoni* mites reproduce exclusively on the long-developing drone bees (Anderson and Trueman 2000). The loss of this exclusive preference for drones was a significant part of the evolutionary leap that produced *Varroa destructor* from *Varroa jacobsoni*. This preference for drones means that drone comb in a hive can pose a risk to the resident colony, potentially breeding many generations of mites. However, as described later in the chapter, this preference can also be exploited by beekeepers either in mite monitoring colonies for mite infestation (the drone comb dissection method) or in the non-chemical control of mite populations (the drone comb trapping method).

Phoretic Phase

In the phoretic phase, mites infest adult bees (typically workers) while the bees live and work within the hive. Though this phase is almost universally called the "phoretic" phase, it is not a true case of phoresy (i.e. non-parasitic riding on the body of another organism) because the mites *do* parasitically feed on the bees they are infesting (Ramsey et al. 2018) and so an alternative term like "dispersal" phase may eventually become the norm. In a colony with a low or moderate degree of mite infestation, the mites infest younger worker bees that are unlikely to leave the hive to forage (Cervo et al. 2014). This keeps the mites in the hive, and on the workers most likely to interact with the developing prepupal larvae on which the mites will reproduce. At high enough levels of mite infestation, however, mites cease to preferentially infest young bees, which shifts a higher proportion of the mite population onto older guard and forager bees (Cervo et al. 2014) (Figure 20.5).

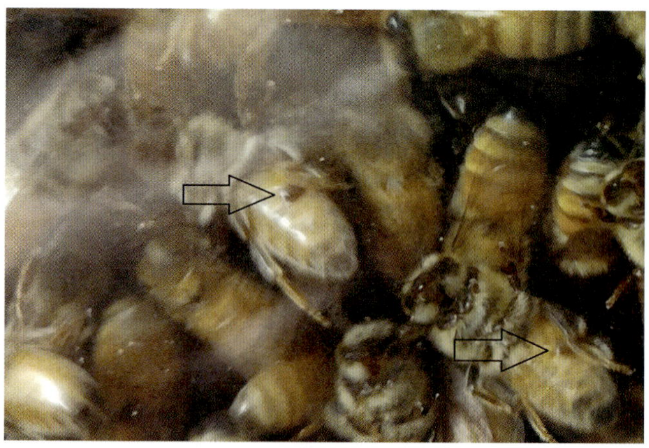

Figure 20.5 Most feeding phoretic mites position themselves between the third and fourth tergites (ventral sclerites) on the abdomen of a bee. The mites show a preference for the left side of the bee, likely because the asymmetries of the bee's abdominal organs make the left side a more efficient site from which to feed (Bowen-Walker et al. 1997). The black arrows each indicate a mite in this sheltered position. *Source:* Photo courtesy of David Peck.

Consequences of Mite Infestation

Uncontrolled mite infestation weakens a colony and can lead to its demise in one or two seasons. The term *varroosis* was coined to describe the diseased state of heavily mite infested colonies, though many beekeepers do not use the term. A more common description is that a colony is "collapsing" due to mites. This may conflate mite-induced colony death with "colony collapse disorder," a phenomenon in which mites are heavily implicated as among the chief causative agents, but which has its own distinct diagnostic criteria. The term "varroosis" is therefore a more precise way to describe symptoms of high mite load, but the term "collapse" is likely to be encountered in many bee yards and is not an incorrect usage as long as it is distinguished from "colony collapse disorder."

Measuring mite population levels (see Monitoring below) is the best way to determine the magnitude of a colony's mite problem, but other symptoms are diagnostically useful. Colonies with heavy mite loads may grow their worker population and also put on honey weight more slowly than healthier colonies nearby. The entrances of heavily infested colonies may have large numbers of small or stunted dead or lethargic adult and pupal bees, as well as live or dead mites, sitting just outside the entrance. High levels of mite infestation may trigger the hygienic removal of infested pupae, which can be detected as uncapped headless brood sitting in cells until the workers completely remove them. Many of the symptoms of high mite load are directly related to the viruses that mites transmit between bees during their parasitism. Seeing bees with the symptoms of these viruses, particularly virus-induced wing deformities, is a clear sign that the mite levels of the colony may be dangerously and even terminally high (Figure 20.6).

Virus Transmission

Parasitism of bees by varroa mites has spread and intensified a number of honey bee viruses (Martin et al. 2012; Kevan et al. 2006). Varroa are significant vectors of a multitude of bee viruses, both between bees in the same colony as well as between bees from different colonies. These varroa-associated viruses, which are readily transmitted during mite feeding, harm individual bees, and threaten entire colonies. All of these viruses are worsened by the presence of varroa mites, and the severity of the mite infestation in a colony is predictive of the viral challenges it will face (Carreck et al. 2010). Since antiviral treatments are not available or forthcoming for beekeepers, the best viral prophylaxis is control of the mite population.

Most of the mite-associated viruses stunt or weaken infected bees particularly during their development, with increased severity as the viral load increases. However, a few viruses produce distinct phenotypes that do not require molecular analysis before a tentative diagnosis can be made. Deformed wing virus (DWV) causes wing deformities in bees by preventing the proper inflation of the wings during the last stages of bee development in the cell. These curled and shriveled wings can arise from non-viral causes but are commonly associated with poorly controlled varroa mite infestation. The bee paralysis viruses, including slow bee paralysis virus (SBPV), chronic bee paralysis virus (CBPV), and acute bee paralysis virus (ABPV) all cause varying degrees of paralysis or lethargy in heavily infected bees, which leads to their early death. Black queen cell virus (BQCV) and sacbrood virus (SBV) cause the conditions for which they are named, while Kashmir bee virus (KBV) weakens bees' ability to cope with other pathogenic microbes (Anderson 1995) (Figure 20.7).

Understanding that much of the harm caused by mite infestation arises from increases in the viral loads throughout the colony offers a crucial insight: If the mite population is able to grow through a summer and is only knocked back by treatment in mid-autumn, it is entirely possible that mite-facilitated transmission and amplification of viruses in the colony may have doomed

Figure 20.6 A normal sized worker bee for scale next to two stunted pupae found outside an experimental colony that was suffering from a terminal mite infestation. *Source:* Photo courtesy of David Peck.

Figure 20.7 This bee developed in a terminally mite-infested colony exhibiting telltale symptoms of varroosis, that likely carried high loads of multiple bee viruses. Her wing deformities rendered this bee incapable of flight and are characteristic of high titers of deformed wing virus (DWV). *Source:* Photo courtesy of Randy Oliver.

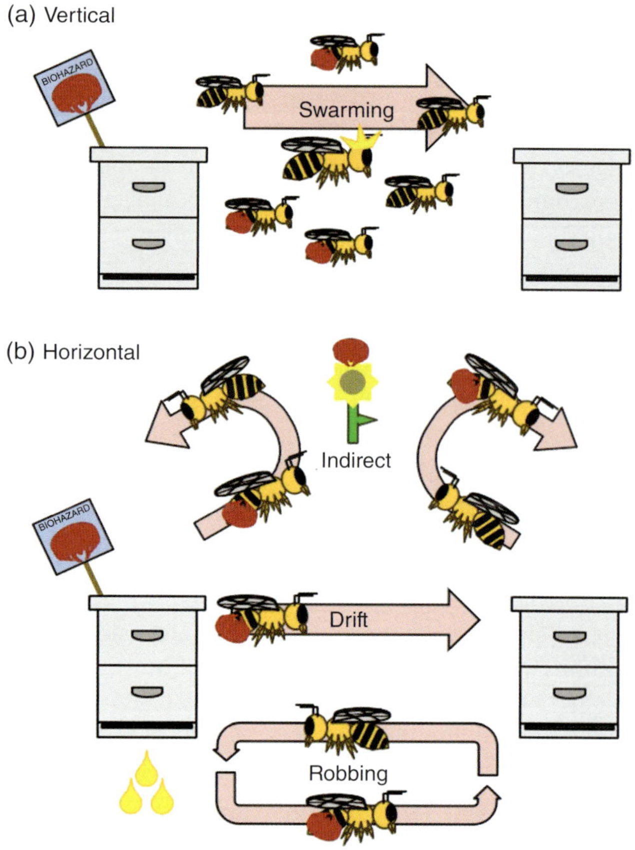

Figure 20.8 Vertical and horizontal mechanisms of mite transmission between colonies. Indirect transmission, exemplified here as transmission via flowers, is hypothesized but as yet unproven, while drift and honey robbing are known drivers of mite transmission between colonies.

the colony to death even if every last mite is killed before winter. This viral amplification over time highlights why mites cannot be considered a seasonal problem to be treated once at the end of the beekeeping season, but instead must be managed and suppressed throughout the year (Carreck et al. 2010). Doing so effectively requires diligent and accurate monitoring of the mite population, as well as timely treatment or intervention.

Transmission Between Colonies

Natural Modes of Transmission

Varroa are spread between colonies attached to bees – though a mite should be capable of walking a few meters from one colony to another, such intercolonial dispersal has not been documented (Figure 20.8). Vertical transmission (between mother and daughter colonies) occurs when an infested colony swarms, with mites infesting both departing bees and bees that remain in the original nest. The two primary modes of natural horizontal transmission (between unrelated colonies) occur via bee drift, and honey robbing. In drift transmission, mites phoretically infesting workers or drones are carried from the bee's natal nest into a new nest. As long as the drifting bee is not repelled by the guards and the entrance to the new hive, the mite is simply carried in and can begin reproducing once she finds a suitable larval bee. In robbing transmission, mites infest robber bees who are stealing honey from the combs of another colony. The mites may originate in the robber bees' nest and climb off in the robbed colony, or they may originate in the robbed colony and be carried back by the robbers along with their stolen honey. This last scenario, in which mite-weakened colonies are robbed by strong colonies situated nearby, represents an ideal opportunity for mite transmission between colonies unless beekeepers intervene.

Additional modes of transmission have been hypothesized, but not conclusively demonstrated. The presence of mites on drone bees captured at drone congregation areas (Mortensen et al. 2018) suggests the possibility of sexual transmission between drones and young queens, though the extremely brief physical contact between drone and queen bees during mating makes this unlikely. Indirect transmission of mites has also been hypothesized: A mite may be transmitted from a bee to a neutral location such as a flower in a field, and then subsequently infest another bee that arrives at the neutral location. Mites may walk off bees at flowers, or be groomed off, and then climb onto foragers from another colony and be airlifted to a new hive. Mites have been found on international shipments of cut flowers (Pettis et al. 2003), mites placed onto flowers will generally quickly orient to the edges of petals from which they could infest arriving bees, and mites placed onto flowers are extremely adept at nimbly climbing onto actively foraging workers (Peck et al. 2016). Despite all this, no definitive evidence of floral transmission from infested bee, to flower, to uninfested bee has been reported. Whether mites are actually transmitted by either of these mechanisms, drift and robbing are sufficient to move many mites between honey bee colonies over large intercolony distances, on the order of hundreds or thousands of meters.

Beekeeper-Assisted Transmission

In addition to the natural mechanisms above, some common beekeeping practices offer mites opportunities to spread between colonies. Mites can survive for a limited period outside of a colony (typically a few hours under most conditions [De Guzman et al. 1993]). This means that poor beekeeper sanitation (failure to clean boots, hive tools, gloves, and other equipment) between colony manipulations can transmit live mites from one colony to the next, especially when a beekeeper visits different colonies or even different apiaries on the same day. In addition, beekeeping practices such as moving bees or transplanting frames of eggs, brood, or honey between colonies of different strengths can easily move mites. The most important beekeeper contribution to mite transmission may simply be the arrangement of hives within crowded apiaries, since small interhive distances of a few meters increase the occurrence of bee drift, and therefore mite transmission (Jay 1966; Peck and Seeley 2019; Seeley and Smith 2015).

Monitoring

Mite infestations *can* be managed *without* rigorous monitoring of mite populations, but this is not advisable for a few key reasons:

- First, without monitoring mite levels it is difficult to know when a colony is approaching danger thresholds and requires immediate treatment. Due to the ease of mite transmission within apiaries, mismanagement of one colony can greatly increase the parasite pressure on neighboring colonies.
- Second, if a mite treatment method begins to lose effectiveness as local mites evolve resistance (see Miticide Resistance in the Treatment section below) it can only be detected if the beekeeper measures mite populations both before and after applying the treatment.
- Third, in the same vein, if any local honey bees begin to demonstrate mite-resistance traits, this highly valuable breeding stock will not be detected by the beekeeper unless mite levels are monitored and the resistant bees identified by their relatively low mite levels prior to any treatments.

However, the time and labor required to repeatedly measure mite levels in each managed colony is often not feasible for beekeepers operating at commercial scales. Thus, despite the benefits of regular and repeated monitoring, it is not difficult to find beekeepers practicing haphazard monitoring or no monitoring at all. Worse, both experienced and beginner beekeepers sometimes struggle to explain precisely how their chosen monitoring method(s) function, and therefore may perform assays incorrectly or may misinterpret the mite population data they obtain. Thus, a bee doctor must have a deep understanding of the mite and bee biology underlying each possible mite population monitoring technique.

Variation in Mite Populations

Mite populations vary between colonies according to colony genotype, colony history (swarming, past miticide treatments), and season. Because mite populations are affected by many colony-level traits, it is easy to miss "problem" colonies if only a subset of the colonies in an apiary are sampled. When feasible, measuring the mite levels in each colony will provide the best guidance on what each colony needs.

Even within a hive, not all bees are equally likely to be infested with mites. Because older forager bees leave the hive, mites infesting them face higher risk than mites on younger bees that spend their time inside the hive. Thus, it is not surprising that mites preferentially infest young bees, particularly the nurse bees that spend much of their time in the broodnest. Because of this preference, if older and younger bees from the same hive are sampled for mites, far more mites will usually be found on the younger bees. However, when mite population levels climb high enough, this apparent preference for younger bees is relaxed, which means that more mites can be found on older bees (Cervo et al. 2014). (This also means that drift by those older foragers is more likely to transfer mites from the heavily infested colony to other colonies nearby.)

On the Seasonality of Mites

A beekeeper diligently monitoring mites in their colonies may be alarmed to suddenly find high mite levels in their colonies as autumn arrives, when measurements only a few weeks before in late summer showed mite levels well below treatment thresholds. This happens when beekeepers forget that most mite monitoring methods measure only the mites in the phoretic phase of their lives. When the colony contains brood, approximately half of the mite population may be reproducing at any time, meaning that measurements of the phoretic mite population may only detect half of the mites in the colony. When brood rearing stops in the autumn (or, indeed, whenever swarming, queen failure, a honeybound broodnest, or any other

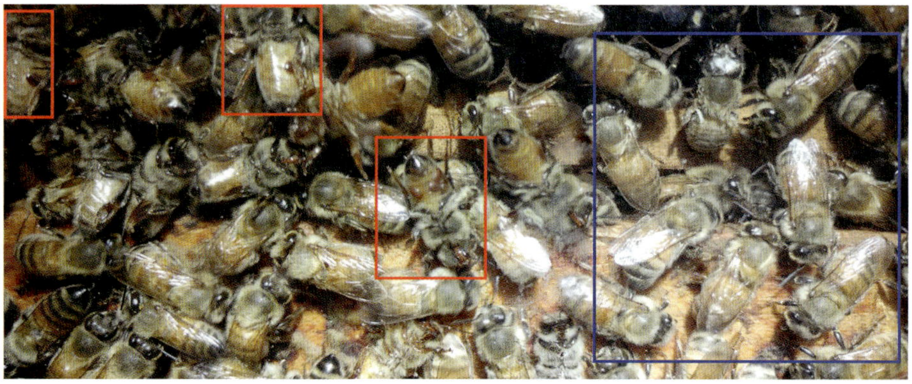

Figure 20.9 An experimentally heavily infested colony in a glass-walled observation hive. The blue box shows the dorsal surfaces of bees with no detectable mites. The red boxes highlight the ventral abdomens of two singly infested bees, and one bee with four mites that are only visually detectable due to the glass hive wall that she is standing on. (Note: This level of hyperinfestation on one bee is likely highly atypical under most beekeeping circumstances). *Source:* Photo courtesy of David Peck.

interruption of bee brood-rearing takes place) most or all of the mites in a colony have no choice but to live phoretically, which produces the alarming late-season "spike" in the apparent mite population that troubles some beekeepers.

Methods not Based on Bee Samples

Visual Inspection

One of the most common "mite monitoring strategies," especially among beginning beekeepers, is the visual inspection or "Keeping an eye out for them" technique, in which a beekeeper visually examines their bees during routine hive inspections and notes whether or not they see any mites on the combs or walking on any of the bees. Figure 20.9 makes clear the shortcomings of this strategy, as these mites were only detectable when the undersides of the bees could be examined and would be completely invisible to someone observing the backs of the bees walking on a frame. Since actively feeding mites wedge themselves underneath the overlapping sclerites on the ventral surface of their host bee, such mites are even harder to spot. Often, by the time mites can be seen walking on the backs of workers a colony's infestation has already progressed well past recommended treatment thresholds.

Visual inspection of the broodnest can also reveal mite-associated problems, but again the damage will be so severe by the time it is visually detected that it will almost certainly be too late for the colony to recover. So-called Parasitic Mite Brood Syndrome, or PMBS, is a visually distinct phenotype in which diseased and frequently partially removed (headless) pupae fill entire frames. Close examination of the exposed pupae often reveals mites actively feeding on the still living pupae (Figure 20.10).

Figure 20.10 "Parasitic Mite Brood Syndrome" in honey bees is characterized by exposed, partially removed, sickly brood with few or no worker bees attending to them. Mites can be seen walking and feeding on the pupae. *Source:* Photo courtesy of David Peck.

Visually inspecting the grass outside a hive can also reveal dangerous levels of mite infestation: When mite levels climb high enough the mite-associated viruses that harm bees begin to kill pupae or cause frequent wing deformities. Young bees with deformed wings are not of value to the hive and are ejected from the colony by healthy workers. These young, flightless bees then crawl through the grass until they are eaten, starve, or die of exposure. Again, by the time these crawlers are seen in the grass, the mite levels in the colony are typically well beyond recommended treatment thresholds. A more rigorous and systematic mite monitoring plan must be used by the responsible beekeeper.

Capped Brood Cell Dissection

A simple method to detect reproducing mites is the drone comb dissection technique, in which capped cells containing developing drone bees of the same approximate age are opened with forceps, the corner of a hive tool, or a cappings scratcher comb. Once the cells are opened, the pale bodies of the drone pupae are examined for bright red mites. Due to the mite preference for reproducing on long-developing drone bees, even rare mites in colonies with low infestations can often be detected by this method. Since drones eat honey but do not make it, beekeepers are often willing to destroy developing drones during inspections. However, the lack of a clear and predictable relationship between the percent of drone cells infested by mites and the overall mite population in the colony means that this method is only modestly informative. It may nonetheless supplement a more rigorous monitoring scheme by showing if any colonies in an apiary have particularly high levels of mite reproduction.

Assessing the infestation level of worker brood can also be informative, though it is rarely necessary when deciding whether or not to apply a miticide treatment to a colony (since knowing the phoretic infestation level is typically sufficient to guide decisions regarding mite treatment). By opening at least 100 cells of capped worker brood, an estimate of the percentage of worker brood cells infested can be made. This can be combined with a measurement of the phoretic mite population (described below) to create an estimate of the complete mite population in a colony. Though these methods are informative, they are prone to high degrees of error unless a very large number of cells are opened, or unless the brood infestation level is at least 2% (Dietemann et al. 2013).

Sticky Boards and Screened Bottom Boards

The sticky board method requires a 2 mm^2 screened bottom board installed beneath the hive, as well as a pale-colored rectangular board, of roughly the dimensions of the hive's footprint, with an adhesive applied to it. Live and dead mites falling down from the combs above are collected on the adhesive and the board can be removed and the mites counted to estimate the mite population (Dietemann et al. 2013). As ants and other detritivorous insects will eagerly eat fallen mites, these creatures must be excluded from the board in order to accurately estimate mite infestation (Dainat et al. 2011). This can be accomplished with either a sufficiently sticky substance on the board, or else hives elevated on wooden legs immersed in cups of oil, water, or some other ant barrier. Even with these controls, the number of mites detected can vary widely due to variations in the patterns of bee brood emergence, the build-up of detritus on the screen, or a number of other factors that may be taking place in the combs above. The natural 24- or 48-hour mite drop onto the sticky board does not allow a direct estimate of the mite infestation per bee in the colony, and so this method is informative but does not give the best data to make treatment decisions.

The sticky board method's great advantage is that it is by far the most sensitive mite sampling method. Since sticky boards sample 100% of the mites falling beneath the entire colony during the sampling period, it will detect mites even at a low level of infestation. This is of value if extremely low levels of mite infestation need to be detected, such as early spring in temperate regions when mite populations are at their natural low, or in mite-free regions attempting to detect the arrival of the first invading mite (in order to prevent their spread.) However, high interday variation in mite drop beneath a moderately or heavily infested colony means that this method is not as helpful as the sample-based methods below (Branco et al. 2006).

An accelerated sticky board sample can also be used to assess mite infestation. Using the same sticky board and screened bottom board as above, mites can be dislodged or killed by a rapid miticidal treatment, and then the number of fallen mites counted. Application of large quantities of powdered sugar to the bees between each frame can prompt grooming and might help to dislodge mites quickly. Application of vaporized oxalic acid (see Treatment below) will kill mites decisively and allow the colony's phoretic mite population to be quickly assessed. (Though importantly, oxalic acid does not kill the mites reproducing inside capped cells and so this method still does not provide a measure of the entire mite population in the colony.)

Bee Sample-Based Methods

The following methods are all based on detaching phoretic mites from a measured sample of bees to determine the percent of bees infested with a mite. These sample-based methods are recommended for beekeepers operating at any scale. Properly noting the units of one's data is critical for these tests, as some beekeepers will report their measurements in mites per 300 bees (that is, total mites counted in the assay) while others will convert their data into mites per 100 bees, or "percent infestation." Treatment thresholds for mite control are likewise reported in either unit, so careful checking is important before numbers are compared.

Both of these methods require a sample of worker bees taken from the brood nest, and particularly taken from frames containing both capped and uncapped cells (Dietemann et al. 2013). Bees from this part of the colony are most likely to be infested with mites throughout the year, and consistently sampling these bees means that assay results will be comparable between colonies and between time points. Sampling bees from the broodnest also reduces the amount of nectar and honey on the frame, which can otherwise drip out of the comb and contaminate the collected sample of bees. Each of the methods described below begin with a sample of 300 worker bees, which are typically collected volumetrically, since 300 worker bees occupy 100 mL of volume. Whether the bees are shaken from the comb or scraped into a measuring container, it is imperative that the queen not be taken in the sample, since even the non-lethal sugar shake method below is capable of killing a fragile queen bee.

Sugar Shake

The sugar shake method (Dietemann et al. 2013) uses powdered sugar (or "icing sugar," or "confectioners' sugar") purchased from a grocery store to dislodge mites from the bees in the sample. Three hundred worker bees (100 mL by volume) are placed in a wide-mouthed glass or plastic jar, with a screw-top lid that has been cut and fitted with a piece of 2 mm^2 hardware cloth that will allow mites and sugar to pass through when inverted, but will not allow bees to escape. Two tablespoons (30 mL) of powdered sugar is added to the bees, and the jar is quickly shaken to coat the bees. The jar is then set aside for two minutes so that the bees can begin grooming. A faint cloud of sugar floating up through the screen indicates that all is working. After two minutes, the jar is inverted over a light-colored plate or other container, and the jar is shaken vigorously for 1.5 minutes to knock both sugar and mites through the screen and onto the plate. The contents of the plate can then be sprayed with water, which will dissolve the sugar and expose the mites for counting. Once the mites are counted, the still-living sugar-coated bees in the jar can be poured onto the hive entrance or the tops of the hive's frames where they will be cleaned by their sisters (Figures 20.11 and 20.12).

No mite-sampling method will dislodge 100% of the mites in the sample 100% of the time, but many criticisms of reduced mite yield from the sugar shake test arise from the technique being only partially understood, and therefore being partially misapplied. Because phoretic mites are often securely wedged between the abdominal sclerites of the bees they are feeding from, application of powdered sugar to the outside of the host bee is not enough to immediately dislodge the mite. The sugar shake method works by covering bees in sugar, *then leaving the bees for enough time that their grooming and wing buzzing increases their body temperature*. This increase in temperature drives the mite out of the relative safety of its feeding site, and from there

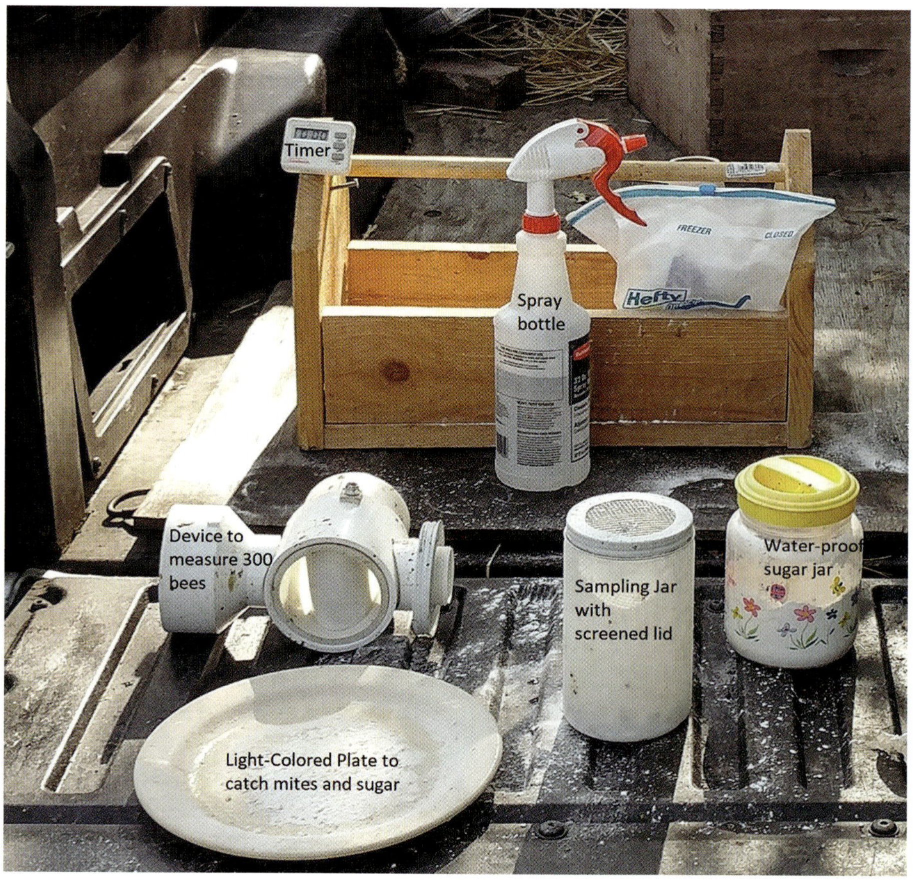

Figure 20.11 The author's sugar shake sampling kit arranged on the tailgate of the field truck. As 300 typical worker bees fill a volume of 100 mL, any device capable of measuring out 100 mL of bees is sufficient to obtain a correctly sized sample. *Source:* Photo courtesy of David Peck.

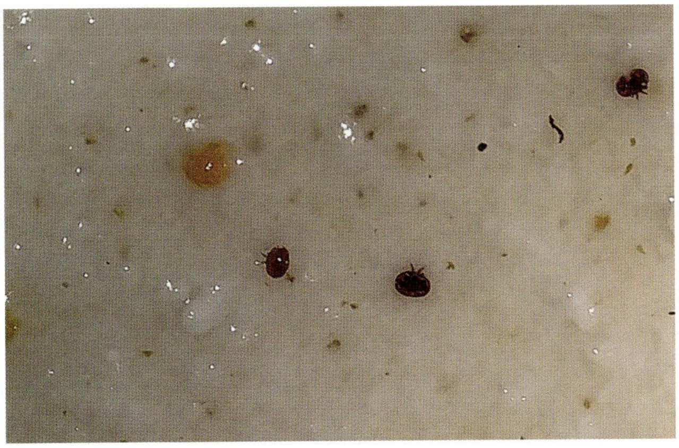

Figure 20.12 A yellow ball of pollen and three red mites floating in dissolved sugar after having been shaken onto the sampling plate. *Source:* Photo courtesy of David Peck.

the slippery powdered sugar and the grooming of the bees will separate the mites from the bees. An impatient sampler who proceeds before this process takes place will detect far fewer mites than one who performs the test properly.

The other common source of error in this method is the sensitivity of the sugar shake to moisture. Rain, high humidity, nectar contaminating the sample of bees, nectar regurgitated by bees in the sample, or residual moisture from cleaning the sampling equipment between colonies can all cause the powdered sugar to form clumps or even a paste. In such cases, the mechanical properties of the sugar change and it is no longer able to dislodge mites (and if extremely wet, may also suffocate some of the bees in the sample). Thus, the test should only be performed in dry weather, and the bees should be sampled from frames with as little nectar as possible.

Alcohol Wash

As in the sugar roll, the alcohol wash performed on a sample of workers, typically 300 bees, taken from the broodnest. In the alcohol wash, the bees can be placed in the same type of screen-topped jar used for the sugar shake, or can be put into any other container that contains screen or perforations that will allow liquid and mites to pass through but will keep bees to one side. The bees are killed when alcohol is added to the container, and then the sample must be agitated slightly for about 20 seconds to dislodge the now-dead mites from the dead bees (Dietemann et al. 2013). Passing the alcohol and mites through the screen, sieve, or perforation will allow the beekeeper to quantify the number of mites on the 300 bees. As with the sugar shake method above, the alcohol was highly accurate, but still sometimes fails to detect 100% of the mites in the sample (Table 20.2).

Problems with Lethal Sampling Methods

Though the alcohol wash sampling procedure offers a fast, accurate, and relatively foolproof method to measure the phoretic mite level in a colony, it carries one significant drawback: it is lethal to the sampled bees. Some beekeepers, especially commercial beekeepers with many hives, are willing to accept the death of a few hundred bees per colony to ensure an accurate mite count. However, other beekeepers express discomfort about killing so many bees, either because doing so depletes the colony's workforce at critical times in the season, or because they simply don't like to kill bees at all. In such cases, recommending the sugar-shake sampling method is ideal. Though it can be slightly more complicated to perform properly, and though it may not dislodge every mite in the sample, because it does not kill bees it is more likely to actually be performed by some beekeepers. A highly accurate monitoring method that a beekeeper is reluctant to use will prove less effective than a slightly less accurate method that the beekeeper will not avoid using.

Treatment

That a colony contains mites is no surprise unless it is in one of the few regions still considered varroa-free at the time of this writing (i.e. Australia, Newfoundland, the Isle of Man, and a few other small islands around the world). Thus, assuming that essentially every colony has mites, the first question any bee doctor must answer is: How many

Table 20.2 Summary of methods for mite monitoring.

Method	Sensitivity	Accuracy	Ease of use	Time required	Recommended use
Visual inspection	Low	Low	High	Moderate	Not recommended
Brood dissection	Moderate	Low	High	Moderate	Supplemental to other methods
Sugar shake	Moderate	High	Moderate	Moderate	Perform repeatedly throughout season
Alcohol wash	Moderate	High	High	Low	Perform repeatedly throughout season
Sticky board mite drop	High	Low	Moderate	High	Supplemental to other methods. Sensitive for low populations or detecting arrival to mite-free regions.

mites is too many? Many treatments or interventions can have negative impacts on the health of a hive or the safety or quality of honey and wax and should not be implemented without considering the costs to colony health and hive productivity. Once a decision is made to treat, the second question is: Which treatment should be used? The availability and costs of treatments, the time required per treatment, beekeeper comfort with treatment methods, whether honey intended for human consumption is currently being produced, and a number of other factors may all influence which treatment method is chosen, and many successful beekeepers make use of multiple methods throughout a season, or year to year.

Treatment Thresholds

Definitive economic treatment thresholds, while often requested by beekeepers, are rarely backed by definitive science, and what data do exist are complicated by factors that make data from one region of questionable use to beekeeping in another region. Some beekeepers treat their colonies regardless of infestation level. Some treat if mite levels are even slightly elevated in the spring (>1% phoretic infestation) but will not apply autumn treatments unless levels are above a higher threshold, typically 3%. Others immediately treat whenever phoretic mite levels climb above 1%, no matter what time of year, while others wait until 2%, 3%, or 4% phoretic infestation before they take action, often basing their decisions on mite management guidelines published in the past. Ongoing research into effective mite control and best practices will likely change the recommended best practices again in the future, so reliable and current sources of information should be sought out, especially those based in bee health data from your region. How many miticide applications will be required each year, with which miticides, applied at what times, depends on the reproductive patterns of the bees and mites over the season, the ability of the bees to suppress mite growth, the local climate, the timing of honey crops meant for human consumption, and most importantly: the beekeeper's relative tolerance for the risk of colony loss due to mites and the risk of bee toxicity or honey contamination from miticide application.

A very healthy colony assayed in early spring with one of the sample-based methods above might not yield a single mite. Colonies with one or two mites (per 300 bees) at the first spring measurement in a temperate climate are also likely reasonably healthy. Because mite populations grow exponentially throughout the summer in temperate climates, controlling mite levels in the spring can be highly protective by knocking the starting mite population lower and reducing the maximum population they will attain. For this reason, some beekeepers will treat their bees in the spring, when mite populations are at their lowest. Currie and Gatien (2006) described increased honey yields when spring treatments were applied to Canadian colonies showing more than 2% phoretic mite infestation. Mite populations reach their peak in mid- to late-autumn in most temperate climates, after a full season of mite reproduction. A mite sample performed in October in New York State may yield a large number of mites, but it may be too late to apply effective treatment at that time, since developing winter bees may already be infected with viruses and some chemical miticides are ineffective at lower ambient temperatures. Modeling of mite population growth suggests that mite levels must be suppressed by mid-August to prevent winter mortality (DeGrandi-Hoffman and Curry 2005).

Chemical

Government regulatory agencies around the world maintain lists of chemical miticides that have been approved for control of *V. destructor* in honey bee colonies. An array of chemical miticides are currently approved in the United States and Canada for control of *V. destructor* in honey bee colonies, each with strict guidelines regarding their application to bee hives, which are beyond the scope of this chapter. However, a brief discussion of the chemical treatments currently available may help to contextualize the different classes of miticide (Table 20.3).

Table 20.3 Miticides approved by the U.S. Environmental Protection Agency for the control of *Varroa destructor* in honey bee colonies.

Product name	Active ingredient
APISTAN STRIPS	Fluvalinate (10.25%)
CHECKMITE+ BEE HIVE PEST CONTROL STRIP	Coumaphos (10%)
AVACHEM SUCROSE OCTANOATE [40.0%]	Sucrose octanoate (40%)
APIGUARD	Thymol (25%)
API LIFE VAR	Thymol (74.09%), oil of eucalyptus (16%), menthol (3.73%)
HOPGUARD II	Hop beta acids resin (16%)
FOR-MITE	Formic acid (65.9%)
MITE-AWAY QUICK STRIPS	Formic acid (46.7%)
FORMIC PRO	Formic acid (42.25%)
OXALIC ACID DIHYDRATE	Oxalic acid (100%)
Apivar	Amitraz (3.33%)

Source: Modified from www.epa.gov.

Synthetic Pesticides

Fluvalinate and coumaphos were both effective compounds for controlling *V. destructor* when they were first introduced, but mites have evolved resistance to both compounds (Baxter et al. 1998; Pettis 2004; Sammataro et al. 2005) and both compounds are capable of contaminating beeswax and equipment for long periods of time, which may cause harm to developing bees (Pettis et al. 2004). Amitraz, meanwhile, is the highly effective miticide of choice for many beekeepers, though reports of evolving amitraz resistant mites (Rinkevich 2020) suggest that it may not be a permanent solution for mite control. New synthetic miticides will continue to be developed, and mite populations will continue to experience selective pressures to evolve resistance to these miticides, so regular reading of the most current literature is required to develop a treatment plan based on best practices.

Organic Acids

Formic acid and oxalic acid are both naturally occurring organic acids capable of killing mites without causing direct mortality to bees (when applied as approved). Usefully, both are found in small quantities in naturally produced honey, which means that minor residual contamination of honey by the treatments is generally allowable, though neither compound is meant to be applied to colonies actively making honey. An added complexity of these treatments is that both are relatively sensitive to the ambient temperature, and misapplication, especially of formic acid, can kill bees and/or cause colonies to replace their queens. The compounds are applied to the colony differently, but are both effective at killing mites. Most applications of formic acid can penetrate the wax caps of sealed brood cells, and can kill mites therein, while oxalic acid is only effective at killing phoretic mites and must be reapplied to catch mites that were within brood cells during the last application. Oxalic acid can be applied by use of a battery powered heat vaporizer, or via other methods which attempt to achieve a slow release of the compound in the hive.

Other Compounds

The plant derivatives in Table 20.3 are known to be effective miticides, though thymol imparts a detectable fragrance to equipment treated with it, and the odor can be detected in honey that is produced in such equipment. Hop beta acids are a relatively recent offering for mite control but are effective (Degrandi-Hoffman et al. 2012). The final compound in the table above, sucrose octanoate, is an effective miticide but requires direct contact with the mites, which makes application difficult and time-consuming, and it is not currently preferred by many beekeepers.

Non-Chemical Mite Interventions

Brood Interruption by Small Hives, Swarming, and Splitting

By studying free-living and experimental colonies, it has become clear that bees in small colonies that swarm during the summer generally show lower mite levels in the summer and autumn compared to large colonies that are prevented from swarming (Loftus et al. 2016). This interruption of the honey bee brood rearing cycle, when the colony swarms or when a small colony fills its brood nest with nectar, prevents the mites from reproducing and allows any mite-resistance traits in the bees such as grooming to work on the mite population while it cannot grow. Simply housing colonies in small hives and allowing them to swarm may reduce (but almost never eliminate) the need for mite treatments. Attempts to replicate these effects by deliberately splitting colonies have yielded positive results. However, such methods produce small colonies, which will field relatively smaller foraging workforces, significantly reducing per-colony honey production and pollination efficiency.

Drone Comb Removal

One method which can be highly effective is the removal of infested drone pupae. By inserting a frame of drone comb into the colony, waiting for it to be filled with drone larvae, then waiting for those drone larvae to be preferentially infested by mites, and then removing and freezing or destroying the comb, a beekeeper can pull mites out of their colony without applying any chemicals. However, the high economic costs to the colony of rearing a frame of drones (in the form of pollen, honey, and labor sacrifices for the drones) means that this may not be efficient. Also, when perfectly executed this technique will lead to a sizeable reduction in the mite population, but when drone comb removal is poorly executed, by neglecting to remove the drone combs on the correct day, it leads to skyrocketing populations. When experiments have required colonies to die from high mite loads, our lab group has been able to achieve this simply by giving large amounts of drone comb to colonies with moderate spring mite loads. In one batch of 10 colonies kept in this way, seven experienced apparently mite-induced death before their first winter.

Screened Bottom Boards

Since live mites are sometimes found on sticky boards placed under sticky bottom boards, it stands to reason that by removing the bottom board beneath the screen, some number of live mites will fall out of the hive and perish in the debris below. Some beekeepers apply powdered sugar to the bees in the colony to promote the dislodging of the mites, using the same principle as the sugar shake sampling method. However, there is no conclusive evidence that any beekeeper can control mite populations using this method alone.

Treatment-Free Beekeeping

In recent years, interest in treatment-free beekeeping practices and philosophies has grown widely and rapidly. Treatment-free beekeepers reason that the only sustainable solution to the problem of *V. destructor* will be the production of mite-resistant bees that can control the mites on their own, and that the best way to obtain such bees is to allow weak colonies and their genes to die and be replaced by only the strongest bees and genes. Though a laudable goal, some treatment-free beekeepers misunderstand elements of host-parasite coevolution, mite transmission biology, and honey bee behavior, and therefore may be doing more harm than good. Some beekeepers use the excuse of treatment-free beekeeping to avoid learning about mite biology or management at all. Therefore, they often misunderstand the impacts that their behaviors may be having on the health of both their colonies and the local honey bees in their region.

Declining to treat colonies for varroa mites imposes a selective pressure on the bees that will kill mite-susceptible colonies and leave behind mite-resistant ones. However, unless the treatment-free beekeeper's bees happen to carry robust mite-resistance, such a selective pressure will simply kill all of the colonies under their care and terminate their evolutionary project. In addition, even if a colony is obtained that is able to suppress the growth of its mite population through a season, if such a colony is placed within drifting or robbing range of sickly colonies it is likely that they will not survive. Mites flowing into resistant hives in the autumn may introduce enough viruses and inflict enough damage on the developing winter bees that the colony will succumb to the mites anyway or will at least be seriously challenged by a small bee population and a high mite load at the beginning of the following spring. Unless a treatment-free beekeeper can ensure that any mite-resistant colonies will be safe from outside mites, it's unlikely that such a program will succeed.

Since mite-weakened colonies can serve as "robber lures," attracting robber bees from healthy colonies and transmitting large numbers of mites to them (Peck and Seeley 2019), if an untreated colony develops a high mite population in mid-summer it may pose a serious disease risk to nearby colonies. Therefore, a beekeeper disinclined to treat their colony may be best served by euthanizing the entire colony and their mites to protect neighboring bees. An alternative is to use a mite treatment to kill the mites and then to replace the queen in that mite-susceptible colony the next spring. In this way the beekeeper is helping to shape the mite-resistance traits in the gene pool, while preventing their infested colonies from being robbed by their bees, their neighbors' bees, or the bees living in wild colonies nearby. Our lab group has found that some bees in unmanaged colonies (living without beekeeper assistance) have been found to carry behavioral mite-resistance traits, making swarms from wild or feral colonies a potential source of mite-resistant bees and genetics.

Mite-Resistant Bees

In the long term, the best way to control *V. destructor* may be by obtaining bees that are less permissive hosts. Bees that can either resist the mites (kill and/or suppress mite populations) or tolerate mites and their viruses (survive despite high parasite load) would obviate the need for chemical treatments and complex bee husbandry. Efforts to obtain naturally or artificially selected mite-resistant bees is ongoing, and a number of lines show promise, though none are currently "mite-proof." One tactic available to a beekeeper trying to control mites would be obtaining bees that demonstrate hygienic behavior, grooming behavior, or other putative mite-resistance traits, but unfortunately none of the commercially available bee lines are currently resistant enough that a beekeeper can neglect mite monitoring and treatment (Table 20.4).

Conclusion

In general, beekeepers accept that they must understand honey bee behavior and biology in order to keep bees well. Many beekeepers are eager to learn new information about their bees if it is made available to them. Beekeepers, like other agriculturalists, are sometimes less interested in learning the biological details of the many organisms that infect and infest their livestock. One important way a bee doctor can assist beekeepers (and their bees) is to ensure that the beekeeper understands *Varroa destructor* as a creature with its own evolutionary history, physiological adaptations, and behavioral repertoire. By understanding the mites as a complex enemy, beekeepers can be better prepared to fight and control the mites in their bee colonies.

Table 20.4 Putative mite-resistance traits identified in different bee populations.

Putative mite-resistance trait	Mechanism of action
Hygienic behavior	Brood hygienic behavior describes the uncapping and removal of diseased, dead, or distressed capped brood through worker sensitivity to chemical cues detected through the caps of brood cells. This trait allows bees to remove brood infected with a number of pathogens and is often assayed using tests that do not directly involve mites. For example, hygienic behavior can be scored by freezing a section of capped brood with liquid nitrogen and assessing how many of the dead pupae are removed when the brood comb is placed back into the colony for 24 hours.
Varroa-sensitive hygienic (VSH) behavior	This term describes a more specific version of the generalized brood hygienic behavior described above. Bees can be assayed specifically for their ability to detect and remove brood containing reproducing mites, and thus demonstrate that their hygienic behavior is sensitive to the reproduction of mites, and not generalized brood hygiene that might be triggered by various infectious or non-infectious problems with brood health.
Suppression of mite reproduction (SMR)	Originally described as an unknown mechanism through which bees somehow stopped mites from reproducing in brood cells. Subsequent research revealed that pupating bees with this trait were not preventing mite reproduction directly but were instead uncapping and hygienically emptying cells that contained reproducing mites, while leaving cells with non-reproducing mites alone. This behavior was detected as large numbers of non-reproductive mites in brood cells, and it was thought that pupating bees were suppressing mite reproduction, but this is now generally considered a form of varroa-sensitive hygienic behavior. However, research continues into the possibility that bee pupae could somehow suppress mite reproduction.
Grooming	Individual bees can groom mites from their bodies and may or may not damage the mites in the process. Breeding programs have successfully increased grooming rates by selectively breeding colonies wherein a larger percent of the mites falling below the colony show signs of damage from the bee's mandibles. This is one of the few resistance traits that targets mites in the phoretic phase and not the reproductive phase.
Uncapping/recapping behavior	Bees with this trait uncap mite-infested cells, which disrupts the reproduction and mating of mites in the cells. Instead of removing the pupal bees, as in hygienic lines, colonies with this trait recap the brood cells and allow the pupal bees to continue development.
Reduced development time	Since mites lay eggs at regular intervals while reproducing in brood cells, if the cell is capped for a shorter duration it can significantly reduce the reproductive success of the mites. Some populations of *Apis mellifera* (e.g. many subspecies native to mainland Africa) produce workers with a shorter post-capping period, which partially explains their relative resistance to the mites compared to bees derived from European subspecies.
"Survivor" bees or "Russian" bees	Populations of bees that persist despite mite infestation, and without any mite treatment, are often called "survivor" bees. Studies of these bees around the world have found elevated levels of one or more known mite-resistance traits, and novel resistance traits may yet be discovered. Known wild-living "survivor" bee populations are summarized by Locke (2016). "Russian" bees are derived from a queen-breeding program that imported bees from Eastern Russia to the United States, reasoning that since bees in Eastern Russia had been exposed to the mites the longest, they would have had the most time to evolve resistance traits. The breeding program documented relatively high mite-resistance in the bees, but beekeepers sometimes complain that the bees also have undesired traits (defensive temperament, proclivity to swarm, poor population growth, low honey productivity, etc.)

The preceding is a non-exhaustive list of traits currently under consideration in breeding programs attempting to produce mite-resistant bees.

References

Anderson, D.L. (1995). Viruses of *Apis cerana* and *Apis mellifera*. In: *The Asiatic Bee Hive: Apiculture, Biology, and Role in Sustainable Development in Tropical and Subtropical Asia* (ed. P.G. Kevan), 161–170. Cambridge, ON, Canada: Enviroquest, Ltd.

Anderson, D.L. and Trueman, J.W.H. (2000). *Varroa jacobsoni* (Acari: Varroidae) is more than one species. *Experimental & Applied Acarology* 24: 165–189. https://doi.org/10.1023/a:1006456720416.

Baxter, J., Eischen, F., Pettis, J. et al. (1998). Detection of fluvalinate-resistant varroa mites in US honey bees. *American Bee Journal* 138: 291.

Bowen-Walker, P.L., Martin, S.J., and Gunn, A. (1997). Preferential distribution of the parasitic mite, *Varroa jacobsoni* Oud. on overwintering honey bee (*Apis mellifera* L.) workers and changes in the level of parasitism. *Parasitology* 114: 151–157. https://doi.org/10.1017/S0031182096008323.

Branco, M.R., Kidd, N.A.C., and Pickard, R.S. (2006). A comparative evaluation of sampling methods for *Varroa destructor* (Acari: Varroidae) population estimation. *Apidologie* 37: 452–461. https://doi.org/10.1051/apido:2006010.

Carreck, N.L., Ball, B.V., and Martin, S.J. (2010). Honey bee colony collapse and changes in viral prevalence associated with *Varroa destructor*. *Journal of Apicultural Research* 49: 93–94. https://doi.org/10.3896/IBRA.1.49.1.13.

Cervo, R., Bruschini, C., Cappa, F. et al. (2014). High Varroa mite abundance influences chemical profiles of worker bees and mite-host preferences. *The Journal of Experimental Biology* 217: 2998–3001. https://doi.org/10.1242/jeb.099978.

Currie, R.W. and Gatien, P. (2006). Timing acaricide treatments to prevent *Varroa destructor* (Acari: Varroidae) from causing economic damage to honey bee colonies. *Canadian Entomologist* 138: 238–252. https://doi.org/10.4039/n05-024.

Dainat, B., Kuhn, R., Cherix, D., and Neumann, P. (2011). A scientific note on the ant pitfall for quantitative diagnosis of *Varroa destructor*. *Apidologie* 42: 740–742. https://doi.org/10.1007/s13592-011-0071-3.

De Guzman, L.I., Rinderer, T.E., and Beaman, L.D. (1993). Survival of *Varroa jacobsoni* Oud. (Acari: Varroidae) away from its living host *Apis mellifera* L. *Experimental and Applied Acarology* 17: 283. https://doi.org/10.1007/BF02337278.

DeGrandi-Hoffman, G. and Curry, R. (2005). The population dynamics of varroa mites in honey bee colonies: part I – the VARROAPOP program. *American Bee Journal* 145 (7): 592–595.

DeGrandi-Hoffman, G., Ahumada, F., Probasco, G., and Schantz, L. (2012). The effects of beta acids from hops (*Humulus lupulus*) on mortality of *Varroa destructor* (Acari: Varroidae). *Experimental & Applied Acarology* 58 (4): 407–421. https://doi.org/10.1007/s10493-012-9593-2.

Dietemann, V., Nazzi, F., Martin, S.J. et al. (2013). Standard methods for varroa research. In: *The COLOSS BEEBOOK, Volume II: Standard Methods for Apis mellifera Pest and Pathogen Research*, vol. 52 (eds. V. Dietemann, J.D. Ellis and P. Neumann), 11–3896. Treforest: International Bee Research Association IBRA.

Guzmán-Novoa, E., Eccles, L., Calvete, Y. et al. (2009). *Varroa destructor* is the main culprit for the death and reduced populations of overwintered honey bee (*Apis mellifera*) colonies in Ontario, Canada. *Apidologie* 41: 443–450. https://doi.org/10.1051/apido/2009076.

Jay, S.C. (1966). Drifting of honeybees in commercial apiaries. III. Effect of apiary layout. *Journal of Apicultural Research* 5: 137–148. https://doi.org/10.1080/00218839.1966.11100147.

Kather, R., Drijfhout, F.P., and Martin, S.J. (2015). Evidence for colony-specific differences in chemical mimicry in the parasitic mite *Varroa destructor*. *Chemoecology* 25: 215. https://doi.org/10.1007/s00049-015-0191-8.

Kevan, P.G., Hannan, M.A., Ostiguy, N., and Guzman-Novoa, E. (2006). A summary of the Varroa-virus disease complex in honey bees. *American Bee Journal* 146 (8): 694–697.

Le Conte, Y., Ellis, M., and Ritter, W. (2010). Varroa mites and honey bee health: can varroa explain part of the colony losses? *Apidologie* 41: 353–363. https://doi.org/10.1051/apido/2010017.

Locke, B. (2016). Natural *Varroa* mite-surviving *Apis mellifera* honeybee populations. *Apidologie* 47: 467–482. https://doi.org/10.1007/s13592-015-0412-8.

Loftus, J.C., Smith, M.L., and Seeley, T.D. (2016). How honey bee colonies survive in the wild: testing the importance of small nests and frequent swarming. *PLoS One* 11 (3): e0150362. https://doi.org/10.1371/journal.pone.0150362.

Martin, S.J. (1994). Ontogenesis of the mite *Varroa jacobsoni* Oud. in worker brood of the honeybee *Apis mellifera* L. under natural conditions. *Experimental & Applied Acarology* 18: 87–100. https://doi.org/10.1007/BF00130823.

Martin, S.J. (2001). Biology and life-history of varroa mites. In: *Mites of the Honey Bee* (eds. T.C. Webster and K.S. Delaplane), 131–148. Hamilton, IL: Dadant & Sons.

Martin, S.J. and Cook, C. (1996). Effect of host brood type on number of offspring produced by the honeybee parasite *Varroa jacobsoni*. *Experimental & Applied Acarology* 20: 387–390. https://doi.org/10.1007/BF00130551.

Martin, S.J., Highfield, A.C., Brettell, L. et al. (2012). Global honey bee viral landscape altered by a parasitic mite. *Science* 336 (6086): 1304–1306. https://doi.org/10.1126/science.1220941.

Mortensen, A.N., Jack, C.J., and Ellis, J.D. (2018). The discovery of *Varroa destructor* on drone honey bees, *Apis mellifera*, at drone congregation areas. *Parasitology Research* 117 (10): 3337–3339. https://doi.org/10.1007/s00436-018-6035-z.

Oldroyd, B.P. (1999). Coevolution while you wait: *Varroa jacobsoni*, a new parasite of Western honeybees. *Trends in Ecology & Evolution* 14 (8): 312–315. https://doi.org/10.1016/s0169-5347(99)01613-4.

Oudemans, A.C. (1904). Note VIII. On a new genus and species of parasitic Acari. *Notes Leyden Museum* 24: 216–222.

Peck, D.T. and Seeley, T.D. (2019). Mite bombs or robber lures? The roles of drifting and robbing in *Varroa destructor* transmission from collapsing honey bee colonies to their neighbors. *PLoS One* 14 (6): e0218392. https://doi.org/10.1371/journal.pone.0218392.

Peck, D.T., Smith, M.L., and Seeley, T.D. (2016). *Varroa destructor* mites can nimbly climb from flowers onto foraging honey bees. *PLoS One* 11 (12): e0167798. https://doi.org/10.1371/journal.pone.0167798.

Pettis, J.S. (2004). A scientific note on *Varroa destructor* resistance to coumaphos in the United States. *Apidologie* 35 (1): 91–92. https://doi.org/10.1051/apido:2003060.

Pettis, J.S., Ochoa, R., and Orr, J. (2003). Interception of a live Varroa mite on imported cut flowers in the United States. *International Journal of Acarology* 29: 291–292. https://doi.org/10.1080/01647950308684342.

Pettis, J.S., Collins, A.M., Wilbanks, R., and Feldlaufer, M.F. (2004). Effects of coumaphos on queen rearing in the honey bee, *Apis mellifera*. *Apidologie* 35: 605–610. https://doi.org/10.1051/apido:2004056.

Ramsey, S.D., Ochoa, R., Bauchan, G. et al. (2018). *Varroa destructor* feeds primarily on honey bee fat body tissue and not hemolymph. *Proceedings of the National Academy of Sciences of the United States of America* 116: 1792–1801. https://doi.org/10.1073/pnas.1818371116.

Rinkevich, F.D. (2020). Detection of amitraz resistance and reduced treatment efficacy in the Varroa Mite, *Varroa destructor*, within commercial beekeeping operations. *PLoS One* 15 (1): e0227264. https://doi.org/10.1371/journal.pone.0227264.

Rosenkranz, P., Aumeier, P., and Ziegelmann, B. (2010). Biology and control of *Varroa destructor*. *Journal of Invertebrate Pathology* 103 (Suppl. 1): S96–S119. https://doi.org/10.1016/j.jip.2009.07.016.

Sammataro, D., Olafson, P., Guerrero, F., and Finley, J. (2005). The resistance of varroa mites (Acari: Varroidae) to acaricides and the presence of esterase. *International Journal of Acarology* 31: 67–74. https://doi.org/10.1080/01647950508684419.

Seeley, T.D. and Smith, M.L. (2015). Crowding honey bee colonies in apiaries raises their vulnerability to the deadly ectoparasite *Varroa destructor*. *Apidologie* 46: 716–727. https://doi.org/10.1007/s13592-015-0361-2.

Staveley, J.P., Law, S.A., Fairbrother, A., and Menzie, C.A. (2014). A causal analysis of observed declines in managed honey bees (*Apis mellifera*). *Human and Ecological Risk Assessment* 20: 566–591. https://doi.org/10.1080/10807039.2013.831263.

Winston, M.L. (1987). *The Biology of the Honey Bee*. Cambridge, MA: Harvard University Press.

21

Honey Bee Viral Diseases

Esmaeil Amiri[1,2], *Olav Rueppell*[1,3], *and David R. Tarpy*[2]

[1] *Department of Biology, University of North Carolina at Greensboro, Greensboro, NC, USA*
[2] *Department of Entomology & Plant Pathology, North Carolina State University, Raleigh, NC, USA*
[3] *Department of Biological Sciences, University of Alberta, Edmonton, Canada*

Introduction

Recently, bees and other pollinators have received considerable attention because of the important ecosystem services they provide through pollination. In agricultural systems, the economic value of the nearly 17 000+ species of native bees (primarily solitary species living in non-managed contexts) is roughly equivalent to that of a single, colonial species managed by humans – the Western honey bee, *Apis mellifera* (Kleijn et al. 2015). While actively managed by beekeepers, honey bees are not domesticated in the strict sense because they are still freely foraging, will readily become feral, and breeding is difficult to control. Beekeepers range from lay hobbyists to professionals with varying levels of expertise. Like most domesticated livestock systems, however, managed honey bees can be subject to increased population densities, genetic bottlenecks, and management intensification, leading to similar challenges in their management.

The common factors most frequently attributed to declining honey bee health include diminished nutrition and forage as a result of habitat loss, environmental contaminants such as various pesticides, and increased management stressors as well as the increased impact and variety of parasites, pathogens, and pests that are consistently ranked among the highest management concerns for beekeepers (vanEngelsdorp and Meixner 2010). There is little doubt that disease plays a critical role in the ecology and management of all domesticated livestock, but it is an especially intriguing problem for managed social insects because their nests provide microhabitats where parasites and pathogens find favorable temperatures, humidity, and high concentrations of hosts (Schmid-Hempel 1998). Indeed, the high densities of managed honey bees greatly facilitate disease transmission within and among different beehives, especially in the US and Canada where large scale migratory commercial beekeeping predominates (Brosi et al. 2017; Dynes et al. 2019). Honey bee parasites weaken colonies and may carry multiple viral pathogens. Hence, controlling parasites will not only reduce their direct impact on honey bees but also diminish their roles as microbial vectors.

The most problematic parasite of honey bees is the parasitic mite *Varroa destructor* (Anderson and Trueman 2000; Sammataro et al. 2000; Boecking and Genersch 2008). This pest shifted hosts from the Eastern honey bee, *Apis cerana*, to the Western honey bee, *A. mellifera* (Oldroyd 1999), and entire feral populations have been decimated as a result (Kraus and Page Jr. 1995). *Varroa* is described in Chapter 20 but is particularly relevant for this chapter because it serves as a vector for numerous viruses. A second parasitic mite, the tracheal mite *Acarapis woodi*, lives in the airway passages of adult bees. Infestation alone by these small arachnids can kill entire colonies. However, they can also compromise the immunodefenses of workers thus allowing secondary infections (Bailey and Perry 1982; Schmid-Hempel 1998). *A. woodi* is increasingly infrequent in the honey bee population and is therefore much less of a major management concern than it once was. Another notable pest is the fungus-causing Nosema disease (*Nosema apis*), which infects the gastrointestinal tract of adults and associates with viruses (Bailey et al. 1983). It has been documented that a new species of Nosema, *Nosema ceranae*, has recently shifted hosts from the Eastern to the Western honey bee (Higes et al. 2006), not unlike the *Varroa* mite. Small hive beetles (or SHBs; *Aethina tumida*) and Tropilaelaps mites (*Tropilaelaps clareae* and *Tropilaelaps mercedesae*) can also spread viruses (Dainat et al. 2009; Eyer et al. 2009).

Honey Bee Medicine for the Veterinary Practitioner, First Edition. Edited by Terry Ryan Kane and Cynthia M. Faux.

Honey bees serve as hosts to a wider range of known parasites and pathogens than all other social insects (Schmid-Hempel 1998). These diseases and parasites weaken colonies and carry multiple viral and bacterial pathogens. Hence, other diseases facilitate viral diseases—these interactions cannot be ignored.

Compared to the wealth of knowledge and actionable management options for parasitic mites and bacterial diseases, few choices exist for beekeepers to mitigate the economically important viral pathogens. Viruses are the largest class of pathogens that infect honey bees, with over 30 different viruses (Chen and Siede 2007; Maori et al. 2007; Runckel et al. 2011; de Miranda et al. 2015; Remnant et al. 2017; McMenamin and Flenniken 2018). Many of the more economically important viruses are associated with parasitic infestations (see below). Furthermore, the viral pathogenicity and infection dynamics are influenced by numerous factors, including agrochemical exposure, synergistic and/or antagonistic pathogenic infections, nutrition, genetic composition of both host and virus, and host immune responses, making viruses very difficult to study and manage for optimal colony health. Here, we review the honey bee viruses that are most germane to honey bee husbandry to advance our understanding of how to diagnose and effectively manage them.

Iflaviridae Family

The *Iflaviridae* is a family of small, non-enveloped viruses within the *Picornavirales* order (Valles et al. 2017b). Virions are roughly spherical, 22–30 nm in diameter, and have no distinctive surface structure (Valles et al. 2017b). Viruses in this family possess a positive, single-stranded, non-segmented RNA genome (+RNA) (Valles et al. 2017b). In order to replicate, these viruses need to enter into the host cell cytoplasm and utilize cell machinery (Valles et al. 2017b). Members of this family are grouped into a single genus, *Iflavirus*, that includes the honey bee viruses Deformed Wing Virus (DWV), Sacbrood Virus (SBV), Slow Bee Paralysis Virus (SBPV), and the newly discovered Moku Virus (MV).

Deformed Wing Virus

DWV was first found in adult honey bees in Egypt and initially named Egypt Bee Virus (EBV) (Bailey et al. 1979). Subsequently, an isolated virus from symptomatic adult bees in Japan had a similar serological relationship and was categorized as a Japanese strain of EBV. The virus has since been designated as DWV (Bailey and Ball 1991; Ribière et al. 2008), because it causes the overt symptom of misshapen and crippled wings of heavily infected worker bees, often associated with *V. destructor* (henceforth *Varroa* mite) (de Miranda and Genersch 2010). DWV is a non-enveloped icosahedral virion, which contains one copy of a positive-sense single-stranded 10 kb RNA genome with a single open reading frame (ORF) that encodes a 2893 amino acid polyprotein that is post-translationally processed (Lanzi et al. 2006; Organtini et al. 2017).

After the development of molecular technologies, other genetically close variants of DWV have been detected, specifically Varroa Destructor Virus 1 (VDV-1) and Kakugo Virus. VDV-1 was initially found in *Varroa* mites, and Kakugo Virus was isolated from the brains of aggressive worker guard bees (Fujiyuki et al. 2004; Ongus et al. 2004). More recently, however, genome sequencing has revealed that these viruses belong to DWV as a quasi-species, where a diversity of viral sequence variants are categorized into at least three distinct genotypes denoted as DWV-A, DWV-B, and DWV-C (Martin et al. 2012; Mordecai et al. 2016a,c). All three strains belong to the DWV clade, but they differ in virulence and distribution (McMahon et al. 2016; Mordecai et al. 2016c; Natsopoulou et al. 2017; Ryabov et al. 2017). DWV type-A is the original virus from this clade that has been studied as the "generic" DWV, to which the Kakugo Virus also belongs with 98% genome homology (Fujiyuki et al. 2004; Lanzi et al. 2006). DWV type-B corresponds to VDV-1 and has 84.4% sequence identity to DWV-A (Mordecai et al. 2016a). DWV type-C is a very recently recognized variant of this complex that is a recombinant of DWV type-A and B and other less frequent recombinants may exist (Mordecai et al. 2016c).

The majority of global surveillance studies indicate that, except in Australia (where *V. destructor* has not been introduced), DWV infection has been reported wherever honey bees exist (Martin et al. 2012; Wilfert et al. 2016; Martin and Brettell 2019), making DWV ubiquitous and one of the most prevalent viral pathogens of honey bees (Martin and Brettell 2019). DWV also can be detected from the Asian honey bee, *A. cerana*, and the parasitic mite *T. mercedesae*, supporting the high prevalence and widespread distribution of the virus in honey bees and associated arthropods (Forsgren et al. 2009; Hassanyar et al. 2019).

DWV has been observed as both overt and covert infections in over 55% of colonies/apiaries in different studies, although detection sensitivity likely influences these reported prevalence levels (Ryabov et al. 2014; Martin and Brettell 2019). The seasonal variation in DWV incidence increases from spring to autumn, which is strongly associated with the *Varroa* mite population in colonies (de Miranda and Genersch 2010; D'Alvise et al. 2019). DWV can be detected in all body parts of bees, as well as all honey bee castes and sexes (queens, workers, and drones) and developmental stages (sperm, eggs, larvae, and pupae) (Fievet et al. 2006; de Miranda and Fries 2008;

de Miranda and Genersch 2010; Amiri et al. 2016; Martin and Brettell 2019). Depending on the route of transmission, the same virus (although not necessarily the same strain) may or may not be symptomatic (de Miranda and Genersch 2010; Martin and Brettell 2019). DWV transmission occurs vertically from queen to offspring via eggs (Yue et al. 2007; de Miranda and Fries 2008; Amiri et al. 2018) and horizontally through drone sperm (de Miranda and Fries 2008; Amiri et al. 2016) and several oral routes including larval food, trophallaxis, or cannibalism of DWV-infected pupae by adult bees engaging in hygienic behavior (Mockel et al. 2011; Mazzei et al. 2014). In the presence of *Varroa* mites, DWV is predominantly horizontally transmitted and actively vectored by *V. destructor*. Varroa-mediated transmission of DWV selects for highly virulent strains and decreases overall virus population diversity (Bowen-Walker et al. 1999; Martin et al. 2012; Ryabov et al. 2014). There is some suggestion that DWV-infection may cause *Varroa* mites to alter their movements (Giuffre et al. 2019), perhaps making them more effective vectors of the virus. Varroa-mediated DWV transmission causes high prevalence and virus load and probably deformity of newly emerged bees and death (Mondet et al. 2014; Di Prisco et al. 2016).

In symptomatic individuals, the virus is prevalent in all body parts but accumulates especially in the epithelial cells of the digestive tract (Fievet et al. 2006), shedding large amount of virus particles into the lumen, and in the basal regions of the antennal epithelium close to the haemolymph. DWV is also particularly prevalent in the ovaries and fat body of queens and the seminal vesicles, mucus glands, and testes of drones (Fievet et al. 2006; Shah et al. 2009). During honey bee development, DWV causes well-characterized symptoms in the presence of mites, including pupal death and shrunken, crumpled wings, bloated abdomen, decreased body size, and discoloration in the resulting newly emerged bees (Figure 21.1). Infected bees are slower to emerge from their capped cells, demonstrate hypoplasia of hypopharyngeal and mandibular glands, and experience increased mortality during and shortly after emergence (Koziy et al. 2019). DWV can compromise sensory and communication systems, resulting in depressed hygienic and other behaviors (Kim et al. 2019). While clinical symptoms such as crippled wings and shortened abdomens have rarely been reported in queens as a consequence of DWV infection (Williams et al. 2009), internally, DWV may cause extreme cases of ovarian degradation (Gauthier et al. 2011). Morphological symptoms are often seen in drones, especially because *Varroa* mites prefer to parasitize drones, likely because of their longer development time enabling higher mite reproduction. In the absence of *Varroa* mites, or when adult stages are infected, DWV normally persists at low levels with no obvious symptoms except a shorter lifespan (Yang and Cox-Foster 2007). Nonetheless, covert DWV infection in forager honey bees can increase responsiveness to low concentrations of sucrose and impair associative olfactory learning (Iqbal and Mueller 2007). Overall, DWV is a major, widespread honey bee pathogen with a range of sublethal and lethal effects that can lead to depopulation and colony mortality (McMenamin and Genersch 2015; Wilfert et al. 2016).

Sacbrood Virus

SBV is the first virus to have been described in *A. mellifera* and is the etiological agent of sacbrood disease (Bailey et al. 1964; Chen and Siede 2007; Ribière et al. 2008). SBV was also the first honey bee virus to have its complete genome sequenced and assembled (Ghosh et al. 1999). SBV

(a) (b)

Figure 21.1 Symptomatic deformed wing virus infection of worker honey bees. (a) Partially crumpled wings, bloated abdomen and discoloration. (b) Fully crumpled wings and decreased body size and shortened abdomens. *Source:* Photos courtesy of Per Kryger.

has a positive-sense, single-stranded RNA genome that is 8832 nucleotides long and contains an additional poly-A sequence at its 3′end (Ghosh et al. 1999; Procházková et al. 2018). Genome sequencing has revealed several lineages of SBV worldwide, including new strains of SBV-like Korean Sacbrood Virus (KSBV; Choe et al. 2012), Chinese Sacbrood Virus (CSBV; Zhang et al. 2001), Thai Sacbrood Virus (TSBV; Verma et al. 1990) and Indian Sacbrood Virus (AcSBV-IndTN1; Aruna et al. 2018). Among all honey bee viruses, SBV has the greatest number of complete genomes identified from both *A. mellifera* and *A. cerana* so far, enhancing our understanding of its evolution and pathogenicity. Phylogenetic analysis has revealed that SBV strains can be split into two distinct lineages (Shen et al. 2005; Li, Zeng et al. 2016; Li, Wang et al. 2019) – SBV strains detected from *A. cerana* (named "AcSBV") and those isolated from *A. mellifera* (named "AmSBV"; Roberts and Anderson 2014, Gong et al. 2016). However, there is no strong evidence that these two groups strictly adhere to their host specificity, geographic origin, or both (Li et al. 2016; Ko et al. 2019); experimental studies have shown that AcSBV is able to infect *A. mellifera* colonies, although with lower pathogenicity (Gong et al. 2016).

SBV is a nearly ubiquitous honey bee virus that is geographically widespread in colonies in all continents where beekeeping is practiced (Allen and Ball 1996; Chen and Siede 2007). It is frequently detected from seemingly healthy adult queens, drones, and worker bees as a latent infection (Anderson and Gibbs 1989; Chen et al. 2005). Large numbers of symptomatic larvae are rarely seen in the colony because adult bees are very efficient at detecting and removing most infected larvae in the early stages of infection. However, this "hygienic behavior" may also be the primary means of SBV transmission within a colony (Grabensteiner et al. 2001; Shen et al. 2005), as virus particles accumulate in the hypopharyngeal glands of infected nurse bees and are spread horizontally throughout the colony by nurses feeding larvae with their glandular secretions or exchanging food with other adult bees (Shen et al. 2005). As a result of infection, worker bees start foraging much earlier than usual and may further spread the virus by passing it from their glandular secretions onto their pollen loads that are stored in the hive for colony consumption, although the pollen stores appear unaltered (Bailey and Fernando 1972; Anderson and Giacon 1992).

Young larvae primarily become infected with the virus by ingesting virus-contaminated food, but there are also other transmission routes. SBV has been detected in *Varroa* mites, but no evidence presented that SBV replicates within mites, suggesting that the mite neither serves as host nor a biological vector of SBV (Shen et al. 2005). Instead, *Varroa* mites serve as mechanical SBV vectors after acquiring virus particles through parasitizing SBV infected pupae or adult bees. This could be the reason that many studies found SBV to be positively associated with *Varroa* mite infestation of honey bee colonies (Meixner et al. 2014; Amiri et al. 2015). Moreover, individual mites that are positive for SBV may have an increased rate of movement (Giuffre et al. 2019), suggesting that SBV may affect their behavior and increase its horizontal spread. SBV has been detected from queens' ovaries, collected sperm, and egg samples in many surveillance studies, which suggests the potential of venereal and vertical transmission, but no study has demonstrated this empirically (Chen et al. 2005; Shen et al. 2005; Ravoet et al. 2015b; Amiri et al. 2018; Prodělalová et al. 2019). Among colonies, SBV infection can be spread through robbing and drifting worker bees.

SBV can multiply in adult bees without causing obvious signs of disease, but honey bee larvae are most susceptible to SBV infection with clear symptoms appearing a few days after capping (Hitchcock 1966). Infected larvae are not able to shed their leathery endocuticle, and a substantial amount of fluid containing millions of virus particles accumulates between the body of the diseased larvae and its unshed cuticle, forming a sac-like appearance that is the characteristic symptom of the disease and the reason for its name. Infected brood fails to pupate, are stretched on their backs with heads lifted up toward the cell opening, and usually die during the last larval stage (Chen and Siede 2007). Cell caps are usually perforated or completely removed with the diseased, headless brood inside. SBV infection can prevent normal larval growth in the colony and thus reduce colony populations. This is especially true in spring when colony size is expanding, large numbers of susceptible larvae can be present, and the temperature fluctuates more than during other seasons (D'Alvise et al. 2019). The effects of SBV in *A. mellifera* colonies are often relatively mild and mostly limited to larval death and colony depopulation, but sacbrood infection in *A. cerana* usually causes serious colony collapse (Blanchard et al. 2014; Gong et al. 2016; Ko et al. 2019).

Slow Bee Paralysis Virus

SBPV was first discovered in 1974 in the United Kingdom fortuitously from an isolation laboratory experiment (Bailey and Woods 1974; Ribière et al. 2008). The virus was so named to differentiate it from the comparatively fast-acting Acute and Chronic Bee Paralysis viruses (ABPV and CBPV) (see below). The complete RNA genomes of two distinct strains of SBPV have been sequenced (de Miranda et al. 2010b), both of which were derived from the original England SBPV isolate, one produced in 1994 in Canada, and the other in 2006 in Sweden. The two strains are named

"Rothamsted" and "Harpenden," respectively, and they are 83% identical at the nucleotide level (additional details in de Miranda et al. 2010b; Kalynych et al. 2016).

Because many pathogen surveys of honey bee populations do not include SBPV, there is relatively little information about its prevalence worldwide. Before the near-global spread of *Varroa* mites, SBPV was rare and recorded only in Britain, Switzerland, Fiji, and Western Soma, where covert infection – most likely facilitated through oral transmission – did not cause any brood or adult mortality (Bailey and Ball 1991; Allen and Ball 1996; Ribière et al. 2008). After the *Varroa* host-shift to *A. mellifera*, however, SBPV has attained a higher prevalence in some honey bee populations because it can be vectored by the mite (Carreck et al. 2010). SBPV can be transmitted by *Varroa* mites directly to adult bees and pupae during parasitic feeding (Santillán-Galicia et al. 2010, 2014). Subsequent studies could not prove *Varroa* as a biological vector, however, and instead showed inefficient SBPV transmission by *V. destructor*. This, together with the high virulence of SBPV, may be the reasons that the virus is not highly prevalent or a significant factor in colony decline (Santillán-Galicia et al. 2014). Reverse transcription polymerase chain reaction (RT-PCR) analysis of samples from England and Switzerland has revealed a background prevalence of <2% in colonies with variable degrees of *V. destructor* infestation (de Miranda et al. 2010b).

While SBPV is mainly a disease of adult honey bees, it can be detected in larvae and pupae without any clear symptoms and is probably transmitted orally or by *Varroa* mites. Overt SBPV infection can also be induced in infected adult bees by injection with inert fluids (Bailey and Woods 1974). Symptoms include the paralysis of the anterior two pairs of legs about 10–12 days after injection into the abdomen of adult bees (Bailey and Woods 1974; Denholm 1999) with high virus accumulation in the head (specifically the hypopharyngeal, mandibular, and salivary glands), the fat body, the crop, and forelegs but less in the hindlegs, midgut, rectum, and thorax (Denholm 1999). SBPV can be lethal at the individual and colony levels (Martin et al. 1998).

Dicistroviridae Family

The *Dicistroviridae* is a family of small, non-enveloped RNA viruses with spherical particles about 30 nm in diameter (Valles et al. 2017a). Virion capsids comprised of 60 copies of three different viral proteins: VP1, VP2, and VP3 and a smaller protein, VP4, that is located inside the virion in contact with the genome (Sánchez-Eugenia et al. 2015; Valles et al. 2017a). The genome is a linear, positive-sense, monopartite RNA that contains two main non-overlapping ORFs that are flanked by untranslated regions (UTR)s and separated by an intergenic region (IRG) (Bonning and Miller 2010; Valles et al. 2017a). Replication of the viruses in this family occurs exclusively in the cytoplasm of infected host cells (Bonning and Miller 2010; Valles et al. 2017a). Different viruses in the *Dicistrovidae* vary considerably in virulence and pathogenicity and the severity of disease ranges from asymptomatic to lethal with some devastating economic consequences (Valles et al. 2017a). Common honey bee viruses in the *Dicistroviridae* family include Black Queen Cell Virus (BQCV), Acute Bee Paralysis Virus (ABPV), Kashmir Bee Virus (KBV), and Israeli Acute Paralysis Virus (IAPV). The latter three viruses form a closely-related complex that is distributed worldwide, with different viruses being more predominant in different geographical regions (de Miranda et al. 2010a).

Black Queen Cell Virus

BQCV is one of the most prevalent, yet least understood, honey bee viruses (Allen and Ball 1996; Chen and Siede 2007; Mondet et al. 2014; Amiri et al. 2015). It was first isolated in 1977 from field-collected prepupae and pupae of dead and partially decomposed queens that had turned dark brown to black within their cells (Bailey and Woods 1977; Leat et al. 2000). The name of the virus was attributed to darkened areas of the walls of the queen cells containing infected pupae. The sequencing of BQCV revealed an 8550 nt polyadenylated genome containing two large ORFs (Leat et al. 2000; Spurny et al. 2017). BQCV has been detected both from European and Asian honey bees and comprises many different strains. Phylogenetic analysis based on the partial capsid protein-coding region gave contradictory results in respect to whether isolates cluster according to geography or according to *Apis* species (Tapaszti et al. 2009; Reddy et al. 2013a,b; Mookhploy et al. 2015; Yang et al. 2016). However, there is a significant association of geography with phylogeny, indicating that physical location is an important factor in the host-host transfer of the virus (Tapaszti et al. 2009; Mookhploy et al. 2015).

There has not been any systematic study on BQCV transmission, although its chronic and asymptomatic persistence in the colony indicates that it is horizontally transmitted through social transmission among adults and from adults to larvae through glandular secretions (Chen et al. 2005, 2006; Chen and Siede 2007). BQCV is largely a gut pathogen that is often associated with *N. apis*, a fungal intestinal parasite of adult honey bees that increases virus susceptibility among worker bees (Bailey et al. 1983;

Tentcheva et al. 2004). BQCV has been detected from queens, drones, collected sperm (used for instrumental insemination), and eggs, all of which suggests the potential for venereal- and vertical transmission from infected drones to the queen and consequently to the offspring (Siede and Buchler 2003; Chen et al. 2005; Prodělalová et al. 2019).

BQCV has been detected during surveillance studies globally and is consistently one of the most prevalent viruses in the honey bee industry (Tentcheva et al. 2004; Chen and Siede 2007; Amiri et al. 2015). BQCV is also reported to be the most common cause of death of queen larvae and thus a known concern for the honey bee queen-rearing industry. Its incidence increases during the swarming season, when queens and drones are reared (Tentcheva et al. 2004; D'Alvise et al. 2019). BQCV has been detected in all honey bee castes, sexes, and developmental stages. In queen larvae, the clinical signs consist of pale-yellow appearance and the presence of a sac-like sink, resulting in the death of queen larvae and pupae in their cells. The virus does not cause overt symptoms in adult queens, even though it can be detected in their guts, feces, and ovaries (Chen et al. 2005). Likewise, infected adult workers are normally asymptomatic. However, BQCV may compromise their orientation, possibly by affecting learning performance. Therefore infected workers may enter into a foreign colony (drifting) due to orientation errors more often, spreading the disease (Retschnig et al. 2019). Interestingly, high titers of BQCV have been detected in collapsed colonies (Cornman et al. 2012). Its co-infection with *N. apis* causes higher virus load and increases worker bee mortality (Bailey et al. 1983; Meixner et al. 2014), although data to support the potential synergistic interaction between BQCV and *N. ceranae* are equivocal (Retschnig et al. 2014; Doublet et al. 2015).

Acute Bee Paralysis Virus

ABPV was first discovered in 1963 as an asymptomatic infection during a laboratory experiment but had not previously been reported as a cause for disease or colony losses (Bailey et al. 1963). It has a single-stranded, positive-sense polyadenylated RNA genome, containing 9470 nt that encodes two open reading frames (Govan et al. 2000). Phylogenetic studies from American and European strains suggested at least two distinct genetic lineages of ABPV, where one major branch is formed by the US strains while the other contains isolates from continental Europe (Bakonyi et al. 2002b).

Until the establishment of *Varroa* mites in Western honey bees, ABPV persisted at relatively low frequencies as an asymptomatic infection of adult bees (Allen and Ball 1996), suggesting a largely horizontal intra-colony transmission and low inter-colony spread (Chen and Siede 2007; Ribière et al. 2008). The virus has been detected in the brains, hypopharyngeal glands, and feces of worker bees, implying oral transmission. It also has been detected in young queens (although not in their ovaries), collected sperm, and eggs, suggesting possible vertical transmission (Chen et al. 2006; Yue et al. 2006; Francis et al. 2013; Prodělalová et al. 2019). High levels of ABPV have been detected from varroa-parasitized individuals and entire honey bee colonies (Faucon et al. 1992; Tentcheva et al. 2004; D'Alvise et al. 2019). *Varroa* mites seem to be only mechanical transmitters for the virus, but this transmission results in higher viral incidence during the late summer when *Varroa* populations peak (Bakonyi et al. 2002a; D'Alvise et al. 2019). As yet, there is no evidence of ABPV replication within mites or an increase of virus virulence caused by *Varroa* transmission (de Miranda et al. 2010a). ABPV has also been detected in *A. cerana* and its ectoparasite, *T. mercedesae* (Chanpanitkitchote et al. 2018).

ABPV is one of the more common honey bee viruses (particularly in Europe) and has a worldwide distribution (Chen and Siede 2007; Ribière et al. 2008; de Miranda et al. 2010a). It normally persists at low titers in covert infections without obvious symptoms at the individual and colony level. Under some circumstances, however, ABPV is a particularly virulent virus; the injection of only 100 virions can cause death of pupae or adults within two to six days (Bailey et al. 1963; Bailey and Woods 1977). ABPV can infect and kill larval and pupal stages, especially in highly infected colonies (de Miranda et al. 2010a). ABPV-killed capped and uncapped brood resemble the symptoms of certain bacterial diseases (specifically American and European foulbrood) (Figure 21.2), making in-hive diagnosis a challenge for beekeepers (Ribière et al. 2008; de Miranda et al. 2010a). When the virus affects adults, it causes paralysis and other obvious symptoms including trembling, inability to fly, and the gradual darkening and loss of hair from the thorax and abdomen (21.2) (Bailey et al. 1963; Ribière et al. 2008; de Miranda et al. 2010a). ABPV has been associated with colony depopulation and overwintering collapse in Europe (McMenamin and Genersch 2015).

Kashmir Bee Virus

KBV was first identified in 1974 in *A. cerana* originating from north India, as a contaminant of Apis Iridescent Virus, a virus that was found from *A. cerana* samples a year before KBV (Bailey and Woods 1977). Initially, it was proposed that KBV had an Asian origin, but subsequent

(a)

(b) (c)

Figure 21.2 Symptomatic larvae and adult bees infected by acute bee paralysis virus. (a) Highly infected larvae. (b) ABPV-killed uncapped brood similar to American and European foulbrood. (c) Typical ABPV symptoms in adult bees are gradual darkening and loss of hair from the thorax and abdomen. *Source:* Photos courtesy of Per Kryger.

identification of virus variants in other *A. mellifera* populations around the world did not substantiate this hypothesis (Anderson 1990; Allen and Ball 1996; Chen and Siede 2007; de Miranda et al. 2010a). Like other Dicistroviruses, the KBV genome is a positive single-stranded RNA containing two ORFs, separated by an IRG and flanked by UTRs (de Miranda et al. 2004). KBV is serologically, genetically, and pathologically related to ABPV (Allen and Ball 1995; de Miranda et al. 2004). It has roughly 70% nucleotide identity with ABPV across the genome, although there are significant differences in several critical areas between the two viruses, for example in the capsid protein profile (de Miranda et al. 2004). Phylogenetic analyses have confirmed that KBV and ABPV are two distinct viruses and fall into separated clades (de Miranda et al. 2004, 2010a). There are few sequenced KBV genomes available to enable investigations of genetic diversity across different geographies. However, analysis of currently available KBV genome sequences distinguishes a Korean genotype from the reference KBV genotype, possibly a result of phylogeographical distance and genetic recombination (Hung et al. 2000; Reddy et al. 2014).

Like ABPV, KBV usually persists as a covert and low-prevalence infection, but environmental stress (especially in late summer into autumn) can trigger the virus to replicate, resulting in a higher prevalence (Dall 1987; de Miranda et al. 2010a; D'Alvise et al. 2019). KBV appears to have a worldwide distribution, although in recent years it has become increasingly rare in the US (Evans 2001; Siede et al. 2005; Pettis et al. 2007). Detection of KBV in workers, honey, pollen, royal jelly, and brood food points to an oral-fecal transmission pathway (Evans 2001). It also has been detected from queens and surface-sterilized eggs, supporting the potential for vertical, transovarial virus transmission (Hung 2000; Chen et al. 2005, 2006). *Varroa* mites are both effective vectors and activators of KBV, and as a result their presence leads to high virulence and colony depopulation (Hung and Shimanuki 1999; Chen et al. 2004; Shen et al. 2005). Comparing varroa-mediated and varroa-independent transmission indicates that the pathology and virulence of KBV depends on the route of transmission (Todd et al. 2007; Roberts et al. 2017); KBV can be highly virulent, with the injection of only a few particles, similar to varroa-mediated transmission route leading to overt infection. In contrast, large virus doses are needed to cause death after inoculation by feeding and the disease typically proceeds in this case without any clear symptom (Allen and Ball 1995; Ribière et al. 2008; de Miranda et al. 2010a).

Israeli Acute Paralysis Virus

IAPV was isolated and characterized in 2007 (Maori et al. 2007). IAPV has a positive-sense, single-stranded RNA genome of 9487 nt containing two ORFs, coated by an icosahedral capsid of 29 nm (Maori et al. 2007; de Miranda et al. 2010a; Mullapudi et al. 2016). Phylogenetic studies from different countries have revealed that multiple strains of IAPV exist in different countries, with three typical genotypes including an Australian genotype, a cluster of China, Israel and Korean genotypes, and the USA genotype, that has two subtypes circulating in the US with a long period of independent evolution (Palacios et al. 2008; Reddy et al. 2013b; Chen et al. 2014).

IAPV was first detected in Israel and has since been detected in many countries around the globe (de Miranda et al. 2010a; Chen et al. 2014). It has not yet been reported from New Zealand, although it should not be presumed to be absent in any area (McFadden et al. 2014). IAPV infection was initially associated with Colony Collapse Disorder (CCD) in the US (Cox-Foster et al. 2007), but subsequent studies have failed to provide an unequivocal association between IAPV infection and colony losses consistent with CCD symptoms (Vanengelsdorp et al. 2009; McMenamin and Genersch 2015; vanEngelsdorp et al. 2017). Instead, once IAPV is acquired and induced to replicate in a colony, it acts as an infectious factor that affects overall bee health and colony productivity (Hou et al. 2014). Many honey bee viral surveys have found IAPV as a prevalent, persistent, but symptomless infection in a population (de Miranda et al. 2010a, Chen et al. 2014). IAPV has been detected from honey, pollen, royal jelly, worker bees, larvae, pupae, and queen feces, which imply food-born transmission pathways (Chen et al. 2014). Experimental studies have also shown that physical contact between infected workers and queens leads to queen infection, suggesting that IAPV infections are also spread through close bodily contact (Amiri et al. 2019). IAPV is readily transmitted to bees by injection and topically applying virus particles after hairs on the surface of the body have been removed (Boncristiani et al. 2013; Amiri et al. 2019). IAPV detected in honey bees after exposure to *Varroa* mites confirms vector-mediated horizontal transmission (Di Prisco et al. 2011; Chen et al. 2014). IAPV quantification from exposed bees is positively correlated with the density of *Varroa* mites and the time period of exposure to *Varroa* mites (Di Prisco et al. 2011). *Varroa* mites are not only a mechanical vector for IAPV but also play a biological vectoring role in supporting the replication of the virus (Di Prisco et al. 2011). Moreover, IAPV has been detected in queens, drone semen, and eggs, suggesting venereal and vertical transmission among reproductive individuals and their offspring (de Miranda et al. 2010a; Chen et al. 2014). In situ hybridization of IAPV indicates the virus in the egg, gut, ovaries, and spermatheca of infected queens (Chen et al. 2014) in support of this hypothesis.

IAPV is highly virulent, where IAPV-injected worker bees die within two to five days post-infection (Maori et al. 2007; Hou et al. 2014). IAPV also replicates readily in pupae, and IAPV injection into white-eyed pupae is lethal, albeit with variable phenotypes (Boncristiani et al. 2013). Symptoms vary from complete cessation of development with no visual evidence of disease to rapid darkening of a part or the entire body. Even injection of naked IAPV RNA (100 μg per larvae) can cause death in over 80% of treated larvae (Maori et al. 2007). Caged worker bees supplied with IAPV-inoculated pollen supplement gradually develop symptoms, showing darkening of the abdomen tips and thorax, constant movement, and reduced flying or eating provided sugar candy. Within a few days, their abdomens and thoraxes become black, the thoraxes become hairless, they completely stop flying, their wings begin to shiver, and they die within 10 days of exposure (Maori et al. 2007). Nearly the same suite of symptoms can be seen in naturally infected bees: shivering wings, followed by progressive paralysis – typical symptoms of nerve-function impairment (Hou et al. 2014). One tissue-tropism study found that IAPV replicates within all bee tissues but tends to concentrate in gut, nerve tissues, and hypopharyngeal glands (Chen et al. 2014). IAPV-infected foragers are more responsive to low sucrose solutions early in the infection. Higher IAPV titer decreases learning, navigation, and orientation of infected forager, which leads to depressed homing ability and increased drifting or death outside of the hive (Li et al. 2013).

Viruses in Unassigned Families

Many honey bee viruses have been identified but not yet assigned to particular families pending further study. Of this group, CBPV and Lake Sinai Virus (LSV) are special cases because they are relatively well-studied and sufficiently distinguished biologically, serologically, and genetically, although they do not fit into either of the virus families described above.

Chronic Bee Paralysis Virus

CBPV is the causative agent of chronic bee paralysis (Rinderer and Green 1976). CBPV was isolated in 1963 in Britain (Bailey et al. 1963). Paralysis symptoms caused by CBPV were probably the first viral disease that was recognized by beekeepers before the identification of its etiological agents. It has been given a variety of names such as "hairless black syndrome," "little black," "black robbers," and "maladie noire" in different countries because of its characteristic symptoms (Ribiere et al. 2010). The morphology of CBPV's particles and RNA genome structural organization are different from other honey bee viruses. CBPV is a positive, single-stranded fragmented RNA virus with a genome composed of two major RNA fragments (Olivier et al. 2008a; Chevin et al. 2015; Li et al. 2017). Full-length CBPV genomic sequence data from different geographic locations are limited, and consequently phylogenetic studies have mostly been based on partial sequences (Blanchard et al. 2009; Morimoto et al. 2012; Li et al. 2017). Several clusters are suggested with incomplete geographical segregation; most European isolates form a group, likewise for American isolates and isolates from southeast Asia (Morimoto et al. 2012; Li et al. 2017).

CBPV has been reported in several different countries and has been found on every continent (Ribiere et al. 2010). The virus has an overall low prevalence in the honey bee populations. However, recent large-scale surveys show CBPV increases in prevalence in apiaries with poor health and in commercial-scale beekeeping operations (Pettis et al. 2007; Traynor et al. 2016). Outbreaks of severe chronic paralysis are sporadic but appear more frequently in spring and summer months when the colony populations are at their peak (Ribière et al. 2008, 2010). In an asymptomatic colony, CBPV is mostly detected from worker bees, but in a symptomatic colony CBPV can be detected from honey, eggs, larvae, pupae, emerging bees, drones, and foragers (Blanchard et al. 2007). Detected virus loads have been shown to be significantly higher in guard, symptomatic, and dead bees outside the hive (Blanchard et al. 2007). CBPV is readily transmitted among adult bees by bodily contact when healthy bees are crowded together with infected individuals. Crowded populations cause body hairs to be broken on healthy bees and provide opportunities for the virus to infect them (Ribiere et al. 2010). Establishing pollen traps onto beehives can increase virus prevalence (Dubois et al. 2018), suggesting a mechanical means to break worker's hair, nutritional stress, or the combination of both may increase infection within colonies. CBPV is also transmitted through trophallaxis and oral-fecal pathways via feces inside the colony (Ribiere et al. 2007; Amiri et al. 2014). CBPV has been detected from eggs, drones, and queens, which also supports the possibility of vertical transmission (Chen et al. 2005, 2006; Blanchard et al. 2007).

While CBPV has been detected in *Varroa* mites, they seem to only play a mechanical role in viral spread, and there is no strong association between mite population and virus load in a colony (Celle et al. 2008; Ribiere et al. 2010). In laboratory experiments, adult workers and queens are susceptible to CBPV infection through topical application and feeding (Amiri et al. 2014), where CBPV show a synergistic effect in combination with *N. ceranae* (Toplak et al. 2013). Even naked CBPV RNAs are infectious after injecting adult bees in laboratory settings, causing symptoms and mortality (Youssef et al. 2015).

Experimentally infected adult bees start showing symptoms after five to six days post-infection (Rinderer and Green 1976; Ribiere et al. 2010). The most frequently observed symptoms are trembling legs, partially spread dislocated wings, and a bloated abdomen (the body cavity is full of hemolymph and the honey sac is dilated). In naturally infected bees, two distinct sets of symptoms have been described. Type 1 symptoms include an abnormal trembling motion of the wings and bodies of paralyzed bees. These bees are unable to fly, frequently huddle together on top of the bee cluster in the hive or in front of the colony entrance, and crawl in front of the hive on the ground or vegetation. Severely affected colonies can suddenly collapse, particularly at the height of summer, and leave the queen with a few workers on neglected combs (Ribiere et al. 2010).

Another set of symptoms, Type 2, include bees that can fly but become almost hairless, appearing dark or almost black, which make them seem smaller than healthy bees, with a relatively broader abdomen. They are shiny, appearing greasy in bright light (Figure 21.3). Within a few days they become flightless, tremble, and die (Ribiere et al. 2010). CBPV can be found in the thoracic and abdominal nerve ganglia, in the hypopharyngeal and mandibular glands, and in the brain (Blanchard et al. 2007). In situ hybridization has shown CBPV is neurotropic and can infect and multiply in neuronal cells of somatic regions and neuropile regions of the protocerebron and deuterocerebron (Olivier et al. 2008b). In the brain, CBPV has been observed in the neurons of the higher integration centers, the mushroom bodies, the central complex, and sensory neuropiles, such as the optic and antennal lobes (Olivier et al. 2008b).

(a)

(b)

Figure 21.3 Type 2 symptoms of chronic bee paralysis virus. (a) Typical symptom of black, hairless infected worker bee, appearing greasy. (b) CBPV symptomatic bees are suffering nibbling by their nestmates. *Source:* Photos courtesy of Per Kryger.

Lake Sinai Viruses

The LSVs were found in 2009 from honey bee samples collected near Lake Sinai in South Dakota, USA (Runckel et al. 2011). LSVs contain monopartite positive sense, single-stranded RNAs (Runckel et al. 2011; Daughenbaugh et al. 2015). The LSV group has remarkable diversity, and since the discovery of LSV1 and LSV2 this group has been expanded to many LSV variants, most supported only by partial sequences, that have been given LSV labels (LSV3–LSV7, LSV-NE, LSV-SA1, LSV-SA2, LSV-Navarra, and LSV-TO). While isolates of LSV1 and LSV2 are clearly distinct sequence clusters, they are embedded within a diverse set of sister lineages and no evidence of geographical pattern could be associated with these clades (Ravoet et al. 2015a,b; Cornman 2019). While LSVs are very diverse compared to other honey bee RNA viruses, they were designated as an individual species within the Sinaivirus genus (Daughenbaugh et al. 2015).

LSVs are among the most abundant honey bee viruses, with a wide distribution across continents (Cornman et al. 2012; Granberg et al. 2013; Cepero et al. 2014; Amakpe et al. 2016; Roberts et al. 2017). In early studies, LSV1 and LSV2 were found to have very different seasonal distribution, with LSV1 common and LSV2 rare during summer months (Runckel et al. 2011; Cepero et al. 2014). However, further findings question if there is a seasonal distribution of LSV1 and LSV2 throughout the year (Daughenbaugh et al. 2015). There is no information about LSVs transmission pathways, although virus detection from the gut of worker bees and collected pollen somewhat support oral–fecal transmission pathways among colony members (Daughenbaugh et al. 2015; Amakpe et al. 2016). LSV1 and LSV2 have been detected from *Varroa* mites, but a positive relation between LSV and mite prevalence has not been observed (Daughenbaugh et al. 2015; Malfroy et al. 2016). In contrast, a positive correlation with the presence of LSV and Nosema microsporidia has been demonstrated, although the cause for this is unclear (Daughenbaugh et al. 2015; Traynor et al. 2016).

Initially, it was thought that LSV is associated with weak colonies, but it is abundant in both weak and healthy colonies (Granberg et al. 2013; Cepero et al. 2014; Traynor et al. 2016). So far, no disease phenotypes or fitness consequences have been demonstrated for LSV, therefore its pathology remains largely unknown in honey bees (Daughenbaugh et al. 2015).

Honey Bee DNA Virus

Apis mellifera Filamentous Virus

Apis mellifera Filamentous Virus (AmFV) was identified in the US in 1978 and later found in several other countries (Clark 1978). AmFV was first described as a double-stranded RNA virus isolated from bees showing the symptoms of bee rickettsiosis but later characterized as a large enveloped DNA virus (Bailey et al. 1981b; Sitaropoulou et al. 1989; Gauthier et al. 2015). There are only a few genome sequences of AmFV available from different geographical regions and phylogenetic analysis shows that AmFV isolates from Chinese *A. mellifera* and *A. cerana* exhibited a high genetic similarity with isolates from Belgium, Switzerland, and the US (Hou et al. 2017).

AmFV is a widespread virus, distributed in many honey bee populations, and it has been detected from different countries (Granberg et al. 2013; Hou et al. 2017). In general, AmFV does not seem to follow seasonal patterns, although higher viral loads might be detected in spring (Hartmann et al. 2015). AmFV has been detected from honey, pollen, gut, larvae, and pupae, indicating possible oral-fecal transmission routes (Hartmann et al. 2015). It can be highly prevalent among honey bee workers and as a result, emerging bees get progressively infected during the first week following trophallaxis with their nest mates (Gauthier et al. 2015; Hartmann et al. 2015). It has also been detected in queen ovaries, eggs, and sperm from drone endophalli, which supports the potential of vertical and venereal transmission routes (Hartmann et al. 2015; Prodělalová et al. 2019).

Earlier studies indicate that virus acute infections of individual honey bees lead to the lysis of bee tissues such as fat body, resulting in characteristic milky-white appearance of worker hemolymph (Sitaropoulou et al. 1989). Acutely infected worker bees are often found crawling at the hive entrance. However, these symptoms are rare and may be linked to other factors (Sitaropoulou et al. 1989; Hartmann et al. 2015). It has been shown that there is a positive correlation between AmFV load and DWV, SBV, and BQCV during spring and summer, but not directly with *Varroa* mite infestation or Nosema spp. (Bailey et al. 1983; Bailey and Ball 1991; Hartmann et al. 2015). Despite possible individual effects, AmFV does not seem to have a pathogenic effect at the colony level, nor has it been associated with colony depopulation or losses in the past (Clark 1978; Sitaropoulou et al. 1989; Gauthier et al. 2015; Hartmann et al. 2015).

Recently Identified Honey Bee Viruses

The development of high-throughput sequencing technologies and accompanying bioinformatic tools have enabled the identification and characterization of many new viruses in honey bees that serological methods had previously overlooked (McMenamin and Flenniken 2018). Because of the immediate data on complete genome sequences, this approach provides a powerful tool to classify new viruses taxonomically (Runckel et al. 2011, Cepero et al. 2014). The newly discovered viruses are from both positive-sense, single-stranded RNA (+ssRNA) and negative-sense, single-stranded RNA (−ssRNA) groups that belong to the *Tymoviridae, Secoviridae, Nodaviridae, Flaviviridae, Bunyaviridae, Iflaviridae*, and *Rhabdoviridae* (Bigot et al. 2017; Remnant et al. 2017; McMenamin and Flenniken 2018; Kraberger et al. 2019). Since these viruses have been identified very recently, information regarding their pathology, mode of transmission, and effects on individuals or colonies is very limited (McMenamin and Flenniken 2018).

Bee Macula-like Virus

One such newly identified virus is the Bee Macula-like Virus (BeeMLV) (de Miranda et al. 2015). Its genome was characterized and demonstrated to be a positive-sense, single-stranded RNA member of the *Tymoviridae* family, a virus family that predominantly infects plants (Granberg et al. 2013; de Miranda et al. 2015; Remnant et al. 2017). BeeMLV was first detected in the US and France, appears to be closely linked to *Varroa* mites, with the highest prevalence found in *Varroa* samples and a clear increase in prevalence in the autumn, at least in French bees (de Miranda et al. 2015).

Tobacco Ringspot Virus

Another plant-associated virus that was detected in honey bees is the Tobacco Ringspot Virus (TRSV) from the *Secoviridae* family (Li et al. 2014). Like the majority of honey bee viruses, TRSV is a positive-sense, single-stranded RNA virus. The virus may be a pollen-borne plant virus that is able to replicate and produce virions in honey bees (Li et al. 2014). TRSV prevalence increases over the course of the active season, correlating with the incidence of colony depopulation and winter mortality, suggesting a potential negative impact on colony health status (Flenniken 2014; Li et al. 2014).

Moku Virus

An additional new virus identified from honey bees by sequencing is the Moku virus (MV), a virus closely related to SBPV and a member of the *Iflaviridae* family (Mordecai et al. 2016b). Moku Virus was first detected in Hawaii in the social wasp *Vespula pensylvanica*, *A. mellifera*, and *V. destructor*, but recently was also detected from historical honey bee samples in Belgium (Mordecai et al. 2016b; Garigliany et al. 2019).

Rhabdovirus-1 and -2

In another effort, seven honey bee-associated viruses were identified using RNA-sequencing (Remnant et al. 2017). These diverse, novel viruses were found from colonies resistant to *Varroa* mites either naturally or because of artificial selection. For the first time, two negative-sense, single-stranded RNA viruses were characterized from honey bees in this study (Remnant et al. 2017). These new viruses belong to the *Rhabdoviridae* and are called *Apis mellifera* Rhabdovirus-1 (ARV-1) and *Apis mellifera* Rhabdovirus-2 (ARV-2) (Remnant et al. 2017). ARV-1 was detected in honey bee samples from North America,

Europe, the Middle East, Africa, and the South Pacific, suggesting a near-global distribution (Levin et al. 2017; Remnant et al. 2017). The complementary positive strand of ARV-1 RNA – a replication form of −ssRNA virus – was also found in *A. mellifera*, which implies active replication of ARV-1 in honey bees (Remnant et al. 2017).

Many of these novel viruses may be benign, but they could equally likely be responsible for some unresolved health problems and deserve further study.

Multi-virus Infection

Surveillance studies analyzing pooled samples of multiple individuals from numerous colonies have revealed that simultaneous virus infections are a common phenomenon in managed honey bee colonies (Ribière et al. 2008; Amiri et al. 2015). Multiple infections tend to be more common toward the end of summer and fall, which is often concurrent with a higher prevalence of *Varroa* mites and colony failure (Tentcheva et al. 2004). Viral co-infection is also observed in individual bees including workers, drones, and queens (Evans 2001; Chen et al. 2005; D'Alvise et al. 2019). Different viruses – or strains of the same virus – can simultaneously persist in a colony or an individual without obvious symptoms (Evans 2001; Tentcheva et al. 2004; McMahon et al. 2016). Multiple virus infections should not be neglected because of their synergistic effects, virus-virus competition, and possibly recombination between strains of the same virus (Carrillo-Tripp et al. 2016; Martin and Brettell 2019).

In a recent study, pathogen co-occurrence including 11 viruses was assessed within more than 1000 single worker bees over the course of one year: on average, an individual bee carried 2.1 different viruses, with 38% of the bees infected with more than three viruses, and 14% infected with four to six different viruses (D'Alvise et al. 2019). In the same study, BQCV showed a significant positive correlation with all other common viruses, but specifically DWV and LSV, which suggests synergistic interactions. BQCV may either increase in the presence of other viruses, it may facilitate infection with or proliferation of other viruses, or both (D'Alvise et al. 2019).

A separate experiment, using serial passaging of DWV by direct injection of crude host extract, indicated that DWV was outcompeted by SBV and BQCV, and as a result DWV levels rapidly declined while the levels of SBV and BQCV increased (Remnant et al. 2019). A similar result showed that ABPV did not replicate when injected into pupae already infected with either SBV or BQCV (Bailey et al. 1981a). These results suggest competitive mutual exclusion among viruses and that the reason that BQCV and SBV are highly competitive is because of their ability to replicate rapidly. A recent study demonstrated that, after experimental infection with different mixed virus inoculates, virus proliferation was strongly influenced by what other viruses were present in the mixed inoculum (Carrillo-Tripp et al. 2016). In that case, IAPV outcompeted SBV while SBV was present in much higher abundance, but IAPV cannot replicate efficiently in combination with KBV. Thus, complex virus–virus interactions, such as competition for cellular resources or modulation of host defense systems, likely affect infection dynamics (Carrillo-Tripp et al. 2016).

Transmission Pathways

Possible transmission pathways have been discussed for each virus in previous sections, but it is crucial to understand transmission routes and directionality because they have direct effects on the prevalence and virulence evolution of viruses (Chen et al. 2006; Martin et al. 2012; Nazzi et al. 2012). Knowing transmission pathways could facilitate the development of strategies to reduce intracolony viral loads and disrupt intercolony transmission, which could significantly alleviate the threats of viruses in honey bee populations.

Viruses are non-living, opportunistic, and obligate intracellular pathogens that require the host-cellular machinery for transcription, translation, and replication. In honey bees, viruses utilize and may even manipulate honey bee behavior to transmit among individuals, colonies, and generations either via horizontal transmission, vertical routes, or both (Chen and Siede 2007) (Figure 21.4).

Horizontal transmission routes occur frequently through trophallactic activities, hygienic behavior, grooming, and larval cannibalism (Chen and Siede 2007; Mockel et al. 2011). Specifically, glandular secretions of nurse bees can transmit viruses to young larvae and cause infection. Viruses can also be transmitted through forager visits to flowers. Infected worker bees can contaminate pollen through salivary secretion while collecting pollen, increasing the virus transmission to nestmates and other colonies foraging on the same resource (Chen and Siede 2007; de Miranda and Genersch 2010). Viruses also can be readily transmitted among workers in a crowded colony via topical or bodily contact (Ribiere et al. 2010).

Among all horizontal transmission routes, vector-mediated transmission is probably most important for honey bee health. Honey bee viruses have been detected from *V. destructor*, *T. mercedesae*, and *A. tumida* (Bowen-Walker et al. 1999; Forsgren et al. 2009; Di Prisco et al. 2011; Chanpanitkitchote et al. 2018). Among them, *V. destructor*

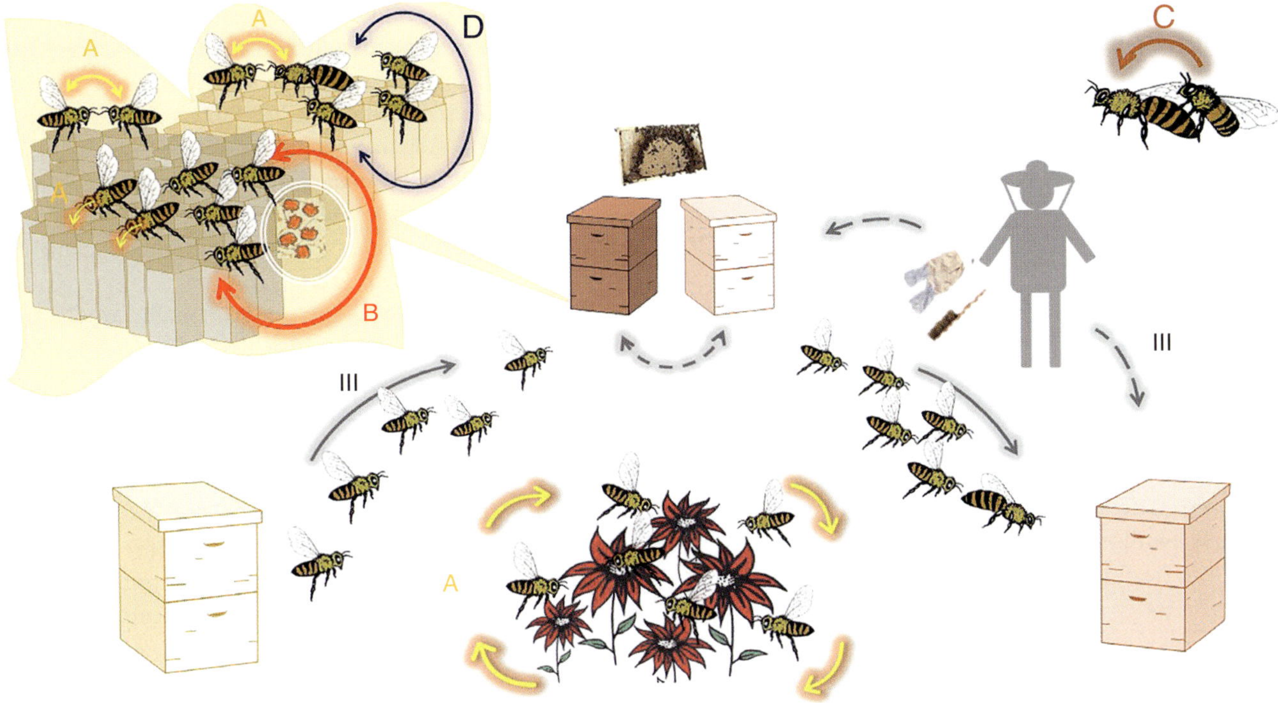

- I Horizontal
  - A. Food-born transmission
    1. Worker-queen trophallaxis
    2. *Worker-worker food sharing*
    3. *Hypopharyngeal and larval food*
    4. *Common yard foraging*
    5. *Hygienic and cannibalism*
  - B. *Vector-born transmission*
    1. *Varroa destructor*
    2. *Tropilaelaps mercedesae*
    3. *Aethina tumida*
  - C. *Venereal Transmission*
  - D. *Topical and body contact*
- II *Vertical*
  1. *Trans-ovarial*
  2. *Trans-ovum*
  3. *Trans-spermal*
- III *Other routes*
  1. *Drifting and robbing*
  2. *Swarming*
  3. *beekeeping activities and tools*

Figure 21.4 Modes of virus transmission. *Source:* Created by Phoebe Snyder and Esmaeil Amiri.

has played a critical role in changing the virus ecology in the honey bee industry after shifting from *A. cerana* to *A. mellifera*, and it is now known, with its associated viruses, to be one of the main drivers for colony mortality (Martin 2001; Carreck et al. 2010). As an example, *Varroa* mites play a mechanical role to transmit the viruses among individuals, and also play a more direct role by selectively vectoring or activating particularly virulent virus strains (Bowen-Walker et al. 1999; Di Prisco et al. 2011).

Venereal transmission is also a form of horizontal transmission since it involves individuals from the same generation. Detailed studies detected honey bee viruses from drones, collected semen, and endophalli (plural term for drone's mating organ, endophallus) from queens returning from mating flights, followed by corresponding virus detection in the queens' ovaries (Yue et al. 2007; de Miranda and Fries 2008; Amiri et al. 2016). Infected queens pass viruses to their progeny through vertical transmission, playing a role for virus persistence in honey bee populations (de Miranda and Fries 2008; Amiri et al. 2018).

Particularly important is inter-colonial virus transmission. High-density beekeeping practices favor virus transmission among colonies because infected adults (workers and drones) can move more readily among hives ("drifting"). The most relevant behaviors that beekeepers need to prevent are drifting and robbing, although intentional transfer of individuals and equipment among colonies are certainly also efficient transmission routes that beekeepers must minimize (Dynes et al. 2019). In addition to the mentioned possible virus transmission routes, international trade in bees and bee products (queens, semen, pollen, honey, royal jelly, and beeswax) is prevalent and arguably represents a main factor for long-distance distribution of most honey bee diseases including viruses (Mutinelli 2011).

Viral Disease Management

Despite their widespread distribution and major implications for honey bee health, there are few direct options for beekeepers to mitigate viral diseases. Antibiotics are not effective against viruses, meaning there are limited therapeutic approaches, none of which can be readily incorporated into honey bee management. Recent genetic and molecular biology developments may provide future directions for limiting virus loads in honey bee population. RNA interference (RNAi) is a post-transcriptional, sequence specific gene regulation mechanism that plays a vital part in insect antiviral immunity. In recent years, RNAi technologies have already been effectively used in some insect-virus systems. Delivery of targeted virus double stranded RNA (dsRNA) by injection or feeding has been demonstrated to be effective in potential honey bees against several viral infections (Maori et al. 2009; Hunter et al. 2010; Desai et al. 2012; Leonard et al. 2020). The potential for silencing virus-specific targets with minimal secondary effects on honey bees is a highly intriguing possibly if the challenges of cost and delivery can be solved (Hunter et al. 2010). Another challenge to this approach is the rapid evolution and genetic diversity of viral strains, making the potential for resistance and adaptation very high. Another possibility would be the use of genetically engineered honey bees that are resistant to viruses, although it deserves careful scrutiny because of the risk of environmental contamination. A long-term and sustainable strategy is to selectively breed honey bees that are naturally more resistant to viruses, and several efforts are under way to promote genetic variation for virus resistance in breeding programs.

Based on the current lack of preventative or curative medication, bee management and breeding are currently the most promising approaches in mitigating viral diseases in honey bees. In addition, understanding transmission routes is a crucial step to managing honey bees and colonies to minimize virus transmission. Hive spacing, visual landmarks, and entrance reducers can reduce drift and robbing. The beekeeper also should refrain from exchanging brood or hive equipment among colonies, employ proper sanitary practices with equipment such as hive tools, and manage viral vectors (Prodělalová et al. 2017; De Guzman et al. 2019). The development and implementation of a biosecurity plan is advisable. Most important among these vectors is the *Varroa* mite, and minimizing its presence will also lighten the viral impact on honey bee health.

Generally, strong hives that are not stressed by nutritional deficits or other factors will tolerate viruses more effectively, which simplifies colony management recommendations. For example, nectar and pollen phytochemicals, as well as extracts from amadou (Fomes) and reishi (Ganoderma) fungi have shown to be effective against DWV and LSV under controlled experiments (Palmer-Young et al. 2017; Stamets et al. 2018). These findings indicate that high-quality diets, that ensure proper nutrition and foraging throughout the active season, can mitigate the impact of viruses (DeGrandi-Hoffman and Chen 2015; Dolezal et al. 2019).

The wax matrix of hives may accumulate viral pathogens over time, thus one management strategy for healthier colonies would be to regularly replace old combs and frames. This is energetically costly for the bees and will lessen honey yields for the beekeeper, but it may be an important change in management tactics for the industry. Finally, collapsing hives should be immediately removed from their apiary and their materials quarantined, replaced, or destroyed from the beekeeping operation. The long-term survival of viruses has been demonstrated in apiaries and bee equipment at the ambient temperature as a possible source of viral infection. Therefore, chemical disinfection of both surfaces and tools using iodopher- and peracetic acid-based sanitizers have been recommended to decrease the honey bee virus load in apiaries and reduce the spread of viruses by beekeepers (Prodělalová et al. 2017).

In local beekeeping communities, the health of colonies in one yard depends partly on the health of the colonies maintained by neighboring beekeepers, because viral diseases can be transmitted by drifting foragers, shared floral resources, or mating. Active beekeeper associations promote education and cooperation to promote best management practices to reduce viral diseases and consequently the viral spread from operation to operation, benefiting the health of everyone's colonies. Additionally, bee breeders should regularly screen for the virus prevalence and diversity in their operations and minimize virus distribution through shipped queens by selecting queen sources and rearing colonies with low viral prevalence. The queen is central for vertical transmission (Amiri et al. 2017) and thus queen breeders – who produce a majority of all honey bee queens used in the US – have a crucial role to play in virus control. Queens that are well-mated with many genetically diverse drones typically produce colonies better able to resist disease (Tarpy 2003; Spivak et al. 2019). Queen replacement has been recommended in CBPV infected colonies with a young, healthy, and productive queen to change the colony genetic composition and maintain colony homeostasis (Ribiere et al. 2010). Therefore, production of high quality and healthy queens and drones will help to increase mating success and colony health (Amiri et al. 2017).

Finally, to prevent virus spread internationally, both importing and exporting countries need to establish epidemiological regulations and health surveillance practices based on standard procedures and recognized tests. Diseases in honey bees caused by viruses are not listed by the World Organization for Animal Health (OIE) and so the *Terrestrial Code* makes no recommendations for worldwide trading of honey bees and their products (Mutinelli 2011).

Virus Isolation, Detection, and Diagnoses

Since the first description of honey bee viruses, there have been numerous attempts to isolate and propagate specific viruses to study symptoms and virus-host interaction. The combination of cell culture and infectious-virus molecular clones has been considered the gold standard for isolation and viral diagnosis, since they are convenient tools to eliminate contamination and environmental variables that are impossible to control when working at organismal or colony levels. Although a few honey bee viruses have been cloned (Lamp et al. 2016; Seitz et al. 2019), there has been less success to develop a non-contaminated immortal honey bee cell line (Carrillo-Tripp et al. 2016) and honey bee viruses are still mostly isolated from naturally or experimentally infected adults or pupae (Boncristiani et al. 2013; de Miranda et al. 2013). These virus-isolation procedures are problematic because co-infections can occur, and the genotypic composition of the inoculum could change between experiments.

The classical diagnostic methods for honey bee viruses have relied on electron microscope detection of viral particles and serological methods. Electron microscopy has been and continues to be one of the most important techniques for diagnosis and virus discovery with the possibility to classify viruses based on their capsid morphology (Roingeard 2008). However, both methods are laborious and require specialist expertise. Thus, symptoms are still one of the principle methods to diagnose virus infections in the apiary, as discussed for each virus in previous sections. It is a simple and fast method but greatly relies on beekeeper expertise. In recent years with the expansion of the Internet and the use of smartphones, there have been several attempts to design and develop diagnostic applications (e.g.: www.thebeemd.com, or the Bee Health app) for use in the field to diagnose honey bee health problems. The major disadvantage of symptom-based diagnoses is that many viral infections remain covert, present with similar symptoms, or involve co-infection. Bioassays, or infectivity tests, were used to discover and diagnose several honey bee viruses. The method proceeds from virus purification from infected bees to injection into apparently healthy individuals to detect amplify symptoms for better diagnosis (Bailey et al. 1963; de Miranda et al. 2013). This is a labor-intensive method, however, and in many cases injection to inoculate with specific virus may trigger latent viruses in the injected bees or pupae confounding the results. The primary purpose of diagnosis at this point is prevention of further infections, but diagnosis by the classic methods is time-consuming, allowing the virus to spread throughout the colony and/or apiary.

Based on the rapid establishment of molecular techniques, continuing method developments have established increasingly sensitive, accurate, reliable, universal, rapid, and cheap approaches to detect and diagnose viral infections. Serological methods including enzyme-linked immune sorbent assays (ELISAs), agar gel immunodiffusion tests (AGIDs), and western blots were the first molecular tools developed for the characterization of viruses (de Miranda et al. 2013). However, the current methods of choice are nucleic acid-based assays. Most targeted assays rely on amplification of specific sequences to quantify and distinguish individual viruses. The various amplification protocols rely on variants of the basic PCR that exponentially amplifies a stretch of nucleic acid based on two specific oligonucleotide primers added to the reaction. Variants include RT-PCR, as well as less common techniques, such as Reverse Transcription Loop-Mediated Isothermal Amplification (RT-LAMP) and Reverse Transcription Multiplex Ligation-dependent Probe Amplification (RT-MLPA). By far the most widely used method is reverse transcriptase quantitative polymerase chain reaction (RT-qPCR), and many individual RT-qPCR protocols have been described for the detection of specific honey bee viruses (de Miranda et al. 2013). The goal of any quantitative assay is to faithfully mirror the biomolecular status of the sample at the time of collection. Therefore, sample management strategy including sample type, transportation time, temperature and chemical stabilization, as well as extraction methods and storage can influence the purity, quantity, and integrity of nucleic acids, which is particularly critical for RNA (Chen et al. 2007; Dainat et al. 2011). Several protocols have been developed to establish standard procedures for sample collection, transport and processing strategies minimize artifactual variation in the results of downstream processes (Dainat et al. 2011; Forsgren et al. 2017). Recently, next-generation sequencing (NGS) has been proposed as a valuable tool for diagnostic virology. It is highly sensitive and has the potential to detect the full spectrum of viruses infecting a sample, including known and even unknown viruses (Runckel et al. 2011; Granberg et al. 2013; Roberts et al. 2017). Indeed, several of the recently discovered honey bee viruses were detected by NGS (Runckel et al. 2011; Remnant et al. 2017; McMenamin and Flenniken 2018). The broad-spectrum and unbiased nature of NGS makes it a valuable tool to detect and diagnose honey bee viruses, but practical hurdles such as high cost and involved data processing need to be solved for NGS to become the standard for near future.

References

Allen, M. and Ball, B. (1995). Characterisation and serological relationships of strains of Kashmir bee virus. *Annals of Applied Biology* 126 (3): 471–484.

Allen, M. and Ball, B. (1996). The incidence and world distribution of honey bee viruses. *Bee World* 77 (3): 141–162.

Amakpe, F., De Smet, L., Brunain, M. et al. (2016). Discovery of Lake Sinai virus and an unusual strain of acute bee paralysis virus in West African apiaries. *Apidologie* 47 (1): 35–47.

Amiri, E., Meixner, M., Büchler, R., and Kryger, P. (2014). Chronic bee paralysis virus in honeybee queens: evaluating susceptibility and infection routes. *Viruses* 6 (3): 1188–1201.

Amiri, E., Meixner, M., Nielsen, S.L., and Kryger, P. (2015). Four categories of viral infection describe the health status of honey bee colonies. *PLoS One* 10 (10): e0140272.

Amiri, E., Meixner, M.D., and Kryger, P. (2016). Deformed wing virus can be transmitted during natural mating in honey bees and infect the queens. *Scientific Reports* 6: 33065.

Amiri, E., Strand, M.K., Rueppell, O., and Tarpy, D.R. (2017). Queen quality and the impact of honey bee diseases on queen health: potential for interactions between two major threats to colony health. *Insects* 8 (2): 48.

Amiri, E., Kryger, P., Meixner, M.D. et al. (2018). Quantitative patterns of vertical transmission of deformed wing virus in honey bees. *PLoS One* 13 (3): e0195283.

Amiri, E., Seddon, G., Zuluaga Smith, W. et al. (2019). Israeli acute paralysis virus: honey bee queen–worker interaction and potential virus transmission pathways. *Insects* 10 (1): 9.

Anderson, D.L. (1990). Pests and pathogens of the honeybee (*Apis mellifera* L) in Fiji. *Journal of Apicultural Research* 29 (1): 53–59.

Anderson, D.L. and Giacon, H. (1992). Reduced pollen collection by honey bee (Hymenoptera: Apidae) colonies infected with *Nosema apis* and sacbrood virus. *Journal of Economic Entomology* 85 (1): 47–51.

Anderson, D.L. and Gibbs, A.J. (1989, 1989). Transpuparial transmission of Kashmir bee virus and sacbrood virus in the honey bee (*Apis mellifera*). *Annals of Applied Biology* 114 (1): 1–7.

Anderson, D.L. and Trueman, J.W. (2000). *Varroa jacobsoni* (Acari: Varroidae) is more than one species. *Experimental and Applied Acarology* 24 (3): 165–189.

Aruna, R., Srinivasan, M.R., Balasubramanian, V., and Selvarajan, R. (2018). Complete genome sequence of sacbrood virus isolated from Asiatic honey bee *Apis cerana* indica in India. *Virusdisease* 29 (4): 453–460.

Bailey, L. and Ball, B.V. (1991). *Honey Bee Pathology*. London: Academic Press.

Bailey, L. and Fernando, E.F.W. (1972). Effects of sacbrood virus on adult honey bees. *Annals of Applied Biology* 72 (1): 27–35.

Bailey, L. and Perry, J. (1982). The diminished incidence of *Acarapis woodi* (Rennie)(Acari: Tarsonemidae) in honey bees, *Apis mellifera* L. (Hymenoptera: Apidae), in Britain. *Bulletin of Entomological Research* 72 (4): 655–662.

Bailey, L. and Woods, R.D. (1974). Three previously undescribed viruses from the honey bee. *Journal of General Virology* 25 (2): 175–186.

Bailey, L. and Woods, R.D. (1977). Two more small RNA viruses from honey bees and further observations on sacbrood and acute bee-paralysis viruses. *Journal of General Virology* 37 (1): 175–182.

Bailey, L., Gibbs, A.J., and Woods, R.D. (1963). Two viruses from adult honey bees (*Apis mellifera* Linnaeus). *Virology* 21: 390–395.

Bailey, L., Gibbs, A.J., and Woods, R.D. (1964). Sacbrood virus of the larval honey bee (*Apis mellifera* Linnaeus). *Virology* 23 (3): 425–429.

Bailey, L., Carpenter, J.M., and Woods, R.D. (1979). Egypt bee virus and Australian isolates of Kashmir bee virus. *Journal of General Virology* 43 (3): 641–647.

Bailey, L., Ball, B.V., and Perry, J.N. (1981a). The prevalence of viruses of honey bees in Britain. *Annals of Applied Biology* 97 (1): 109–118.

Bailey, L., Carpenter, J.M., and Woods, R.D. (1981b). Properties of a filamentous virus of the honey bee (*Apis mellifera*). *Virology* 114 (1): 1–7.

Bailey, L., Ball, B.V., and Perry, J.N. (1983). Association of viruses with two protozoal pathogens of the honey bee. *Annals of Applied Biology* 103 (1): 13–20.

Bakonyi, T., Farkas, R., Szendröi, A. et al. (2002a). Detection of acute bee paralysis virus by RT-PCR in honey bee and *Varroa destructor* field samples: rapid screening of representative Hungarian apiaries. *Apidologie* 33 (1): 63–74.

Bakonyi, T., Grabensteiner, E., Kolodziejek, J. et al. (2002b). Phylogenetic analysis of acute bee paralysis virus strains. *Applied and Environmental Microbiology* 68 (12): 6446–6450.

Bigot, D., Dalmon, A., Roy, B. et al. (2017). The discovery of Halictivirus resolves the Sinaivirus phylogeny. *Journal of General Virology* 98 (11): 2864–2875.

Blanchard, P., Ribiere, M., Celle, O. et al. (2007). Evaluation of a real-time two-step RT-PCR assay for quantitation of chronic bee paralysis virus (CBPV) genome in experimentally-infected bee tissues and in life stages of a symptomatic colony. *Journal of Virological Methods* 141 (1): 7–13.

Blanchard, P., Schurr, F., Olivier, V. et al. (2009). Phylogenetic analysis of the RNA-dependent RNA polymerase (RdRp) and a predicted structural protein (pSP) of the chronic bee paralysis virus (CBPV) isolated from various geographic regions. *Virus Research* 144 (1): 334–338.

Blanchard, P., Guillot, S., Antùnez, K. et al. (2014). Development and validation of a real-time two-step RT-qPCR TaqMan® assay for quantitation of sacbrood virus (SBV) and its application to a field survey of symptomatic honey bee colonies. *Journal of Virological Methods* 197: 7–13.

Boecking, O. and Genersch, E. (2008). Varroosis – the ongoing crisis in bee keeping. *Journal für Verbraucherschutz und Lebensmittelsicherheit* 3 (2): 221–228.

Boncristiani, H.F., Evans, J.D., Chen, Y. et al. (2013). *In vitro* infection of pupae with Israeli acute paralysis virus suggests disturbance of transcriptional homeostasis in honey bees (*Apis mellifera*). *PLoS One* 8 (9): e73429.

Bonning, B.C. and Miller, W.A. (2010). Dicistroviruses. *Annual Review of Entomology* 55 (1): 129–150.

Bowen-Walker, P.L., Martin, S.J., and Gunn, A. (1999). The transmission of deformed wing virus between honeybees (*Apis mellifera* L.) by the ectoparasitic mite *Varroa jacobsoni* Oud. *Journal of Invertebrate Pathology* 73 (1): 101–106.

Brosi, B.J., Delaplane, K.S., Boots, M., and de Roode, J.C. (2017). Ecological and evolutionary approaches to managing honeybee disease. *Nature Ecology & Evolution* 1 (9): 1250–1262.

Carreck, N.L., Ball, B.V., and Martin, S.J. (2010). Honey bee colony collapse and changes in viral prevalence associated with *Varroa destructor*. *Journal of Apicultural Research* 49 (1): 93–94.

Carrillo-Tripp, J., Dolezal, A.G., Goblirsch, M.J. et al. (2016). *In vivo* and *in vitro* infection dynamics of honey bee viruses. *Scientific Reports* 6: 22265.

Celle, O., Blanchard, P., Olivier, V. et al. (2008). Detection of chronic bee paralysis virus (CBPV) genome and its replicative RNA form in various hosts and possible ways of spread. *Virus Research* 133 (2): 280–284.

Cepero, A., Ravoet, J., Gómez-Moracho, T. et al. (2014). Holistic screening of collapsing honey bee colonies in Spain: a case study. *BMC Research Notes* 7 (1): 649.

Chanpanitkitchote, P., Chen, Y., Evans, J.D. et al. (2018). Acute bee paralysis virus occurs in the Asian honey bee *Apis cerana* and parasitic mite *Tropilaelaps mercedesae*. *Journal of Invertebrate Pathology* 151: 131–136.

Chen, Y. and Siede, R. (2007). Honey bee viruses. *Advances in Virus Research* K. Maramorosch, S.A. Shabalina and F.A. Murphy. San Diego, Elsevier Academic Press Inc. 70: 33–80.

Chen, Y., Pettis, J.S., Evans, J.D. et al. (2004). Transmission of Kashmir bee virus by the ectoparasitic mite *Varroa destructor*. *Apidologie* 35 (4): 441–448.

Chen, Y., Pettis, J.S., and Feldlaufer, M.F. (2005). Detection of multiple viruses in queens of the honey bee *Apis mellifera* L. *Journal of Invertebrate Pathology* 90 (2): 118–121.

Chen, Y., Pettis, J.S., Collins, A., and Feldlaufer, M.F. (2006). Prevalence and transmission of honeybee viruses. *Applied and Environmental Microbiology* 72 (1): 606–611.

Chen, Y., Evans, J., Hamilton, M., and Feldlaufer, M. (2007). The influence of RNA integrity on the detection of honey bee viruses: molecular assessment of different sample storage methods. *Journal of Apicultural Research* 46 (2): 81–87.

Chen, Y., Pettis, J.S., Corona, M. et al. (2014). Israeli acute paralysis virus: epidemiology, pathogenesis and implications for honey bee health. *PLoS Pathogens* 10 (7): e1004261.

Chevin, A., Coutard, B., Blanchard, P. et al. (2015). Characterisation of structural proteins from chronic bee paralysis virus (CBPV) using mass spectrometry. *Viruses* 7 (6): 3329–3344.

Choe, S.E., Nguyen, L.T.K., Noh, J.H. et al. (2012). Analysis of the complete genome sequence of two Korean sacbrood viruses in the Honey bee, *Apis mellifera*. *Virology* 432 (1): 155–161.

Clark, T.B. (1978). A filamentous virus of the honey bee. *Journal of Invertebrate Pathology* 32 (3): 332–340.

Cornman, R.S. (2019). Relative abundance and molecular evolution of Lake Sinai virus (Sinaivirus) clades. *PeerJ* 7: e6305.

Cornman, R.S., Tarpy, D.R., Chen, Y. et al. (2012). Pathogen webs in collapsing honey bee colonies. *PLoS One* 7 (8): e43562.

Cox-Foster, D.L., Conlan, S., Holmes, E.C. et al. (2007). A metagenomic survey of microbes in honey bee colony collapse disorder. *Science* 318 (5848): 283–287.

Dainat, B., Ken, T., Berthoud, H., and Neumann, P. (2009). The ectoparasitic mite *Tropilaelaps mercedesae* (Acari, Laelapidae) as a vector of honeybee viruses. *Insectes Sociaux* 56 (1): 40–43.

Dainat, B., Evans, J.D., Chen, Y.P., and Neumann, P. (2011). Sampling and RNA quality for diagnosis of honey bee viruses using quantitative PCR. *Journal of Virological Methods* 174 (1): 150–152.

Dall, D.J. (1987). Multiplication of Kashmir bee virus in pupae of the honeybee, *Apis-mellifera*. *Journal of Invertebrate Pathology* 49 (3): 279–290.

D'Alvise, P., Seeburger, V., Gihring, K. et al. (2019). Seasonal dynamics and co-occurrence patterns of honey bee pathogens revealed by high-throughput RT-qPCR analysis. *Ecology and Evolution* 9 (18): 10241–10252.

Daughenbaugh, K.F., Martin, M., Brutscher, L.M. et al. (2015). Honey bee infecting Lake Sinai viruses. *Viruses* 7 (6): 3285–3309.

De Guzman, L.I., Simone-Finstrom, M., Frake, A.M., and Tokarz, P. (2019). Comb irradiation has limited, interactive effects on colony performance or pathogens in bees, *Varroa destructor* and wax based on two honey bee stocks. *Insects* 10 (1): 15.

DeGrandi-Hoffman, G. and Chen, Y. (2015). Nutrition, immunity and viral infections in honey bees. *Current Opinion in Insect Science* 10: 170–176.

Denholm, C.H. (1999). *Inducible Honey Bee Viruses Associated with Varroa jacobsoni*. University of Keele.

Desai, S.D., Eu, Y.-J., Whyard, S., and Currie, R.W. (2012). Reduction in deformed wing virus infection in larval and adult honey bees (*Apis mellifera* L.) by double-stranded RNA ingestion. *Insect Molecular Biology* 21 (4): 446–455.

Di Prisco, G., Pennacchio, F., Caprio, E. et al. (2011). *Varroa destructor* is an effective vector of Israeli acute paralysis virus in the honeybee, *Apis mellifera*. *Journal of General Virology* 92 (1): 151–155.

Di Prisco, G., Annoscia, D., Margiotta, M. et al. (2016). A mutualistic symbiosis between a parasitic mite and a pathogenic virus undermines honey bee immunity and health. *Proceedings of the National Academy of Sciences of the United States of America* 113 (12): 3203–3208.

Dolezal, A.G., Carrillo-Tripp, J., Judd, T.M. et al. (2019). Interacting stressors matter: diet quality and virus infection in honeybee health. *Royal Society Open Science* 6 (2): 181803.

Doublet, V., Labarussias, M., de Miranda, J.R. et al. (2015). Bees under stress: sublethal doses of a neonicotinoid pesticide and pathogens interact to elevate honey bee mortality across the life cycle. *Environmental Microbiology* 17 (4): 969–983.

Dubois, E., Reis, C., Schurr, F. et al. (2018). Effect of pollen traps on the relapse of chronic bee paralysis virus in honeybee (*Apis mellifera*) colonies. *Apidologie* 49 (2): 235–242.

Dynes, T.L., Berry, J.A., Delaplane, K.S. et al. (2019). Reduced density and visually complex apiaries reduce parasite load and promote honey production and overwintering survival in honey bees. *PLoS One* 14 (5): e0216286.

Evans, J.D. (2001). Genetic evidence for coinfection of honey bees by acute bee paralysis and Kashmir bee viruses. *Journal of Invertebrate Pathology* 78 (4): 189–193.

Eyer, M., Chen, Y., Ping, M.S. et al. (2009). Small hive beetle, *Aethina tumida*, as a potential biological vector of honeybee viruses. *Apidologie* 40 (4): 419–428.

Faucon, J.P., Vitu, C., Russo, P., and Vignoni, M. (1992). Diagnosis of acute paralysis – application to epidemic honeybee diseases in France during 1990. *Apidologie* 23 (2): 139–146.

Fievet, J., Tentcheva, D., Gauthier, L. et al. (2006). Localization of deformed wing virus infection in queen and drone *Apis mellifera* L. *Virology Journal* 3: 16.

Flenniken, M.L. (2014). Honey bee-infecting plant virus with implications on honey bee colony health. *mBio* 5 (2): e00877-00814.

Forsgren, E., de Miranda, J.R., Isaksson, M. et al. (2009). Deformed wing virus associated with *Tropilaelaps mercedesae* infesting European honey bees (*Apis mellifera*). *Experimental and Applied Acarology* 47 (2): 87–97.

Forsgren, E., Locke, B., Semberg, E. et al. (2017). Sample preservation, transport and processing strategies for honeybee RNA extraction: influence on RNA yield, quality, target quantification and data normalization. *Journal of Virological Methods* 246: 81–89.

Francis, R.M., Nielsen, S.L., and Kryger, P. (2013). Patterns of viral infection in honey bee queens. *Journal of General Virology* 94 (Pt 3): 668–676.

Fujiyuki, T., Takeuchi, H., Ono, M. et al. (2004). Novel insect picorna-like virus identified in the brains of aggressive worker honeybees. *Journal of Virology* 78 (3): 1093–1100.

Garigliany, M., El Agrebi, N., Franssen, M. et al. (2019). Moku virus detection in honey bees, Belgium, 2018. *Transboundary and Emerging Diseases* 66 (1): 43–46.

Gauthier, L., Ravallec, M., Tournaire, M. et al. (2011). Viruses associated with ovarian degeneration in *Apis mellifera* L. queens. *PLoS One* 6 (1): e16217.

Gauthier, L., Cornman, S., Hartmann, U. et al. (2015). The *Apis mellifera* filamentous virus genome. *Viruses* 7 (7): 3798–3815.

Ghosh, R.C., Ball, B.V., Willcocks, M.M., and Carter, M.J. (1999). The nucleotide sequence of sacbrood virus of the honey bee: an insect picorna-like virus. *Journal of General Virology* 80 (6): 1541–1549.

Giuffre, C., Lubkin, S.R., and Tarpy, D.R. (2019). Does viral load alter behavior of the bee parasite *Varroa destructor*? *PLoS One* 14 (6): e0217975.

Gong, H.-R., Chen, X.-X., Chen, Y.P. et al. (2016). Evidence of *Apis cerana Sacbrood virus* infection in *Apis mellifera*. *Applied and Environmental Microbiology* 82 (8): 2256–2262.

Govan, V.A., Leat, N., Allsopp, M., and Davison, S. (2000). Analysis of the complete genome sequence of acute bee paralysis virus shows that it belongs to the novel group of insect-infecting RNA viruses. *Virology* 277 (2): 457–463.

Grabensteiner, E., Ritter, W., Carter, M.J. et al. (2001). Sacbrood virus of the honeybee (*Apis mellifera*): rapid identification and phylogenetic analysis using reverse transcription-PCR. *Clinical and Diagnostic Laboratory Immunology* 8 (1): 93–104.

Granberg, F., Vicente-Rubiano, M., Rubio-Guerri, C. et al. (2013). Metagenomic detection of viral pathogens in Spanish honeybees: co-infection by aphid lethal paralysis, Israel acute paralysis and Lake Sinai viruses. *PLoS One* 8 (2): e57459.

Hartmann, U., Forsgren, E., Charrière, J.-D. et al. (2015). Dynamics of *Apis mellifera* filamentous virus (AmFV) infections in honey bees and relationships with other parasites. *Viruses* 7 (5): 2654–2667.

Hassanyar, A.K., Huang, S., Li, Z. et al. (2019). Prevalence of bee viruses in *Apis cerana* cerana populations from different locations in the Fujian province of China. *MicrobiologyOpen* 8 (9): e00830.

Higes, M., Martin, R., and Meana, A. (2006). *Nosema ceranae*, a new microsporidian parasite in honeybees in Europe. *Journal of Invertebrate Pathology* 92 (2): 93–95.

Hitchcock, J.D. (1966). Transmission of sacbrood disease to individual honey bee larvae. *Journal of Economic Entomology* 59 (5): 1154–1156.

Hou, C., Rivkin, H., Slabezki, Y., and Chejanovsky, N. (2014). Dynamics of the presence of Israeli acute paralysis virus in honey bee colonies with colony collapse disorder. *Viruses* 6 (5): 2012–2027.

Hou, C., Li, B., Deng, S. et al. (2017). Diagnosis and distribution of the *Apis mellifera* filamentous virus (AmFV) in honey bees (*Apis mellifera*) in China. *Insectes Sociaux* 64 (4): 597–603.

Hung, A.C.F. (2000). PCR detection of Kashmir bee virus in honey bee excreta. *Journal of Apicultural Research* 39 (3–4): 103–106.

Hung, A.C. and Shimanuki, H. (1999). A scientific note on the detection of Kashmir bee virus in individual honeybees and *Varroa jacobsoni* mites. *Apidologie* 30 (4): 353–354.

Hung, A.C.F., Peng, C.Y.S., and Shimanuki, H. (2000). Nucleotide sequence variations in Kashmir bee virus isolated from *Apis mellifera* L. and *Varroa jacobsoni* Oud. *Apidologie* 31 (1): 17–23.

Hunter, W., Ellis, J., vanEngelsdorp, D. et al. (2010). Large-scale field application of RNAi technology reducing Israeli acute paralysis virus disease in honey bees (*Apis mellifera*, Hymenoptera: Apidae). *PLoS Pathogens* 6 (12): e1001160 - e1001160.

Iqbal, J. and Mueller, U. (2007). Virus infection causes specific learning deficits in honeybee foragers. *Proceedings of the Royal Society B: Biological Sciences* 274 (1617): 1517–1521.

Kalynych, S., Přidal, A., Pálková, L. et al. (2016). Virion structure of iflavirus slow bee paralysis virus at 2.6-Angstrom resolution. *Journal of Virology* 90 (16): 7444–7455.

Kim, S.H., Mercer, A., Mitchell, A. et al. (2019). Viral infections alter antennal epithelium ultrastructure in honey bees. *Journal of Invertebrate Pathology* 168: 107252.

Kleijn, D., Winfree, R., Bartomeus, I. et al. (2015). Delivery of crop pollination services is an insufficient argument for wild pollinator conservation. *Nature Communications* 6 (1): 7414.

Ko, C.-Y., Chiang, Z.-L., Liao, R.-J. et al. (2019). Dynamics of *Apis cerana* sacbrood virus (AcSBV) prevalence in *Apis cerana* (Hymenoptera: Apidae) in Northern Taiwan and demonstration of its infection in *Apis mellifera* (Hymenoptera: Apidae). *Journal of Economic Entomology* 112 (5): 2055–2066.

Koziy, R.V., Wood, S.C., Kozii, I.V. et al. (2019). Deformed wing virus infection in honey bees (*Apis mellifera* L.). *Veterinary Pathology* 56 (4): 636–641.

Kraberger, S., Cook, C.N., Schmidlin, K. et al. (2019). Diverse single-stranded DNA viruses associated with honey bees (*Apis mellifera*). *Infection, Genetics and Evolution* 71: 179–188.

Kraus, B. and Page, R.E. Jr. (1995). Effect of *Varroa jacobsoni* (Mesostigmata: Varroidae) on feral *Apis mellifera* (Hymenoptera: Apidae) in California. *Environmental Entomology* 24 (6): 1473–1480.

Lamp, B., Url, A., Seitz, K. et al. (2016). Construction and rescue of a molecular clone of deformed wing virus (DWV). *PLoS One* 11 (11): e0164639-e0164639.

Lanzi, G., de Miranda, J.R., Boniotti, M.B. et al. (2006). Molecular and biological characterization of deformed wing virus of honeybees (*Apis mellifera* L.). *Journal of Virology* 80 (10): 4998–5009.

Leat, N., Ball, B., Govan, V., and Davison, S. (2000). Analysis of the complete genome sequence of black queen-cell virus, a picorna-like virus of honey bees. *Journal of General Virology* 81 (8): 2111–2119.

Leonard, S.P., Powell, J.E., Perutka, J. et al. (2020). Engineered symbionts activate honey bee immunity and limit pathogens. *Science* 367 (6477): 573–576.

Levin, S., Galbraith, D., Sela, N. et al. (2017). Presence of apis rhabdovirus-1 in populations of pollinators and their parasites from two continents. *Frontiers in Microbiology* 8 (2482).

Li, Z., Chen, Y., Zhang, S. et al. (2013). Viral infection affects sucrose responsiveness and homing ability of forager honey bees, *Apis mellifera* L. *PLoS One* 8 (10): e77354.

Li, J.L., Cornman, R.S., Evans, J.D. et al. (2014). Systemic spread and propagation of a plant-pathogenic virus in European honeybees, *Apis mellifera*. *mBio* 5 (1): e00898-00813.

Li, Y., Zeng, Z.J., and Wang, Z.L. (2016). Phylogenetic analysis of the honeybee sacbrood virus. *Journal of Apicultural Science* 60 (1): 31–38.

Li, B., Deng, S., Yang, D. et al. (2017). Complete sequences of the RNA 1 and RNA 2 segments of chronic bee paralysis virus strain CBPV-BJ detected in China. *Archives of Virology* 162 (8): 2451–2456.

Li, J., Wang, T., Evans, J.D. et al. (2019). The phylogeny and pathogenesis of sacbrood virus (SBV) infection in European honey bees, *Apis mellifera*. *Viruses* 11 (1): 61.

Malfroy, S.F., Roberts, J.M.K., Perrone, S. et al. (2016). A pest and disease survey of the isolated Norfolk Island honey bee (*Apis mellifera*) population. *Journal of Apicultural Research* 55 (2): 202–211.

Maori, E., Lavi, S., Mozes-Koch, R. et al. (2007). Isolation and characterization of Israeli acute paralysis virus, a dicistrovirus affecting honeybees in Israel: evidence for diversity due to intra- and inter-species recombination. *Journal of General Virology* 88 (12): 3428–3438.

Maori, E., Paldi, N., Shafir, S. et al. (2009). IAPV, a bee-affecting virus associated with colony collapse disorder can be silenced by dsRNA ingestion. *Insect Molecular Biology* 18 (1): 55–60.

Martin, S.J. (2001). The role of *Varroa* and viral pathogens in the collapse of honeybee colonies: a modelling approach. *Journal of Applied Ecology* 38 (5): 1082–1093.

Martin, S.J. and Brettell, L.E. (2019). Deformed wing virus in honeybees and other insects. *Annual Review of Virology* 6 (1): 49–69.

Martin, S., Hogarth, A., Van Breda, J., and Perrett, J. (1998). A scientific note on *Varroa jacobsoni* Oudemans and the collapse of *Apis mellifera* L. colonies in the United Kingdom. *Apidologie* 29 (4): 369–370.

Martin, S.J., Highfield, A.C., Brettell, L. et al. (2012). Global honey bee viral landscape altered by a parasitic mite. *Science* 336 (6086): 1304–1306.

Mazzei, M., Carrozza, M.L., Luisi, E. et al. (2014). Infectivity of DWV associated to flower pollen: experimental evidence of a horizontal transmission route. *PLoS One* 9 (11): e113448.

McFadden, A.M.J., Tham, K.-M., Stevenson, M. et al. (2014). Israeli acute paralysis virus not detected in *Apis mellifera* in New Zealand in a national survey. *Journal of Apicultural Research* 53 (5): 520–527.

McMahon, D.P., Natsopoulou, M.E., Doublet, V. et al. (2016). Elevated virulence of an emerging viral genotype as a driver of honeybee loss. *Proceedings of the Royal Society B: Biological Sciences* 283 (1833): 20160811.

McMenamin, A.J. and Flenniken, M.L. (2018). Recently identified bee viruses and their impact on bee pollinators. *Current Opinion in Insect Science* 26: 120–129.

McMenamin, A.J. and Genersch, E. (2015). Honey bee colony losses and associated viruses. *Current Opinion in Insect Science* 8: 121–129.

Meixner, M.D., Francis, R.M., Gajda, A. et al. (2014). Occurrence of parasites and pathogens in honey bee colonies used in a European genotype–environment interactions experiment. *Journal of Apicultural Research* 53 (2): 215–219.

de Miranda, J.R. and Fries, I. (2008). Venereal and vertical transmission of deformed wing virus in honeybees (*Apis mellifera* L.). *Journal of Invertebrate Pathology* 98 (2): 184–189.

de Miranda, J.R. and Genersch, E. (2010). Deformed wing virus. *Journal of Invertebrate Pathology* 103: S48–S61.

de Miranda, J.R., Drebot, M., Tyler, S. et al. (2004). Complete nucleotide sequence of Kashmir bee virus and comparison with acute bee paralysis virus. *Journal of General Virology* 85 (8): 2263–2270.

de Miranda, J.R., Cordoni, G., and Budge, G. (2010a). The acute bee paralysis virus–Kashmir bee virus–Israeli acute paralysis virus complex. *Journal of Invertebrate Pathology* 103: S30–S47.

de Miranda, J.R., Dainat, B., Locke, B. et al. (2010b). Genetic characterization of slow bee paralysis virus of the honeybee (*Apis mellifera* L.). *Journal of General Virology* 91 (10): 2524–2530.

de Miranda, J.R., Bailey, L., Ball, B.V. et al. (2013). Standard methods for virus research in *Apis mellifera*. *Journal of Apicultural Research* 52 (4): 1–56.

de Miranda, J.R., Cornman, R.S., Evans, J.D. et al. (2015). Genome characterization, prevalence and distribution of a macula-like virus from *Apis mellifera* and *Varroa destructor*. *Viruses* 7 (7): 3586–3602.

Mockel, N., Gisder, S., and Genersch, E. (2011). Horizontal transmission of deformed wing virus: pathological consequences in adult bees (*Apis mellifera*) depend on the transmission route. *Journal of General Virology* 92 (Pt 2): 370–377.

Mondet, F., de Miranda, J.R., Kretzschmar, A. et al. (2014). On the front line: quantitative virus dynamics in honeybee (*Apis mellifera* L.) colonies along a new expansion front of the parasite *Varroa destructor*. *PLoS Pathogens* 10 (8): e1004323.

Mookhploy, W., Kimura, K., Disayathanoowat, T. et al. (2015). Capsid gene divergence of black queen cell virus isolates in Thailand and Japan honey bee species. *Journal of Economic Entomology* 108 (3): 1460–1464.

Mordecai, G.J., Brettell, L.E., Martin, S.J. et al. (2016a). Superinfection exclusion and the long-term survival of honey bees in *Varroa*-infested colonies. *The ISME Journal* 10 (5): 1182–1191.

Mordecai, G.J., Brettell, L.E., Pachori, P. et al. (2016b). Moku virus; a new iflavirus found in wasps, honey bees and *Varroa*. *Scientific Reports* 6: 34983.

Mordecai, G.J., Wilfert, L., Martin, S.J. et al. (2016c). Diversity in a honey bee pathogen: first report of a third master variant of the deformed wing virus quasispecies. *The ISME Journal* 10 (5): 1264–1273.

Morimoto, T., Kojima, Y., Yoshiyama, M. et al. (2012). Molecular identification of chronic bee paralysis virus infection in *Apis mellifera* colonies in Japan. *Viruses* 4 (7): 1093–1103.

Mullapudi, E., Přidal, A., Pálková, L. et al. (2016). Virion structure of Israeli acute bee paralysis virus. *Journal of Virology* 90 (18): 8150–8159.

Mutinelli, F. (2011). The spread of pathogens through trade in honey bees and their products (including queen bees and semen): overview and recent developments. *Revue Scientifique et Technique de l'OIE* 30 (1): 257–271.

Natsopoulou, M.E., McMahon, D.P., Doublet, V. et al. (2017). The virulent, emerging genotype B of deformed wing virus is closely linked to overwinter honeybee worker loss. *Scientific Reports* 7 (1): 5242.

Nazzi, F., Brown, S.P., Annoscia, D. et al. (2012). Synergistic parasite-pathogen interactions mediated by host immunity can drive the collapse of honeybee colonies. *PLoS Pathogens* 8 (6): e1002735.

Oldroyd, B.P. (1999). Coevolution while you wait: *Varroa jacobsoni*, a new parasite of western honeybees. *Trends in Ecology & Evolution* 14 (8): 312–315.

Olivier, V., Blanchard, P., Chaouch, S. et al. (2008a). Molecular characterisation and phylogenetic analysis of chronic bee paralysis virus, a honey bee virus. *Virus Research* 132 (1–2): 59–68.

Olivier, V., Massou, I., Celle, O. et al. (2008b). *In situ* hybridization assays for localization of the chronic bee paralysis virus in the honey bee (*Apis mellifera*) brain. *Journal of Virological Methods* 153 (2): 232–237.

Ongus, J.R., Peters, D., Bonmatin, J.-M. et al. (2004). Complete sequence of a picorna-like virus of the genus iflavirus replicating in the mite *Varroa destructor*. *Journal of General Virology* 85 (12): 3747–3755.

Organtini, L.J., Shingler, K.L., Ashley, R.E. et al. (2017). Honey bee deformed wing virus structures reveal that conformational changes accompany genome release. *Journal of Virology* 91 (2) e01795-01716.

Palacios, G., Hui, J., Quan, P.L. et al. (2008). Genetic analysis of Israel acute paralysis virus: distinct clusters are circulating in the United States. *Journal of Virology* 82 (13): 6209–6217.

Palmer-Young, E.C., Tozkar, C.Ö., Schwarz, R.S. et al. (2017). Nectar and pollen phytochemicals stimulate honey bee (Hymenoptera: Apidae) immunity to viral infection. *Journal of Economic Entomology* 110 (5): 1959–1972.

Pettis, J., Vanengelsdorp, D., and Cox-Foster, D. (2007). Colony collapse disorder working group pathogen sub-group progress report. *American Bee Journal* 147 (7): 595–597.

Procházková, M., Füzik, T., Škubník, K. et al. (2018). Virion structure and genome delivery mechanism of sacbrood honeybee virus. *Proceedings of the National Academy of Sciences of the United States of America* 115 (30): 7759–7764.

Prodělalová, J., Malenovská, H., Moutelíková, R., and Titěra, D. (2017). Virucides in apiculture: persistence of surrogate enterovirus under simulated field conditions. *Pest Management Science* 73 (12): 2544–2549.

Prodělalová, J., Moutelíková, R., and Titěra, D. (2019). Multiple virus infections in western honeybee (*Apis mellifera* L.) ejaculate used for instrumental insemination. *Viruses* 11 (4): 306.

Ravoet, J., De Smet, L., Wenseleers, T., and de Graaf, D.C. (2015a). Genome sequence heterogeneity of Lake Sinai virus found in honey bees and Orf1/RdRP-based polymorphisms in a single host. *Virus Research* 201: 67–72.

Ravoet, J., De Smet, L., Wenseleers, T., and de Graaf, D.C. (2015b). Vertical transmission of honey bee viruses in a Belgian queen breeding program. *BMC Veterinary Research* 11 (1): 1–6.

Reddy, K.E., Noh, J.H., Choe, S.E. et al. (2013a). Analysis of the complete genome sequence and capsid region of black queen cell viruses from infected honeybees (*Apis mellifera*) in Korea. *Virus Genes* 47 (1): 126–132.

Reddy, K.E., Noh, J.H., Kim, Y.-H. et al. (2013b). Analysis of the nonstructural and structural polyprotein regions, and complete genome sequences of Israel acute paralysis viruses identified from honeybees (*Apis mellifera*) in Korea. *Virology* 444 (1): 211–217.

Reddy, K.E., Yoo, M.-S., Kim, Y.-H. et al. (2014). Analysis of the RdRp, intergenic and structural polyprotein regions, and the complete genome sequence of Kashmir bee virus from infected honeybees (*Apis mellifera*) in Korea. *Virus Genes* 49 (1): 137–144.

Remnant, E.J., Shi, M., Buchmann, G. et al. (2017). A diverse range of novel RNA viruses in geographically distinct honey bee populations. *Journal of Virology* 91 (16) e00158-00117.

Remnant, E.J., Mather, N., Gillard, T.L. et al. (2019). Direct transmission by injection affects competition among RNA viruses in honeybees. *Proceedings of the Royal Society B: Biological Sciences* 286 (1895): 20182452.

Retschnig, G., Williams, G.R., Mehmann, M.M. et al. (2014). Sex-specific differences in pathogen susceptibility in honey bees (*Apis mellifera*). *PLoS One* 9 (1): e85261.

Retschnig, G., Kellermann, L.A., Mehmann, M.M. et al. (2019). Black queen cell virus and drifting of honey bee workers (*Apis mellifera*). *Journal of Apicultural Research* 58 (5): 754–755.

Ribiere, M., Lallemand, P., Iscache, A.L. et al. (2007). Spread of infectious chronic bee paralysis virus by honeybee (*Apis mellifera* L.) feces. *Applied and Environmental Microbiology* 73 (23): 7711–7716.

Ribière, M., Ball, B., and Aubert, M. (2008). Natural history and geographical distribution of honey bee viruses. In: *Virology and the Honey Bee* (eds. M. Aubert, B.V. Ball, I. Fries, et al.), 15–84, 460 pp. Luxembourg: Office for Official Publications of the European Communities. ISBN: 92-79-00586-3.

Ribière, M., Olivier, V., and Blanchard, P. (2010). Chronic bee paralysis: a disease and a virus like no other? *Journal of Invertebrate Pathology* 103: 120–131.

Rinderer, T.E. and Green, T.J. (1976). Serological relationship between chronic bee paralysis virus and virus causing hairless-black syndrome in honeybee. *Journal of Invertebrate Pathology* 27 (3): 403–405.

Roberts, J.M.K. and Anderson, D.L. (2014). A novel strain of sacbrood virus of interest to world apiculture. *Journal of Invertebrate Pathology* 118: 71–74.

Roberts, J.M.K., Anderson, D.L., and Durr, P.A. (2017). Absence of deformed wing virus and *Varroa destructor* in Australia provides unique perspectives on honeybee viral landscapes and colony losses. *Scientific Reports* 7 (1): 6925.

Roingeard, P. (2008). Viral detection by electron microscopy: past, present and future. *Biology of the Cell* 100 (8): 491–501.

Runckel, C., Flenniken, M.L., Engel, J.C. et al. (2011). Temporal analysis of the honey bee microbiome reveals four novel viruses and seasonal prevalence of known viruses, *Nosema*, and *Crithidia*. *PLoS One* 6 (6): e20656.

Ryabov, E.V., Wood, G.R., Fannon, J.M. et al. (2014). A virulent strain of deformed wing virus (DWV) of honeybees (*Apis mellifera*) prevails after *Varroa destructor* -mediated, or *in vitro*, transmission. *PLoS Pathogens* 10 (6): e1004230.

Ryabov, E.V., Childers, A.K., Chen, Y. et al. (2017). Recent spread of *Varroa destructor* virus-1, a honey bee pathogen, in the United States. *Scientific Reports* 7 (1): 17447.

Sammataro, D., Gerson, U., and Needham, G. (2000). Parasitic mites of honey bees: life history, implications, and impact. *Annual Review of Entomology* 45 (1): 519–548.

Sánchez-Eugenia, R., Goikolea, J., Gil-Cartón, D. et al. (2015). Triatoma virus recombinant VP4 protein induces membrane permeability through dynamic pores. *Journal of Virology* 89 (8): 4645–4654.

Santillán-Galicia, M.T., Ball, B.V., Clark, S.J., and Alderson, P.G. (2010). Transmission of deformed wing virus and slow paralysis virus to adult bees (*Apis mellifera* L.) by *Varroa destructor*. *Journal of Apicultural Research* 49 (2): 141–148.

Santillán-Galicia, M.T., Ball, B.V., Clark, S.J., and Alderson, P.G. (2014). Slow bee paralysis virus and its transmission in honey bee pupae by *Varroa destructor*. *Journal of Apicultural Research* 53 (1): 146–154.

Schmid-Hempel, P. (1998). *Parasites in Social Insects*. Princeton University Press.

Seitz, K., Buczolich, K., Dikunová, A. et al. (2019). A molecular clone of chronic bee paralysis virus (CBPV) causes mortality in honey bee pupae (*Apis mellifera*). *Scientific Reports* 9 (1): 16274.

Shah, K.S., Evans, E.C., and Pizzorno, M.C. (2009). Localization of deformed wing virus (DWV) in the brains of the honeybee, *Apis mellifera* Linnaeus. *Virology Journal* 6: 182.

Shen, M.Q., Cui, L.W., Ostiguy, N., and Cox-Foster, D. (2005). Intricate transmission routes and interactions between picorna-like viruses (Kashmir bee virus and sacbrood virus) with the honeybee host and the parasitic *varroa* mite. *Journal of General Virology* 86 (pt 8): 2281–2289.

Siede, R. and Buchler, R. (2003). Symptomatic black queen cell virus infection of drone brood in Hessian apiaries. *Berliner und Münchener Tierärztliche Wochenschrift* 116 (3–4): 130–133.

Siede, R., Derakhshifar, I., Otten, C. et al. (2005). Prevalence of Kashmir bee virus in central Europe. *Journal of Apicultural Research* 44 (3): 129–129.

Sitaropoulou, N., Neophytou, E.P., and Thomopoulos, G.N. (1989). Structure of the nucleocapsid of a filamentous virus of the honey bee (*Apis mellifera*). *Journal of Invertebrate Pathology* 53 (3): 354–357.

Spivak, M., Goblirsch, M., and Simone-Finstrom, M. (2019). Social-medication in bees: the line between individual and social regulation. *Current Opinion in Insect Science* 33: 49–55.

Spurny, R., Přidal, A., Pálková, L. et al. (2017). Virion structure of black queen cell virus, a common honeybee pathogen. *Journal of Virology* 91 (6): e02100-02116.

Stamets, P.E., Naeger, N.L., Evans, J.D. et al. (2018). Extracts of polypore mushroom mycelia reduce viruses in honey bees. *Scientific Reports* 8 (1): 13936.

Tapaszti, Z., Forgách, P., Kővágó, C. et al. (2009). Genetic analysis and phylogenetic comparison of black queen cell virus genotypes. *Veterinary Microbiology* 139 (3): 227–234.

Tarpy, D.R. (2003). Genetic diversity within honeybee colonies prevents severe infections and promotes colony growth. *Proceedings of the Royal Society of London. Series B: Biological Sciences* 270 (1510): 99–103.

Tentcheva, D., Gauthier, L., Zappulla, N. et al. (2004). Prevalence and seasonal variations of six bee viruses in *Apis mellifera* L. and *Varroa destructor* mite populations in France. *Applied and Environmental Microbiology* 70 (12): 7185–7191.

Todd, J.H., De Miranda, J.R., and Ball, B.V. (2007). Incidence and molecular characterization of viruses found in dying New Zealand honey bee (*Apis mellifera*) colonies infested with *Varroa destructor*. *Apidologie* 38 (4): 354–367.

Toplak, I., Jamnikar Ciglenečki, U., Aronstein, K., and Gregorc, A. (2013). Chronic bee paralysis virus and *Nosema ceranae* experimental co-infection of winter honey bee workers (*Apis mellifera* L.). *Viruses* 5 (9): 2282–2297.

Traynor, K.S., Rennich, K., Forsgren, E. et al. (2016). Multiyear survey targeting disease incidence in US honey bees. *Apidologie* 47 (3): 325–347.

Valles, S.M., Chen, Y., Firth, A.E. et al. (2017a). ICTV virus taxonomy profile: Dicistroviridae. *Journal of General Virology* 98 (3): 355–356.

Valles, S.M., Chen, Y., Firth, A.E. et al. (2017b). ICTV virus taxonomy profile: Iflaviridae. *Journal of General Virology* 98 (4): 527–528.

vanEngelsdorp, D. and Meixner, M.D. (2010). A historical review of managed honey bee populations in Europe and the United States and the factors that may affect them. *Journal of Invertebrate Pathology* 103: S80–S95.

Vanengelsdorp, D., Evans, J.D., Saegerman, C. et al. (2009). Colony collapse disorder: a descriptive study. *PLoS One* 4 (8): e6481.

vanEngelsdorp, D., Traynor, K.S., Andree, M. et al. (2017). Colony collapse disorder (CCD) and bee age impact honey bee pathophysiology. *PLoS One* 12 (7): e0179535.

Verma, L.R., Rana, B.S., and Verma, S. (1990). Observations on *Apis cerana* colonies surviving from Thai sacbrood virus infestation. *Apidologie* 21 (3): 169–174.

Wilfert, L., Long, G., Leggett, H. et al. (2016). Deformed wing virus is a recent global epidemic in honeybees driven by *Varroa* mites. *Science* 351 (6273): 594–597.

Williams, G.R., Rogers, R.E.L., Kalkstein, A.L. et al. (2009). Deformed wing virus in western honey bees (*Apis mellifera*) from Atlantic Canada and the first description of an overtly-infected emerging queen. *Journal of Invertebrate Pathology* 101 (1): 77–79.

Yang, X. and Cox-Foster, D. (2007). Effects of parasitization by *Varroa destructor* on survivorship and physiological traits of *Apis mellifera* in correlation with viral incidence and microbial challenge. *Parasitology* 134 (Pt 3): 405–412.

Yang, Q., Song, Z.Y., Feng, X. et al. (2016). Analysis of the complete genome sequence of black queen cell virus JL1 from infected honeybees in China. *Bulletin of Entomological Research* 106 (5): 561–568.

Youssef, I., Schurr, F., Goulet, A. et al. (2015). RNA 1 and RNA 2 genomic segments of chronic bee paralysis virus are infectious and induce chronic bee paralysis disease. *Journal of Immunology Research* 2015: 8.

Yue, C., Schroder, M., Bienefeld, K., and Genersch, E. (2006). Detection of viral sequences in semen of honeybees (*Apis mellifera*): evidence for vertical transmission of viruses through drones. *Journal of Invertebrate Pathology* 92 (2): 105–108.

Yue, C., Schroder, M., Gisder, S., and Genersch, E. (2007). Vertical-transmission routes for deformed wing virus of honeybees (*Apis mellifera*). *Journal of General Virology* 88 (Pt 8): 2329–2336.

Zhang, J., Feng, J., Liang, Y. et al. (2001). Three-dimensional structure of the Chinese sacbrood bee virus. *Science in China Series C: Life Sciences* 44 (4): 443–448.

22

Honey Bee Bacterial Diseases

Meghan Milbrath

Michigan Pollinator Initiative, Department of Entomology, Michigan State University, East Lansing, MI, USA

Bacterial diseases in European honey bees (*Apis mellifera)* are widely distributed, causing severe economic damage worldwide (Matheson 1993). Unlike many emerging pests and viral diseases today, bacterial diseases have a long-known association with honey bees; they were first described as "foulbrood" in the eighteenth century (Schirach 1769), named by the strong smell associated with diseased hives. In the early twentieth century, scientists determined there were two separate foulbrood diseases: European foulbrood (EFB) and American foulbrood (AFB) (White 1912). While both foulbrood diseases affect honey bee brood and are bacterial in origin, they have different causes, health outcomes, and treatment recommendations.

Distribution

The names "European foulbrood" and "American foulbrood" have nothing to do with the origins or distributions of these diseases; they were named after the regions where the first scientific investigations of these diseases occurred. Both AFB and EFB have a worldwide distribution, with cases found almost everywhere honey bees are kept. The distribution of EFB is worldwide but non-ubiquitous (Matheson 1993), and cases of AFB can be found everywhere except regions of sub-Saharan Africa (Raina and Fries 2004; OIE [World Organisation for Animal Health] 2009). Though AFB and EFB are known as diseases of *Apis mellifera*, there is some evidence of spread to other *Apis* species – the etiologic agent for AFB has the capability to cause disease in *Apis cerana* (Krongdang et al. 2019), and the etiologic agent for EFB has been isolated from other *Apis* species including *A. cerana* (Bailey 1974) and *Apis laborosa* (Allen et al. 1990).

Economic Importance in North America

Bacterial disease is one of the most serious economic threats to American beekeeping. AFB is generally considered the most devastating disease of honey bees, as clinical infection typically results in both the death of the colony and the destruction of the hive equipment. It is estimated that in the year 2000, AFB resulted in an economic loss of US$5 million to beekeepers in the United States (Eischen et al. 2005). The outcome of EFB at the colony level is more variable: some colonies may die, while other colonies may show signs of complete recovery. Even when colonies appear to recover completely, the remaining population is generally too weak to make a honey crop or to meet a pollination contract, and the beekeeper generally loses the income from that colony for that season. Furthermore, EFB disease will often reappear when the colony is exposed to subsequent stressors. Both AFB and EFB are responsible for high numbers of colony loss, honey crop loss, and colonies that are too weak to meet pollination contracts.

High background levels of bacterial diseases in honey bee populations have the potential to cause exceptional devastation in North American beekeeping. In contrast to Europe and most other parts of the world, commercial beekeeping operations in Canada and in the United States are often much larger and more mobile, containing thousands of colonies that are moved across the country to meet crop pollination contracts. The best example of this large-scale migration is almond pollination, where over 1.5 million colonies are brought to the central valley of California in February (USDA National Agricultural Statistics Service 2017). A survey of colonies in almond pollination found that while only a small number (<5%) of

Honey Bee Medicine for the Veterinary Practitioner, First Edition. Edited by Terry Ryan Kane and Cynthia M. Faux.

the colonies had AFB disease (based on high bacterial colony forming units), 70–80% were positive for the etiologic agent that causes AFB (Eischen et al. 2005). A study on the East Coast of the United States found a similar trend: few colonies (<10%) had visible disease, but infectious spores were found in about half of the registered apiaries (Dingman 2012). In a study in Manitoba, about 15% of the colonies had signs of disease, but spores were found in the honey of almost all of the colonies (Pernal and Melathopoulos 2006). With AFB, the actual cases are low, but the disease potential remains quite high.

The trend in EFB is different from AFB and is not as clearly documented. EFB generally has large seasonal outbreaks in late spring and early summer, with infection rates of up to 60% (Milbrath unpublished). EFB severity varies geographically; it is relatively benign in some regions, and severe in others (Forsgren et al. 2013a). Similar to AFB, the infectious agent for EFB can be present in the absence of visual signs of disease – adult worker bees from seemingly healthy colonies can have high loads of the bacteria that causes EFB (Roetschi et al. 2008). Almost all colonies within an apiary with a clinical case of EFB will contain adults that carry the bacterium (Belloy et al. 2007; Forsgren 2009; Erban et al. 2017), including colonies with no signs of disease (Forsgren et al. 2005; Belloy et al. 2007; Budge et al. 2010; Erban et al. 2017). Infected adult worker bees likely carry the bacteria within the colony and between colonies and apiaries (Belloy et al. 2007; Erban et al. 2017). During pollination, hives are placed in close proximity, with thousands of colonies within flight distance and therefore within contact with one another. With such a high prevalence of colonies containing infectious material for EFB and AFB, the risk of widespread and devastating epidemics of bacterial disease is enormous. Historically, this risk has been managed through antibiotic use and state and provincial level monitoring and control programs. It remains essential to control bacterial diseases in honey bees to preserve the health of honey bees as well as the security of the food systems that depend on honey bees for pollination.

European Foulbrood (EFB)

Etiologic Agent

The etiologic agent of EFB is the gram-positive, microaerophilic, and lanceolate bacterium *Melissococcus plutonius* (Bailey 1983). This bacterium was originally described in 1912 by White (1912), and was first cultured and characterized in detail by Bailey (1957). It was classified into the genus *Melissococcus* in 1982 (Bailey and Collins 1982), but was referred to by other names before that time, namely *Bacillus alvei*, *Bacillus pluton*, and *Streptococcus pluton*. Early studies assumed homogeneity within *M. plutonius*, but recent investigations using multilocus sequence typing (MLST) have identified about 30 different sequence types, divided into three clonal complexes: CC3, CC12, and CC13 (Haynes et al. 2013; Budge et al. 2014; Takamatsu et al. 2014; Lewkowski and Erler 2019) (http://pubmlst.org/mplutonius). Different strains appear to differ in host virulence, with CC3 and CC12 considered more virulent than CC13 in both field (Budge et al. 2014) and *in vitro* (Nakamura et al. 2016) larval infection assays. While there is variability in strain distribution, it is possible to detect multiple strains from the same larval sample (Arai et al. 2014) and for several strains to be responsible for a single outbreak (Takamatsu et al. 2014).

Melissococcus plutonius is often found in association with other bacteria (Forsgren 2010), but the role of these other bacteria in infection and disease progression is not entirely understood. The presence of these other bacteria may influence the color, odor, and consistency of the dead and dying larvae and therefore may be helpful in diagnosis (Tarr 1937a; Gaggìa 2015), but co-infection with other bacteria may (Giersch et al. 2010) or may not (Lewkowski and Erler 2019) lead to greater mortality. Commonly associated bacteria include *Achromobacter eurydice/Bacterium eurydice/Lactobacillus kunkeei/Fructobacillus fructosus*, *Bacillus pumilus*, *Enterococcus faecalis*, *Paenibacillus alvei/Bacillus alvei*, *Brevibacillus laterosporus*, and *Paenibacillus dendritformus*.

Transmission

Melissococcus plutonius is spread between colonies by adult bees. Infected adult worker bees carry the bacteria within a colony and between colonies and apiaries (Belloy et al. 2007; Erban et al. 2017). Transmission can also occur through the robbing of honey, where bees enter a weakened hive and take contaminated honey back to their own hives, or the movement of hive materials as part of apiary management, as *M. plutonius* has been detected in honey as well as bees (Hornitzky and Smith 1998; Mckee et al. 2003; Forsgren et al. 2005). In colonies that were clinically diseased, 38% of the brood nest honey contained detectable levels of *M. plutonius* (Forsgren et al. 2005). The pathogen load in colonies with active disease is generally quite high, as workers from colonies with clinical signs have been found to have 100-fold higher pathogen load of *M. plutonius* than colonies without clinical signs (Erban et al. 2017), and workers in the brood nest have higher pathogen loads than workers at the hive entrance (Roetschi et al. 2008). In colonies that are located close to EFB-symptomatic hives, *M. plutonius* is highly prevalent in adult bees and in larvae, even in colonies without EFB symptoms (Mckee et al. 2003; Forsgren et al. 2005; Belloy et al. 2007; Roetschi et al. 2008; Budge et al. 2010; Erban et al. 2017). In apiaries with a clinical case of EFB, almost all colonies will have adults that are carrying the bacterium (Belloy et al. 2007; Forsgren 2009; Erban et al. 2017). It is therefore important to understand the high levels of within-apiary and between-apiary transmission in areas where disease is present.

EFB disease may stay in an enzootic state in apparently healthy colonies (Pinnock and Featherstone 1984), as *M. plutonius* can be present in hives that are apparently healthy and show no signs of disease (Pinnock and Featherstone 1984; Belloy et al. 2007; Budge et al. 2010). *M. plutonius* is not detected in adults in areas where there is no history of clinical disease (Belloy et al. 2007), but it can be found in apiaries with no sign of current infection and far away from current EFB outbreaks (Budge et al. 2010; Erban et al. 2017). Clinical disease can be triggered in these healthy-looking colonies that contain the pathogen after they experience stress. This stress is often related to abrupt changes in larval feeding (CAPA [Canadian Association of Professional Apiculturists] 2013). Changes in larval feeding occur several times over the season, as the adult bees shift tasks in response to environmental change. For example, the quick onset of a heavy pollen flow will cause the colony to rapidly expand the brood nest. A heavy nectar flow can cause bees that would normally be engaged in nursing tasks to temporarily devote their time to nectar and honey processing. This growth of the brood nest at the same time as a reduction in bees engaged in nursing duties can result in underfeeding of larvae, leading to larval stress, and subsequent EFB disease. The effect of nursing care loss may be exacerbated or amplified by other colony stressors that lead to early forager mortality (and therefore precocious foraging of nurse bees) including pesticides, viruses, and transportation stress.

Colonies that are used for crop pollination are especially susceptible to EFB infection and disease. Colonies in pollination are often kept in areas near many other colonies, increasing the chance of pathogen transmission through drifting, management, or robbing, and they are subjected to dramatic changes in food availability when they are transported long distances to be placed near crops that are in bloom. Many beekeepers who participate in the pollination of early-blooming crops, such as blueberries, report high rates of visible EFB disease every year. Colonies that are kept in regions with a history of EFB outbreaks and colonies that are expected to experience dramatic shifts in food availability should be considered as high risk for EFB disease.

Infection

EFB disease can affect worker, drone, and queen larvae. Infection occurs when a young larva consumes *M. plutonius* in food. The bacteria can enter the brood food through residual bacteria in the hive or indirectly from nurse bees that have previously cleaned a cell in which an infected larva has defecated. While the bacteria can be found throughout the hive, multiplication of *M. plutonius* has only been found to occur in the larval midgut (Forsgren 2009). The susceptibility of the larva to infection depends on larval nutrition, age, and genetics, as well as pathogen strain and dose. One study demonstrated that fewer than 100 bacterial cells are sufficient to cause infection in larvae (Bailey 1960), while infectivity tests in artificially reared honey bee larvae have shown a minimum of 200 *M. plutonius* organisms /ml are required to elicit a disease response (Mckee et al. 2004). A strong dose response with *M. plutonius* and larval death has been observed in lab trials (Mckee et al. 2003), and a dose response with *M. plutonius* and disease has been shown in field trials (Budge et al. 2010). There was also a strong linear relationship between the time duration of feeding an infected diet and larval mortality (Mckee et al. 2003).

The rate of disease progression in an individual larva is highly variable, ranging from quick larval death to successful pupation and emergence as an adult (Bailey 1959b). *M. plutonius* can be detected in larvae without any clinical signs of disease (Forsgren et al. 2005; Gaggìa et al. 2015), and it can be found in healthy-looking pupae, indicating that infected larvae can survive, pupate, and emerge as adults while carrying the bacteria (Bailey and Ball 1991; Forsgren et al. 2005). Bees surviving to pupation can contribute to the spread of the disease within the hive – immediately prior to pupation, larvae defecate, at which time they can discharge large numbers of bacteria at the base of the cell. *M. plutonius* remains viable in these deposits and can survive in this state through long periods of desiccation (Bailey 1959a), remaining viable to infect other larvae. This defecation is generally cleaned by nurse bees who will later feed other susceptible larvae. If an infected larva dies before the defecation/pupation stage, most of the bacteria is removed from the colony when workers remove the infected larvae. For this reason, infections that induce *greater* larval mortality may be resolved more quickly at the colony level, as the infectious material is removed from the hive more quickly and safely.

Most larvae infected with *M. plutonius* die before they reach the defecation or pupation stage (Mckee et al. 2004), with death generally occurring in four to five day-old larvae (Forsgren 2009). In larvae that develop disease, *M. plutonius* rapidly multiplies in the favorable conditions of the larval midgut after ingestion. As the infection continues, the larvae generally move in the cells, adopting a twisted, "corkscrew" shape, or becoming generally misaligned. As the disease progresses, the larva will lose its shape and become discolored, with colors ranging from yellow to brown.

Many basic aspects of the pathogenesis of EFB remain unknown (Forsgren 2009; Djukic et al. 2018). Early studies by Bailey (1983) posit that larval death occurs from starvation after competition for food resources between *M. plutonius* and the larva, and much of the subsequent literature echoes this theory (Forsgren 2010). When larvae are reared in the presence of consistent excess food, however, the larvae still become ill, indicating that additional pathogenic mechanisms are likely (Mckee et al. 2003). It may be the case that EFB is an intestinal infection that causes death through resource competition, or through resource competition along with other causes. Other proposed pathways include the proliferation of opportunistic bacteria such as *P. alvei* and *E. faecalis* (Bailey and Ball 1991), or the invasion of the peritrophic membrane and penetration of host tissue (Mckee et al. 2004; Takamatsu et al. 2014). Recent genetic work aimed at identifying virulence genes in *M. plutonius* provides a variety of potential pathogenic pathways (Djukic et al. 2018). For example, many strains of *M. plutonius* retained the ability to degrade the pectin backbone of the pollen cell wall, which might result in pollen perforation and the release of nutrients (Djukic et al. 2018). *M. plutonius* may also produce products that are toxic to larvae, including tyramine, which has been shown to lead to changes in larval color – a classic sign of EFB disease (Kanbar et al. 2004). More research is necessary to identify strain-specific models of larval pathogenicity of *M. plutonius*.

Prognosis

There is high diversity in colony-level outcomes after EFB infection, which is likely due to a variety of factors including host genetic background (Lewkowski and Erler 2019), host behavior (e.g. hygienic behavior), environmental factors (e.g. nutrition), co-infections, and strain virulence. Even factors such as the diversity of anti-microbial properties in honey from different floral sources can affect bacterial growth (Erler et al. 2014). Some colonies can overcome infection, while in other colonies, brood loss will lead to sufficient weakening to cause eventual collapse of the colony (Forsgren et al. 2013a). Many colonies that do recover from EFB disease are at risk for reinfection during later times of colony stress (Figure 22.1).

Figure 22.1 Michigan State University researchers study colonies affected with European foulbrood. Beekeepers often report high cases of EFB after blueberry pollination, and researchers are working to understand the association of the disease with spring pollination. *Source:* Photo courtesy of Charlotte Hubbard.

American Foulbrood

Etiologic Agent

The causative agent of AFB is the gram positive, facultative anaerobic, endospore-forming bacterium *P. larvae*. Members of this genus were originally included in *Bacillus*, until reclassification into *Paenibacillus* in 1993 (Ash et al. 1993). Shortly after, *P. larvae* was split into two subspecies: *P. larvae* subsp. *larvae* and *P. larvae* subsp. *pulvifaciens*, with the latter labeled as the causative agent of "Powder scale disease" (Heyndrickx et al. 1996). In 2006, it was determined that *P. larvae* subsp. *pulvifaciens*, should be reclassified as the same species as *P. larvae*, but that *P. larvae* should be broken down into four genotypes (ERIC (Enterobacterial Repetitive Intergenic Consensus) I-IV) (Genersch et al. 2006). These four types differ in their phenotypic characteristics including colony morphology, metabolism of carbon sources, and virulence. ERIC I and II are the most epidemiologically important, with ERIC I being the most common genotype worldwide. Genotypes III and IV (previously *P. larvae* subsp. *pulvifaciens*) have low prevalence (Morrissey et al. 2015). ERIC I is the main genotype in the Americas, though ERIC II has also been detected (Alippi et al. 2004; Antúnez et al. 2007; Morrissey et al. 2015; Krongdang et al. 2017). There is diversity within the genotypes as well; these four ERIC strains are further separated into about 20 sequence types (STs) using MLST (Heyndrickx et al. 1996) (http://pubmlst.org/plarvae). Only the spores of the *P. larvae* are capable of causing disease (Tarr 1937b). These spores can remain viable for years, surviving under extreme conditions.

Transmission

AFB incidence varies both temporally and spatially (Otten and Otto 2005; Wilkins et al. 2007). Colonies diagnosed with AFB are usually destroyed, so management generally leads to localized extinction of disease. Disease clusters are present, however, providing evidence of local spread facilitated by honey bees and beekeepers (Mill et al. 2014). Spread of *P. larvae* spores often occurs during robbing (De Graaf et al. 2001; Lindström et al. 2008a). The honey may be stored in the colony and can act as a disease reservoir and spores can build up in adult bees over winter (Lindström et al. 2008b). Transmission can also occur vertically at the colony level through swarming, as spores have been found in daughter colonies emitted from diseased mother hives (Fries et al. 2006). Recently, it has been shown that the hive pest *Aethina tumida* (Small hive beetle) can act as a vector of *P. larvae* (Schäfer et al. 2010). The most significant transmission route of AFB, however, is likely movement of equipment by beekeepers, as *P. larvae* spores have been detected from colonies with no visible signs or history of disease (Hornitzky and Karlovskis 1989; Riessberger-Gallé et al. 2001), and beekeepers may not know they are moving infectious material during general management. The spores can contaminate tools used for colony management and honey processing equipment. Since the spores can remain viable on equipment for decades, there is a high risk of transmission through the use of old or previously used equipment (Hasemann 1961).

Because of the severity of this disease, many states and provinces have implemented programs to control its spread. Starting in the 1920s, many states enacted laws that required reporting, inspection, registration, and destruction of infected hives. Many countries do not allow the transportation of equipment across borders, and many states require health certificates certifying that a seller's operation is free from AFB. It is important to note that control programs for AFB have been in place in most countries where bees are kept, but they have failed to eradicate AFB from any region where it is present.

Infection

Paenibacillus larvae infects first instar larvae when they ingest brood food that has been provisioned by a nurse bee carrying spores. The ingested spores pass through the larval foregut and germinate in the midgut about one day post-ingestion (Yue et al. 2008). The vegetative bacteria multiply prolifically in the midgut, synthesizing carbohydrates in the larval brood food. The bacteria produce a variety of highly active extracellular proteases that allow for penetration of the epithelium (Dancer and Chantawannakul 1997; Hrabák and Martínek 2007). *P. larvae* breaches the damaged epithelium and passes from the midgut, invading the hemocoel (Davidson 1970; Bailey and Ball 1991; Yue et al. 2008). The infection spreads through the whole body, and the larva ultimately dies from a septicemic condition.

As the larvae degrade, they are replaced by a bacterial biomass, entering a phase known as the "ropy stage." They become a brownish, semi-fluid, glue-like colloid, releasing a putrid smell. The smell and the ropy characteristics of this stage are primary clinical signs for field diagnosis of the disease. Finally, the larval remains dry down to a hard scale, which adheres tightly to the lower cell wall. The scales contain millions of spores, which can distribute the infection for many years within and between colonies (Bailey and Ball 1991).

Infection outcome is driven by genotypic differences in *P. larvae*; different strains vary greatly in virulence (Genersch et al. 2005). The most virulent genotypes (ERIC II-IV) kill larvae about 7 days, whereas the least virulent genotype (ERIC I) takes around 12 days to kill all infected larvae (Genersch et al. 2005, 2006).

AFB only affects bees in the young larval stages (Hoage and Rothenbuhler 1966; Bamrick 1967), though adult honey bees can physically act as carriers of the spores. AFB has an incredibly low infectious dose; less than 10 spores are sufficient to initiate a fatal infection, with younger larvae requiring fewer spores for infection (Woodrow n.d.; Brødsgaard et al. 1998). The spores do not germinate in older larvae (>48 hours old) or in adult honey bees (Wilson 1971; Hitchcock et al. 2019), which is likely due to the thickening and development of the peritrophic matrix, increasing the retention of the bacteria to the midgut lumen (Davidson 1970).

Prognosis

AFB is generally fatal to the colony if left untreated (Hansen and Brødsgaard 1999).

Diagnosis of Bacterial Diseases

Colony Evaluation for Bacterial Disease

The first indication of foulbrood infection is timing. AFB can occur throughout the active beekeeping season, while EFB has a strong seasonality toward late spring and early summer. Because of the current epidemic of varroa mites, most brood disease observed during late summer through fall is varroa-associated and will not be caused by foulbrood. If brood disease is observed during this time, first sample for varroa mites to rule out this likely cause of late-season brood disease.

Visual Identification

Diagnosis of EFB and AFB in the field is generally based on the visual identification of diseased larvae. Bacterial diseases can often be visually confused with other diseases, including the virally-caused parasitic mite brood syndrome (PMBS), so care must be taken when identifying disease based solely on visual signs. To determine if a colony is infected with bacterial disease, inspect multiple frames that contain both open larvae and capped brood cells. In healthy colonies, larvae should be pearly white and should lay flat in the bottom of the cells, and cappings should have a solid distribution and uniform coloration – ranging from yellowish tan to brown. The first sign of bacterial infection is often a scattered or "shotgun" appearance to the capped brood, indicating that brood died before reaching adulthood. Any signs of inconsistency in the brood – spotty patterns of capped cells, misaligned or discolored larvae, or punctures in cappings – indicate that the colony is in a disrupted or ill state, but these signs are not unique to AFB and EFB. As larvae die, they are removed from the hive by worker bees. In hives that are highly hygienic, the bees may remove the infected larvae quickly, and diseased or dead larvae may not be visible. In cases of highly hygienic colonies, when bees are quick to remove diseased or damaged brood, a spotty brood pattern may be the only visible sign of bacterial disease (Figure 22.2).

Visual Signs of EFB

Very early EFB infection may be indicated by a yellowing of the brood food, but this sign is subtle and rarely observed. As the infection progresses in the hive, disease signs are generally observed in third to fifth instar larvae. An EFB-infected larva loses its well-rounded, opaque appearance, and it can take a variety of forms, highly variable in both color and shape. The younger infected larvae may become lightly translucent so that the trachea may become prominent, giving the larva a skeletal-like appearance. Older infected larvae quickly lose their segmentation, and dead larvae appear as a moist, somewhat collapsed mass, giving the appearance of being melted. Diseased larvae are often discolored, changing from healthy pearly white to yellow to brown, and a grayish black as they decompose (Bailey 1961; Bailey and Ball 1991). Many other diseases cause the larvae to turn brown as they decay, but the yellowing color and visible trachea are unique to EFB.

Larval displacement is also an indication of EFB disease – healthy larvae retain a coiled position, lying flat at the bottom of the brood cell. Younger larvae that die from EFB tend to cover the bottom of the cell, but older infected larvae may move uneasily, forming a classic "corkscrew position," stretching out flat reaching out of the cell, or a "blob position" turning so that both ends of the larvae are to the rear of the cell (Figure 22.3).

As the remains of dead larvae dry, the trachea sometimes become conspicuous, retaining their shape while the rest of the body content dries around them, forming a small, brownish transparent scale at the base of the cell, with strong white lines. For larvae that did not retain their normal position, the mass often dries along the lower cell wall forming a dark brown scale. EFB scales are malleable and are not adhesive; they are easily removed and may be seen on the bottom boards and entrances as the bees attempt to remove them from the hive. Infected brood generally die and decompose in the older larval stage, but some will survive until cell capping. If the larvae die after the cell is capped, the capping

Figure 22.2 A brood frame from a colony infected with EFB exhibiting a typical "shotgun" or spotty brood pattern. The scattered appearance of capped brood cells signifies that many young bees died before emergence and indicates that there is an issue with the colony. Spotty brood is always visible in colonies with severe bacterial disease, but it is also often seen in colonies with other issues as well. *Source:* Photo courtesy of Meghan Milbrath.

Figure 22.3 A frame from a colony with severe EFB disease. Note the multiple discolored and twisted larvae. *Source:* Photo courtesy of Emily Noordyke.

of the cell will become sunken and may develop a hole, strongly resembling AFB. It is common to have many different types of diseased larvae present in the same colony. A single frame may contain healthy looking larvae, punctured cappings, melted looking larvae, and larvae with visible trachea.

Visual Signs of AFB

The visual signs of AFB infection are fairly consistent and are less diverse than those of EFB. As larvae die from AFB, they darken and flatten against the lower cell wall, looking like they have melted into a slimy biofilm. Newly dead larvae appear wet and are light tan or caramel-colored. They will continue to desiccate, eventually drying into a dark scale at the lower cell wall (the side that is parallel to the ground when the frame is in the normal position). The scale of AFB will be hard and stuck to the cell wall. Many infected brood live a bit after they are capped, dying after they have begun the pupation process. Brood that die at this stage often leave what is called a "pupal tongue" – the soft body melts, but the embryonic proboscis structure, which is firmer, can be retained as a visible point. This sign is less commonly seen but is unique to AFB. As the brood continues to die, many of the cappings will darken, sink, and develop a hole. The holes are generally jagged and off-center. The cappings may appear dry or may take on a "greasy" or wet appearance (Figure 22.4).

Odor

Both bacterial diseases are named and known for their foul smells, but an odor may or may not be observed in an infected hive. Not all individuals are able to detect the odor of the diseased larvae, so lack of an observable odor does not indicate lack of disease. EFB-diseased brood may have a foul or sour odor that may smell like vinegar, yeast, or cheese. The odor varies with the presence of saprophytes and other associated bacteria. The odor associated with AFB has been described as dead fish, chicken house, or rotting meat. It is important to note that brood that die from any cause (e.g. chilling, pesticides, viruses) will have an unpleasant smell. Just as lack of an odor does not exclude foulbrood disease, presence of a foul odor does not immediately signify foulbrood disease (Table 22.1).

Figure 22.4 Perforated cappings in a colony infected with AFB disease. As the disease progresses, the cappings will often take on a greasy or wet appearance, as caramel color biofilms leak out of the cells. *Source:* Photo courtesy of Sarah Scott.

Differentiation from Other Diseases

It is not always easy to differentiate EFB from AFB, as both diseases can result in melted larvae and sunken cappings. Visually, EFB infections will likely be more variable than AFB infections – larvae with EFB can take on a variety of colors and positions in the cells, whereas larvae with AFB will generally liquefy and will turn brown. Yellow-colored larvae and the corkscrew position are unique to EFB, whereas pupal tongue and caramel color are unique to AFB. Not all the signs of disease will be present in all infections. It is possible to have AFB with no pupal tongue, and early AFB infections may not yet have developed scales. EFB may have only the classic signs of yellow larvae, visible trachea, and corkscrew larvae, or they may show only

Table 22.1 Observable colony signs of bacterial disease.

	EFB	AFB
Diseased larvae	• Highly variable, colors range from yellow[a] to gray to chocolate brown, • Often misaligned into corkscrew or other displaced positions • Visible trachea[a] often present • Many larvae appear melted	• Quickly melt to a liquid/slimy appearing biofilm • Light to dark brown, or caramel color[a] • Pupal tongue may be present[a]
Capped brood	Can have sunken capped brood with holes	Often has sunken cappings and holes in cappings, cappings may appear moist
Scales	Dark brown, rubbery, detaches from cell, not always present	Dark brown to black, thin and adhered to cell, present in later infection
Odor	None to yeasty, vinegar, cheese-like.	Rotting meat or dead fish

[a] Denotes signs that are unique to that particular disease.

(a) (b)

Figure 22.5 a (left) and Figure 22-5b (right) Comparison of matchstick test from AFB (left) and EFB (right) infected colonies. Note that the EFB infected larvae may be sticky and clump together but will not be ropy. *Source:* Photo courtesy of Sarah B. Scott.

melted larvae, sunken cappings, and scales. This latter description is commonly described by researchers as "atypical EFB" or "idiopathic brood disease" and by beekeepers as "snotbrood" or "the crud." This disease presentation is not well characterized yet, and researchers are working to identify if alternative presentations of disease are due to different associative bacteria, different virulence, or other factors.

Other diseases may be confused with bacterial diseases on visual inspection as well. Early infection with the fungal disease, chalkbrood, may closely resemble EFB. As the disease progresses, the obvious identifying signs of chalkbrood will become more apparent, though it is also common to find co-infection with EFB and chalkbrood. The other disease that is often mistaken for foulbrood disease is PMBS, parasitic might brood syndrome. PMBS is virally caused and can result in a variety of signs, including melted-looking larvae. The visual similarity of PMBS and foulbrood could be caused by secondary bacterial infections during PMBS. Generally, in PMBS, brood will die at many stages – from young larvae up until the moment of emergence. If a diseased frame contains bees that are dying at emergence from the cell (almost adults), the disease is likely PMBS, or pesticide damage and not foulbrood.

Field Tests

Matchstick Test

AFB can be identified in the field through the use of a "matchstick test." Macerate the suspected brood with a matchstick or twig (or any small object with a rough surface). Carefully withdraw the stick from the cell, trying to draw out the diseased brood. EFB infection will be goopy, but AFB infection can draw out like a ropy thread, stretching to >2 cm (Figure 22.5).

Holst Milk Test

Field diagnosis of AFB can be performed by testing the ability of a larval sample to break down milk proteins, because sporulating *P. larvae* produce high levels of proteolytic enzymes. To perform a Holst Milk Test (Holst 1946), add a larval sample to a small tube of milk that is diluted to be just barely cloudy (1% powdered milk in water or highly diluted skim milk). Incubate in a warm area for 10–20 minutes. A positive result occurs when the cloudiness is removed from the milk, indicating that milk proteins have been broken down by AFB-associated peptides.

Commercial Field Tests

Commercially available lateral flow devices are available for the detection of antigens for both EFB and AFB (Tomkies et al. 2009). Kits are available for purchase at most bee supply companies and provide a method for field screening of bacterial diseases.

All of these field tests (matchstick [AFB], Holst [AFB], and Commercial ELISA (Enzyme Linked Immunosorbent Assay) [AFB and EFB]) are highly specific. A positive field test indicates a positive test result, but a colony may still be infected with foulbrood after a negative test result. The ability to produce a positive result will depend on the age of the disease. For example, if the disease has long progressed, the larvae may have all dried and will not be ropy. A black light can also be used to visualize AFB on old equipment, as scales will grow a greenish-blue color. Positive identification of EFB or AFB is sufficient to take action for treatment, though laboratory confirmation is often desired to substantiate the diagnosis or to identify the presence of antibiotic resistance.

Laboratory Diagnosis

Laboratory confirmation of bacterial disease can be done through a microscopic preparation, by cultivation on selective culture media, or using immunological or molecular methods (Shimanuki and Knox 2000; Bailey 1968; Pinnock and Featherstone 1984; Hornitzky and Karlovskis 1989; Dancer and Barnes 1995; Hornitzky and Smith 1998; Mckee et al. 2003; *Honey bee diseases and pests: a practical guide* 2006; Roetschi et al. 2008; Forsgren et al. 2013b). Positive diagnosis of AFB is generally done through identification of spores, and EFB is done by identification of either the causative organism or its associated bacteria. Correct sampling is key if laboratory diagnosis is to be used. With AFB, staining for spores can be done on samples that are >10 days old. Before this point, only vegetative bacteria are present. Conversely, microscopic observation of *M. plutonius* is generally possible only early in the infection cycle, before the appearance of the secondary bacteria. It is important to sample multiple diseased larvae in a variety of disease stages, as all larvae in a diseased hive will not contain *M. plutonius* (Forsgren et al. 2005), and it is quite possible to get a false negative.

Submission of Samples to the USDA

In the United States, samples from hives suspected to have EFB or AFB may be sent to the USDA Bee Research Laboratory in Beltsville, MD. This lab can confirm a field diagnosis, and can screen for antibiotic resistance. The following is taken directly from the USDA lab website, but check with the lab for updated instructions:

General Instructions

- Beekeepers, bee businesses, and regulatory officials may submit samples.
- Samples are accepted from the United States and its territories; samples are NOT accepted from other countries.
- Include a short description of the problem along with your name, address, phone number, or e-mail address.
- There is no charge for this service.
- For additional information, contact Sam Abban by phone at (301) 504-8821 or e-mail: samuel.abban@ars.usda.gov

How to Send Adult Honey Bees

- Send at least 100 bees and if possible, select bees that are dying or that died recently. Decayed bees are not satisfactory for examination.
- Bees should be placed in and soaked with 70% ethyl, methyl, or isopropyl alcohol as soon as possible after collection and packed in leak-proof containers.
- USPS, UPS, and FedEx do not accept shipments containing alcohol. Just prior to mailing samples, pour off all excess alcohol to meet shipping requirements.
- Do NOT send bees dry (without alcohol).

How to Send Brood Samples

- A comb sample should be at least 2 × 2 in. and contain as much of the dead or discolored brood as possible. NO HONEY SHOULD BE PRESENT IN THE SAMPLE.
- The comb can be sent in a paper bag or loosely wrapped in a paper towel, newspaper, etc. and sent in a heavy cardboard box. AVOID wrappings such as plastic, aluminum foil, waxed paper, tin, glass, etc. because they promote decomposition and the growth of mold.
- If a comb cannot be sent, the probe used to examine a diseased larva in the cell may contain enough material for tests. The probe can be wrapped in paper and sent to the laboratory in an envelope.

Send samples to:

Bee Disease Diagnosis
Bee Research Laboratory
10300 Baltimore Ave. BARC-East
Bldg. 306 Room 316
Beltsville Agricultural Research Center - East
Beltsville, MD 20705

Reporting

In response to the risk of AFB, the Honeybee Act of 1922 was enacted by the United States Congress to restrict the importation of living adult honey bees into the country. In the United States and Canada, honey bee disease reporting requirements are at the state or the provincial level. It is important to contact the state or provincial apiarist and to know the reporting and control requirements in your area. A list of state and provincial apiary inspectors is maintained by the Apiary Inspectors of America (https://apiaryinspectors.org\). While many states have no regulation regarding foulbrood diseases, others are quite strict and require immediate destruction of the hive under the guidance of the state inspector. EFB is not a notifiable disease in the United States, though it may be in other countries. While not required, in the United States it is useful to send a sample to the USDA laboratory in Beltsville, MD, which does keep track of positive samples over time as well as samples that are positive for antibiotic resistance.

Treatment and Control

Disease management in honey bees is different from disease management in other animals. In other animals, the point of treatment is generally to cure an individual.

In honey bees, it is usually impossible to cure the individual – larvae that are infected to the point of displaying visual signs will always die. Rather, the point of treatment is disease control – to prevent transmission from the diseased individual to other individuals in the same colony and to prevent transmission from one colony to other colonies. Honey bees are also different from many other animals in terms of the animal group. In most other animals, the herd or animal group is clearly defined, and groups often do not come into contact with each other. Honey bees, however fly for miles, and are often in contact with many other colonies. During agricultural pollination, honey bees may be in a "herd" of tens of thousands of other colonies, managed by many different beekeepers. The final unique consideration for the treatment of honey bees is the dependence on them by our food industry. The security of the vast majority of the fruits and vegetables in Canada and the United States depends on strong colonies at certain times of the year. In order to provide pollination services for food production, honey bees are often put in stressful situations that are known to be high risk for developing bacterial diseases or creating epidemics. Diseased and dying colonies not only affect beekeepers, but they also affect food prices and food availability. It is crucial to take bacterial disease control seriously to protect the health of honey bees as well as national food security.

In dealing with both AFB and EFB, it is important to keep in mind that both diseases can be spread through robbing of honey from weak hives, drifting bees, and movement of contaminated equipment. Disease management must consider the larger context, with attention to the health of the hive, the apiary, and other operations in the area. When controlling bacterial diseases in honey bees, it is essential to take action in three areas:

1) Address colonies that are identified as sick with bacterial disease,
2) Protect colonies in contact with diseased hives, and
3) Prevent future infection from contaminated equipment.

When dealing with sick or suspect colonies, proper measures must be taken to prevent the spread of disease as a result of handling.

- Wear nitrile gloves when dealing with sick or suspect hives, and make sure that they are disposed of properly. If leather gloves were used when handling a diseased hive, they should be disposed of with the infected hive equipment.
- Take extreme care to prevent robbing from sick or suspect hives. Bring extra disposable covers to minimize the time that honey is exposed to other bees. Keep diseased equipment contained at all times until it has been sterilized or disposed of so bees cannot rob honey from it during any stage of the process.
- Sterilize or dispose of other tools and equipment that can become contaminated while working hives, including hive tools and smokers. Focus on removing all of the wax, honey, and propolis, as bacteria and spores can remain in all of these products.

Colonies Identified as Sick with American Foulbrood

Colonies that have AFB must be dealt with immediately because of its highly infectious nature, taking every precaution to avoid the spread of the highly stable spores. Colonies with even a single cell of AFB disease should be dealt with on the day that the disease was discovered. In many states and provinces, regulations require that a colony positive for AFB should be destroyed through burning. Even if not required by law, burning AFB-infected colonies is considered best practice because it removes all of the spores in the bees and in the hive. In areas where burning is not required by law, and there is sufficient time in the season, beekeepers can deal with the colony using the "shook swarm method," outlined below. If the beekeeper lives in an area where burning is not permitted, the colony should be euthanized, and the hive should be double-bagged and disposed of in a landfill or incinerated.

Colonies Identified as Sick with European Foulbrood

The treatment for colonies with EFB is much more context dependent. Appropriate strategies include burning, shook swarms, watchful waiting, and antibiotic treatment. In areas where EFB is rare and there is very little overlap with other beekeeping operations, it is often desirable to destroy hives with EFB, treating it like AFB. Destruction is ideal in these cases, because it may be possible in these areas to completely eradicate EFB from the operation, reducing risk to future colonies. In areas where risk is not likely to be significantly reduced through the removal of a few colonies – either the infection is widespread in the region, or there is frequent migration and high density of commercial hives – other measures can be taken to manage disease spread and mitigate losses.

Colonies in the Same Yard as an Infected Colony

It is important to remember that there is a high prevalence of inapparent infection with both AFB and EFB, and that a high proportion of forager bees will drift from the infected hive into neighboring colonies within the same yard. Therefore, it should be assumed that all colonies in the yard of an infected colony will contain some infectious material, even though the colonies may not test positive or show visual signs of disease. It is usually appropriate to treat bacterial diseases at the herd level; the entire yard should be provided with antibiotics to prevent re-infection, regardless of the action taken with the diseased colony. For example, if a colony with AFB is burned, the remaining colonies in that yard should be provided with a full antibiotic treatment. If antibiotic treatment is not possible due to timing (the colonies are in honey production), then these colonies should be considered at risk and closely monitored. When possible, the yard should be quarantined, with no hive equipment leaving the yard until a year has passed with no signs of disease, and the colonies should be frequently and carefully examined for signs of disease, so that any re-emergence can be dealt with quickly.

Colonies in High Risk Situations

For colonies that are high risk for developing disease, protective action may be taken before signs of disease are noted in the larva. High-risk situations include scenarios where re-emergence is expected, colony stress is high, or there is increased contact with high-risk colonies. In cases of agricultural pollination, colonies with a history of EFB infection may have to be moved at a stressful time, putting the bees at risk for disease re-emergence. For example, beekeepers may move hives in the early spring for blueberry pollination, where they have experienced high rates of EFB for the past few years. Another high risk scenario is when a beekeeper will be moving their hives in close proximity to other beekeepers that may be using antibiotics to suppress AFB infection. Since antibiotics have been used regularly for years in some operations, bacterial replication may be suppressed, even though a large amount of infectious material is present. Bees brought into close range of these colonies may be at risk. Colonies may be especially at risk if the bees are in a state that requires automatic burning, as the economic loss to the beekeeper could be devastating. In high risk scenarios, judicious antibiotic use and frequent monitoring are necessary to prevent widespread outbreaks.

Burning a Honey Bee Hive

Burning a honey bee colony should take place in the evening of the day the disease is identified to capture all of the field bees from that hive that could potentially be carrying spores. The bees may be euthanized before burning, but it should be burned with the hive contents. First, dig a hole that is big enough to hold the hive. The hole is necessary to capture the infected honey that will run out of the hive as the beeswax comb melts. The hive should be placed in the hole and completely and carefully burned. After all of the equipment is destroyed, bury the ashes in the hole, making sure that bees are not able to access any remaining honey or equipment. If the colony contains plastic foundation, or if the region is too dry for open fires, it is best to completely bag the equipment and take it to a disposal site where it can be incinerated. Take care during transport and destruction that the equipment is never left open and exposed in a manner that would allow for robbing and that truck beds and tools are sufficiently cleaned of honey after transport. Equipment such as lids, bottom boards, and boxes may be saved in some cases, if they are sterilized as described below, but frames from diseased colonies should always be destroyed.

When Burning/Hive Destruction Should be Used

- In states or provinces where it is required by law.
- In colonies that have been diagnosed with AFB, when the beekeeper wants to take the strongest precautions.
- In colonies that are too small to prevent robbing or survive.
- In colonies that have been diagnosed with EFB, when it is economically feasible for the beekeeper and burning would likely significantly reduce disease pressure in the area.

Shook Swarm Method

If the beekeeper does not want to kill the bees in an infected colony, they can shake the bees onto new equipment using a process called a "shook swarm" (Phillips 1911; Waite et al. 2003). The focus of the shook swarm method is to remove infectious material from the hive, replacing all the comb with foundation. Removing the comb withholds space for the bees to deposit any infectious food in their crops, compelling them to process any food in their crops, draw fresh comb, and gather new stores. Because the bees need to draw out the comb for a new hive, the shook swarm method can only be used in cases where there is sufficient time in the season and the colony is of sufficient strength to draw new comb. Unlike burning, the shaking treatment should occur during the middle of the day, as bees will return to the original hive. A new hive with clean equipment should be set up in the original location. The bees should be shaken into the new equipment, taking great care to cover the frames from the diseased hive once the bees have been removed. The shook swarm method can be stressful to the colony and may be deadly if performed too late in the season for the colony to draw enough wax and fill the hive with honey. It is important to provide supplemental feed with the shook swarm method. Colonies that were shook without supplemental feed had reduced honey yield and high colony mortality (Guler 2008). Generally, antibiotics are provided to the colony to prevent the infection of newly laid larvae from any spores remaining on the bees. The shook swarm method can be used without antibiotics, but the shook swarm method with oxytetracycline has been reported as an effective treatment for the control of EFB, with low (5%) levels of colony disease recurrence. This method has similar effectiveness when used with other bacterial diseases, including AFB (Shimanuki and Knox 2000). The frames removed from the diseased hive should be burned or destroyed as explained above.

When Shook Swarm Can Be Used

- Shook swarms can be used for colonies with AFB in states where burning is not required by law, with colonies with EFB, or in colonies with unidentified brood disease.
- Colonies must be strong enough to sufficiently rebuild and collect stores.
- There must be sufficient time left in the season for the colony to draw wax and gather sufficient stores.
- The beekeeper must be able to perform frequent checks to identify re-emergence of disease, apply antibiotics, and feed the colony.

Watchful Waiting

Watchful waiting is not an option for colonies with diagnosed AFB, but it can be used in colonies with EFB. With EFB, sudden outbreaks are often followed by spontaneous recovery (Bailey 1961), and recovery is also often observed after severely infected colonies are moved from endemically infected areas to areas free of disease (Bailey and Locher 1968). The beekeeper should take clear notes on the severity of the disease including size of the cluster of honey bees (recorded in terms of frames covered with bees), number of brood frames that are affected, and number of cells infected (e.g. >100 / frame). The beekeeper should check the colony frequently to ensure that the disease is not progressing and the colony is strong enough to prevent robbing. If the disease worsens, antibiotic treatment should be used. If the colony becomes too weak to prevent robbing, then the colony should be euthanized and the frames destroyed. Colonies that have had EFB should be marked and closely monitored the following year, as threatened apiaries tend to have high recurrence rates of the disease (Thompson and Brown 2001).

When Watchful Waiting Can Be Used

- Watchful waiting cannot be used for colonies with AFB, since spontaneous recovery cannot be expected.
- It can be used for colonies with EFB as long as the colony is strong enough that survival is expected and that it can guard against robbing from neighboring colonies.
- Watchful waiting can only be used by beekeepers with sufficient time to closely monitor the hive.

Antibiotics

In the United States, three antibiotics are currently labeled for use in honey bees: oxytetracycline/Terramycin (OTC), tylosin/tylan, and lincomycin hydrochloride/Lincomix. Many other antibiotics are effective against foulbrood diseases, but are not labeled for honey bee use due to their importance in human health applications (Lodesani and Costa n.d.; Kochansky et al. 2001). All three antibiotics are labeled for the control of AFB, while only OTC is labeled for use with EFB. Extra label drug use (ELDU) of tylosin and lincomycin is allowed with a prescription under ELDU regulations, but effectiveness of these drugs for EFB is not well-documented, and OTC remains effective so their use for EFB is not warranted at this time. No resistance to OTC has been identified in *M. plutonius* (Hornitzky and Smith 1999; Waite et al. 2003), but some resistance has been seen to OTC in *P. larvae* (Miyagi and al 2000; Evans 2003). However, these resistant populations remain sensitive to tylosin and lincomycin (Alippi et al. 1999; Feldlaufer et al. 2001), and resistance is not universal. Therefore, OTC is still often an appropriate choice for a first line of treatment for most bacterial disease, but samples should be tested for resistance and the beekeeper should be prepared to switch treatments if necessary.

Antibiotics must be used in a manner that reduces the further development of resistance and that eliminates the potential for human exposure through contamination of honey. The timing and method of antibiotic delivery are important to reduce these risks. Antibiotics should only be applied to the brood chambers of a colony and never when honey supers are present on the hive. Different antibiotic product labels will list specific requirements, but in general, treatment should be concluded at least four weeks before the start of a honey flow. Tylosin is at higher risk for honey contamination, and it is generally only used in the fall, when there is no expectation of a honey flow. Some historical application methods such as grease ("extender") patties increase the length of time that the treatment is in the hive, increasing the risk to honey contamination and development of resistance. For that reason, application in powdered sugar is preferred, at the rate specified on the label.

All of the antibiotics labeled for honey bees work by inhibiting the multiplication of the bacteria. It is important to note that they do not kill spores or non-replicating bacteria. Therefore, in cases where the disease is present, antibiotics should be used in conjunction with the removal of infectious material. Colonies treated with the shook swarm method were less likely than colonies treated with OTC alone to have detectable levels of *M. plutonius* and lower disease recurrence in the spring following treatment (Waite et al. 2003; Budge et al. 2010).

There have been many recent studies investigating alternative strategies for the control of foulbrood diseases including propolis, essential oils, indoles, probiotics, prebiotics, fatty acids, bacteria, and phages (Lodesani and Costa n.d.; Genersch 2010; Alonso-Salces et al. 2017; Alvarado et al. 2017; Lamei et al. 2019). At present, none of these treatments have demonstrated to be as effective as antibiotics in managing bacterial disease, though many have shown promise in laboratory trials.

Sterilization of Equipment

Managing equipment is essential to the management of foulbrood disease. The bacteria for both types of foulbrood proliferate dramatically in the larval guts, and diseased larva or larval scales can contain billions of bacteria or spores. The scales and diseased larvae are cleaned by young nurse bees who will distribute infectious material throughout the hive. This perpetual reinfection by bees drives the severity of the disease at the colony level (Bailey and Ball 1991; Genersch 2010). Between-colony transmission is driven by movement of infected equipment as well as robbing and drifting (Fries et al. 2006; Lindström et al. 2008b). The most conservative method to manage disease risk from equipment is to burn the entire hive and hive contents. While it is essential in all cases to destroy the frames from colonies with AFB and severe EFB, the boxes, lids, and bottom boards provide much lower risk of infection and can be reused after sterilization. Both EFB and AFB can be managed by gamma irradiation, and many beekeepers may live near a facility that offers this service for beekeeping equipment. If gamma irradiation cannot be used, then woodenware can be dipped in hot wax for a minimum of 10 minutes at a minimum temperature of 150 °C. Both gamma irradiation and hot wax are considered sufficient for reducing the risk of reinfection from AFB spores. Other methods such as ethylene oxide, scorching, or boiling in lye will not completely remove the risk of AFB spores, but these methods can reduce the risk of infectious material to a point where reinfection is unlikely.

Prevention and Control

- Never feed honey or pollen from other operations to honey bees. Both honey and pollen are commonly infected with *M. plutonius* and *P. larvae*.
- When possible, limit colony stress through the provision of supplemental feed and timing of movement and splits.
- Practice proper sanitation, ensuring that hive tools, smokers, and gloves are frequently sterilized and that equipment movement is minimized.
- Promote frequent monitoring for early recognition of disease signs.
- When possible, practice equipment quarantine at the apiary level.

Other Bacterial Pathogens

Other bacteria have been identified in honey bees, including *Pseudomonas aeruginosa* (Shimanuki and Knox 2000; Papadopoulou-Karabela 1992) and *Serratia marcescens* (Burritt et al. 2016; Rayman et al. 2018). These and other bacteria may cause disease opportunistically after microbiome disruption, immune disturbance, or as secondary infection after foulbrood of viral infections. It has often been suspected that secondary bacterial infections are present during severe cases of parasitic mite syndrome, and some beekeepers report improvement after treatment with antibiotics in conjunction with mite control. Much more research is needed before treatment recommendations can be made for non-foulbrood bacterial disease.

Conclusion

Bacterial diseases continue to remain a large problem in beekeeping because of high background levels and high infectivity. In areas where disease risk is low, such as small operations in remote locations, a combination of early detection and burning may be sufficient to minimize risk of future infection. In migratory operations, disease risk may be near constant, and great care must be taken to minimize the negative outcomes of bacterial disease.

References

Alippi, A.M. et al. (1999). Comparative study of tylosin, erythromycin and oxytetracycline to control American foulbrood of honey bees. *Journal of Apicultural Research* 38 (3–4): 149–158.

Alippi, A.M. et al. (2004). Molecular epidemiology of *Paenibacillus larvae larvae* and incidence of American foulbrood in Argentinean honeys from Buenos Aires province. *Journal of Apicultural Research* 43 (3):

135–143. https://doi.org/10.1080/00218839.2004.11101124.

Allen, M., Ball, B., and Underwood, B. (1990). An isolate of *Melissococcus pluton* from Apis laboriosa. *Journal of Invertebrate Pathology* 55: 439–440.

Alonso-Salces, R.M. et al. (2017). Natural strategies for the control of *Paenibacillus larvae*, the causative agent of American foulbrood in honey bees: a review. *Apidologie* 48 (3): 387–400. https://doi.org/10.1007/s13592-016-0483-1.

Alvarado, I. et al. (2017). Inhibitory effect of indole analogs against *Paenibacillus larvae*, the causal agent of American foulbrood disease. *Journal of Insect Science (Online)* 17 (5) https://doi.org/10.1093/jisesa/iex080.

Antúnez, K. et al. (2007). Phenotypic and genotypic characterization of *Paenibacillus larvae* isolates. *Veterinary Microbiology* 124 (1–2): 178–183. https://doi.org/10.1016/J.VETMIC.2007.04.012.

Arai, R. et al. (2014). Development of duplex PCR assay for detection and differentiation of typical and atypical *Melissococcus plutonius* strains. *Journal of Veterinary Medical Science* 76 (4): 491–498. https://doi.org/10.1292/jvms.13-0386.

Ash, C., Priest, F.G., and Collins, M.D. (1993). Molecular identification of rRNA group 3 bacilli (Ash, Farrow, Wallbanks and Collins) using a PCR probe test. *Antonie van Leeuwenhoek* 64 (3): 253–260. https://doi.org/10.1007/BF00873085.

Bailey, L. (1957). The isolation and cultural characteristics of *Streptococcus pluton* and further observations on bacterium eurydice. *Journal of General Microbiology* 17: 39–48.

Bailey, L. (1959a). An improved method for the isolation of *Streptococcus pluton*, and observations on its distribution and ecology. *Journal of Insect Pathology* 1: 80–85.

Bailey, L. (1959b). Recent research on the natural history of European foulbrood. *Bee World* 40: 66–70.

Bailey, L. (1960). The epizootiology of European foulbrood of the larval honey bee, *Apis mellifera linneaus*. *Journal of Insect Pathology* 2: 67–83.

Bailey, L. (1961). European foulbrood. *American Bee Journal* 101: 89–92.

Bailey, L. (1968). Honey bee pathology1. *Annual Review of Entomology* 13: 191–212.

Bailey, L. (1974). An unusual type of *Streptococcus pluton* from the eastern hive bee. *Journal of Invertebrate Pathology* 23: 246–247.

Bailey, L. (1983). *Melissococcus pluton*, the cause of European foulbrood of honey bees (*Apis spp*). *Applied Bacteriology* 55: 65–69.

Bailey, L. and Ball, B. (1991). *Honey Bee Pathology*. London: Academic Press.

Bailey, L. and Collins, M. (1982). Reclassification of *Streptococcus pluton* (White) in a new genus *Melissococcus pluton*. *Journal of Applied Bacteriology* 53: 209–213.

Bailey, L. and Locher, N. (1968). Experiments on the etiology of European foulbrood of the honeybee. *Journal of Apicultural Research* 7: 103–107.

Bamrick, J.F. (1967). Resistance to American foulbrood in honey bees: VI. Spore germination in larvae of different ages. *Journal of Invertebrate Pathology* 9 (1): 30–34. https://doi.org/10.1016/0022-2011(67)90039-0.

Belloy, L. et al. (2007). Spatial distribution of *Melissococcus plutonius* in adult honey bees collected from apiaries and colonies with and without symptoms of European foulbrood*. *Apidologie* 38: 136–140. https://doi.org/10.1051/apido:2006069.

Brødsgaard, C.J., Ritter, W., and Hansen, H. (1998). Response of in vitro reared honey bee larvae to various doses of *Paenibacillus larvae larvae* spores. 29: 569–578.

Budge, G.E. et al. (2010). The occurrence of *Melissococcus plutonius* in healthy colonies of *Apis mellifera* and the efficacy of European foulbrood control measures. *Journal of Invertebrate Pathology* 105: 164–170. https://doi.org/10.1016/j.jip.2010.06.004.

Budge, G.E. et al. (2014). Molecular epidemiology and population structure of the honey bee brood pathogen *Melissococcus plutonius*. *The ISME Journal* 8 (8): 1588–1597. https://doi.org/10.1038/ismej.2014.20.

Burritt, N.L. et al. (2016). Sepsis and hemocyte loss in honey bees (*Apis mellifera*) infected with *Serratia marcescens* strain sicaria. *PLoS One* 11 (12): 1–26. https://doi.org/10.1371/journal.pone.0167752.

CAPA (Canadian Association of Professional Apiculturists) (2013). *Honey Bee Diseases and Pests*, 3ee (eds. S. Pernal and H. Clay). Beaverlodge, AB.

Dancer, B. and Barnes, M. (1995). *Detection of European Foulbrood by DNA Hybridisation Analysis*. Central Association of Bee-Keepers.

Dancer, B.N. and Chantawannakul, P. (1997). The proteases of American foulbrood scales. *Journal of Invertebrate Pathology* 70 (2): 79–87. https://doi.org/10.1006/jipa.1997.4672.

Davidson, E.W. (1970). Ultrastructure of peritrophic membrane development in larvae of the worker honey bee (*Apis mellifera*). *Journal of Invertebrate Pathology* 15 (3): 451–454. https://doi.org/10.1016/0022-2011(70)90190-4.

De Graaf, D.C. et al. (2001). Influence of the proximity of American foulbrood cases and apicultural management on the prevalence of *Paenibacillus larvae* spores in Belgian honey. *Apidologie* 32 (6): 587–599. https://doi.org/10.1051/apido:2001146.

Dingman, D. W. (2012) American foulbrood in honey bees: prevalence and geographic distribution of *Paenibacillus larvae* in connecticut. Available at: https://reeis.usda.gov/web/crisprojectpages/0218205-american-foulbrood-in-honey-bees-prevalence-and-geographic-distribution-of-paenibacillus-larvae-in-connecticut.html (Accessed: 9 December 2019).

Djukic, M. et al. (2018). Comparative genomics and description of putative virulence factors of *Melissococcus plutonius*, the causative agent of European foulbrood disease in honey bees. *Genes* 9 (8): 1–20. https://doi.org/10.3390/genes9080419.

Eischen, F.A., Graham, R.H., and Cox, R. (2005). Regional distribution of *Paenibacillus larvae* subspecies larvae, the causative organism of American foulbrood, in honey bee colonies of the Western United States. *Journal of Economic Entomology* 98 (4): 1087–1093. https://doi.org/10.1603/0022-0493-98.4.1087.

Erban, T. et al. (2017). European foulbrood in Czechia after 40 years: application of next-generation sequencing to analyze *Melissococcus plutonius* transmission and influence on the bacteriome of *Apis Mellifera*. *PeerJPreprints* 5: e3816.

Erler, S. et al. (2014). Diversity of honey stores and their impact on pathogenic bacteria of the honeybee, *Apis mellifera*. *Ecology and Evolution* 4 (20): 3960–3967. https://doi.org/10.1002/ece3.1252.

Evans, J.D. (2003). Diverse origins of tetracycline resistance in the honey bee bacterial pathogen *Paenibacillus larvae*. *Journal of Invertebrate Pathology* 83 (1): 46–50. https://doi.org/10.1016/S0022-2011(03)00039-9.

Feldlaufer, M. et al. (2001). Lincomycin hydrochloride for the control of American foulbrood disease of honey bees. *Apidologie* 32 (6): 547–554. https://doi.org/10.1051/apido:2001100ï.

Forsgren, E. (2009). European foulbrood in honey bees. *Journal of Invertebrate Pathology* 103: S5–S9. https://doi.org/10.1016/j.jip.2009.06.016.

Forsgren, E. (2010). European foulbrood in honey bees. *Journal of Invertebrate Pathology* 103: S5–S9.

Forsgren, E. et al. (2005). Distribution of *Melissococcus plutonius* in honeybee colonies with and without symptoms of European foulbrood. *Microbial Ecology* 50 (3): 369–374. https://doi.org/10.1007/s00248-004-0188-2.

Forsgren, E. et al. (2013a). Standard methods for European foulbrood research. *Journal of Apicultural Research* 52 (1) https://doi.org/10.3896/IBRA.1.52.1.12.

Forsgren, E. et al. (2013b). Standard methods for European foulbrood research. *Journal of Apicultural Research* 52 (1): 1–14. https://doi.org/10.3896/IBRA.1.52.1.12.

Fries, I., Lindström, A., and Korpela, S. (2006). Vertical transmission of American foulbrood (*Paenibacillus larvae*) in honey bees (*Apis mellifera*). *Veterinary Microbiology* 114 (3–4): 269–274. https://doi.org/10.1016/J.VETMIC.2005.11.068.

Gaggìa, F. (2015). Microbial investigation on honey bee larvae showing atypical symptoms of European foulbrood. *Bulletin of Insectology* 68 (2): 321–327.

Gaggìa, F. et al. (2015). Microbial investigation on honey bee larvae showing atypical symptoms of European foulbrood. *Bulletin of Insectology* 68 (2): 321–327.

Genersch, E. (2010). American foulbrood in honeybees and its causative agent, *Paenibacillus larvae*. *Journal of Invertebrate Pathology* 103: S10–S19.

Genersch, E., Ashiralieva, A., and Fries, I. (2005). Strain-and genotype-specific differences in virulence of *Paenibacillus larvae* subsp. larvae, a bacterial pathogen causing American foulbrood disease in honeybees. *Applied and Environmental Microbiology* 71 (11): 7551–7555. https://doi.org/10.1128/AEM.71.11.7551-7555.2005.

Genersch, E. et al. (2006). Reclassification of *Paenibacillus larvae* subsp. *pulvifaciens* and *Paenibacillus larvae* subsp. *larvae* as *Paenibacillus larvae* without subspecies differentiation. *International Journal of Systematic and Evolutionary Microbiology* 56 (Pt 3): 501–511. https://doi.org/10.1099/ijs.0.63928-0.

Giersch, T., Barchia, I., and Hornitzky, M. (2010). Can fatty acids and oxytetracycline protect artificially raised larvae from developing European foulbrood? *Apidologie* 41 (2): 151–159. https://doi.org/10.1051/apido/2009066.

Guler, A. (2008). The effects of the shook swarm technique on honey bee (*Apis mellifera L.*) colony productivity and honey quality. *Journal of Apicultural Research* 47 (1): 27–34. https://doi.org/10.1080/00218839.2008.11101420.

Hansen, H. and Brødsgaard, C.J. (1999). American foulbrood: a review of its biology, diagnosis and control. *Bee World* 80 (1): 5–23. https://doi.org/10.1080/0005772X.1999.11099415.

Hasemann, L. (1961). How long can spores of American foulbrood live? *American Bee Journal* 101: 298–299.

Haynes, E. et al. (2013). A typing scheme for the honeybee pathogen *Melissococcus plutonius* allows detection of disease transmission events and a study of the distribution of variants. *Environmental Microbiology Reports* 5 (4): 525–529. https://doi.org/10.1111/1758-2229.12057.

Heyndrickx, M. et al. (1996). Reclassification of *Paenibacillus* (formerly *Bacillus*) *pulvifaciens* (Nakamura 1984) Ash et al. 1994, a later subjective synonym of *Paenibacillus* (formerly *Bacillus*) *larvae* (White 1906) Ash et al. 1994, as a subspecies of *P. larvae*, with emended description. *International Journal of Systematic Bacteriology* 46 (1): 270–279. https://doi.org/10.1099/00207713-46-1-270.

Hitchcock, J.D. et al. (2019). Pathogenicity of bacillus pulvifaciens to honey bee larvae of various ages (Hymenoptera: Apidae). *Journal of the Kansas Entomological Society* 52 (2): 238–246.

Hoage, T.R. and Rothenbuhler, W.C. (1966). Larval honey bee response to various doses of *Bacillus larvae* spores. *Journal of Economic Entomology* 59 (1): 42–45. https://doi.org/10.1093/jee/59.1.42.

Holst, E. (1946). A simple field test for American foulbrood. *American Bee Journal* 86: 14–34.

Honey bee diseases and pests: a practical guide (2006). Available at: http://www.fao.org/3/a-a0849e.pdf (Accessed: 4 November 2016).

Hornitzky, M.A.Z. and Karlovskis, S. (1989). A culture technique for the detection of *Bacillus larvae* in honeybees. *Journal of Apicultural Research* 28 (2): 118–120. https://doi.org/10.1080/00218839.1989.11100831.

Hornitzky, M.A.Z. and Smith, L. (1998). Procedures for the culture of *Melissococcus pluton* from diseased brood and bulk honey samples. *Journal of Apicultural Research* 37 (4): 293–294. https://doi.org/10.1080/00218839.1998.11100987.

Hornitzky, M.A.Z. and Smith, L.A. (1999). Sensitivity of Australian *Melissococcus pluton* isolates to oxytetracycline hydrochloride. *Australian Journal of Experimental Agriculture* 39 (7): 881–883.

Hrabák, J. and Martínek, K. (2007). Screening of secreted proteases of *Paenibacillus larvae* by using substrate-SDS-polyacrylamide gel electrophoresis. *Journal of Apicultural Research* 46 (3): 160–164. https://doi.org/10.1080/00218839.2007.11101388.

Kanbar, G. et al. (2004). Tyramine functions as a toxin in honey bee larvae during *Varroa* -transmitted infection by *Melissococcus pluton*. *FEMS Microbiology Letters* 234 (1): 149–154. https://doi.org/10.1111/j.1574-6968.2004.tb09526.x.

Kochansky, J. et al. (2001). Screening alternative antibiotics against oxytetracycline-susceptible and -resistant *Paenibacillus larvae*. *Apidologie* 32 (3): 215–222. https://doi.org/10.1051/apido:2001123.

Krongdang, S. et al. (2017). Multilocus sequence typing, biochemical and antibiotic resistance characterizations reveal diversity of North American strains of the honey bee pathogen *Paenibacillus larvae*. *PLoS One* 12 (5): e0176831–e0176831. https://doi.org/10.1371/journal.pone.0176831.

Krongdang, S. et al. (2019). Comparative susceptibility and immune responses of Asian and European honey bees to the American foulbrood pathogen, *Paenibacillus larvae*. *Insect Science* 26 (5): 831–842. https://doi.org/10.1111/1744-7917.12593.

Lamei, S. et al. (2019). Feeding honeybee colonies with honeybee-specific lactic acid bacteria (Hbs-LAB) does not affect colony-level Hbs-LAB composition or *Paenibacillus larvae* spore levels, although American foulbrood affected Colonies Harbor a more diverse Hbs-LAB community. *Microbial Ecology* https://doi.org/10.1007/s00248-019-01434-3.

Lewkowski, O. and Erler, S. (2019). Virulence of *Melissococcus plutonius* and secondary invaders associated with European foulbrood disease of the honey bee. *MicrobiologyOpen* 8 (3): 1–9. https://doi.org/10.1002/mbo3.649.

Lindström, A., Korpela, S., and Fries, I. (2008a). Horizontal transmission of *Paenibacillus larvae* spores between honey bee (*Apis mellifera*) colonies through robbing. *Apidologie* 39 (5): 515–522. https://doi.org/10.1051/apido:2008032.

Lindström, A., Korpela, S., and Fries, I. (2008b). The distribution of *Paenibacillus larvae* spores in adult bees and honey and larval mortality, following the addition of American foulbrood diseased brood or spore-contaminated honey in honey bee (*Apis mellifera*) colonies. *Journal of Invertebrate Pathology* 99 (1): 82–86. https://doi.org/10.1016/J.JIP.2008.06.010.

Lodesani, M. and Costa, C. (2005) 'Bee World Limits of chemotherapy in beekeeping: wdevelopment of resistance and the problem of residues'. doi: https://doi.org/10.1080/0005772X.2005.11417324.s

Matheson, A. (1993). World bee health report. *Bee World* 74 (4): 176–212. https://doi.org/10.1080/0005772X.1993.11099183.

Mckee, B.A. et al. (2003). The detection of *Melissococcus pluton* in honey bees (*Apis mellifera*) and their products using a hemi-nested PCR. *Apidologie* 34: 19–27. https://doi.org/10.1051/apido:2002047.

Mckee, B. A. et al. (2004) 'The transmission of European foulbrood (*Melissococcus plutonius*) to artificially reared honey bee larvae (*Apis mellifera*)'. doi: https://doi.org/10.1080/00218839.2004.11101117.

Mill, A.C. et al. (2014). Clustering, persistence and control of a pollinator brood disease: epidemiology of American foulbrood. *Environmental Microbiology* 16 (12): 3753–3763. https://doi.org/10.1111/1462-2920.12292.

Miyagi and al (2000) NOTE Verification of Oxytetracycline-Resistant American Foulbrood Pathogen Paenibacillus larvae in the United States. Available at: www.idealibrary.com (Accessed: 16 December 2019).

Morrissey, B.J. et al. (2015). Biogeography of *Paenibacillus larvae*, the causative agent of American foulbrood, using a new multilocus sequence typing scheme. *Environmental Microbiology* 17 (4): 1414–1424. https://doi.org/10.1111/1462-2920.12625.

Nakamura, K. et al. (2016). Virulence differences among *Melissococcus plutonius* strains with different genetic backgrounds in *Apis mellifera larvae* under an improved experimental condition. *Scientific Reports* 6: 1–12. https://doi.org/10.1038/srep33329.

OIE (World Organisation for Animal Health) (2009). *World Animal Health Information Database*. Paris, France. Available at: www.oie.int.

Otten, C. and Otto, A. (2005). Epidemiology and control of the American foul- brood in Germany. *Journal of Apicultural Research* 40: 16–22.

Papadopoulou-Karabela, K. (1992, 1992). Experimental infection of honeybees by *Pseudomonas aeruginosa* Papadopoulou-Karabela N Iliadis. *Journal of Economic Entomology* 23 (5): 393–397.

Pernal, S.F. and Melathopoulos, A.P. (2006). Monitoring for American foulbrood spores from honey and bee samples in Canada. *Apiacta* 41: 99–109.

Phillips, E. F. (1911) The Treatment of Bee Diseases. Available at: https://naldc.nal.usda.gov/download/5420774/PDF (Accessed: 19 January 2019).

Pinnock, D.E. and Featherstone, N.E. (1984). Detection and quantification of *Melissococcus pluton* infection in honeybee colonies by means of enzyme-linked immunosorbent assay. *Journal of Apicultural Research* 23 (3): 168–170. https://doi.org/10.1080/00218839.1984.11100627.

Raina SK and Fries I (2004) 'American foulbrood and African honey bees (Hymenoptera: Apidae)', doi: https://doi.org/10.1603/0022-0493-96.6.1641.

Rayman, K. et al. (2018). Pathogenicity of *Serratia marcescens* strains in honey bees. *MBio* 9: e01649–e01618. https://doi.org/10.1128/microbiolspec.gpp3-0053-2018.

Riessberger-Gallé, U., von der Ohe, W., and Crailsheim, K. (2001). Adult Honeybee's resistance against *Paenibacillus larvae larvae*, the causative agent of the American foulbrood. *Journal of Invertebrate Pathology* 77 (4): 231–236. https://doi.org/10.1006/JIPA.2001.5032.

Roetschi, A. et al. (2008). Infection rate based on quantitative real-time PCR of *Melissococcus plutonius*, the causal agent of European foulbrood, in honeybee colonies before and after apiary sanitation*. *Apidologie* 39: 362–371. https://doi.org/10.1051/apido:200819.

Schäfer, M.O. et al. (2010). Small hive beetles, Aethina tumida, are vectors of *Paenibacillus larvae*. *Apidologie* 41 (1): 14–20. https://doi.org/10.1051/apido/2009037.

Schirach, G. A. (1769) Histoire des Abeilles, p. 56 (Chapter 3).

Shimanuki, H. and Knox, D. A. (2000) Diagnosis of Honey Bee Diseases United States Department of Agriculture Agricultural Research Service Agriculture Handbook Number 690, 61 pp. Available at: https://www.ars.usda.gov/is/np/honeybeediseases/honeybeediseases.pdf (Accessed: 19 January 2019).

Takamatsu, D. et al. (2014). Typing of *Melissococcus plutonius* isolated from European and Japanese honeybees suggests spread of sequence types across borders and between different apis species. *Veterinary Microbiology* 171 (1–2): 221–226. https://doi.org/10.1016/j.vetmic.2014.03.036.

Tarr, H. (1937a). Studies on Eurpoean foul brood of bees: further experiments on the production of the disease. *Annals of Applied Biology* 24: 614–626.

Tarr, H.L.A. (1937b). Studies on American foul brood of bees: the relative pathogenicity of vegetative cells and endospores of bacillus larvae for the brood of the bee. *Annals of Applied Biology* 24 (2): 377–384. https://doi.org/10.1111/j.1744-7348.1937.tb05040.x.

Thompson, H.M. and Brown, M.A. (2001). Is contact colony treatment with antibiotics an effective control for European foulbrood? *Bee World* 82 (3): 130–138. https://doi.org/10.1080/0005772X.2001.11099515.

Tomkies, V. et al. (2009). Development and validation of a novel field test kit for European foulbrood*. *Apidologie* 40: 63–72. https://doi.org/10.1051/apido:2008060.

USDA National Agricultural Statistics Service (2017) 'Cost of Pollination', pp. 1–13. Available at: http://usda.mannlib.cornell.edu/usda/current/CostPoll/CostPoll-12-21-2017.pdf.

Waite, R.J. et al. (2003). Controlling European foulbrood with the shook swarm method and oxytetracycline in the UK. *Apidologie* 34: 569–575.

Waite, R., Jackson, S., and Thompson, H. (2003). Preliminary investigations into possible resistance to oxytetracycline in *Melissococcus plutonius*, a pathogen of honeybee larvae. *Letters in Applied Microbiology* 36 (1): 20–24. https://doi.org/10.1046/j.1472-765X.2003.01254.x.

White, G. (1912) The Cause of European Foulbrood.

Wilkins, S., Brown, M.A., and Cuthbertson, A.G.S. (2007). The incidence of honey bee pests and diseases in England and Wales. *Pest Management Science* 63: 1062–1068. https://doi.org/10.1002/ps.1461.

Wilson, W.T. (1971). Resistance to American foulbrood in honey bees. XI. Fate of *Bacillus larvae* spores ingested by adults. *Journal of Invertebrate Pathology* 17 (2): 247–255. https://doi.org/10.1016/0022-2011(71)90099-1.

Woodrow, A. W. (n.d.) Susceptibility of Honeybee Larvae to Individual Inoculations with Spores of Bacillus larvae 1. Available at: https://academic.oup.com/jee/article-abstract/35/6/892/2202832 (Accessed: 16 October 2019).

Yue, D. et al. (2008). Fluorescence in situ hybridization (FISH) analysis of the interactions between honeybee larvae and *Paenibacillus larvae*, the causative agent of American foulbrood of honeybees (*Apis mellifera*). *Environmental Microbiology* 10 (6): 1612–1620. https://doi.org/10.1111/j.1462-2920.2008.01579.x.

23

Honey Bee Fungal Diseases

Yanping (Judy) Chen and Jay D. Evans

USDA-ARS Beltsville Bee Research Laboratory, Beltsville, MD, USA

Nosema Disease

Causative Agents

Nosemosis, or *Nosema* disease, caused by microsporidia of the genus *Nosema,* is the most serious and widespread fungal disease among honey bees. Microsporidia are spore-forming intracellular parasites originally considered to be primitive protozoa; however, they have been reclassified as specialized fungi based on their molecular characteristics and evolutionary phylogeny (Capella-Gutiérrez et al. 2012; James et al. 2006). For decades, *Nosema* disease was exclusively attributed to a single species *Nosema apis*, which was first described in European honey bees *Apis mellifera* (Zander 1909).

Nosema ceranae, a species of *Nosema* originally found among Asian honey bees, *Apis cerana* (Fries et al. 1996) has emerged as a disease-causing agent in *A. mellifera* (Huang et al. 2005, 2007). Since its identification in 2005 in *A. mellifera*, of the two *Nosema* species, *N. ceranae* infection has predominated among European honey bees in the majority of the world. *Nosema* infections are associated with bee colony losses worldwide (Chen et al. 2007; Higes et al. 2006; Klee et al. 2007; Giersch et al. 2009; Paxton et al. 2007). Recently, a new species of Nosema, *N. neumanni* was found to be more common than the two other *Nosema* species in *A. mellifera* in Uganda (Chemurot et al. 2017). However, its pathological effects on the host, as well as its distribution and prevalence, have yet to be determined.

Biology

Like other microsporidia, the infective form of *Nosema* is the spore which is highly resistant to environmental conditions and capable of surviving outside the host for up to several years.

The spore is protected by an outer layer comprised of an electron-dense exospore and an electron-lucent chitinous endospore layer, separated from the cell by a thin plasma membrane. A long, thread-like polar tubule, or polar filament, is coiled inside the spore, which is a characteristic structure of all microsporidia. The polar filament is attached to the anterior end of the spore by an anchoring disc followed by lamellae-type polaroplast. The sporoplasm, which contains a single nucleus and the posterior vacuole, is surrounded by the polar filament coils at the posterior region of the spore (Figure 23.1A).

Nosema spores are oval or rod shaped, with the *N. ceranae* spores roughly 5–7 μm in length and 3–4 μm in width (Chen et al., 2009), while the spores of *N. apis* are roughly 6.0 μm in length and 3.0 μm in width (Fries et al. 1996). The number of polar filament coils inside *N. ceranae* spores ranges from 18 to 21 (Chen et al. 2009) (Figure 23.1B), compared to *N. apis* which has more than 30 coils (Fries 1989; Liu 1984). Thus, *N. ceranae* has a fewer number of coils in the polar filament. Both the size difference and the number of polar filament coils inside a spore provide evidence of a morphological differences between these two *Nosema* species (Figure 23.1).

The life cycle of a spore consists of a *proliferative phase* (merogony), a *spore production* phase (sporogony), and an *infective phase* (mature spore). Honey bees become infected when they ingest spore-contaminated food or water and clean spore-contaminated combs. While the germination mechanism of spores has not been definitively elucidated, the physical and chemical conditions of the honey bee midgut are believed to create osmotic pressure which stimulates the germination of spores, involving the rapid expulsion of a polar tubule from a spore. The spore then infects the host cell by piercing a cell membrane and injecting through the polar tubule the infective spore contents, sporoplasm, into an epithelial cell that lines the

Honey Bee Medicine for the Veterinary Practitioner, First Edition. Edited by Terry Ryan Kane and Cynthia M. Faux.

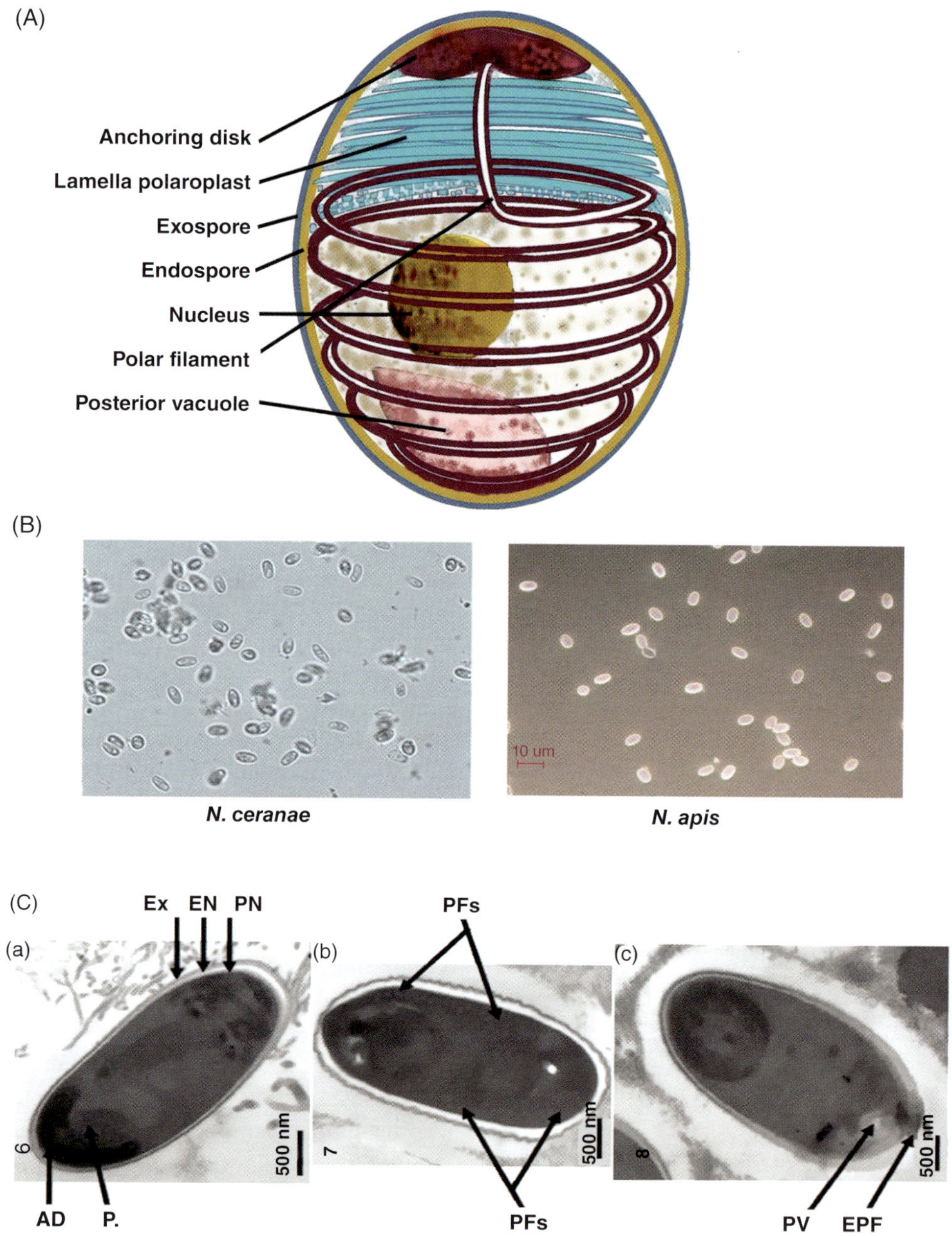

Figure 23.1 (A) Schematic representation of the microsporidia spore. (B) Light micrograph of *Nosema apis and N. ceranae* spores. (C) Electron-micrograph of longitudinal section of *Nosema ceranae* spore showing (a) anchoring disk (AD), polaroplast (P), posterior vacuole (PV), polar filament (PF); (b) endospore (EN), exospore (EX), plasma membrane (PM), nucleus (N), 20–22 isofilar coils of the polar filament (PFs); and (c) extruded polar filament (EPF). *Source:* Image (B) and (C) From Chen et al. (2009).

midgut. Inside the host cell cytoplasm, the sporoplasm undergoes multiple divisions to produce proliferative meronts which undergo binary division and differentiate into sporonts (merogony). Each sporont divides into two sporoblasts which mature into spores by the formation of a thick wall around the spore (sporogony). Repeated multiplication results in the host cell cytoplasm becoming completely filled with spores. Roughly 30–50 million spores can be found inside a bee's midgut within two weeks post infection (Bailey and Ball 1991). Mature spores either germinate with the midgut to infect adjacent cells or are released into the midgut lumen via cell lysis and excreted in feces into the hive environment. The spore-contaminated food, water, combs, etc. provide new sources of infection through feeding and cleaning activities in the colonies (Figure 23.2).

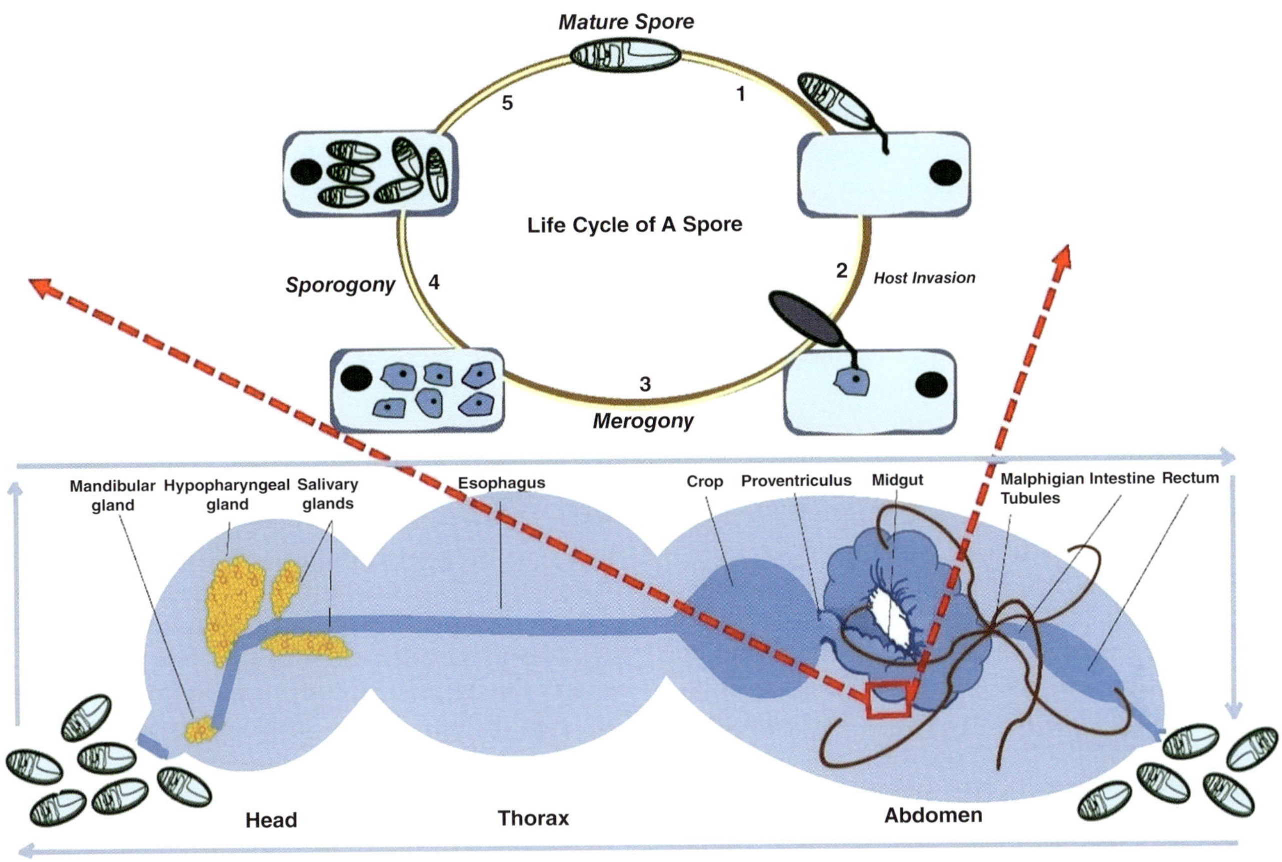

Figure 23.2 Life cycle of a Nosema spore. Infection begins when a bee ingests *Nosema* spore contaminated food/water. Within the midgut, the spore germinates and injects its infectious content into a gut epithelial cell via polar filament or polar tubule. Following a proliferative vegetative phase, mature spores are formed and released into the gut and passed out of the body in feces.

Pathology and Epidemiology

Nosema infection impacts bee health and performance in multiple ways (reviewed in Fries 2010; Goblirsch 2018; Holt and Grozinger 2016; Martín-Hernández et al. 2018). Colony members, including adult worker bees, drones, and queens, are all susceptible to *Nosema* infections. Recent laboratory experiments demonstrated that larvae fed with *N. ceranae* spores had a significantly higher number of spores once they reached a prepupae stage and subsequently had a decreased adult longevity compared to a negative control (Eiri et al. 2015). However, the pathological effects of *Nosema* infection on honey bees, to date, have been primarily documented for adult bees.

At the individual honey bee level, infection with either *N. apis* or *N. ceranae* is associated with the behavior and physiological impairments of adult workers, including suppressed immunity, metabolic abnormality, pronounced hunger, energetic stress, and cognitive decline, and the eventual reduction of life expectancy (Mayack et al. 2015; Mayack and Naug 2009; Martín-Hernández et al. 2017; Vidau et al. 2014; Goblirsch et al. 2013; Li et al. 2018). When young nurse bees become infected, their hypopharyngeal glands become atrophied, resulting in a significant reduction, to the complete loss, of their ability to produce royal jelly. This triggers the infected nurse bees to skip the brood rearing stage of their life and engage in precocious foraging. As a result, normal bee polyethism within the colony is altered, ultimately leading to shortened adult lifespan (Hassanein 1953; Wang and Moeller 1970).

Drones, like other members of the colony, can be infected by *Nosema* (Traver and Fell 2011). Drones infected by *N. ceranae* were found to have a higher mortality and a lower body mass than infected workers, suggesting that drones are more susceptible to the *Nosema* infection compared to female workers (Retschnig et al. 2014). A decrease in energy level, flight activity, and survival has also been observed in drones infected with *Nosema* (Holt et al. 2018; Peng et al. 2015). *Nosema* was also detected in the ejaculate of drones and old drones infected with *N. apis* were found to have decreased sperm viability and lifespan. However, drones were found to have the innate immunity to minimize the spread of *Nosema* infection in their reproductive tissue (Peng et al. 2015).

Nosema not only directly causes serious disease in honey bees but can also compromise the physical and immunological barriers of honey bees toward disease, leaving bees more susceptible to other pathogens (Antúnez et al. 2009; Dussaubat et al. 2012). *N. ceranae* was reported to interact with viral pathogens resulting in a more complex disease that diminished the vitality of a honey bee colony (Costa et al. 2011). The synergistic interaction between *Nosema* and neonicotinoid pesticide exposure is also incriminated in increased bee mortality and health decline worldwide (Alaux et al. 2010; Aufauvre et al. 2012; Pettis et al. 2012; Wu et al. 2012). The disease caused by *N. ceranae* has been implicated in colony declines in the United States and Europe and poses a serious threat to the health of honey bee colonies (Cox-Foster et al. 2007; vanEngelsdorp et al. 2009; Goblirsch 2018; Higes et al. 2008, 2009; Paxton 2010).

At the colony level, natural infection by *Nosema* exerts a direct negative impact on adult population size, brood rearing, honey production, queen supersedure, and the survival of bee colonies (Bailey and Ball 1991). Historically, late winter and early spring dwindling of adult populations are often associated with *Nosema* infection in colonies. During the winter months, infected bees trapped in their hives by cold weather may defecate inside the hive. Winter workers performing comb cleaning and colony maintenance become infected as they pick up spore-laden fecal matter and distribute spores throughout the inside of the hive (Bailey 1953). The spores existing in the hive serve as new sources of infection in colonies through the oral-fecal transmission. In spring and summer, infected bees live half as long as non-infected bees, which in turn leads to a significant decrease in colony performance including nectar collection, honey production, royal jelly secretion, and brood production. Both species of *Nosema* not only infect worker bees but also queens of colonies. When queens become infected, they decrease or completely stop laying eggs. Alaux et al. (2011) showed that vitellogenin titer, antioxidant capacity, and mandibular pheromone production were significantly altered in *Nosema*-infected queens which in turn seriously affected queens' fertility, longevity, and performance, thereby inducing queen supersedure or replacement. Natural supersedure often occurs at the end of the summer or early autumn, and the queenless colonies will eventually dwindle away and die (Loskotova et al. 1980).

The seasonality and disease dynamics of *N. apis* and *N. ceranae* differ considerably. Incidence of *N. apis* infection generally varies during the year and the highest level of infection is typically observed in spring. Bee colonies generally display the lowest level of infection during the summer months, low prevalence but with a small detectable peak of infection in the fall, notably increased infection during winter, and peak infection in spring (Bailey 1955). With *N. ceranae*, the infection can be detected in samples throughout the year with no seasonal pattern of prevalence. The epidemiological and pathological differences between the two *Nosema* species in field conditions are likely influenced by climate factors, primarily high and low temperature extremes, which can exert a direct impact on the growth and multiplication of the pathogen. The two *Nosema* species exhibit marked differences in sensitivity to temperatures. *N. ceranae* is more vulnerable to cool temperatures compared to *N. apis* as evidenced by the fact that *N. ceranae* spores lost roughly 90% infectivity at freezing temperature for one week, while *N. apis* spores retained 100% of their activity (Fenoy et al. 2009). Further cell culture experiments confirmed that *N. ceranae* has a higher proliferative activity than *N. apis* at 27° and 33 °C (Gisder et al. 2017). The results from these studies provide a potential explanation for the observed tendency that *N. apis* is more prevalent in temperate, northern countries while *N. ceranae* is more prevalent in sub-tropical, southern countries and has a higher infection potential than *N. apis* in warm climates (Fries 2010; Malone et al. 2001; Martín-Hernández et al. 2009).

Symptoms and Diagnosis

Nosema disease is often referred to as a "silent killer" as *Nosema* infected honey bees typically do not exhibit outward symptoms until the colony is significantly diminished. However, COLOSS (Prevention of honey bee COlony LOSSes), an international research network for better understanding bee health at a global level, defined two disease patterns associated with *Nosema* infection: nosemosis type A (caused by *N. apis*) and nosemosis type C (caused by *N. ceranae*) (COLOSS 2009). The acute manifestations of nosemosis type A include the following: crawling bees with disjointed wings and swollen abdomens, and milky-white gut coloration, dead bees at hive entrances, and brownish-yellow fecal streaks present on combs and hive exterior (Bailey 1955; Bailey and Ball 1991; Higes et al. 2010; Martín-Hernández et al. 2018).

By contrast, the signs and symptoms of nosemosis type C do not include dysentery or crawling. Rather, the *N. ceranae* caused disease is characterized by suppressed immunity, energetic stress, decreased foraging activity, diminished honey production and poor colony growth, particularly during spring (Higes et al. 2008, 2009, 2010; Paxton 2010; Botías et al. 2013) (Figure 23.3).

Early diagnosis of *Nosema* infection is difficult because the signs or symptoms of the disease are lacking, especially with infection by *N. ceranae*. The definitive diagnosis of

Figure 23.3 Brownish-yellow fecal spotting on the exterior of a beehive.

Nosema infection in honey bees relies on the presence of spores and species determination. Traditionally, *Nosema* infection has been diagnosed via microscopic examination. A common method of microscopic testing for *Nosema* spores in bees involves collecting foraging bees at the at the hive entrance, grinding the abdomens of suspect bees in water, placing one drop of homogenate in the center of a microscope slide, and examining the slide under a light microscope with a magnification of 400×. Bees of a foraging age are more likely to suffer from an *N. ceranae* infection when compared to nurse-aged bees. This makes them a reliable sample for *Nosema* detection (Higes et al. 2008; Jack et al. 2016). *Nosema* infection levels can be quantitatively determined using a special type of microscope slide, hemocytometer, which allows estimating of the total number of spores per bee (Cantwell 1970). Over the years, molecular techniques have been increasingly incorporated in the detection, quantification, species differentiation, and phylogenetic analysis of *Nosema* infection in honey bees. A polymerase chain reaction (PCR), a powerful and effective method that selectively amplifies a target sequence of interest following the purification of deoxyribonucleic acids (DNAs) from infected bee samples, has become a popular means of diagnosing of *Nosema* infection (Chen et al. 2007). The availability of complete genome sequences for both *N. ceranae* and *N. apis* (Cornman et al. 2009; Chen et al. 2013) has provided ample opportunity for the development of sensitive and specific PCR diagnostic techniques for the *Nosema* disease. Other methods such as scanning electron microscopy, histological staining, and the Enzyme-Linked Immunosorbent Assay (ELISA) test have been also deployed for diagnosing and characterizing *Nosema* infection in honey bees (Aronstein et al. 2013; Maiolino et al. 2014; Ptaszyńska et al. 2014; Wang and Moeller 1969).

Treatment

Of the numerous potential *Nosema* treatments that have been explored and tested, fumagillin (Fumidil-B®), a naturally secreted antibiotic of the fungus *Aspergillus fumigatus*, has performed the best in the treatment of *Nosema* infections in honey bees (Katznelson and Jamieson 1952; Williams et al. 2011). While fumagillin has been used to control nosemosis in managed honey bee colonies in North America for decades, the use of the compound is forbidden within the European Union due to the absence of an established maximum residue level (MRL) which is required for medicines used to treat the food producing animals (FPA). Additionally, there are several problems associated with the use of fumagillin, including serious toxic effects on mammals (Stanimirovica et al. 2007).

Moreover, although fumagillin inhibits the reproduction of spores, it does not kill pre-existing spores. Therefore, treatment using fumagillin does not eliminate *Nosema* from the colony, and the infection may return after the chemical therapy ends (Johnya et al. 2008). Furthermore, there is evidence to suggest that honey bee *Nosema* parasites have developed a resistance to fumagillin (Williams et al. 2008; Huang et al. 2013). Fumagillin has recently come off the market due to production problems. As a result, the field of veterinary medicine is in urgent need of new, diverse, and robust drugs and therapies that effectively treat *Nosema* infections in honey bees.

Over the past decade, significant progress has occurred in finding and testing novel therapeutic agents to treat honey bee nosemosis. For example, RNA interference (RNAi) is a biological process used by a wide range of organisms to silence genes and has been explored for developing a new class of therapies for various diseases. RNAi has been employed in suppression of *N. ceranae* virulence factors that facilitate the invasion of hosts as well as in the enhancement of the ability of honey bees to fight off *Nosema* infection by silencing the expression of a host immune suppressor triggered by the *Nosema* infection. The silencing of the parasite virulence factors and the host immune suppressor was positively linked to an increase in the expression level of host genes encoding antimicrobial peptides, a decrease in *Nosema* spore load, and an extension of infected bee lifespan (Li et al. 2016; Paldi et al. 2010; Rodríguez-Garcíaa et al. 2018). These findings suggest that RNAi-based therapies hold promise for the effective treatment of *Nosema* disease in honey bees and warrant further investigation.

Recent laboratory and field studies report many compounds including oxalic acid, formic acid, thymol, and resveratrol, all of which are typically used to combat parasitic *Varroa* mite infestation, significantly decrease the multiplication and infectivity of *Nosema* and lengthen the lifespan of infected honey bees (Costa et al. 2010; Maistrello et al. 2008; Nanetti et al. 2015; Underwood and Currie 2009). These dual-function compounds are a promising development for *Nosema* disease treatment and deserve further investigation.

Natural products can exhibit antimicrobial properties and boost or support immunity and have provided a rich source of potential treatments for bee and hive health. Many natural products have been identified as possessing anti-microsporidian activity and are effective in reducing *Nosema* spore loads as well as improving survival of infected bees following oral treatment (Arismendi et al. 2018; Bravo et al. 2017; Damiani et al. 2014; Porrini et al. 2011; Roussel et al. 2015; Yemor et al. 2015).

Prevention

An important element in disease management is prevention. The decontamination of combs and hive tools and equipment that contain Nosema spores is a crucial element of the disease prevention process. Clorox® bleach (5.25% sodium hypochlorite as the active ingredient) diluted to 10–20% of the original concentration, is an effective disinfectant for surface sterilization. Fumigation with acetic acid and ethylene oxide can deactivate both *N. apis* and *N. ceranae* spores. Placing combs contaminated by cold susceptible *N. ceranae* spores in a freezer is also an effective and safe method of deactivation. As ever, good beekeeping practices help control infection of *Nosema* and other bee diseases as well. The best bee health management practices, including replacing contaminated combs, introducing new queens, providing good colony nutrition, and selective breeding of disease-resistant bees, will help to sustain healthy bee populations and reduce the risk of diseases.

Chalkbrood Disease

Agent and Biology

Chalkbrood disease is caused by *Ascosphaera apis*, a fungal pathogen that affects sealed and unsealed brood and causes "chalkbrood" in honey bee larvae (Gilliam et al. 1988). Spores of the fungus enter the gut of a larva through the ingestion of contaminated food and germinate when conditions are favorable (Gilliam et al. 1988). Inside the sealed brood cell, the fungus hyphae extract nutrients from the larva and consume the rest of the host's body, and ultimately form a false skin. The dead larva is covered by a fluffy white mold that eventually dries to become a chalky white mass referred to as a white mummy. When spores form, the mummified larva will become mottled black or completely black and is referred to as a black mummy (Jensen et al. 2013). Each black mummy may contain as many as 100 million spores which can remain infective for many years in the environment and can be spread between hives via robbing bees, drifting bees, or the moving of infected equipment and hive tools (Aronstein and Murray 2010).

Symptoms and Diagnosis

The symptoms vary depending on the infection stages of larvae. Young infected larvae do not usually show signs of the disease. Fungus-killed larvae shrink and dry to form a white or gray-black chalk-like mummy. The field diagnosis of chalkbrood is based on the visual detection of these mummies. In an infected colony, brood pattern in the comb is scattered. The wax cell capping may also have small holes as the nurse bees cut inspection holes in the capping and try to remove the mummy. Chalkbrood mummies can often be seen in the combs, at the hive entrance and on the bottom board (Figure 23.4). The mummies can easily be removed from the brood cells by tapping the comb against a solid surface. This easy removal of infected larval remains also differentiates chalkbrood from other brood diseases such as American foulbrood (AFB), Sacbrood, Stonebrood, and European foulbrood (EFB).

Following field diagnosis, a microscopic examination can confirm the presence of spore cysts in the infected samples (Jensen et al. 2013). The spherical spore cysts measure 47–140 μm in diameter. Spore balls enclosed within the cysts range from 9 to 19 μm in diameter, while individual hyaline spores are 2.7–3.5 μm by 1.4–1.8 μm (Skou 1972; Bissett 1988). PCR-based tests using DNA sequence data from the nuclear ribosomal internal transcribed spacer (ITS) region and other target areas and various bioassays have also been developed and employed for the detection, quantification, and characterization of chalkbrood infection in honey bees (Jensen et al. 2013).

Pathology and Treatment

Chalkbrood is considered a relatively minor disease in honey bees and chemotherapy is not recommended. High moisture in a hive and cold temperature in spring promote the growth of the chalkbrood pathogen. While chalkbrood weakens a colony's health, it rarely kills a colony. Worker

Figure 23.4 Honey bee Chalkbrood disease. (A) Chalkbrood mummy. *Source*: Photo courtesy of Virginia Williams. (B) Brood cell filled with a chalkbrood mummy. *Source*: Photo courtesy of Bart Smith Jr. (C) White and dark chalkbrood mummies found on the bottom board of a hive. *Source:* Photo courtesy of Bart Smith Jr.

bees with hygienic traits detect, uncap, and remove the infected brood and strong, established bee colonies with low levels of diseased brood may recover from the chalkbrood disease (Gilliam et al. 1988). The effective approach to reduce the impact of the chalkbrood disease is to maintain strong, healthy, populous colonies, requeen affected colonies with young mated queens, use bee stocks selected for hygienic behavior, and keep beehives well ventilated.

Stonebrood Disease

Agent and Biology

Stonebrood is usually attributed to *Aspergillus* spp. including *A. flavus*, *A. fumigatus*, and *A. niger* and causes the mummification of the brood. These *Aspergillus* species are cosmopolitan filamentous fungi often found in soil and are also pathogenic to adult bees and pupae. They produce airborne conidia which can cause respiratory damage to humans and other animals (Gilliam and Vandenberg 1997). Larvae become infected by ingesting spore-contaminated brood food. Spores germinate in the gut, and the mycelium grows rapidly to form a whitish-yellow collar-like ring near the larval heads. The larvae often die in the capped cell from toxins (aflatoxins) produced by the fungus. After death, the larvae turn black and become difficult to crush, hence the name stonebrood. Eventually, the fungus erupts from the integument of the larva and forms a false skin (mummy). In this stage, the larvae are covered with powdery fungal spores. The spores of the different species have different colors. The surfaces of the larvae are covered by yellow-green powdery spores when infected with *A. flavus*, and gray-green powdery spores when infested by *A. fumigatus*.

Symptom and Diagnosis

Like chalkbrood disease, stonebrood can usually be diagnosed from its gross symptoms – mummies in combs, irregular capping of the brood, and holes in infected cell capping. Compared to the sponge-like chalkbrood mummies, stonebrood mummies turn hard and resemble small stones which are very difficult to remove from the cells. Definitive identification of the species requires a combination of cultivation, microscopic examination of specialized fungal structures, conidiophores and conidia (400× magnification), and molecular assay (Jensen et al. 2013).

Pathology and Treatment

Stonebrood is a rare honey bee brood disease, with only slight effects on a bee colony. As with chalkbrood, no chemotherapy is recommended. Any beekeeping practices that maximize bee population strength will help to clear the stonebrood infection from bee colonies.

Conclusion

In summary, the importance of fungal infections in honey bees has increased over the last few decades, especially with the new emergence of *N. ceranae*, a highly virulent and specialized fungal pathogen of the European honey bee, *A. mellifera*, thereby underscoring the need for the innovative and effective strategies for prevention and control of the disease infections. While most fungal infections rarely kill bee colonies, they can significantly diminish the overall robustness and foraging efficiency of colonies and weaken bees, making them more susceptible to other diseases and pests. Selection for hygienic behavior, in which worker bees detect, uncap, and remove diseased or dead brood from sealed cells, holds the most promise against outbreaks of diseases. Good sanitation practices such as providing good colony ventilation, maintaining a warm, dry hive interior; and periodically renewing or replacing comb and hive materials can be expected to reduce the chance of spreading the diseases to healthy bees.

References

Alaux, C., Brunet, J.L., Dussaubat, C. et al. (2010). Interactions between Nosema microspores and a neonicotinoid weaken honeybees (*Apis mellifera*). *Environmental Microbiology* 12 (3): 774–782.

Alaux, C., Folschweiller, M., McDonnell, C. et al. (2011). Pathological effects of the microsporidium *Nosema ceranae* on honey bee queen physiology (*Apis mellifera*). *Journal of Invertebrate Pathology* 106: 380–385.

Antúnez, K., Martín-Hernández, R., Prieto, L. et al. (2009). Immune suppression in the honey bee (*Apis mellifera*) following infection by *Nosema ceranae* (Microsporidia). *Environmental Microbiology* 11 (9): 2284–2290.

Arismendi, N., Vargas, M., López, M.D. et al. (2018). Promising antimicrobial activity against the honey bee parasite *Nosema ceranae* by methanolic extracts from Chilean native plants and propolis. *Journal of Apicultural Research* 57: 522–535.

Aronstein, K. and Murray, K. (2010). Chalkbrood disease in honey bees. *Journal of Invertebrate Pathology* 103: S20–S29.

Aronstein, K.A., Webster, T.C., and Saldivar, E. (2013). A serological method for detection of *Nosema ceranae*. *Journal of Applied Microbiology* 114: 621–625.

Aufauvre, J., Biron, D.G., Vidau, C. et al. (2012). Parasite-insecticide interactions: a case study of *Nosema ceranae* and fipronil synergy on honeybee. *Scientific Reports* 2 (326): 1–7.

Bailey, L. (1953). The transmission of *nosema* disease. *Bee World* 34: 171–172.

Bailey, L. (1955). The epidemiology and control of *Nosema* disease of the honeybee. *The Annals of Applied Biology* 43: 379–389.

Bailey, L. and Ball, B.V. (1991). *Honey Bee Pathology*, 2e. London: Academic Press.

Bissett, J. (1988). Contribution toward a monograph of the genus *Ascosphaera*. *Canadian Journal of Botany* 66: 2541–2560.

Botías, C., Martín-Hernández, R., Barrios, L. et al. (2013). *Nosema* spp. infection and its negative effects on honey bees (*Apis mellifera* iberiensis) at the colony level. *Veterinary Research* 44: 25–14.

Bravo, J., Carbonell, V., Sepúlveda, B. et al. (2017). Antifungal activity of the essential oil obtained from *Cryptocarya alba* against infection in honey bees by *Nosema ceranae*. *Journal of Invertebrate Pathology* 49: 141–147.

Cantwell, G.E. (1970). Standard methods for counting nosema spores. *American Bee Journal* 110 (6): 222–223.

Capella-Gutiérrez, S., Marcet-Houben, M., and Gabaldón, T. (2012). Phylogenomics supports microsporidia as the earliest diverging clade of sequenced fungi. *BMC Biology* 10: 47.

Chemurot, M., De Smet, L., Brunain, M. et al. (2017). *Nosema neumanni* n. sp. (Microsporidia, Nosematidae), a new microsporidian parasite of honeybees, *Apis mellifera* in Uganda. *European Journal of Protistology* 61: 13–19.

Chen, Y.P., Evans, J.D., Smith, J.B., and Pettis, J.S. (2007). *Nosema ceranae* is a long-present and wide-spread microsporidian infection of the European honey bee (*Apis mellifera*) in the United States. *Journal of Invertebrate Pathology* 92: 152–159.

Chen, Y.P., Evans, J.D., Murphy, C.A. et al. (2009). Morphological, molecular, and phylogenetic characterization of *Nosema ceranae*, a microsporidian parasite isolated from the European honey bee, *Apis mellifera*. *Journal of Eukaryotic Microbiology* 56: 142–147.

Chen, Y.P., Pettis, J.S., Zhao, Y. et al. (2013). Genome sequencing and comparative genomics of honey bee microsporidia parasite, *Nosema apis* reveal novel insights into host-parasite interactions. *BMC Genomics* 14: 451.

Cornman, R.S., Chen, Y.P., Schatz, M.C. et al. (2009). Genomic analyses of the microsporidian *Nosema ceranae*, an emergent pathogen of honey bees. *PLoS Pathogens* 5: e1000466.

Costa, C., Lodesani, M., and Maistrello, L. (2010). Effect of thymol and resveratrol administered with candy or syrup on the development of *Nosema ceranae* and on the longevity of honeybees (*Apis mellifera* L.) in laboratory conditions. *Apidologie* 41: 141–150.

Costa, C., Tanner, G., Lodesani, M. et al. (2011). Negative correlation between *Nosema ceranae* spore loads and deformed wing virus infection levels in adult honey bee workers. *Journal of Invertebrate Pathology* 108: 224–225.

Cox-Foster, D.L., Conlan, S., Holmes, E.C. et al. (2007). A metagenomic survey of microbes in honey bee colony collapse disorder. *Science* 318: 283–287.

Damiani, N., Fernández, N.J., Porrini, M.P. et al. (2014). Laurel leaf extracts for honeybee pest and disease management: antimicrobial, microsporicidal, and acaricidal activity. *Parasitology Research* 113: 701–709.

Dussaubat, C., Brunet, J.-L., Cousin, M. et al. (2012). Gut pathology and responses to the microsporidium *Nosema ceranae* in the honey bee *Apis mellifera*. *PLoS One* 7 (5): e37017.

Eiri, D.M., Suwannapong, G., Endler, M., and Nieh, J.C. (2015). *Nosema ceranae* can infect honey bee larvae and reduces subsequent adult longevity. *PLoS One* (5): 10, e0126330.

Fenoy, S., Rueda, C., Higes, M. et al. (2009). High-level resistance of *Nosema ceranae*, a parasite of the honeybee, to temperature and desiccation. *Applied and Environmental Microbiology* 75: 6886–6889.

Fries, I. (1989). Observation on the development and transmission of *Nosema apis* Z. in the ventriculus of the honey bee. *Journal of Apicultural Research* 28: 107–117.

Fries, I. (2010). *Nosema ceranae* in European honey bees (*Apis mellifera*). *Journal of Invertebrate Pathology* 103: S73–S79.

Fries, I., Feng, F., Silva, A.D. et al. (1996). *Nosema ceranae* n. sp. (Microspora, Nosematidae), morphological and molecular characterization of a microsporidian parasite of the Asian honey bee *Apis cerana* (Hymenoptera, Apidae). *European Journal of Protistology* 32: 356–365.

Giersch, T., Berg, T., Galea, F., and Hornitzky, M. (2009). *Nosema ceranae* infects honey bees (*Apis mellifera*) and contaminates honey in Australia. *Apidologie* 40: 117–123.

Gilliam, M. and Vandenberg, J. (1997). Fungi. In: *Honey Bee Pests, Predators, & Diseases*, 3e (eds. R.A. Morse and K. Flottum), 81–110. Medina, OH: A I Root Co.

Gilliam, M., Taber, S. III, Lorenz, B.J., and Prest, D.B. (1988). Factors affecting development of chalkbrood disease in colonies of honey bees, *Apis mellifera*, fed pollen contaminated with *Ascosphaera apis*. *Journal of Invertebrate Pathology* 52: 314–325.

Gisder, S., Schüler, V., Horchler, L.L. et al. (2017). Long-term temporal trends of *Nosema* spp. infection prevalence in Northeast Germany: continuous spread of *Nosema ceranae*, an emerging pathogen of honey bees (*Apis mellifera*), but no general replacement of *Nosema apis*. *Frontiers in Cellular and Infection Microbiology* 7: 301.

Goblirsch, M. (2018). *Nosema ceranae* diseases of honey bee (*Apis mellifera*). *Apidologie* 49: 131–150.

Goblirsch, M., Huang, Z.Y., and Spivak, M. (2013). Physiological and behavioral changes in honey bees (*Apis mellifera*) induced by *Nosema ceranae* infection. *PLoS One* 8 (3): e58165.

Hassanein, M.H. (1953). The influence of infection with *Nosema apis* on the activities and longevity of the worker honeybee. *The Annals of Applied Biology* 40: 418–423.

Higes, M., Martín, R., and Meana, A. (2006). Nosema ceranae, a new microsporidian parasite in honey bees in Europe. *Journal of Invertebrate Pathology* 92: 93–95.

Higes, M., Martín-Hernández, R., Botías, C. et al. (2008). How natural infection by *Nosema ceranae* causes honey bee colony collapse. *Environmental Microbiology* 10: 2659–2669.

Higes, M., Martín-Hernández, R., Garrido-Bailón, E. et al. (2009). Honeybee colony collapse due to *Nosema ceranae* in professional apiaries. *Environmental Microbiology Reports* 1: 110–113.

Higes, M., Martín-Hernández, R., and Meana, A. (2010). *Nosema ceranae* in Europe: an emergent type C nosemosis. *Apidologie* 41: 375–392.

Holt, H. and Grozinger, C.M. (2016). Approaches and challenges to managing *Nosema* (Microspora: Nosematidae) parasites in honey bee (Hymenoptera: Apidae) colonies. *Journal of Economic Entomology* 109 (4): 1487–1503.

Holt, H.L., Villar, G., Cheng, W. et al. (2018). Molecular, physiological and behavioral responses of honey bee (*Apis mellifera*) drones to infection with microsporidian parasites. *Journal of Invertebrate Pathology* 155: 14–24.

Huang, W.F., Jiang, J.H., and Wang, C.H. (2005). *Nosema ceranae* infection in *Apis mellifera*. 38th Annual Meeting of Society for Invertebrate Pathology. Anchorage, Alaska.

Huang, W.-F., Jiang, J.-H., Chen, Y.-W., and Wang, C.-H. (2007). A *Nosema ceranae* isolate from the honeybee *Apis mellifera*. *Apidologie* 38: 30–37.

Huang, W., Solter, L.F., Yau, P.M., and Imai, B.S. (2013). *Nosema ceranae* escapes fumagillin control in honey bees. *PLoS Pathogens* 9: e1003185.

Jack, C.J., Lucas, H.M., Webster, T.C., and Sagili, R.R. (2016). Colony level prevalence and intensity of *Nosema ceranae* in honey bees (*Apis mellifera* L.). *PLoS One* 11 (9): e0163522.

James, T.Y., Kauff, F., Schoch, C.L. et al. (2006). Reconstructing the early evolution of fungi using a six-gene phylogeny. *Nature* 443: 818–822.

Jensen, A.B., Aronstein, K., Flores, J.M. et al. (2013). Standard methods for fungal brood disease research. *Journal of Apicultural Research* 52 (1): 1–20. https://doi.org/10.3896/IBRA.1.52.1.13.

Johnya, S., Whitman, D.W., and Bridge study group (2008). Effect of four antimicrobials against an *Encephalitozoon* sp. (Microsporidia) in a grasshopper host. *Parasitology International* 57: 362–367.

Katznelson, H. and Jamieson, C.A. (1952). Control of nosema disease of honey bees with fumagillin. *Science* 115: 70–71.

Klee, J., Besana, A.M., Genersch, E. et al. (2007). Widespread dispersal of the microsporidian *Nosema ceranae*, an emergent pathogen of the western honey bee, *Apis mellifera*. *Journal of Invertebrate Pathology* 96: 1–10.

Li, W.F., Evans, J.D., Huang, Q. et al. (2016). Silencing honey bee (*Apis mellifera*) naked cuticle (nkd) improves host immune function and reduces *Nosema ceranae* infections. *Applied and Environmental Microbiology* 82: 6779–6787.

Li, W., Chen, Y., and Cook, S.C. (2018). Chronic *Nosema ceranae* infection inflicts comprehensive and persistent immunosuppression and accelerated lipid loss in host *Apis mellifera* honey bees. *International Journal for Parasitology* 48 (6): 433–444.

Liu, T.P. (1984). Ultrastructure of the midgut of the worker honey *Apis mellifera* heavily infected with *Nosema apis*. *Journal of Invertebrate Pathology* 44: 103–105.

Loskotova, J., Peroutka, M., and Vesely, V. (1980). *Nosema* disease of honey bee queens (*Apis mellifera*). *Apidologie* 12: 53–61.

Maiolino, P., Iafigliola, L., Rinaldi, L. et al. (2014). Histopathological findings of the midgut in European honey bee (*Apis Mellifera* L.) naturally infected by *Nosema* spp. *Veterinary Medicine & Animal Sciences* 2: 4. https://doi.org/10.7243/2054-3425-2-4.

Maistrello, L., Lodesani, M., Costa, C. et al. (2008). Screening of natural compounds for the control of nosema disease in honeybees (*Apis mellifera*). *Apidologie* 39: 436–445.

Malone, L.A., Gatehouse, H.S., and Tregidga, E.L. (2001). Effects of time, temperature, and honey on *Nosema apis* (Microsporidia: Nosematidae), a parasite of the honeybee, *Apis mellifera* (Hymenoptera: Apidae). *Journal of Invertebrate Pathology* 77: 258–268.

Martín-Hernández, R., Meana, A., García-Palencia, P. et al. (2009). Effect of temperature on the biotic potential of honeybee microsporidia. *Applied and Environmental Microbiology* 75: 2554–2557.

Martín-Hernández, R., Higes, M., Sagastume, S. et al. (2017). Microsporidia infection impacts the host cell's cycle and reduces host cell apoptosis. *PLoS One* 12 (2): e0170183.

Martín-Hernández, R., Bartolomé, C., Chejanovsky, N. et al. (2018). *Nosema ceranae* in *Apis mellifera*: a 12 years post detection perspective. *Environmental Microbiology* 20: 1302–1329.

Mayack, C. and Naug, D. (2009). Energetic stress in the honeybee *Apis mellifera* from *Nosema ceranae* infection. *Journal of Invertebrate Pathology* 100 (3): 185–188.

Mayack, C., Natsopoulou, M.E., and McMahon, D.P. (2015). *Nosema ceranae* alters a highly conserved hormonal stress pathway in honeybees. *Insect Molecular Biology* 24 (6): 662–670.

Nanetti, A., Rodriguez-García, C., Meana, A. et al. (2015). Effect of oxalic acid on *Nosema ceranae* infection. *Research in Veterinary Science* 102: 167–172.

Paldi, N., Glick, E., Oliva, M. et al. (2010). Effective gene silencing of a microsporidian parasite associated with honey bee (*Apis mellifera*) colony declines. *Applied and Environmental Microbiology* 76 (17): 5960–5964.

Paxton, R.J. (2010). Does infection by *Nosema ceranae* cause "Colony Collapse Disorder" in honey bees (*Apis mellifera*)? *Journal of Apicultural Research* 49: 80–84.

Paxton, R., Klee, J., Korpela, S., and Fries, I. (2007). *Nosema ceranae* has infected *Apis mellifera* in Europe since at least 1998 and may be more virulent than *Nosema apis*. *Apidologie* 38: 558–565.

Peng, Y., Baer-Imhoof, B., Millar, A.H., and Baer, B. (2015). Consequences of *Nosema apis* infection for male honey bees and their fertility. *Scientific Reports* 5: 10565.

Pettis, J.S., vanEngelsdorp, D., Johnson, J., and Dively, G. (2012). Pesticide exposure in honey bees results in increased levels of the gut pathogen *Nosema*. *Naturwissenschaften* 99: 153–158.

Porrini, M.P., Fernández, N.J., Garrido, P.M. et al. (2011). *In vivo* evaluation of antiparasitic activity of plant extracts on *Nosema ceranae* (Microsporidia). *Apidologie* 42: 700–707.

Ptaszyńska, A.A., Borsuk, G., Mułenko, W., and Demetraki-Paleolog, J. (2014). Differentiation of *Nosema apis* and *Nosema ceranae* spores under scanning electron microscopy (SEM). *Journal of Apicultural Research* 53: 537–544.

Retschnig, G., Williams, G.R., Mehmann, M.M. et al. (2014). Sex-specific differences in pathogen susceptibility in honey bees (*Apis mellifera*). *PLoS One* 9 (1): e85261.

Rodríguez-Garcíaa, C., Evans, J.D., Li, W.F. et al. (2018). Nosemosis control in European honey bees *Apis mellifera* by silencing the gene encoding *Nosema ceranae* polar tube protein 3. *The Journal of Experimental Biology* 5 (Pt 19): 221.

Roussel, M., Villay, A., Delbac, F. et al. (2015). Antimicrosporidian activity of sulphated polysaccharides from algae and their potential to control honeybee nosemosis. *Carbohydrate Polymers* 133: 213–220.

Skou, J.P. (1972). Ascosphaerales. *Friesia* 10 (1): 1–24.

Stanimirovica, Z., Stevanovica, J., Bajicb, V., and Radovicc, I. (2007). Evaluation of genotoxic effects of fumagillin by cytogenetic tests in vivo. *Mutation Research* 628: 1–10.

Traver, B.E. and Fell, R.D. (2011). *Nosema ceranae* in drone honey bees (*Apis mellifera*). *Journal of Invertebrate Pathology* 107: 234–236.

Underwood, R.M. and Currie, R.W. (2009). Indoor winter fumigation with formic acid for control of *Acarapis woodi* (Acari: Tarsonemidae) and *Nosema* disease, *Nosema* sp. *Journal of Economic Entomology* 102: 1729–1736.

vanEngelsdorp, D., Evans, J.D., Saegerman, C. et al. (2009). Colony collapse disorder: a descriptive study. *PLoS One* 4: e6481.

Vidau, C., Panek, J., Texier, C. et al. (2014). Differential proteomic analysis of midguts from *Nosema ceranae*-infected honeybees reveals manipulation of key host functions. *Journal of Invertebrate Pathology* 121: 89–96.

Wang, D.I. and Moeller, F.E. (1969). Histological comparisons of the development of hypopharyngeal glands in healthy and *Nosema*-infected worker honey bees. *Journal of Invertebrate Pathology* 14: 135–142.

Wang, D.I. and Moeller, F.E. (1970). The division of labor and queen attendance behavior of *Nosema* infected worker honeybees. *Journal of Economic Entomology* 63: 1540–1541.

Williams, G.R., Shafer, A.B.A., Rogers, R.E.L. et al. (2008). First detection of *Nosema ceranae*, a microsporidian parasite of European honey bees (*Apis mellifera*), in Canada and central USA. *Journal of Invertebrate Pathology* 97: 189–192.

Williams, G.R., Shutler, D., Little, C.M. et al. (2011). The microsporidian *Nosema ceranae,* the antibiotic Fumagilin-B®, and Western honey bee (*Apis mellifera*) colony strength. *Apidologie* 42: 15–22.

Wu, J.Y., Smart, M.D., Anelli, C.M., and Sheppard, W.S. (2012). Honey bees (*Apis mellifera*) reared in brood combs containing high levels of pesticide residues exhibit increased susceptibility to *Nosema* (Microsporidia) infection. *Journal of Invertebrate Pathology* 109 (3): 326–329.

Yemor, T., Phiancharoen, M., Benbow, E.M., and Suwannapong, G. (2015). Effects of stingless bee propolis on *Nosema ceranae* infected Asian honey bees, *Apis cerana*. *Journal of Apicultural Research* 54: 468–473.

Zander, E. (1909). Tierische Parasiten als Krankenheitserreger bei der Biene. *Münchener Bienenzeitung* 31: 196–204.

24

Honey Bee Parasites and Pests

Britteny Kyle

Department of Population Medicine, Ontario Veterinary College, University of Guelph, Ontario, Canada

Tracheal Mites

The tracheal mite, or *Acarapis woodi*, is a microscopic obligatory endoparasite that spends almost its entire lifecycle within the trachea of the adult honey bee (Sammataro et al. 2013). The mites have an oval, white to pearly white body with females measuring 140–175 µm in length, and males measuring 125–136 µm in length (Ritter 2014). Mated females have a short phoretic phase in which they leave the trachea and grab onto the hairs of their bee host waiting for direct contact to transfer to a young adult bee (Pernal and Clay 2013; Sammataro et al. 2013; Ritter 2014; Downey and Winston 2001). The female mites enter the trachea through the spiracles where they lay eggs, and all parts of the life cycle – eggs, larva and adult – occur within the tracheal network of the host bee. The mites pierce the tracheal walls to feed on hemolymph (Sammataro et al. 2013). Although all honey bees in a colony can be infested, tracheal mites are most attracted to, and more likely to heavily infest, drones (Ritter 2014; Pernal and Clay 2013; OIE 2019; Sammataro et al. 2013). The majority of mites will be found in the large prothoracic trachea, though they can also be found in air sacs of the head, thorax and abdomen (OIE 2019; Ritter 2014).

Tracheal mites were first identified in 1919 in the Isle of Wight following devastating colony losses for which they were originally blamed (Vidal-Naquet 2015; Pernal and Clay 2013). The first appearance in North America was in the early 1980s and they have now spread across the continent (Sammataro et al. 2013; Pernal and Clay 2013; Ritter 2014). Tracheal mites have an almost worldwide distribution, with only a few countries in Northern Europe, Australia, New Zealand, and Hawaii being free of them (Ritter 2014).

Tracheal mites impact their host bee in a number of ways. First and foremost, they can cause a partial or complete obstruction of the air ducts by occluding the trachea. This obstruction leads to reduced oxygenation of the tissues supplied by the prothoracic trachea – namely the brain and the flight muscles. Furthermore, bees can suffer a significant loss of hemolymph due to mite feeding (Vidal-Naquet 2015; OIE 2019). These result in a reduced lifespan of the individual bee and, when infestations are severe, particularly in winter and early spring, can result in loss of the colony (Ellis 2016; Pernal and Clay 2013).

Clinical signs include dead bees, bees that are unable to fly, crawling bees, and bees with K-wing, all of which are most often noted at the entrance of the hive. During the overwintering period, the ability of the cluster to thermoregulate can be impaired as the bees' thoracic muscles may be unable to generate sufficient heat. Dysentry in affected hives has also been noted. Clinical signs of tracheal mites are not specific and, therefore, may be suggestive but cannot be considered diagnostic (Vidal-Naquet 2015; Sammataro et al. 2013; OIE 2019; Ritter 2014).

The life cycle of a tracheal mite is almost as long as the entire lifespan of the individual bees during the active beekeeping season and this, along with the high population of young adult bees at this time, leads to subclinical infestation during the summer months. However, in the winter (diutinus) bees, which can live for five–six months, the mites can multiply through five or six generations and develop a large population within individual bees. Consequently, the winter infestation can be much more severe and can result in colony losses (Vidal-Naquet 2015; Pernal and Clay 2013; Sammataro et al. 2013). Tracheal mites are more serious in colder climates and the colder the temperature in the winter, the more likely the mites are to cause mortality of the colony (Sammataro et al. 2013; Downey and Winston 2001; Pernal and Clay 2013).

Though at one time tracheal mites were considered a major honey bee colony problem, they are presently

Honey Bee Medicine for the Veterinary Practitioner, First Edition. Edited by Terry Ryan Kane and Cynthia M. Faux.

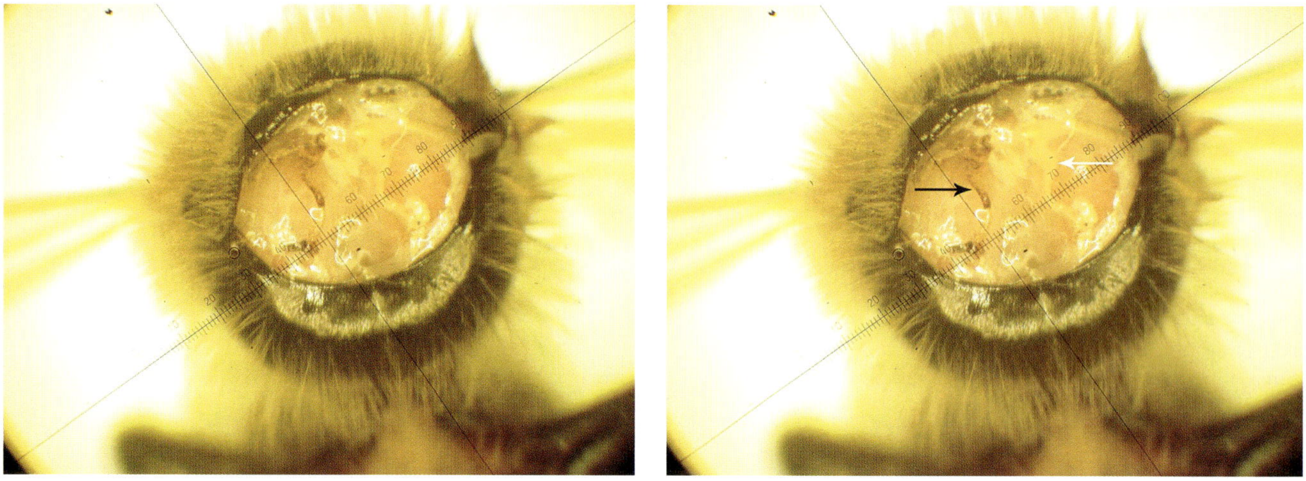

Figure 24.1 A dissected honey bee showing a blotchy, darkened discolouration to the trachea on the left (black arrow). The trachea on the right (white arrow) appears normal. *Source:* Photo courtesy of Don Hopkins.

considered a moderate problem (Ellis 2016). However, in colonies that are dually infested with both varroa and tracheal mites there appears to be a synergistic interaction resulting in significantly greater colony mortality when compared to colonies infested with only one of the mite species (Downey and Winston 2001).

Differential diagnoses include viral infections (Chronic Bee Paralysis Virus and Acute Bee Paralysis Virus), nosemosis and poisoning (Vidal-Naquet 2015). A diagnostic sample of 50 adult bees should be collected and, ideally, should consist of older bees and those showing clinical signs. The best time of year to collect samples for diagnosis of tracheal mites is in winter or early spring when the mite population is expected to be at its highest (Sammataro et al. 2013; OIE 2019; Vidal-Naquet 2015). Diagnosis is made by dissecting the thorax to reveal the tracheae which are then examined under a dissecting microscope (OIE 2019; Vidal-Naquet 2015). The mites can be found unilaterally or bilaterally, and since the tracheal networks on each side do not communicate, each bee represents two tracheal samples. Dissection is easiest on fresh or frozen bees, but the sample can also be preserved in 70% alcohol (Sammataro et al. 2013). To dissect a bee for examination, the bee is pinned in two places through the thorax. It is a good idea to remove the abdomen first so that digestive contents do not contaminate the field. Next, the head and the first pair of legs are removed, followed by the collar. The tracheae can then be visualized (Sammataro et al. 2013). Normal trachea are white and transparent as shown by the white arrow in Figure 24.1. The trachea turn opaque and a acquire a darkened, blotchy discoloration with infestation, as shown by the black arrow in Figure 24.1. The mites themselves can also be visualized within the trachea (Figure 24.2) (OIE 2019; Pernal and Clay 2013). This examination can be performed in the field or at a laboratory. Molecular tests are available for diagnosis of tracheal mites as well (Sammataro et al. 2013).

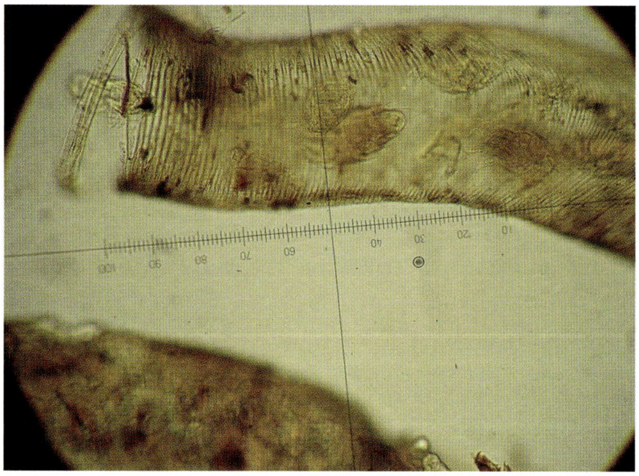

Figure 24.2 Tracheal mites can be visualized within the trachea of a honey bee viewed under the microscope. *Source:* Photo courtesy of Don Hopkins.

A good method of control of tracheal mites is to keep resistant bee stock, such as the Buckfast, Russian, or Ontario-mite resistant bees (Pernal and Clay 2013; Sammataro et al. 2013; Vidal-Naquet 2015). Another technique is the use of grease patties, made by mixing two parts granulated sugar into one part vegetable shortening. These are then placed onto the top bars of the center of the brood nest, ideally in the fall or early spring. The grease patties keep the prevalence of mites below the typical economical threshold by disrupting the ability of the mite to transfer from one host to another (Sammataro et al. 2013;

Pernal and Clay 2013; OIE 2019; Ellis 2016). Alternatively, methanol crystals can be used which work by producing vapors that enter the trachea of the bee and kill the mites. Temperature is important if using methanol crystals as it must be warm enough (>20 °C/68 F) to cause a sufficient rate of evaporation but not so hot (>30 °C/84 F) that the bees are driven outside of their hive by the amount of vapor produced (Pernal and Clay 2013; Sammataro et al. 2013; OIE 2019). Oftentimes methods, such as synthetic acaricides, that are meant to control varroa also have the effect of controlling tracheal mites (Ellis 2016; Vidal-Naquet 2015; Sammataro et al. 2013). Furthermore, the location of the apiary can influence the development of tracheal mite infestation, though the environmental factors responsible for this effect are largely unknown. It is important to keep good records, and apiary locations that experience heavy infestations should be avoided in the future (Pernal and Clay 2013).

Wax Moths

There are two wax moths that affect honey bee colonies – the Greater Wax moth (*Galleria mellonella*) and the Lesser Wax moth (*Achrola grisella*). Both have a worldwide distribution; however, the Greater Wax moth is more common and causes more destruction than the Lesser Wax moth (Ellis et al. 2013; Vidal-Naquet 2015). Adult wax moths are a dull grayish color (Figure 24.3) and are nocturnal, with mated females gaining entrance to the hive at night. Adult moths do not cause any damage. The female will lay hundreds of eggs in cracks or crevices within the hive. The egg hatches into an off-white 1–3 mm long larva which grows up to 20 mm long (Figure 24.4). It is in this stage of the life cycle that the wax moth causes destruction. The larvae feed, not so much on wax, which has a low nutritive value, but on deposits of pollen, cocoons remaining in the cell from bee brood, and feces from the bee larvae. Larvae thus prefer dark brood comb over comb used for honey storage (Ellis et al. 2013; Ritter 2014; Pernal and Clay 2013; Weast 2016). The greater wax moth larvae feed along the midrib – or base of the honeycomb cells. As they tunnel along, they leave a trail of silk webbing that protects them from being removed by worker bees (Pernal and Clay 2013; Ellis et al. 2013). This webbing can result in galleriasis – a condition in which the newly developed adult bees will become trapped by the silken threads on eclosure and be unable to emerge (Ellis et al. 2013). In contrast, the lesser wax moth tunnels just beneath the cappings, causing the cells to become uncapped. This condition is known as bald brood and can be differentiated from worker hygienic behavior as it has a linear pattern that follows the path of the larvae as it tunnels (Ellis et al. 2013). Greater wax moths can also cause damage to the wooden ware of the hives as they chew boat-shaped gouges in which to spin their cocoons (Figure 24.5) (Ellis et al. 2013; Pernal and Clay 2013). The cocoons of the wax moths are hard, white and have a leather consistency (Figure 24.5) (Weast 2016).

Figure 24.3 Adult wax moth and cocoons on a frame. *Source:* Photo courtesy of Jeffrey R. Applegate, Jr.

Figure 24.4 Wax moth larvae. *Source:* Photo courtesy of Jeffrey R. Applegate, Jr.

Figure 24.5 Wax moth cocoons on the top frames of a hive. Note the gouges in the wooden ware. *Source:* Photo courtesy of Jeffrey R. Applegate, Jr.

Following the pupal stage, adults emerge and leave the hive to start the life cycle again (Pernal and Clay 2013).

Both the greater and lesser wax moth can cause extensive damage (Figure 24.6) in colonies weakened by a primary stressor such as: being queenless, viral, bacterial or fungal disease, parasitism, or exposure to pesticides (Pernal and Clay 2013; Delaplane 2018; Ellis et al. 2013; Vidal-Naquet 2015).

The best way to control wax moths in active hives is to have strong, healthy colonies (Ellis 2016; Vidal-Naquet 2015). Hygienic bees may have a further advantage as they are more resistant to the primary stressors and also remove more of the debris that is attractive to the wax moths (Pernal and Clay 2013). Wax debris and scrapings should be removed by the beekeeper from the floor of the hive and the surrounding area regularly (Vidal-Naquet 2015; Pernal and Clay 2013).

Control measures against wax moths must be taken with stored equipment. Before storage, all honey and pollen should be removed and dark brood comb frames should be separated from other comb. There are two options for physical control of wax moth infestation on stored equipment. The first option is to use air flow and light, which are conditions the wax moth will avoid. The stored combs are placed into supers that are stacked one on top of the other, at 90° angles, to a maximum height of 2 m. This allows air and light to penetrate the boxes and can be aided by the addition of an oscillating fan and/or constant light. This method can work well for comb that has not contained brood (Ellis et al. 2013; Vidal-Naquet 2015). The second method for physical control is to freeze supers containing comb or to freeze individual combs. Freezing at −7 °C (20 F) for 4.5 hours, −12 °C (10 F) for 3 hours, or − 15 °C (5 F) for 2 hours is sufficient to eliminate all life stages of the wax moth (Pernal and Clay 2013; Ellis et al. 2013; Weast 2016). Following the freezing treatment, it is important to ensure all comb and equipment is dry prior to storage, to prevent mold growth (Ellis et al. 2013).

Figure 24.6 A heavy wax moth infestation leads to extensive damage to the frames and woodenware. *Source:* Photo courtesy of Jeffrey R. Applegate, Jr.

Other methods of control for stored equipment includes biological control methods, such as with *Bacillus thuringiensis,* which produces a biopesticide that kills the wax moths. This is not available commercially in North America at the time of publication (Pernal and Clay 2013; Ritter 2014; Vidal-Naquet 2015). Chemical control methods should not be used so as to avoid the risk of residues in the wax as well as health risks for the applicant (Ritter 2014; Vidal-Naquet 2015).

Hive Beetles

Small Hive Beetle

The small hive beetle (SHB), *Aethina tumida,* is a pest of honey bee colonies native to sub-Saharan Africa (Neumann et al. 2013; OIE 2019). In 1996 small hive beetle was detected in the United States, followed a few years later by Australia. It has now also been found in Canada, parts of Central and South America, Egypt, Italy, Korea and the Philippines (OIE 2019; Ellis and Ellis 2010; Neumann et al. 2013).

Eggs are about 2/3 the size of honey-bee eggs with a similar pearly-white coloration. These are typically laid in clutches in cracks or crevices but can be found throughout the hive, including on the comb or within capped brood (Neumann et al. 2013; OIE 2019). Larvae are a creamy-white color and freely move on, and within, the combs, feeding on honey, pollen and bee brood (Figure 24.7). Once mature, the larvae – now called wandering larvae – leave the hive to search for suitable substrate in which to pupate (Neumann et al. 2013; OIE 2019). The pupae begin as pearly-white but darken during development within

Figure 24.7 Small Hive Beetle larvae. *Source:* Photo courtesy of Don Hopkins.

individual pupation chambers in soil (Neumann et al. 2013). Adult beetles, upon emergence, will fly to infest new colonies, and are attracted by volatiles released by worker honey bees (Torto et al. 2010).

Adult small hive beetles are oval in shape and smaller than adult honey bees, measuring 5–7 mm in length and 3–4.5 mm in width. They are a reddish-brown color upon emergence but darken to a dark brown or black color as they mature. The antennae have a club shape and the elytra, or hardened forewings, are short, so that some of the abdomen is visible (Figure 24.8). Adults can be found anywhere within the hive but show a preference for the bottom board. They tend to avoid sunlight and will hide in corners and beneath material (Neumann et al. 2013). Honey bees will aggressively chase small hive beetles into confinement sites or "prisons," with Cape honeybee colonies encapsulating the beetles within propolis. These prisons are then strategically guarded by worker bees to prevent escape (Neumann et al. 2001, 2013). Interestingly, the small hive beetle will mimic honey bee behavior to induce the trophallactic feeding from bees during their confinement (Ellis et al. 2002; OIE 2019).

The damage caused by small hive beetle can be extensive and is due in large part to the feeding of adults and larvae on honey, pollen and bee brood. During feeding, the wax comb is destroyed and honey will ferment, likely due to the presence of particular yeasts associated with small hive beetles, making it unfit for human consumption (Ellis and Ellis 2010). Colonies with infestations may abscond due to the presence of a large number of adult beetles in the hive (Ellis et al. 2003). The amount of sealed brood within an infested colony will be reduced. This is from the feeding behavior of the larvae, but also potentially because of brood

Figure 24.8 Adult small hive beetle. Note the club shaped antennae and the shortened elytra. *Source:* Photo courtesy of Paul van Westendorp.

abortion, in which the larvae and pupae are pulled from the comb and discarded outside of the colony by the bees as they prepare to abscond (Ellis et al. 2003). Warm climates with high humidity are more likely to have problems with small hive beetles such that the adults will infest even strong colonies (OIE 2019). Small hive beetles can also be a vector for honeybee pathogens such as *Paenibacillus larvae* (the etiological agent of American Foulbrood), Deformed Wing Virus (DWV) and Sacbrood virus (Schafer et al. 2010; Eyer et al. 2009a,b). Small hive beetles can also be problematic in honey houses where they breed and feed on the comb, honey and wax stored in these facilities (OIE 2019; Ellis and Ellis 2010).

Diagnosis of small hive beetles begins in the field with direct visualization of the adults, eggs, larvae or pupae. The adults can be difficult to spot as they avoid sunlight and will hide during inspection. They may be seen running for cover when the hive is opened. There may be a rotten smell to the hive from fermentation of the honey as well as death of the brood. Often there are smear trails or slime left within or outside the hive by the wandering larvae. There may be perforated cappings from the adult female ovipositing within the capped cells – these can be opened to examine for eggs. The soil around the colony can be sifted to look for pupae or pupal chambers. Traps can also aid in the detection of small hive beetles (see below). Any suspect specimens should be killed in either 70% ethanol or by storing at −20 °C overnight, then sent to a diagnostic laboratory for confirmation by morphological identification and possibly polymerase chain reaction (PCR). This is particularly important in areas where small hive beetles are not endemic (OIE 2019).

It is important to inspect shipments of imported queens, worker bees and colonies for small hive beetles in order to prevent the establishment of this pest in non-endemic areas (Neumann et al. 2013). Furthermore, sentinel apiaries can be used in areas at risk for introduction of this invasive pest (OIE 2019).

One method of control is to use traps, of which several devices exist. Which trap to use, and hence where to place it, depends on climactic conditions, as small hive beetles are more likely to be on the bottom board or periphery during warm weather and with the cluster of bees during cold weather. Due to the size difference between the small hive beetle and honey bees, the entrances of the traps will allow the beetles to enter while excluding honey bees. Most of the traps contain mineral or vegetable oil which will kill small hive beetles; some traps will also include bait to increase the efficacy. A trap to capture the wandering larvae is also available. Diagnostic strips made of corrugated plastic, which provides tunnels for the beetles to hide, can be placed directly on the bottom board to either diagnose an infestation or to screen for small hive beetle in areas not known to have this pest (OIE 2019; Neumann et al. 2013). One non-chemical method showing great promise for monitoring, and potentially as a method of control, is to place unscented non-woven dry-cleaning cloths within the hive. The tarsi of the adult beetles become trapped within the fibers of the cloth, allowing for easy detection and removal (R. Johnson, personal communication).

In addition to traps, some chemical control products are available – ones used as a ground drench around colonies to target the pupae and ones used in the colony to target the adults. Important cultural control methods for the honey house include keeping the area clean of honey, comb and cappings, extracting supers quickly, and reducing the relative humidity to 50% or less to prevent the small hive beetle eggs from hatching. In the apiary, keeping strong colonies of hygienic bees can help reduce the damage caused by SHB (Ellis and Ellis 2010).

Large Hive Beetle

There are two species of large hive beetle that are a known pest of honey bee colonies – *Oplostomus fuligineus* and *Oplostomus haroldi*. Both are scarab beetles found throughout sub-Saharan Africa (Oldroyd and Allsopp 2017). The beetles are, as the common name suggests, large in size – measuring 20–23 mm in length (BeeAware 2019). *O. fuligineus* has an all-black body with orange ends of the antennae. *O. haroldi* is similar in appearance but the body coloration can vary between all-black, or black with either reddish-brown or orange stripes (Oldroyd and Allsopp 2017). Eggs are laid in cattle or horse dung, where they hatch, become larvae and pupate (Oldroyd and Allsopp 2017). The emerging adults are attracted by volatiles from worker honey bees and will enter hives to feed on brood, pollen, and honey. Adults will remain in the hive, feeding for 30 or more days, with the females leaving after mating to search for suitable sites for oviposition (Torto et al. 2010; Oldroyd and Allsopp 2017). Large hive beetles can cause significant damage to a hive by consuming the brood and destroying the comb. The large size of these pests – being larger than *Apis mellifera* – makes control possible by using a beetle barrier across the entrance to restrict access to the hive (Oldroyd and Allsopp 2017; BeeAware 2019). Although this pest is currently restricted to countries in Southern Africa, there is potential for spread to other countries, particularly on agricultural equipment and contaminated soil (Oldroyd and Allsopp 2017). Therefore, it is imperative that beekeepers and those working with honey bees, such as veterinarians, be on the lookout for this pest and notify the appropriate authorities immediately if they suspect a case.

Tropilaelaps

There are four species of this mite – *Tropilaelaps clareae, Tropilaelaps mercedesae, Tropilaelaps koenigerum, Tropilaelaps thaii* – all of which are natural brood parasites of giant Asian honey bees. Two of these mites, *T. clareae* and *T. mercedesae*, have been able to parasitize *A. mellifera* (OIE 2019; Pernal and Clay 2013; Vidal-Naquet 2015; Chantawannakul et al. 2018). Currently, all species are confined to Asia, however, there is concern this parasite, and in particular *T. mercedesae*, could spread worldwide due to beekeeping practices and international trade (Chantawannakul et al. 2018; Anderson and Roberts 2013; Pernal and Clay 2013).

These mites, like *Varroa destructor*, are small, reddish-brown in color and feed on developing bees (Figure 24.9). However, there are a number of differences. Tropilaelaps are smaller than *V. destructor*, measuring 1.0 mm long by 0.6 mm wide, and have an elongated shape as opposed to the crab-like shape of *Varroa* (Ritter 2014; OIE 2019; Anderson and Roberts 2013). In addition, they are capable of very quick movement and can be seen darting about the comb, in contrast to the slow-moving nature of *Varroa* (Anderson and Roberts 2013; Vidal-Naquet 2015; OIE 2019). Furthermore, *Tropilaelaps* mouth parts are

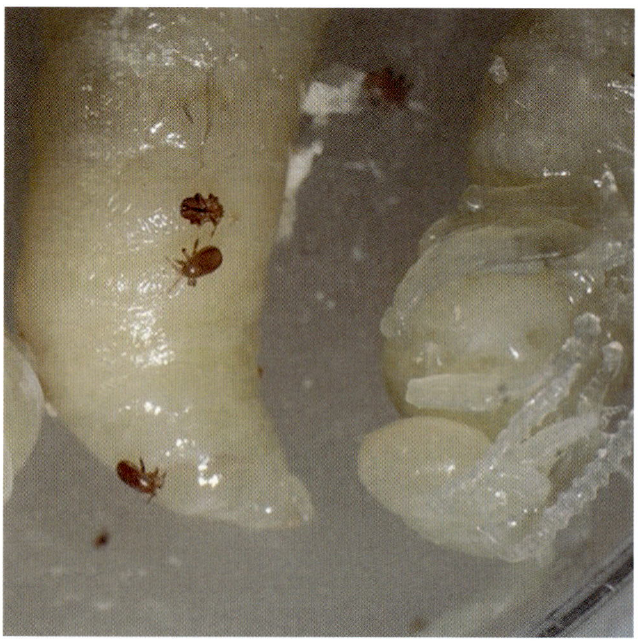

Figure 24.9 *Tropilaelaps mites* on a developing bee. Source: Photo courtesy of Samuel Ramsey.

unable to pierce the adult bee integument to feed. Therefore, the phoretic phase (if it is indeed truly phoretic) is very short and consists of climbing onto adult bees for the purpose of transportation within the colony and, potentially, to other colonies through drifting and robbing behaviors (Anderson and Roberts 2013; Vidal-Naquet 2015; Ritter 2014; OIE 2019; Chantawannakul et al. 2018; Pernal and Clay 2013).

In some ways, *Tropilaelaps* life cycle is similar to *Varroa* in that a mated female will enter a cell prior to capping and lay eggs, which then develop into adults beneath the cappings (Chantawannakul et al. 2018; Anderson and Roberts 2013; Vidal-Naquet 2015; Ritter 2014). Each mother founder will typically lay three to four eggs, with a ratio of several females to one male (OIE 2019; Ritter 2014; Vidal-Naquet 2015; Anderson and Roberts 2013). The total period of development to the adult stage is six days, which is significantly shorter than *Varroa* (Ritter 2014; OIE 2019; Anderson and Roberts 2013; Chantawannakul et al. 2018; Vidal-Naquet 2015). Once the adult bee ecloses (emerges from the cell), all of the adult mites exit the cell and can run across the comb to enter other cells, continuing their reproduction (Anderson and Roberts 2013; Ritter 2014; Vidal-Naquet 2015). These mites are also able to mate outside of the brood cell (Chantawannakul et al. 2018; Anderson and Roberts 2013; Vidal-Naquet 2015).

It is the short phoretic phase coupled with the faster development that makes *Tropilaelaps* more virulent than *Varroa* to a colony (OIE 2019; Pernal and Clay 2013; Chantawannakul et al. 2018).

Tropilaelaps mites feed not only on the capped brood, as *Varroa* does, but also on pre-capped brood. This ability to feed on the developing larvae may increase the survival time of the individual mites, which is potentially important in colder environments that experience limited brood rearing in winter (Phokasem et al. 2019). Also, in contrast to the feeding behavior of *Varroa*, *Tropilaelaps* mites will inflict multiple small wounds through which they feed (Phokasem et al. 2019; De Guzman et al. 2017). Consequences of this indiscriminate feeding behavior can be observed on the adult bees, which may develop with injured antennae, wings, abdomens, proboscis, and legs arising from the feeding sites. These injuries may have significant effects on the adult bees' ability to carry out tasks both inside and outside the hive (Phokasem et al. 2019).

The clinical signs of *Tropilaelaps* infestation at the colony level include an irregular brood pattern, adult bees with deformities and shrunken abdomens (Vidal-Naquet 2015; Ritter 2014; OIE 2019; Chantawannakul et al. 2018). *Tropilaelaps* parasitism has been linked to infection with DWV (Forsgren et al. 2009; Phokasem et al. 2019). In addition to DWV, Black queen cell virus has also been identified in honey bees infested with these mites (De Guzman et al. 2017; Phokasem et al. 2019). Adults that do emerge after being parasitized during development can be significantly lighter in weight and have shortened lifespans (Phokasem et al. 2019; Vidal-Naquet 2015; Ritter 2014; OIE 2019; Chantawannakul et al. 2018).

The diagnosis of this parasite can be done by direct visualization of the mites running across the comb or by inspecting the capped brood. An estimation of infestation rate can be determined by examining a predetermined number of capped brood cells and calculating the percent of brood cells that contained live mites (OIE 2019; Ritter 2014; De Guzman et al. 2017). Early infestation can be detected using the bump method, in which a frame of capped brood is bumped four times over a light colored paper or pan to knock the mites loose. This can be aided by perforating the cappings with a capping scratcher before bumping onto a dusting of powdered sugar to keep the mites from escaping (De Guzman et al. 2017). Mite collection and quantification can be done using the same techniques as are used for *Varroa*, such as the alcohol wash or sugar roll (OIE 2019; Anderson and Roberts 2013; De Guzman et al. 2017). A sticky board placed at the bottom of the hive and covered by a mesh or screen can also be used to check for the presence of dead mites (Ritter 2014; OIE 2019; De Guzman et al. 2017). In countries where *Tropilaelaps* species are not known to inhabit, it is important to get a confirmed diagnosis quickly and to alert the proper authorities in order to implement the appropriate measures for foreign animal pests. Diagnosis is best done by morphological identification at an official laboratory and can be confirmed by molecular identification by PCR. Samples that are collected should be killed prior to submission by 70% ethanol or stored overnight at −20 °C (OIE 2019).

The similarities between *Tropilaelaps* species and *Varroa* have led many to adapt management and treatment regimens for *Tropilaelaps* based on those for *Varroa*. The differences in their life cycle, however, may be significant when it comes to effectiveness of treatment (Chantawannakul et al. 2018). For example, a study done comparing the efficacy of different treatments traditionally used on *Varroa* found that while formic acid, powdered sulfur and the cultural control method of making a colony division worked well, hops acids and amitraz did not. It is proposed that the latter two treatments need contact between adult bees and the mite in order to be effective and this does not happen on a significant enough level to produce a positive clinical outcome due to the short phoretic phase of *Tropilaelaps* (Pettis et al. 2017).

Breaking the brood cycle works well to control this mite as the mite only feeds on brood. When no brood is available, the mites will starve to death (Ritter 2014; De Guzman et al. 2017). This attribute is thought to be beneficial in climates with cold winters that have an extended period of no brood rearing, as this may prevent this parasite from becoming established in these regions (Pernal and Clay 2013). However, *Tropilaelaps* has become established in South Korea, which has a temperate climate with limited brood rearing in the winter (De Guzman et al. 2017). Furthermore, it could become a problem in any country where brood rearing is year-round (Anderson and Roberts 2013). Since *Tropilaelaps* is a potential threat not only to the keeping of *A. mellifera* in Asia, but to the global beekeeping world, and because it has the potential to be much more devastating than *Varroa*, some experts are calling for more focused research to be done on *Tropilaelaps* (Anderson and Roberts 2013; Chantawannakul et al. 2018).

Asian Hornets

Asian Yellow-Legged Hornet

The Asian yellow-legged hornet, or *Vespa velutina,* is a predator of honey bees native to tropical and subtropical regions of Asia. One subspecies, *Vespa velutina nigrithorax*, has spread outside of its native range, being first observed in the Republic of Korea in 2003, and in France in 2004 where it appears to have been accidentally introduced through international trade of goods that harbored overwintering mated queens from China. Subsequently, these hornets have been found in Spain in 2010; Portugal and Belgium in 2011; Italy in 2012; Japan in 2012; Germany in 2014; Great Britain in 2016; Switzerland, the Netherlands, and Scotland in 2017. The introduction into Japan appears to be by trade with the Republic of Korea, however entry into the other European countries is believed to have occurred naturally by spread from France (Espinosa et al. 2019; Laurino et al. 2020). It is believed that it is no longer possible to eradicate this species from Europe (Chauzat et al. 2015).

Vespa velutina is noticeably larger than *A. mellifera*, measuring 17–32 mm in length. It has a yellow–orange face on a black head, with a dark brown to black thorax. The abdominal segments are brown and are bordered across the dorsum with a narrow band of yellow, however the fourth segment is a bright yellow–orange color. The legs are brown and yellow, giving rise to the common name of yellow-legged hornet (Espinosa et al. 2019).

In the temperate regions of the world where it is found, this hornet has an annual life cycle which begins every spring with a mated foundress queen building a small primary nest where she lays eggs that develop into female workers. As the colony population increases, the primary nest will be abandoned and a secondary nest with a spherical to pear shape will be built in an aerial location, most commonly in a tree canopy. At its peak population in early autumn, a single colony can contain a couple thousand workers and produce hundreds of new queens and drones. The new queens and drones will leave the nest to mate. The newly mated queens will then find a protected location to hibernate over the winter, until they can emerge in the spring to start the life cycle again. The remainder of the colony will die in the autumn or early winter (Espinosa et al. 2019; Laurino et al. 2020).

The developing brood requires a protein source, which is provided by the workers preying upon other arthropods, including honey bees, and carrion. When hunting honey bees, *V. velutina* will hover 20–40 cm from the entrance of a honey bee hive, in a behavior known as hawking, and will catch foraging bees in flight. It will then carry its prey to a nearby tree branch, where the hornet will hang down by its back legs and remove the honey bee's head, legs, wings and abdomen, leaving only the thorax which is carried back to the nest to feed to the larvae. A single hornet has the potential to catch up to 25–50 bees per day which, over time, can have a significant impact on the honey bee population. In colonies with a decreased population, the hornets can invade the hive to rob larvae and honey stores (Espinosa et al. 2019; Chauzat et al. 2015). Furthermore, the appearance of these hornets close to the entrance of a hive may cause stress and decrease foraging activity, which could in turn lead to insufficient stores with which to overwinter. *V. velutina nigrithorax* predation can lead to the death of a honey bee colony (Espinosa et al. 2019; Laurino et al. 2020).

Another potential concern with this hornet is that a number of the honey bee viruses, namely sacbrood virus, black queen cell virus, and DWV, and to a lesser extent chronic bee paralysis virus and acute bee paralysis virus, have been detected in *V. velutina*. The significance of this is unknown; however, these hornets could be a potential reservoir, or may even play a role in the dissemination, of these honey bee viruses (Chauzat et al. 2015). *Apis cerena*, which has co-evolved with *V. velutina*, appears to have a variety of defense mechanisms, such as heat-balling where the bees will engulf a hornet and raise the temperature through vibration of their bodies to a point that is lethal to the hornet but not the honey bees. *A. cerena* will also use stop signaling which keeps the honey bees inside the nest, will increase guarding, and will modify flying behavior. These mechanisms are not observed, at least not effectively, in *A. mellifera* (Chauzat et al. 2015; Espinosa et al. 2019;

Laurino et al. 2020; Tan et al. 2016). These hornets not only pose a significant threat to the beekeeping industry, but also to pollination and biodiversity within an ecosystem (Espinosa et al. 2019; Laurino et al. 2020).

Due to globalization, international trade, and climate change, this honey bee predator could spread to other parts of the world. North America, in particular regions along both the Eastern and Western coasts, could be suitable environments. Monitoring and early detection of this pest is very important. Although traps can be used for monitoring purposes, they are not selective for *V. velutina* and may have a negative impact on native insect populations. Unfortunately, there are currently no useful control measures to prevent the spread of this invasive species; however, multiple control strategies should be combined to limit the impact. The most effective method of control is to destroy the nests, particularly if done before the next generation of queens is produced. However, locating the nests can be difficult as they are often hidden within the tree canopy. There are no current organisms that havc been identified as suitable and effective candidates for biological control (Laurino et al. 2020; Espinosa et al. 2019).

Asian Giant Hornet

Another species of *Vespa* that poses a significant threat to the beekeeping industry is the Asian Giant Hornet *Vespa mandarinia*. This hornet rightfully deserves its common name as it is the largest hornet in the world, measuring 50 mm in length with a wingspan of 76 mm and a 6 mm long sting. The orange colored head is also wide by comparison to other hornets (Figure 24.10) (McCaffrey and Walker 2012; Kozak and Otis 2020). The life cycle of *V. mandarinia* is very similar to that of *V. velutina* discussed above, with the main difference being the nesting habit. *V. mandarinia* nests in subterranean cavities and the primary nest site selected by the foundress will be used throughout the season. The workers will extend the nest cavity by removing small balls of soil with their mouths and dropping these just outside of the entrance, creating a platform of soil that may be used as an indicator that a nest is present (Matsuura and Sakagami 1973).

Figure 24.10 *Vespa mandarinia. Source:* Illustration by Patrick D. Wilson.

The Asian Giant Hornet preys on honey bees in late summer and autumn as a source of protein for workers to feed the developing larvae. Unlike *V. velutina*, these hornets are not as skilled at catching bees in flight and will tend to sit near the hive entrance or on the hive itself. Similarly to *V. velutina*, *V. mandarinia* will prepare a meatball from the bee's thorax to take back to the hive. Sometimes these hornets will change their behavior from the typical hunting to slaughtering a hive. During this later behavior, an attack will be focused on one hive by anywhere from 3 to 50 hornets and the attack will not stop until the hornets occupy the hive. Tens of thousands of bees may be killed in a matter of hours, with the hornets biting the bees to death and leaving the corpses on the ground in front of the hive rather than carrying them back to the nest as in the hunting phase. The slaughter will stop when there are no more bees defending the hive, at which point the hornets enter and will remove the pupa first, and later the larvae, to take back to their nest. Some of the hornets that occupy the hive will guard the entrance and display territorially defensive behavior (Matsurra and Sakagami 1973; Kozak and Otis 2020).

Vespa mandarinia is native to temperate Asia, however it was found for the first time outside of its native range in 2019 where it was confirmed in British Columbia, Canada. In September of that year, two independent sightings were reported on Vancouver Island and a subsequent search of the area located a nest which was destroyed. Two months later, in November, a specimen was confirmed to be *V. mandarinia* on the coastal mainland, close to the United States border (Kozak and Otis 2020). Shortly thereafter, in early December of 2019, a specimen was collected in Blaine, Washington and confirmed to be *V. mandarinia* (Penner 2019).

Ants, Wasps and Hornets, and Robbing

Ants

Ants are very common and, although most of the time they are of no consequence to the bees, at times they can become persistent pests within apiaries and honey houses. Depending on the species of ant present they may eat dead bees and brood, eat nectar and honey stores, and make

Figure 24.11 Ants feeding on a honey bee colony. *Source:* Photo courtesy of Don Hopkins.

Figure 24.12 Entrance reducers will decrease the area that the colony needs to defend itself against and are particularly useful in the fall when there is increased activity by wasps and hornets at honey bee colonies. *Source:* Photo courtesy of Britteny Kyle.

nests within hives and stored equipment (Figure 24.11). Weak hives in particular are susceptible to predation by some ant species and keeping colonies strong is the best defense. Regardless of the species and the type of damage they can cause, exclusion of ants is the preferred method of control. This can be achieved by using commercially available, or home-made, ant stands that use either a moat or a sticky substance as a barrier to trap the ants, thus preventing them from entering the hive. It should be noted that the use of axle grease as the sticky substance is not recommended because of concerns regarding environmental contamination. It is also very important to keep vegetation below the barrier, through mowing or other weed control methods, as otherwise the ants will simply use the vegetation as a bridge to bypass the barrier (Kern 2017).

Wasps and Hornets

Social wasps of the family *Vespidae*, which include yellow jackets, European hornets, Bald-faced hornets and paper wasps, can be a threat to honey bee colonies, particularly in late summer and early autumn (Repasky 2018). In contrast, solitary wasps are not problematic for beekeepers (Magnini 2015). Social wasps are annual nesters with the queen finding a suitable nest site in early spring, where she begins construction of the paper nest and starts to lay eggs. Once the eggs have developed into adults, they are the workers and will take over the nest building, foraging and feeding of larvae (Repasky 2018). Larvae are fed meat protein, while workers sustain themselves on carbohydrates in the form of sugar. As summer dwindles down, the colony begins to decline and the worker wasps need only to forage for carbohydrates and will often do so aggressively. It is at this time that honey bee colonies are most at risk as the wasps will attack in order to rob honey and nectar from the hives. As is true with many other pests of honey bee colonies, maintaining strong colonies that can defend themselves is the best defense. Further protection can be offered by reducing the hive entrance and therefore the space the guard bees need to defend (Figure 24.12) (Repasky 2018; Magnini 2015). Furthermore, although wasps are mainly a problem toward the end of the beekeeping season, one should start looking to prevent them in early spring before the wasp queen has established a nest. Commercial traps are available that use a pheromone to attract the social wasps. If large nests are found near an apiary it may be beneficial to contact a licensed pest control operator to have the nest removed (Repasky 2018).

Robbing

Robbing behavior is the collection of nectar and honey from another colony. This behavior can be between colonies within an apiary, or between apiaries, and can quickly escalate into a robbing frenzy in which multiple colonies in an apiary are being robbed. When resources are good, bees are not likely to engage in robbing behavior, however, during times of nectar dearth strong colonies will invade weaker colonies to steal the nectar and honey stores. Robbing in itself, as well as significant drifting that results from robbing, can transfer pests and diseases between colonies. When robbing is occurring, there will be an increase in the number of bees flying around and attacking at the hive entrance, as well as bees searching for other ways into

Figure 24.13 Note the increased activity around the hive entrance with bees fighting in the foreground during this robbing event. Entrances, cracks and crevices are closed as much as possible in an attempt to stop the robbing. *Source:* Photo courtesy of Britteny Kyle.

the hive – clustering around cracks and joints of the hive (Figure 24.13). Robbing will continue until the resources are depleted, which can occur over hours or days. Newly installed and weak colonies are especially at risk. Beekeeper activity can also trigger robbing behavior by working colonies during dearths and having hives opened for prolonged periods of time. During a robbing event, any hives being worked should be closed, openings in vulnerable hives should be plugged, entrances of colonies being robbed should be reduced, and there have been anecdotal reports of stopping robbing behavior by running an overhead sprinkler to encourage bees to return to their hives. If all else fails, a colony being attacked should be moved to a new location. To prevent robbing from occurring in the first place, ensure all colonies within an apiary are spaced apart, are equal in strength, entrances are reduced during dearths, colonies are worked quickly and supers kept covered, only one colony is worked at a time, the apiary is kept clean of wax debris, honey and sugar spills are cleaned up quickly, all cracks are sealed, and equipment is in good condition. Furthermore, at times when robbing behavior is likely to occur, colonies should be worked later in the day to reduce the amount of time the bees have to carry out robbing behavior (Willingham et al. 2014; Ellis 2015). Colonies that have been robbed may abscond or starve to death. Dead Outs should be examined for signs of robbing, which can include empty cells that have jagged edges and discarded cappings. It is important to note that while the robbing may have resulted in the death of the colony, there is likely an underlying cause that made the colony weak and thus susceptible to robbing behavior (Sebestyen 2019).

Bears, Raccoons, Skunks and Mice

Bears

Bears can cause a large amount of destruction to an apiary in a short amount of time, and although most species of bear will attack beehives, it is the American Black Bear (*Euarctos americanus*) that does most of the damage (Ritter 2014; Pernal and Clay 2013). Despite popular belief, it is not the honey they are after – though they will eat it – but the bee brood which provides a rich source of protein (Ellis 2016). The attack usually happens in the evening or at night, typically destroying one to three colonies each night and returning until the entire apiary has been destroyed (University of Georgia n.d.). Broken frames and equipment strewn about the apiary are suggestive of predation by bears. In areas known to be inhabited by bears, apiaries should be located away from woodlots, berry patches, and bodies of water and should be kept clean of debris (Pernal and Clay 2013). The best defense is an electric fence put in place before bears have visited the apiary. It is very important to maintain the fence in good condition, making sure there is no vegetation touching the wires and grounding it. The fence should be inspected and the voltage checked every time the apiary is visited (Spivak and Reuter 2016; Pernal and Clay 2013). Once a bear has visited an apiary, the electric fence will unlikely be enough to deter them and it is advised to move remaining colonies to another location (Spivak and Reuter 2016; University of Georgia n.d.). It is important to be aware of governing regulations regarding bear control and removal which can differ by state or province. Furthermore, some state or provincial programs may offer compensation to beekeepers who have experienced damage due to bears (Pernal and Clay 2013; Ontario Ministry of Agriculture, Food and Rural Affairs 2019).

Raccoons

Raccoons are similar to bears in that they visit during the night to eat honey and brood. They can also cause a large amount of destruction as large adults can topple over a small hive in order to access the contents. But, unlike bears, raccoons tend to be a problem for beekeepers in urban and suburban environments (Pernal and Clay 2013; Sollenberger 2011). It is important to ensure apiaries are kept clean and stored equipment is housed in secure buildings to reduce the attractiveness to raccoons. Bricks or heavy stones can be placed on top of hives to make them more difficult to push over (Figure 24.14). Electric fencing can also be used (Pernal and Clay 2013).

Skunks

Skunks will visit apiaries during the night where they will scratch at the bottom board of a hive, catching and eating the guard bees that respond (Spivak and Reuter 2016; Mangum 2018; Sollenberger 2011). The population of the colony can be negatively affected from a skunk feeding night after night (Spivak and Reuter 2016; Sollenberger 2011). Evidence of a skunk problem can include agitated bees, scratches in the wooden equipment at the entrance of the hive, and raked grass around the entrance (Spivak and Reuter 2016). There may also be small, shiny, and dark teaspoon sized balls on the ground that contain the exoskeletons of the bees which the skunk has chewed and spat out (Mangum 2018). The easiest prevention is to raise hives off the ground by at least 12–14 in. (Figure 24.14) (Sollenberger 2011; Mangum 2018). Alternatively, an electric fence, with the lowest wire about 3 in. off the ground, can be used (Spivak and Reuter 2016).

Mice

Mice enter hives or stored equipment in late fall and winter to build nests and to eat pollen, honey and dead bees. In order to make room for themselves, they will chew through frames and wax comb, causing considerable destruction (Figure 24.15) (Spivak and Reuter 2016; University of Georgia n.d.; Ellis 2016). Furthermore, their urine and droppings can make the hive unsuitable to the bees, who will in turn abandon the hive. It is recommended to reduce the size of the entrance to the hive starting in early fall and continuing throughout winter to 3/8–1/2 an inch. Stored equipment can be protected by sealing cracks, closing entrances, and placing boxes on top of queen excluders or closed bottom boards (Spivak and Reuter 2016; University of Georgia n.d.).

Non-Harmful Hive Commensals

From time to time other arthropods, such as the bee louse, spiders, cockroaches, and earwigs will be seen inside the hive. These insects have little to no impact on the health or productivity of the colony and do not cause economically significant damage. Therefore, no steps need to be taken for prevention or control (PennState Extension 2017).

Figure 24.14 A heavy rock is placed on top of the lid to prevent raccoons from opening or tipping the hive. The hive is placed on top of a hive stand to raise it off the ground and prevent predation by skunks. *Source:* Photo courtesy of Britteny Kyle.

Figure 24.15 Mice have chewed through the wax frames to make room for a nest. *Source:* Photo courtesy of Jeffrey R. Applegate, Jr.

References

Anderson, D.L. and Roberts, J.M.K. (2013). Standard methods for *Tropilaelaps* mites research. *Journal of Apiculture Research* 52 (4): 1–6. https://doi.org/10.3896/IBRA.1.52.4.21.

BeeAware (2019). *Large Hive Beetle*. http://beeaware.org.au/archive-pest/large-hive-beetle/#ad-image-0 (accessed June 8, 2019).

Chantawannakul, P., Ramsey, S., vanEngelsdorp, D. et al. (2018). *Tropilaelaps mite*: an emerging threat to European honey bee. *Current Opinion in Insect Science* 26: 69–75. https://doi.org/10.1016/j.cois.2018.01.012.

Chauzat, M.P., Ribière-Chabert M., Schurr F., et al. (2015). First detection of honey bee pathogens in nest of the Asian hornet (*Vespa velutina*) collected in France. *Watch letter No. 33,* CIHEAM.

De Guzman, L.I., Williams, G.R., Khongphinitbunjong, K. et al. (2017). Ecology, life history, and management of *Tropilaelaps mites*. *Journal of Economic Entomology* 110 (2): 319–332. https://doi.org/10.1093/jee/tow304.

Delaplane, K. (2018). For the love of bees and beekeeping. *American Bee Journal* 158 (7): 823–827.

Downey, D.L. and Winston, M.L. (2001). Honey bee colony mortality and productivity with single and dual infestations of parasitic mite species. *Apidologie* 32: 567–575.

Ellis, J. (2015). Field guide to beekeeping – inspecting your newly installed colonies for the first time. *American Bee Journal* 155 (5): 509–512.

Ellis, J. (2016). Field guide to beekeeping – biotic stressors of honey bee colonies. *American Bee Journal* 156 (7): 761–766.

Ellis, J.D. and Ellis, A (2010). Featured Creatures: small hive beetle. http://entnemdept.ufl.edu/creatures/misc/bees/small_hive_beetle.htm (accessed 8 June 2019).

Ellis, J.D., Pirk, C.W.W., Hepburn, H.R. et al. (2002). Small hive beetles survive in honeybee prisons by behavioural mimicry. *Naturwissenschaften* 89: 326–328. https://doi.org/10.1007/s00114-002-0326-y.

Ellis, J.D., Hepburn, R., Delaplane, K.S. et al. (2003). The effects of adult small hive beetles, *Aethina tumida* (Coleoptera: Nitidulidae), on nests and flight activity of Cape and European honey bees (*Apis mellifera*). *Apidologie* 34: 399–408. https://doi.org/10.1051/apido:2003038.

Ellis, J.D., Graham, J.R., and Mortensen, A. (2013). Standard methods for wax moth research. In: *The COLOSS BEEBOOK*, Volume II: Standard Methods for *Apis mellifera* Pest and Pathogen Research (eds. V. Dietemann, J.D. Ellis and P. Neuman). *Journal of Apicultural Research* 52(1): https://doi.org/10.3896/IBRA.1.52.1.10.

Espinosa, L., Franco, S., and Chauzat, M.P. (2019). Could *Vespa velutina nigrithorax* be included in the world organisation for animal health list of diesases, infections and infestations? *Scientific and Technical Review* 38 (3).

Eyer, M., Chen, Y.P., Schafer, M.O. et al. (2009a). Small hive beetle, *Aethina tumida*, as a potential biological vector of honeybee viruses. *Apidologie* 40: 419–428. https://doi.org/10.1051/apido:2008051.

Eyer, M., Chen, Y.P., Schafer, M.O. et al. (2009b). Honey bee sacbrood virus infects adult small hive beetles, *Aethina tumida* (Coleoptera: Nitidulidae). *Journal of Apiculture Research and Bee World* 48 (4): 296–297. https://doi.org/10.3896/IBRA.1.48.4.11.

Forsgren, E., de Miranda, J.R., Isaksson, M. et al. (2009). Deformed wing virus associated with *Tropilaelaps mercedesae* infesting European honey bees (*Apis mellifera*). *Experimental and Applied Acarology* 47: 87–97. https://doi.org/10.1007/s10493-008-9204-4.

Kern, W.H (2017). *Ant Control in the Apiary*. http://edis.ifas.ufl.edu/in1181 (accessed 10 June 2019).

Kozak, P. and Otis, G.W. (2020). From the province new honey bee pests in North America: Asian hornets reported and confirmed in British Columbia. *Ontario Bee Journal* 39 (1): 20–21.

Laurino, D., Lioy, S., Carisio, L. et al. (2020). *Vespa velutina*: an alien driver of honey bee colony losses. *Diversity* 12 (1) https://doi.org/10.3390/d12010005.

Magnini, R.M. (2015). My vespa war or the predatory nature of wasps. *American Bee Journal* 155 (5): 547–549.

Mangum, W.A. (2018). Bees and beekeeping present and past: some interesting signs of spring. *American Bee Journal* 158 (3): 315–317.

McCaffrey, S., Walker, K (2012) *Asian Giant Hornet (Vespa mandarinia)* Available online: PaDIL - www.padil.gov.au (accessed 26 January 2020).

Neumann, P., Pirk, C.W.W., Hepburn, H.R. et al. (2001). Social encapsulation of beetle parasites by Cape honeybee colonies (*Apis mellifera capensis* Esch.). *Naturwissenschaften* 88: 214–216. https://doi.org/10.1007/s001140100224.

Neumann, P., Evans, J.D., Pettis, J.S. et al. (2013). Standard methods for small hive beetle research. In: *The COLOSS BEEBOOK*, Volume II: Standard Methods for *Apis mellifera* Pest and Pathogen Research (eds. V. Dietemann, J.D. Ellis and P. Neumann). Journal of Apicultural Research 52(4): https://doi.org/10.3896/IBRA.1.52.4.19.

OIE (2019). *Manual of Diagnostic Tests and Vaccines for Terrestrial Animals 2018. Section 3.2.* http://www.oie.int/standard-setting/terrestrial-manual/access-online (accessed May 12, 2109).

Oldroyd, B.P. and Allsopp, M.H. (2017). Risk assessment for large African hive beetles (*Oplostomus* spp.) – a review.

Apidologie 48: 495–503. https://doi.org/10.1007/s13592-017-0493-7.

Ontario Ministry of Agriculture, Food, and Rural Affairs (2019). *Wildlife damage to Bee Colonies, Bee Hives & Bee Hive Related Equipment*. www.omafra.gov.on.ca/english/food/inspection/bees/wildlifedamage.htm (accessed 9 June 2019).

Penner, D (2019). Asian giant hornet spotted near Blaine, close to B.C. border. *Vancouver Sun* (23 December).

PennState Extension (2017). *A Quick Reference Guide to Honey Bee Parasites, Pests, Predators, and Diseases*. https://extension.psu.edu/a-quick-reference-guide-to-honey-bee-parasites-pests-predators-and-diseases (accessed 27 June 2019).

Pernal, S.F. and Clay, H. (eds.) (2013). *Honey Bee Diseases and Pests*, 3e. Alberta, Canada: Canadian Association Professional Apiculturists.

Pettis, J.S., Rose, R., and Chaimanee, V. (2017). Chemical and cultural control of *Tropilaelaps mercedease* mites in honeybee (*Apis mellifera*) colonies in northern Thailand. *PLoS ONE* 12 (11): e0188063. https://doi.org/10.1371/journal.pone.0188063.

Phokasem, P., de Guzman, L.I., Khongphinitbunjong, K. et al. (2019). Feeding by *Tropilaelaps mercedesae* on pre- and post-capped brood increases damage to *Apis mellifera* colonies. *Scientific Reports* 9: 13044. https://doi.org/10.1038/s41598-019-49662-4.

Repasky, S. (2018). When yellow jackets attack! *American Bee Journal* 158 (12): 1377–1380.

Ritter, W. (ed.) (2014). *Bee Health and Veterinarians*. Paris, France: World Organisation for Animal Health.

Sammataro, D., de Guzman, L., George, S. et al. (2013). Standard methods for tracheal mite research. *Journal of Apiculture Research* 52 (4): 1–20. https://doi.org/10.3896/IBRA.1.52.4.20.

Schafer, M.O., Ritter, W., Pettis, J. et al. (2010). Small hive beetles, *Aethina tumida*, are vectors of *Paenibacillus larvae*. *Apidologie* 41: 14–20. https://doi.org/10.1051/apido/2009037.

Sebestyen, T. (2019). Beekeeping basics – diagnosing a dead-out. *American Bee Journal* 159 (5): 269–272.

Sollenberger, T. (2011). Urban beekeeping survival guide. *American Bee Journal* 151 (5): 455–458.

Spivak, M., Reuter, G.S (2016). *Honey Bee Diseases and Pests*. https://www.beelab.umn.edu/sites/beelab.umn.edu/files/_2016_disease_pdf_version_s.pdf (accessed 10 June 2019).

Tan, K., Dong, S., Li, X. et al. (2016). Honey bee inhibitory signaling is tuned to threat severity and can act as a colony alarm signal. *PLOS Biology* 14 (3): e1002423. https://doi.org/10.1371/journal.pbio.1002423.

Torto, B., Fombong, A.T., Mutyambai, D.M. et al. (2010). *Aethina tumida* (Coleoptera: Nitidulidae) and *Oplostomus haroldi* (Coleoptera: Scarabaeidae): occurrence in Kenya, distribution within honey bee colonies, and responses to host Odors. *Annals of the Entomological Society of America* 103 (3): 389–396. https://doi.org/10.1603/AN09136.

University of Georgia (n.d.). *Bees, Beekeeping & Pollination – Non-infectious Diseases and Pests*. https://bees.caes.uga.edu/bees-beekeeping-pollination/honey-bee-disorders/honey-bee-disorders-non-infectious-diseases-and-pests.html (accessed 10 June 2019).

Vidal-Naquet, N. (2015). *Honeybee Veterinary Medicine: Apis mellifera L*. Sheffield, UK: 5m Publishing.

Weast, R. (2016). The wax moth: the bane of beekeepers. *American Bee Journal* 156 (1): 51–54.

Willingham, R., Klopchin, J., Ellis, J (2014). *Robbing Behaviour in Honey Bees*. https://edis.ifas.ufl.edu/in1064 (accessed 16 June 2019).

Matsuura, M. and Sakagami, S.F. (1973). A Bionomic Sketch of the Giant Hornet, Vespa mandarinia, a Serious Pest for Japanese Apiculture. *Journal of the Faculty of Science, Hokkaido University. Series 6, Zoology* 19 (1): 125–162.

25

Pesticides

Reed M. Johnson

Department of Entomology, The Ohio State University, Wooster, OH, USA

Categorizing Pesticides

In the United States the term pesticide is defined in law by the Federal Insecticide, Fungicide and Rodenticide Act (FIFRA) and refers to any substance or combination of substances deployed with the intention of "preventing, destroying, repelling, or mitigating any pest" (FIFRA Section 2(u), 7 U.S.C. Section 136(u)). This is a very broad definition and can include synthetic chemicals, substances derived from natural sources, living biological control agents and genes capable of controlling pests that are incorporated into genetically engineered organisms. The broad range of pesticides is subdivided into categories depending on the type of organism the pesticide is intended to control: insecticides are applied for controlling insect pests, fungicides for controlling fungal pathogens and herbicides for controlling weeds (Table 25.1).

Within each category of pesticides there have been a range of specific chemical compounds, the active ingredients, formulated with specific activity against a target pest. Early in the history of pesticide development, these substances were chosen solely for their effectiveness; the physiological mechanism through which they interact with target organisms was generally unknown. However, long-term repeated use of a single active ingredient to control the same pest resulted in strong selection pressure for the evolution of resistance against that particular pesticide. Pesticide resistance spurred the development of new active ingredients, but new compounds were not necessarily more effective against existing resistant pest populations. The observation of cross-resistance spurred research into the specific enzymes and receptors perturbed by pesticides, i.e. the mode of action. To address the issue of cross-resistance, all active ingredients are now categorized by their modes of action by industry trade groups (Insecticide Resistance Action Committee, https://www.irac-online.org; Fungicide Resistance Action Committee; https://www.frac.info). Posters and materials published by these resistance action committees are the best resource in determining the mode of action for a particular pesticide.

Active ingredients are not marketed for direct use but are, instead, incorporated into formulations containing inert ingredients added to improve shelf stability, solubility in water, and other handling and application characteristics. Additionally, multiple active ingredients may be combined into a single formulation. These formulated products are given a trade name for marketing purposes.

Regulation of Pesticides

Each formulated product carries a label listing the recommended application rate for particular crops as well as health and safety information to protect the applicator, consumers, and the environment. Labels are written by the companies marketing the product but the label must be approved by the United States Environmental Protection Agency (EPA). Application of a pesticide in violation of the label guidelines is illegal and pesticide applicators may be prosecuted if they are found to have violated label precautions. While the federal EPA performs the risk assessment and approves pesticide labels and regulations, enforcement of pesticide laws is delegated to the states and legal action following a violation will be conducted by state-level departments of agriculture (http://npic.orst.edu/reg/state_agencies.html).

Protections for honey bees are included on pesticide labels under the "Environmental Hazards" statement (Table 25.1, Figure 25.1). These statements serve the function of risk mitigation to reduce the harm that a pesticide application will have on bees and often include

Honey Bee Medicine for the Veterinary Practitioner, First Edition. Edited by Terry Ryan Kane and Cynthia M. Faux.

Table 25.1 Example pests and pesticides, bee toxicity and label language.

Pesticide category	Intended target	Class	Mode of action	Active indredient	Topical LD_{50} (μg/bee)[a]	Trade name	Bee guidance from environmental hazard statement on the product label
Insecticides	Insects	Pyrethroids	Sodium channel modulators	Bifenthrin	0.0146	Brigade 2EC	This product is highly toxic to bees exposed to direct treatment or residues on blooming crops or weeds. Do not apply this product or allow it to drift to blooming crops or weeds while bees are actively visiting the treatment area.
				Lambda-cyhalothrin	0.098	Warrior II	This product is highly toxic to bees exposed to direct treatment or residues on blooming crops or weeds. Do not apply this product or allow it to drift to blooming crops or weeds if bees are visiting the treatment area.
				Tau-fluvalinate	6.75	Mavrik Aquaflow	This product is toxic to honey bees if bees are exposed to direct application.However, dried residues of this product are non-toxic to honey bees. Treat during non-foraging periods to minimize adverse effects.
		Neonicotinoids	Nicotinic acetylcholine receptor competitive modulators	Imidacloprid	0.0439	Admire Pro	This product is highly toxic to bees exposed to direct treatment or residues on blooming crops or weeds. Do not apply this product or allow it to drift to blooming crops or weeds if bees are foraging the treatment area.
				Thimethoxam	0.024	Actara	This pesticide is highly toxic to bees exposed to direct treatment on blooming crops/plants or weeds. Do not apply this product or allow it to drift to blooming crops/plants or weeds while bees are foraging in/or adjacent to the treatment area.
				Acetamiprid	7.07	Assail 70WP	This product is toxic to bees exposed to direct treatment. Do not apply this product while bees are foraging in the treatment area.
		Organo-phosphates		Chlorpyrifos	0.114	Lorsban 50 W	Highly TOXIC to bees exposed to direct treatment, drift, or residues on blooming plants. Do not use on flowering crops or weeds.
Fungicides	Fungi	Demethylation inhibitors	Sterol biosynthesis in membranes	Propiconazole	>25	Tilt	None
				Fenbuconazole	>292	Indar 2F	None
		Pyridine-carboxamides	Respiration (succinate dehydrogenase inhibitors)	Boscalid	>166	Pristine	None
		Methoxy-carbamates	Respiration (quinone outside inhibitors)	Pyraclostrobin	>100	Pristine	None
Herbicides	Weeds	Glycines	Inhibition of 5-enolpyruvylshikimate-3-phosphate synthase	Glyposate	>100	Roundup Pro	None
		Benzoic acids	Synthetic auxins	Dicamba	>90.65	Clarity	None
		Phenoxy-carboxylic acids	Synthetic auxins	2,4-D	32.26	Freelexx	None

[a] Median lethal dose data taken from EPA ECOTOX Knowledgebase (https://cfpub.epa.gov/ecotox).

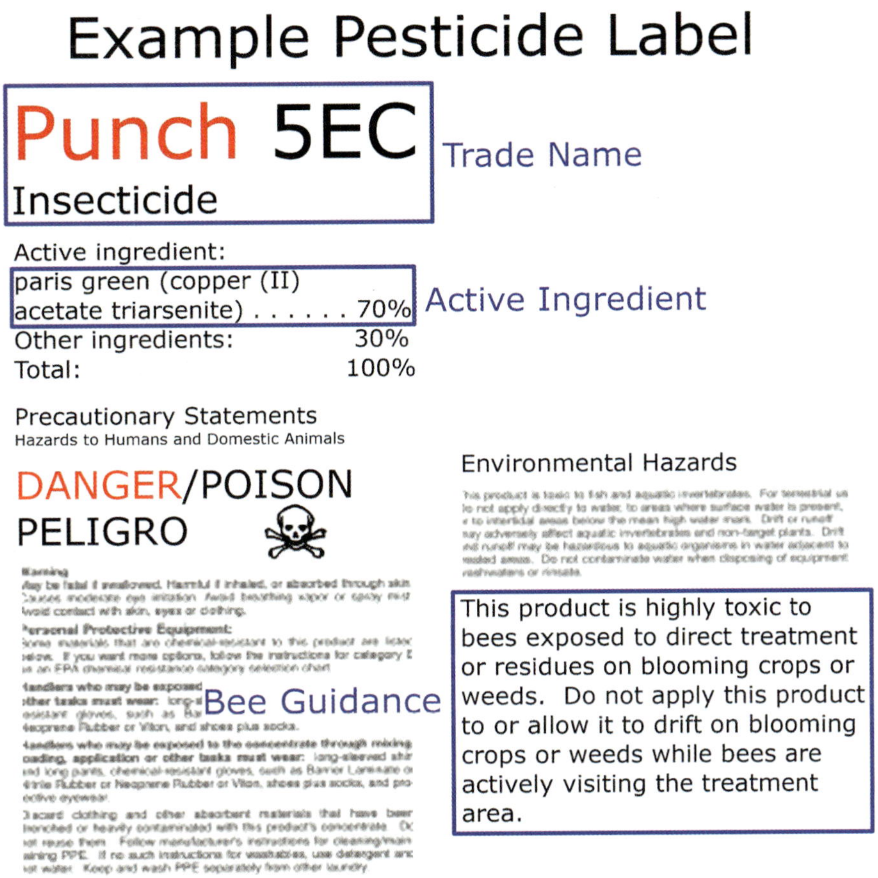

Figure 25.1 Mock pesticide label with bee guidance in the Environmental Hazards statement.

specific prohibitions against application during the daytime hours when bees are actively foraging, or to bee-attractive blooming crops or weeds. Beekeepers often criticize these statements because their interpretation depends on the applicator's subjective assessment of bee foraging activity. Labels for formulated products and the protections they afford bees can be found in compendia of labels, such as the Crop Data Management System's Label Database (http://www.cdms.net/Label-Database).

The EPA derives its authority to regulate pesticide use to protect bees based on the FIFRA's statement that pesticide applications should not have "unreasonable adverse effects on the environment." However, this protection for bees is not all-encompassing and requires accounting for both the "economic, social and environmental costs and benefits" of pesticide use. This implies that economic benefits derived from pesticide use may, in certain circumstances, outweigh harm to bees. In fact, depending on state laws, pesticide applications with known toxicity to bees may be applied to blooming bee-attractive crops if beekeepers are notified in advance of the application.

Determining the Toxicity of Pesticides to Bees

Many peer-reviewed studies have been published by scientists addressing the potential harm pesticides cause to individual bees and whole colonies. Effects studied range from outright mortality of individual bees, (traditionally determined using median lethal dose [LD_{50}] studies), to sublethal effects on bee behavior, learning, and susceptibility to disease (Desneux et al. 2007). Synergistic effects, the combinatorial effect of exposure to multiple pesticides applied together, or the effects of nutritional or disease status on pesticide susceptibility, have also been addressed in these studies. Experiments on individual bees are relatively straightforward to conduct and have been accomplished with a variety of compounds and combinations on adult worker bees, queens, drones, and larvae (Johnson 2015). However, individual-level experiments do not consider the complex social interactions between bees in a colony. Studies using whole colonies are much more challenging to conduct given the high variability observed between colonies and the potential for landscape, disease, and management practices to also affect colony success. Nonetheless, researchers have conducted field-level experiments where pesticides are applied to a bee-attractive crop and the effects on colony success are determined – though in many cases the effects of field-application of an insecticide are less than would have been expected based on individual-level testing (Woodcock et al. 2017; Osterman et al. 2019). The apparent resiliency of the honey bee colony may be due to the dilution of pesticides with uncontaminated food collected over their extensive foraging range, the partitioning of exposure away from the most vulnerable life stages, or through other unknown mechanisms (Sponsler and Johnson 2017).

While the EPA may take into account published scientific studies during the registration or re-registration of pesticides, the variability in experimental design and methods often make these studies difficult to use for risk assessment and, ultimately, in determining the risk mitigation measures that appear on the pesticide label. Instead, the pesticide manufacturers must generate standardized toxicological data to support the risk assessment process. Since the 1970's bee toxicity has been primarily determined based on tests of lethality using varying concentrations of an active ingredient applied to the thoracic notum of adult worker bees (Figure 25.2). Mortality data, collected up to 96 hours after application, is used to generate a dose-response curve and derive a topical or oral LD_{50} (Table 25.1). These LD_{50} values for all pesticide active ingredients are available through the EPA's ECOTOX Knowledgebase (https://cfpub.epa.gov/ecotox).

These tests are simple to conduct and provide useful quantitative data that can be used to compare the sensitivity of bees to different active ingredients and categorize their toxicity: highly toxic ($LD_{50} \leq 2$ µg/bee), toxic ($LD_{50} > 2$ and ≤ 12 µg/bee) and relatively non-toxic ($LD_{50} > 12$ µg/bee). While it is no surprise that most insecticides are

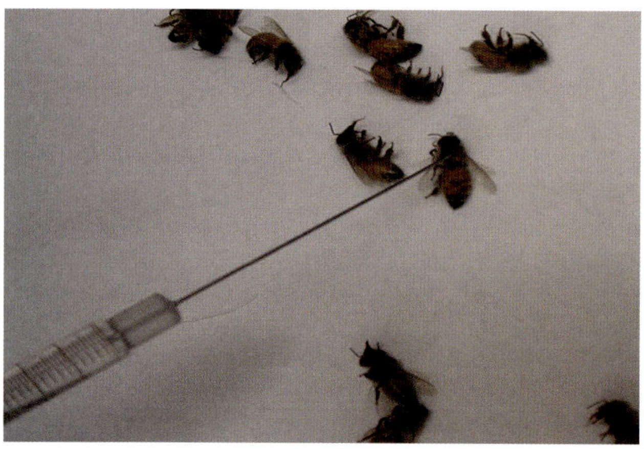

Figure 25.2 Topical application of pesticides to individual honey bees to determine median lethal dose (LD_{50}). *Source:* Photo courtesy of Reed Johnson.

categorized as toxic or highly toxic, there is considerable variability among active ingredients within each insecticide class. Even among the neonicotinoid class of insecticides, with their reputation for bee toxicity, the active ingredient acetamiprid is of relatively low toxicity. Fungicides and herbicides, which are not intended to kill insects, are almost always categorized as relatively non-toxic. Products with active ingredients deemed relatively non-toxic generally do not carry a cautionary statement limiting bee exposure on the pesticide label. Inert ingredients, which are added to improve the handling and application characteristics of a formulated product, are not commonly tested for bee safety. However, some adjuvants, which are added to the tank and mixed with formulated products for application, have been associated with toxic effects on bees (Fine et al. 2017).

Following reports of Colony Collapse Disorder in 2007, and the suspected role of pesticides in causing colony failures, the EPA revised regulatory testing requirements to formalize a tiered testing approach. Tier I consists of individual-level tests, including oral and topical LD_{50}'s as well as a 10 day chronic feeding test. Methods have been developed to rear honey bee larvae *in vitro* and test the effect of active ingredients delivered through the larval diet (Schmehl et al. 2016) (Figure 25.3). Bee exposure is determined by estimating the concentration of active ingredients in pollen and nectar collected from treated plants. Colony-level exposure is predicted from residue studies using a mathematical model named BEEREX (Office of Pesticide Programs 2015). If uncertainties remain following individual-level testing, then a pesticide may undergo Tier II or semi-field testing where a formulated pesticide is applied to a blooming crop inside a tent and effects are determined on the functioning of a small colony restricted to foraging within the tent. Full field studies, where a pesticide is applied to an attractive blooming crop and full-sized colonies forage freely, constitutes Tier III testing. While newly registered or re-registered pesticides have undergone this updated testing regime, many pesticides on the market were registered when only limited testing on adult bees was required.

Many pesticides are applied through a spray application and have relatively short residual toxicity, generally hours or a few days. Residual toxicity is measured by spraying alfalfa and determining the number of hours after application that less than 25% of bees exposed to treated foliage (RT_{25}) will die (https://www.epa.gov/pollinator-protection/residual-time-25-bee-mortality-rt25-data) However, some pesticides are water soluble and have systemic activity, most notably the neonicotinoids. Systemic pesticides may be applied to soil, taken up by plant roots and may appear in pollen and nectar. This presents a drastically different route of exposure and bees

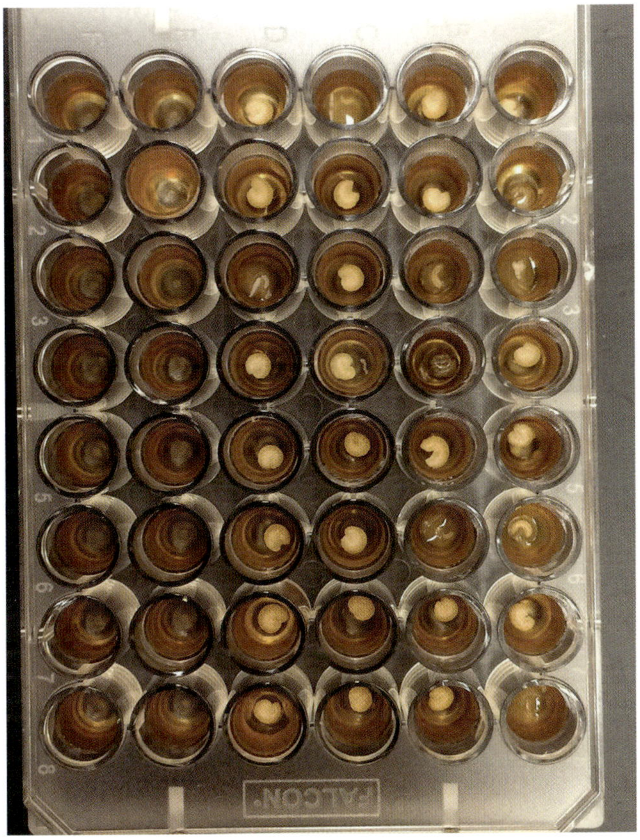

Figure 25.3 Honey bee larvae reared *in vitro* for pesticide testing. *Source:* Photo courtesy of Andrea Wade.

may encounter systemic pesticides well after the application event occurs. The updated regulatory testing addresses oral exposure to systemics, but there continue to be unknown factors related to the movement of systemics into pollen and nectar in different plant species.

Preparing for Expected Pesticide Exposure

Pesticide applications with the potential to harm bees may occur legally if the beekeeper is notified in advance. While laws vary by state, pesticide applicators planning on applying an insecticide that carries cautionary language regarding bees on the label must generally find and notify beekeepers within a radius of ½–1 mile of the treatment site at least 24 hours before application to a bee-attractive crop in bloom. However, identifying apiaries and beekeepers can be a challenge for applicators. The best advice for beekeepers is to be proactive and to talk to neighbors to let them know about the presence of their bees. Beekeepers should also post their contact information prominently in apiaries. Applicators can contact their state departments of agriculture, which often keep lists of apiary locations, to identify bees in the area. Beekeepers can also register the locations of their apiaries with online tools such as BeeCheck (https://beecheck.org), which is currently available in 22 states, where pesticide applicators can search online for apiaries within foraging range of their planned pesticide application.

While there is some tension between pesticide applicators and beekeepers, the majority of applicators are willing to contact and work with beekeepers to protect bee colonies beyond what is required by law. States have developed Managed Pollinator Protection Plans (https://honeybeehealthcoalition.org/managed-pollinator-protection-plan-mp3-resources) to facilitate communication between farmers, pesticide applicators, and beekeepers. Mitigating risk to colonies can be as easy as following label guidance and making pesticide applications late in the day, or even at night, after nectar and pollen have been harvested from flowers and the bees have returned to the colony. Early morning applications may also be protective for honey bees but are likely to harm wild bees that may begin foraging at or before dawn.

If application timing cannot be arranged to protect bees, then a beekeeper must decide what measures, if any, to take to reduce bee exposure (Atkins 1992). Relocating colonies is the most protective action, but it is often logistically challenging to move a large number of colonies on short notice. Where applications occur with predictability, it may be possible for a beekeeper to simply keep their bees out of an area during the period when applications are likely. Alternatively, beekeepers may confine their bees to the colony during and immediately following the pesticide application through the use of an entrance screen, affixed at night or in the early morning, to keep all foragers within the hive. However, confinement has the potential to be much more damaging to a colony than pesticide exposure if the colony suffers overheating from lack of ventilation. A screened lid should be used to provide ventilation and colonies can be confined with a wet burlap sack or sheet to provide some cooling during confinement. Bees can generally be released from confinement at night or the morning following pesticide application since, for many crops and weeds, new uncontaminated flowers will open the next day.

Identifying and Reporting a Suspected Pesticide-Related Bee-Kill

The classic signs of a pesticide-related bee-kill are an accumulation of dead and dying bees around the hive entrance and on the bottom board. Dead bees may be foragers that perished after returning to the colony, but it is more likely that the dead bees observed are younger nurse

Figure 25.4 Dead bees at the entrance of colonies following a moderate bee-kill event observed in May 2016 resulting from exposure to neonicotinoid-laden corn seed treatment dust. All colonies in the apiary were affected and dead bees were found to contain seed-treatment insecticides clothianidin and thiamethoxam. *Source:* Photo courtesy of Reed Johnson.

bees that were poisoned through ingestion of contaminated pollen and nectar. In many cases all colonies in an apiary will be showing similar signs of distress, though it is possible that only a subset of colonies that were foraging on treated flowers will display an obvious accumulation of dead bees. Moderate cases of poisoning may only result in an increased number of dead and dying bees found on the ground in front of colonies (Figure 25.4). If hives are placed in grass it is likely that even a substantial increase in the number of dead bees in front of a colony will go unnoticed as the bees are easily hidden in the vegetation. To better identify bee-kill events it is recommended that beekeepers place colonies on gravel or concrete or, use plywood, cardboard or old carpeting in front of their hives to improve the visibility of dead bees. Dead bee traps, constructed from 2″ ×6″ lumber with window screen stapled on the bottom and ½″ hardware cloth on the top, may be used for quantitative dead bee counts (Figure 25.5).

Figure 25.5 Dead bee traps used for quantifying dead bees resulting from a bee-kill event. *Source:* Photo courtesy of Reed Johnson.

Different classes of insecticides may produce different effects in the dying bees observed around the entrance, though these cannot be relied upon as diagnostic. Pyrethroids are generally fast-acting and may cause excitation and twichiness in affected bees. Organophosphates are slower-acting and dead bees may have a greasy appearance. Neonicotinoid poisoning can cause lethargy and disorientation. It should be noted that foragers exposed to a toxic pesticide away from the hive may simply not return to the colony. While this can result in depopulation of the foraging force for a colony, it is preferable to having these foragers return to the hive where the pesticide will be distributed to younger nestmates. This is one reason why the pyrethroids, with their fast action, may not be as harmful to colonies as their very low LD_{50}'s would suggest.

Varroa infestations, and the associated viruses, are also capable of causing sudden mass die-offs of adult bees in colonies and may also produce greasy-looking or twitching dead bees that can easily be confused with a pesticide-related kill. Dead bees from colonies with severe deformed wing virus (DWV) infections will often have characteristic wing deformations. Thus, varroa monitoring and management must be conducted throughout the season to ensure that varroa cannot explain sudden losses. In addition, varroa-related kills generally occur later in the season, often after pesticide use in crops has ceased.

While most insecticides are neurotoxic and result in relatively quick mortality of adult insects, others have modes of action which target insect development. Other pesticides, particularly some fungicides and adjuvants, have also been associated with effects on larval development (Mussen et al. 2004). Regulatory testing conducted prior to 2014 only examined adult bees and developmental effects were generally not identified. Recognizing larval poisoning in colonies can be extremely challenging – particularly given the bees' behavioral propensity for quickly removing sick larvae. Larval effects are most obvious in colonies where the adult population has been drastically reduced through poisoning and there are insufficient nurse bees to feed healthy larvae or clean out dying larvae from their cells (Figure 25.6).

Whenever a pesticide-related bee-kill is suspected, and other signs cannot immediately rule out pesticides as a cause, the first thing to recommend to beekeepers is they contact their state department of agriculture or appropriate state lead agency responsible for enforcement of pesticide laws (http://npic.orst.edu/reg/state_agencies.html). Typically, a state investigator will be dispatched to document the bee-kill, take samples for pesticide residue analysis and to inspect logs of pesticide use for applicators in the surrounding area. Investigators will also ask the beekeeper about their feeding and varroa management practices. In many cases it may take days for an investigator to arrive at the apiary and, in the meantime, it is important to avoid manipulating affected colonies or doing anything that may compromise the integrity of the investigation. However, the beekeeper should take photos or videos of the situation and make detailed notes on the state of the colonies. Following the inspection, the state will issue a report, often weeks or months later, that is sent to the beekeeper and the state may, based on that report, pursue legal action against the pesticide applicator. However, it is relatively rare that pesticide applicators are subject to fines or other penalties – largely because of the vague and subjective nature of the label statements intended to protect bees.

Apart from the state investigation, beekeepers can pursue their own investigation and legal recourse. Contract labs regularly conduct comprehensive pesticide residue screening for over 200 pesticides in bees, wax, pollen, or honey (https://www.ams.usda.gov/services/lab-testing/nsl). However, this service can be prohibitively expensive – up to $500 per sample in the United States, depending on the lab. Standard protocols for sampling are available from the Bee Informed Partnership (https://beeinformed.org/wp-content/uploads/2019/11/Pesticide-sampling-protocol-for-ERK_DTK.pdf). A minimum of 3 g of bees, pollen, honey, or wax must be submitted for pesticide residue analysis. At least 30 dead bees (~3 g) should be collected from the bottom board or 30 live bees from the brood area. Thirty pollen or honey cells should be scraped out of the comb using a flat wooden stick or clean metal spatula. Wax should be scraped from empty cells using a clean hive tool. All materials should be collected into labeled plastic bags or 50 ml centrifuge tubes and stored in darkness, ideally at −20 °C, before overnight shipment to the testing lab.

Finally, beekeepers should report any suspected pesticide incidents directly to the federal EPA (beekill@epa.gov). Since states are responsible for enforcement of pesticide laws, the federal EPA will not conduct an investigation. However, the EPA does highly value incident reports from beekeepers. These reports help improve the risk assessment and risk mitigation process. The Honey Bee Health Coalition maintains a one-page guide to responding to a bee-kill incident that summarizes the key steps and provides contact information relevant for each state (https://honeybeehealthcoalition.org/quick-guide)

(a)

(b)

Figure 25.6 Dead and dying larvae observed in a colony fed the organophosphate insecticide dimethoate (a). Compare with larvae of a similar age from a healthy colony (b). *Source:* Photo courtesy of Reed Johnsons.

Recovering from a Bee-Kill

Experiencing a loss of bees can be devastating for beekeepers, regardless of the cause, but when pesticides are involved, and the loss could have been prevented, the hit comes especially hard. Beekeepers should not become so discouraged that they give up and allow their bees to die from mismanagement. While colonies can be severely affected, and there may be a substantial financial impact if bees were needed to make honey or for pollination, honey bee colonies often have the capacity to recover and should not be immediately given up for lost. There is generally little risk of pesticide exposure for the beekeepers themselves in a bee-kill incident unless the apiary itself has been directly sprayed. If the apiary has been directly sprayed, then the beekeepers should not enter the area within the Re-Entry Interval or REI period, typically 4–24 hours, as listed on the pesticide label.

Without pesticide residue testing it is impossible to know the extent of contamination of wax and stored pollen and nectar or to determine whether harmful levels of residues remain. Many pesticides are lipophilic and will partition into the wax component of the hive where they can remain for an extended period of time (Traynor et al. 2016), though the bioavailability of pesticides in wax is not well understood. Contaminated pollen may be a greater concern than nectar because it may be more highly contaminated and will be consumed more quickly.

Surviving colonies should be fed supplemental sugar syrup and pollen or pollen substitute to stimulate brood rearing and compensate for the loss of bees. Supplemental feeding may also dilute any remaining pesticide in the hives. If colonies survive the bee-kill event and appear to be recovering they should generally be left in their hives with existing equipment. In situations where colonies are severely affected or completely dead, and pesticide exposure is confirmed as the cause, then contaminated frames should be double-bagged and discarded before robber bees from other colonies can find them. Woodenware can generally be salvaged and reused with new frames. Queens and their brood pattern should be watched over the subsequent weeks and queens should be replaced if they appear to have been harmed by the exposure.

References

Atkins, E. (1992). Injury to honey bees by poisoning. In: *The Hive and the Honey Bee*, 1153–1208. Hamilton, IL: Dadant & Sons, Inc.

Desneux, N., Decourtye, A., and Delpuech, J.M. (2007). The sublethal effects of pesticides on beneficial arthropods. *The Annual Review of Entomology* 52: 81–106.

Fine, J.D., Cox-Foster, D.L., and Mullin, C.A. (2017). An inert pesticide adjuvant synergizes viral pathogenicity and mortality in honey bee larvae. *Science Reports* 7: 40499.

Johnson, R.M. (2015). Honey bee toxicology. *The Annual Review of Entomology* 60: 415–434.

Mussen, E.C., Lopez, J.E., and Peng, C.Y.S. (2004). Effects of selected fungicides on growth and development of larval honey bees, Apis mellifera L. (hymenoptera: Apidae). *Environmental Entomology* 33 (5): 1151–1154.

Office of Pesticide Programs US EPA (2015) *BeeREX*.

Osterman, J., Wintermantel, D., Locke, B. et al. (2019). Clothianidin seed-treatment has no detectable negative impact on honeybee colonies and their pathogens. *Nature Communications* 10 (1): 692.

Schmehl, D.R., Tomé, H.V.V., Mortensen, A.N. et al. (2016). Protocol for the in vitro rearing of honey bee (*Apis mellifera* L.) workers. *Journal of Apicultsural Research*.

Sponsler, D.B. and Johnson, R.M. (2017). Mechanistic modeling of pesticide exposure: the missing keystone of honey bee toxicology. *Environmental Toxicology and Chemistry* 36 (4): 871–881.

Traynor, K.S., Pettis, J.S., Tarpy, D.R. et al. (2016). In-hive pesticide exposome: assessing risks to migratory honey bees from in-hive pesticide contamination in the eastern United States. *Scientific Reports* 6: 33207.

Woodcock, B.A., Bullock, J.M., Shore, R.F. et al. (2017). Country-specific effects of neonicotinoid pesticides on honey bees and wild bees. *Science* 356 (6345): 1393–1395.

26

Diagnostic Sampling

Dan Wyns

Department of Entomology, Michigan State University, East Lansing, MI, USA

Inspection for honey bee diseases and pests is a critical part of maintaining healthy colonies. There are many conditions of honey bee colonies that can be diagnosed through visual inspection alone, but others require sampling to quantify the severity of a condition or confirm visual diagnosis. Sampling methods should be standardized in order to provide consistent and comparable results. Most of the sampling methods are lethal to the bees collected in the sample. Killing bees for sampling purposes can be disturbing to many beginning beekeepers, but it is important to remember that the subject being evaluated is the colony as a whole. By sacrificing a very small percentage of the individual bees in the colony, important diagnostic decisions can be made for the entire colony.

Some samples can be collected and processed in the field for an immediate diagnosis. These processes include quantification of varroa mite levels and determination of foulbrood. Other conditions require laboratory work and samples must be collected, stored, and transported under appropriate conditions to remain viable for analysis. Appropriate protocols for collecting, handling, shipping, and processing samples should be followed to ensure sample viability does not deteriorate between collection and analysis. This chapter will provide guidance on sampling for parasitic mites, nosema, and bacterial diseases.

Sampling for Field Analysis

American Foulbrood: Ropy Test

Cells containing prepupae/pupae infected with the bacteria causing American Foulbrood (AFB) will take on a coffee or caramel color and develop a slimy texture, often under sunken, moist, or perforated cappings eventually drying to a hard scale. The ropy test, while not always providing a definitive diagnosis, is a good place to start when encountering a suspected case of AFB. Dead and dying brood from many causes will exhibit discoloration and decomposition, but only brood infected with AFB will rope out as described below.

To perform the ropy test, use a wooden matchstick, coffee stirrer, toothpick, or small twig. The content of individual cells selected for the ropy test should be in a liquid state with partial cappings as described above. Affected brood in a later stage of deterioration will dry to a scale that does not have suitable moisture content for testing. Insert the stick into a cell containing suspect brood and stir gently. Slowly remove the stick from the cell. If the pupa is infected with AFB, it will typically adhere to the stick and be drawn out into a thread in excess of 2 cm (3/4 in.). Several suspected cells should be checked in this manner. A prepupa/pupa that ropes is considered a positive indication of AFB, but failure to rope does not definitively mean AFB is not present.

While the ropy test is a quick and easy field diagnostic, it does have some limitations. Brood in the earliest stage of decomposition may not contain enough of the bacteria to rope. If the contents of the cell have progressed toward the stage of scale formation, there may not be sufficient moisture to allow for roping. It is important to perform the test using a stick that is both dry and somewhat rough so that the material in the cell may adhere to it. A smooth or moist surface, like a grass stem, may not allow the material to stick well and prevent or limit the roping characteristic (Figure 26.1).

American Foulbrood: Holst Milk Test

The Holst milk test (Holst 1946) is a quick field test for diagnosing AFB. This test can be performed using prepupae/pupae that have reached the ropy stage, described above, or

Honey Bee Medicine for the Veterinary Practitioner, First Edition. Edited by Terry Ryan Kane and Cynthia M. Faux.

Figure 26.1 The contents of a cell roping out at least 2 cm indicates a positive diagnosis of American Foulbrood. *Source:* Photo courtesy of Dan Wyns.

progressed further to form scale. This test is based on the proteolytic enzymes created by the bacteria that causes AFB. These enzymes break down the proteins found in milk. In the Holst milk test, suspected decomposing brood or scales are placed in a diluted milk solution. If the sample is positive for AFB, the enzymes will break down the proteins, and the solution will change from opaque to clear brown.

To perform the Holst milk test you will need two transparent sample tubes, match sticks or coffee stirrers, powdered skim milk, and water. Prepare a dilute solution using powdered skim milk and water, shaking to create a uniform solution of approximately 1% concentration. The precise concentration is not critical; it should have a cloudy appearance. Fill each tube with the milk solution, labeling one "control" and one "test." Add suspect material to the tube marked "test." Larvae from the ropy test above may be used as well as scale removed from the bottom of cells. The more infected material added to the test tube, the faster the reaction will occur. The reaction will not be immediate and requires a brief period of incubation. The speed of the reaction may take 10–20 minutes, depending on the concentration of milk solution, amount of infected

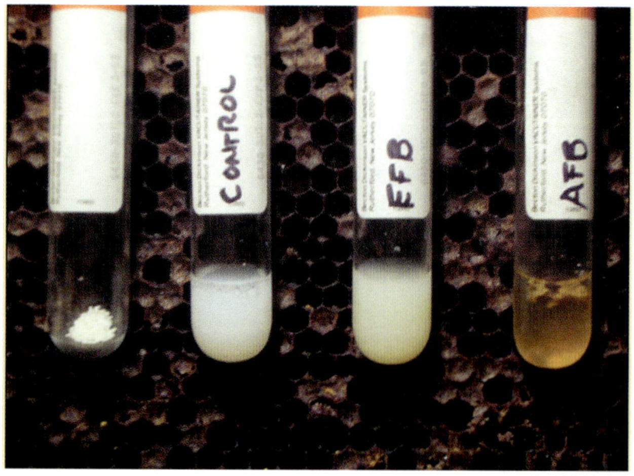

Figure 26.2 Holst Milk Test. *Source:* Photo courtesy of Randy Oliver.

material added, and temperature. Observe the appearance of the test tube compared to the control tube. A positive result for AFB is indicated by a brown transparent solution, while samples without AFB bacteria will remain fully opaque. The "control" tube can be helpful as a point of visual comparison while becoming familiar with the test. As you become experienced performing this test, it may not be necessary to use a control tube for reference (Figure 26.2).

American Foulbrood and European Foulbrood: Diagnostic Test Kits

Diagnostic test kits are commercially available for testing AFB and European foulbrood (EFB). These kits allow for testing of an individual colony for either AFB or EFB in the field and provide results in a matter of minutes. The kits are available through most beekeeping supply companies, and more information can be found at: vita-europe.com/beehealth/products/.

A kit includes instructions and all materials needed for the test. To complete the test:

1) Unpack the contents of the kit and remove a frame with suspect larvae/pupae.
2) Use the small spatula provided with the kit to remove a larva/pupa.
3) Place affected brood into the jar and screw the lid on tightly. Be careful not to spill the jar, as contents are toxic to humans. Do not overload the jar with too much affected brood material, as they may clog test device in subsequent steps.
4) Shake the jar for 20 seconds to mix thoroughly.
5) Remove the test device from the foil packet. Do not touch the rectangular test window or circular sample well.

Figure 26.3 Remove the contents of a suspect cell with the spatula provided in the test kit. *Source:* Photo courtesy of Dan Wyns.

6) Remove a small amount of liquid from the jar using the provided pipette.
7) Gently squeeze several drops from the pipette into a sample well of test device.
8) Read the result of the test. The appearance of the control line (C) indicates the test is working. The appearance of a test line (T) indicates a positive result (Figure 26.3–26.6).

Varroa Mites: Alcohol Wash and Sugar Roll Tests

Varroa mites are a persistent threat to honey bee colony health, and controlling them is fundamental to successful beekeeping. In order to understand when intervention is necessary and to mitigate the negative impacts caused by varroa mites, it is critical that beekeepers monitor mite levels in their colonies. Mite infestation levels are typically quantified as a percentage representing the number of mites per 100 bees. Both the alcohol wash and sugar shake methods below were developed to be performed in the field and provide immediate results so that beekeepers can provide prompt response when needed. For both methods, it is standard to collect a sample of approximately 300 adult bees. This number of bees represents 102 ml, or a little less than ½ cup in volume.

It is generally recommended that at least eight colonies are sampled per apiary (Lee et al. 2010) and management decisions are based on the highest infestation rate in the apiary.

Alcohol Wash Test The alcohol wash is the field method with the most reliable and consistent results, but it is lethal to the bees collected in the sample.

Figure 26.4 Place the test material into the solution container from the kit. *Source:* Photo courtesy of Dan Wyns.

Figure 26.5 After shaking sample bottle for 20 seconds, use the provided pipette to put a drop of solution in the round well on the test strip. *Source:* Photo courtesy of Dan Wyns.

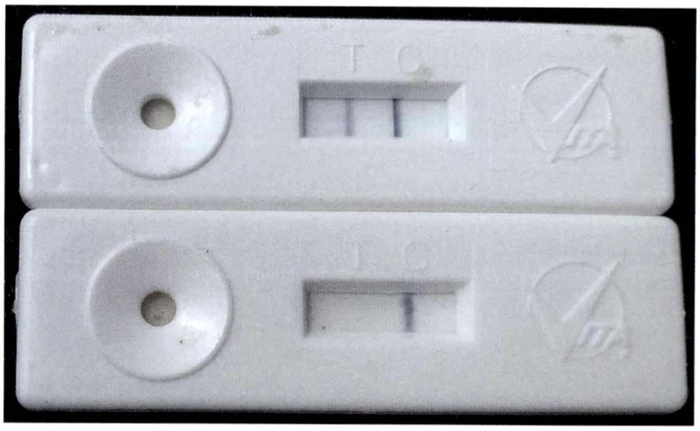

Figure 26.6 The appearance of only the control line (bottom) indicates the test kit works but the sample is negative for target bacteria (AFB or EFB). The appearance of both the treatment and control lines (top) indicates the sample is positive for the target bacteria. *Source:* Photo courtesy of Dan Wyns.

Equipment List:

- Isopropyl alcohol 35%:70% isopropyl alcohol is widely available at grocery and pharmacies and can be mixed 1:1 with water for a 35% solution.
- Digital timer
- 1-quart glass jar with lid and ring, such as a wide-mouth Mason jar, marked at 400 ml
- A measuring scoop, preferably with a flat edge, to collect bees, marked at 102 ml
- A large mesh strainer/sieve with a size such that mites can pass through, but bees cannot
- A fine mesh sieve that mites cannot pass through for alcohol recovery
- Additional jars with lids to store recovered alcohol
- Funnel to get bees into saline jars
- Plastic tubs for mite recovery
- Sealable plastic storage container for collecting washed bees for disposal

Warning:
Working with alcohol around live bees and a lit smoker comes with a couple of hazards. Alcohol kills bees, so make sure to place open jars where they won't get spilled onto bees or into hives. Alcohol is flammable, so be vigilant where your smoker is relative to alcohol and try to minimize spills/drips onto a surface (hive lid) where you may set your smoker.

Instructions:

1) Remove lid from glass jar and fill to pre-marked line with 400 ml 35% isopropyl alcohol.
2) Perform regular colony inspection (frame count, queen check, etc.)
3) As you are collecting bees off of brood frame, be absolutely sure the queen is not within your sample! Use measuring scoop to collect bees off a frame of open brood by gently rolling the scoop Downwards, against the bees, until reaching the fill line pre-marked at 102 ml. Gently tap scoop to knock down bees crawling up the sides, but do not aggressively knock them down in an attempt to "pack them down."
4) Secure lid on jar and set timer for one minute.
5) Shake jar with moderate intensity to dislodge mites from bees for one minute.
6) Place coarse sieve in plastic tub and remove lid from shaker jar.
7) Pour bees, mites, and alcohol across the sieve screen. Pour across sieve in a line so bees are somewhat spread out. Pour quickly so bees and liquid come out together; a slow pour will drain all liquid first and increase the "straining effect" of the bees themselves which will catch mites on their bodies.
8) Check empty jar and lid for any mites adhered to the sides.
9) Remove sieve from plastic tub and count mites in tub.
10) Pour mites and alcohol through the fine sieve into a reservoir container for alcohol reuse. (Discard alcohol in a separate container when it becomes too cloudy to easily see mites.)
11) Gently shake coarse sieve to roll bees around and distribute them in a single layer.
12) Return sieve with bees to the plastic tub and pour another 400 ml over bees so they all get rinsed. The bees will be more spread out than the first pour out of the jar.
13) Remove sieve from tub and count mites in the tub.
14) Add 1st and 2nd mite count together and record the total on the data sheet. The number recorded is the number of mites per 300 bees. Divide the mite count by 3 to calculate the standard mite count per 100 bees.
15) Return alcohol to the reservoir container as in step 10.

This process completes the alcohol wash process. Collect bees in a sealable container for disposal (Figure 26.7–26.13).

Sugar Roll Test This method of assessing varroa mite infestation levels is not immediately lethal to the bees sampled, but it is subject to some potential inaccuracies when performed incorrectly.

Equipment:

- Digital timer
- One-quart glass or plastic jar with ring, such as a wide-mouth Mason jar
- Wire mesh, typically 8 mm gauge (with holes large enough for mites to pass through but small enough that bees cannot fit through) cut to cover the opening of the jar
- A measuring scoop, preferably with a flat edge, to collect bees, marked at 102 ml

Figure 26.7 A simple kit for performing alcohol was to assess varroa mite infestation levels includes: Measuring cup (102 ml), quart jar with Isopropyl alcohol, coarse sieve, fine sieve, was tub, and funnel. *Source:* Photo courtesy of Dan Wyns.

Figure 26.9 After shaking for one minute, pour bees and alcohol through coarse sieve into the wash tub. *Source:* Photo courtesy of Dan Wyns.

Figure 26.8 Quart jar with isopropyl alcohol and approximately 300 bees ready for shaking. *Source:* Photo courtesy of Bee Informed Partnership.

Figure 26.10 Dislodged *Varroa* mites are visible against a light-colored wash tub. *Source:* Photo courtesy of Dan Wyns.

Figure 26.11 After counting mites in the wash tub, the alcohol should be poured through fine sieve to remove mites prior to reusing alcohol for second rinse of bees. *Source:* Photo courtesy of Dan Wyns.

Figure 26.12 The strained alcohol is poured over the previously washed bees to remove any remaining mites. *Source:* Photo courtesy of Dan Wyns.

- Powdered sugar (newly opened bag/box, store-bought powdered sugar that contains corn-starch is fine)
- Plastic tub for mite recovery
- Water

Instructions:

1) Put a few tablespoons (the exact measurement is not important) of powdered sugar in a jar.
2) Use a measuring scoop to collect bees off a frame of open brood by gently rolling downwards, against the bees, until reaching the fill line pre-marked at 102 ml. Gently tap scoop to knock down bees crawling up the sides. Do not collect the queen in the sample of bees.
3) Dump the sample of bees into the jar. Use the jar's ring to secure the wire mesh to the top of the jar.
4) Set the jar in a shady location for two minutes.
5) For one minute, shake the jar upside down so that powdered sugar and mites fall into the plastic bucket. Since bees may release alarm pheromones when shaken, it is best to shake the sample away from the hive. The person sampling must shake the jar quickly and with force to dislodge the mites.
6) Use water to dissolve the powdered sugar so that mites are easy to see and count.
7) Return the bees to the hive.

If at the end of the test the bees look sticky or wet, the test should be considered a failed test. Bees may look wet or sticky because the jar was left in the sun before shaking, because nectar is in the sample, or because of humid weather. Since old or exposed powdered sugar is less effective at dislodging mites, the person sampling should use relatively new or newly opened sugar (Figure 26.14).

Sampling for Laboratory Analysis

The USDA Bee Research Lab (BRL) in Beltsville, MD provides bee disease diagnostic services for bacterial, fungal, and microsporidian diseases in addition to varroa mites and tracheal mites. Analysis is performed on samples of adult bees or wax comb. This service is provided free of charge to beekeepers in the United States. Results typically take several weeks, and reports of findings are sent to the beekeeper, sample submitter, and appropriate apiary

Figure 26.13 After completing a field alcohol wash, bees can be collected in saline bottle for further laboratory analysis (nosema or tracheal mites) or discarded. *Source:* Photo courtesy of Dan Wyns.

(a)

(b)

Figure 26.14 Bees covered in powdered sugar in the sugar roll jar and the sugar-coated bees returned to the hive (ghost bees). *Source:* Photo courtesy of Emma Walters.

inspector when applicable. For more information, see Shimanuki and Knox (2000).

Send samples to:

Bee Disease Diagnosis
Bee Research Laboratory
10300 Baltimore Ave. BARC-East
Bldg. 306 Room 316
Beltsville Agricultural Research Center – East
Beltsville, MD 20705
(301) 504-8205

It is best to contact the lab prior to sending any samples as they will be able to answer specific questions regarding sample collection and shipment as well as advising on current turnaround time for samples.

American Foulbrood (AFB)

Laboratory diagnosis of AFB can be done by microscopy, culture, or molecular methods. In some states AFB is a disease that must be reported when found and infected colonies may need to be destroyed. Regulations vary by state, but a laboratory test may be required to support a field diagnosis before any actions are enforced. In addition to testing for the presence of AFB, the Beltsville lab can also determine if there is any resistance to the antibiotics used in the control of AFB.

To send a comb sample for AFB diagnosis follow the directions below. Any updates to sampling procedures will be available at from the Beltsville lab website at: ars.usda.gov/northeast-area/beltsville-md/beltsville-agricultural-research-center/bee-research-laboratory/docs/how-to-submit-samples/.

How to send brood samples:

- A comb sample should be at least 2 in. by 2 in. and contain as much of the dead or discolored brood as possible. No honey or nectar should be present in the sample.
- The comb can be sent in a paper bag or loosely wrapped in a paper towel, newspaper, etc. and sent in a heavy cardboard box. Avoid wrappings such as plastic, aluminum foil, waxed paper, tin, glass, etc. because they promote decomposition and the growth of mold.

- If a comb sample cannot be sent, the probe used to examine a diseased larva in the cell may contain enough material for tests. The probe can be wrapped in paper and sent to the laboratory in an envelope. When these options aren't available, a diseased pupa can be smeared on an index card, folded, and sent for sampling.

Tracheal Mites

Tracheal mites are internal parasites that infest the respiratory systems of adult bees of all castes. Tracheal mites are microscopic, and their detection requires laboratory dissection. Tracheal mite infestation inhibits flight ability and may result in a large number of crawling bees in front of the hive. These crawling bees may also exhibit a condition called K-wing in which their wings are disjointed. If a tracheal mite infestation is suspected, a sample of bees in alcohol or saline can be sent for dissection. Typically, at least 20 individual bees are dissected to determine presence or absence of tracheal mites at the colony level.

Nosema

Nosema is a microsporidian parasite that lives in the gut of adult bees. Detection and quantification of nosema requires a microscope and hemocytometer to count spores. There are two species found in honey bees, *Nosema ceranae* and *Nosema apis*. When examining gut contents for nosema spores, *N. apis* and *N. ceranae* are usually not differentiated.

There are two common ways of quantifying nosema infestation, generally referred to as prevalence or abundance. A test of prevalence examines a series of individual bees for presence/absence of nosema spores with results being expressed as a percentage of bees infested. Testing for nosema abundance involves processing multiple bees (typically 100) in a single pooled sample and counting spores with results reported as an average spore count in the unit of million spores per bee. For further discussion of nosema prevalence versus abundance, see Jack et al. (2016). Sending bees to the lab for nosema analysis can be done with either a sample of live bees or bees in alcohol.

Pesticide Analysis

There have been incidents of acute pesticide exposure experienced by colonies in addition to a growing awareness of the chronic exposure by accumulation of contaminants in comb. Samples may be collected if colonies are noticeably dwindling or crashing during or immediately after a pollination event or if there is interest in knowing the general level of contamination present in hives. Pesticide samples are relatively expensive to process; however, an effort should be made to collect samples from healthy and sick colonies so comparisons can be made. Mark clearly on the sample and data sheet which samples are from healthy colonies and which samples are from sick colonies.

The following hive components can be sampled for pesticide screening: honey/nectar, trapped pollen, bee bread, brood, adult bees (alive, dead, or dying), and wax. If you are worried about a pesticide kill or exposure, decide which hive component is best to test. Below are a few examples:

If you are concerned about a very recent pesticide kill, collect dying or twitching bees at the entrance of the hive, around the hive, or in the hive (such as on the bottom board). Second best would be the very recently dead bees, meaning the bees that are still soft and have their hair. Try to collect at least 100 dying or recently dead bees.

If you are concerned about a recent pesticide spray but don't see bees dying, collect pollen. Err on collecting fresh pollen stored in cells, not bee bread. Bee bread is shiny, and fresh pollen is matte. Try to sample only the very top pollen in a cell to get the more recent pollen. If possible, use a pollen trap to catch and sample the pollen the bees are bringing into the hive if the event is ongoing.

Collect bee bread if you are interested in knowing which pesticides the colony has been exposed to in the past and levels of accumulated chemicals in the hive.

Sample wax if you are interested in seeing which chemicals the colony has historically been exposed to. Sampling wax will tell you about pesticides used from longer ago than the bee bread. You may also sample wax if you are worried about what bees are exposed to during development.

Instructions specifically for sampling pollen:

1) Find a frame with fresh pollen. Freshly stored pollen will appear to have a matte finish, while older pollen that has been transformed into bee bread will have a shinier appearance.
2) Collect at least 16 cells (or 3+ grams) of fresh pollen. A wooden coffee stir stick works best for scooping pollen out of cells.
3) Deposit the pollen into a plastic conical centrifuge tube with screw top lid or other small sealable container.
4) Label collection tube with the beekeeper's name, date of sampling, and colony ID information.
5) Wrap tube in foil to prevent sample degradation from UV exposure.

Figure 26.15 Bee bread samples should be collected from cells using a small coffee stir stick and centrifuge tube. *Source:* Photo courtesy of Bee Informed Partnership.

6) Fill out a data sheet and record all hive grading measurements for the sampled colony (Figure 26.15).

Instructions specifically for sampling live bees:

Live bees should be shipped via the United States Postal Service. Instructions and regulations for shipping live bees are covered under Postal Regulations Section 526, available here: http://pe.usps.com/text/pub52/pub52c5_008.htm.

To ship live honey bees, the sampler should acquire a cardboard box with screens typically used for shipping honey bee queens. The kit contains enough sugar candy for food and a wet sponge for water to sustain bees for at least two days provided they are not overloaded. Live bees should only be collected and shipped Monday, Tuesday, or Wednesday to ensure there are personnel in the lab to receive them when they arrive.

These instructions assume a pooled sample from 8 colonies with 1/2 cup of bees from each yielding 4 cup total bees per live bee box. If more or fewer colonies are to be aggregated, adjust the quantity of bees per colony so the total volume of bees will remain at 4 cups.

1) Prepare a live bee box used for shipping bees by removing the wax paper covering the queen candy, open the tube of water and secure it to the bottom of the box.
2) Close the box making sure the two sides of the box with metal screens match up to ensure proper ventilation. Insert a funnel into the hole in the top of the box, and place it in a location out of direct sunlight.
3) Perform basic colony inspection and record appropriate data on a field data sheet. Be sure to use the apiary notes section to indicate colony or yard "status" (e.g. healthy, weak, etc.).
4) Find a frame containing uncapped brood. If uncapped brood is not available, find a frame from the center of the cluster.
5) Shake bees from the brood frame into the original plastic tub and scoop two, ¼ cup of adult bees (about 300) into the funnel of the live bee shipping box. Avoid collecting the queen!
6) Repeat steps 3 through 5 until 8 colonies have been sampled. Remove the funnel and close box with packing tape, ensuring top hole is well covered but ventilation screens remain functional.
7) Repeat steps 1 through 6 on the next set of colonies (or apiary) to be sampled.
8) Ensure field data sheets are completely filled out and label each live bee box with yard identifying information (weak, crashing, healthy, apiary name, etc.) (Figure 26.16).

Instructions for sending pesticide samples:

1) Label collection tube with your name, date of sampling, and colony number.
2) Collect the substance of interest (pollen, adult bees, brood, etc.). A minimum of 3 g of each hive matrix (pollen, wax, honey, bees) is requested for pesticide analysis, but amounts as small as 0.5 g can be processed. Samples less than 3 g may result in less accurate results.
3) Deposit samples into plastic conical centrifuge tubes with screw top lids or other small sealable containers.
4) Keep sample(s) on ice, preferably dry ice, and wrap in foil to reduce UV degradation.
5) Fill out a field data sheet and record all hive grading measurements and notes about colony condition and behavior for the sampled colonies (Figure 26.17).
6) Frozen samples should be shipped to the diagnostic lab overnight via FedEx. Guidelines for packing and shipping with dry ice are available at: http://fedex.com/downloads/jp_english/packagingtips/dryice_jobaid.pdf

Figure 26.16 When bees need to be collected kept alive in transit to the laboratory a live bee box should be used. *Source:* Photo courtesy of Bee Informed Partnership.

Figure 26.17 A popsicle stick can be used to scrape wax from the foundation in an approximately 3 in. circle to provide at least 3 g of wax for analysis. *Source:* Photo courtesy of Bee Informed Partnership.

References

Holst, E.C. (1946). A simple field test for American foulbrood. American Bee Journal 86 (1): 14–34.

Jack, C.J., Lucas, H.M., Webster, T.C., and Sagili, R.R. (2016). Colony level prevalence and intensity of Nosema ceranae in honey bees (*Apis mellifera* L.). PLoS One 11 (9): e0163522.

Lee, K. et al. (2010). Standardized sampling plan to detect Varroa density in colonies and apiaries. American Bee Journal 150: 1151–1155.

Shimanuki, H., & Knox, D. (2000). Diagnosis of Honey Bee Diseases. United States Department of Agriculture, Agricultural Research Service, *Agriculture Handbook Number* 690.

27

Necropsy of a Hive

Dewey M. Caron

Western Apiculture Society, University of Delaware, Affiliate Faculty Oregon State University, Portland, OR, USA

Examination of a dead colony can help determine what might have been the reason for colony death. Bee hives are an ideal environment for pathogens, with abundant, highly concentrated stored carbohydrates (honey) and proteins (bee bread), plus social organization behaviors such as food exchange and thermoregulation which aid in rapid pathogen growth and, likewise, can result in rapid colony decline.

Bee colony death can be due to environmental extremes, specific bee pathogens or pest activity, pesticide exposure (external as well as beekeeper internally applied pesticides) or internal colony social dysfunctions, especially related to queen replacement. By and large, colony death can be attributed to the beekeeper doing something incorrectly or, more frequently, missing the signs/symptoms, and/or not properly intervening in a timely fashion (Caron 2018).

Each year, small scale (backyard/hobbyist) beekeepers (individuals who keep fewer than 50 colonies, 50% of whom average three or fewer colonies) lose in excess of 40% of their overwintering colonies. Losses among sideliner (50–500 colonies) or commercial beekeepers owning 500+ colonies losses are not as high (Caron and Fitzpatrick 2019). Considering additional losses during the active year, approximately **one of every two colonies** do not survive over a full year period. Colony losses are generally understood to be multiple and interrelated (vanEngelsdorp et al. 2013).

Colony losses of this magnitude may not be unprecedented. The Reverend L.L. Langstroth, developer of the modern movable frame hive bearing his name, wrote in early editions of *The Hive and the Honey Bee* that losses of 45% overwinter might be expected in some seasons so the beekeeper should plan to take losses in the fall by combining and uniting to insure only strong, well provisioned colonies are overwintered. Unusually heavy losses in some locations and years, given names like Isle of Wight disease, spring dwindling, autumn collapse, etc. have been noted in the bee literature (Underwood and vanEngelsdorp 2007).

Prior to introduction and spread of tracheal and varroa mites, annual losses of 5–10% were considered "normal" in North America. As documented in 2006 and 2007 national surveys, losses have elevated to 20–30%. Present losses are apparently higher. Based on the 2018–2019 Bee Informed Partnership survey year, the annual overwinter loss across the US was 37.7% (Bruckner et al. 2019) and 48% in the Pacific Northwest (Caron and Fitzpatrick 2019, Figure 27.1), at least for small scale beekeepers.

A dead colony analysis, a bee hive necropsy, is an imprecise science and determination of the reason for colony failure may not always be possible. Examination of a dead colony might help eliminate some possible causes, resulting in a probable diagnosis or a shorter list of differential diagnoses.

Colony Phase Dynamics

The colony necropsy must consider the annual colony development cycle. (Figure 27.2; further explanation from Honey Bee Health Coalition [HBHC] *Tools for Varroa Management* http://www.honeybeehealthcoalition.org/varroa).

Honey bee colonies exhibit distinct seasonality, and annually passing through four phases. In the spring, colonies develop rapidly as forage opportunities improve. The further north one keeps bees, the more rapid this spring growth. As colonies approach peak size, and forage opportunities are extensive, a colony may divide, a behavior termed swarming.

Honey Bee Medicine for the Veterinary Practitioner, First Edition. Edited by Terry Ryan Kane and Cynthia M. Faux.

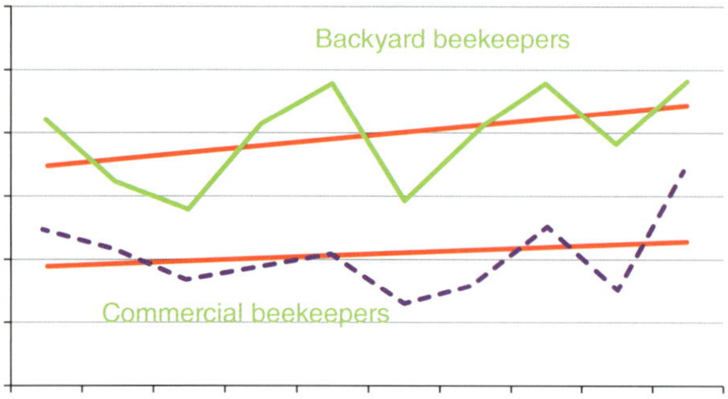

Figure 27.1 Loss of colonies. *Source:* Data DM Caron http://www.pnwhoneybeecurvey.com/survey-reports.

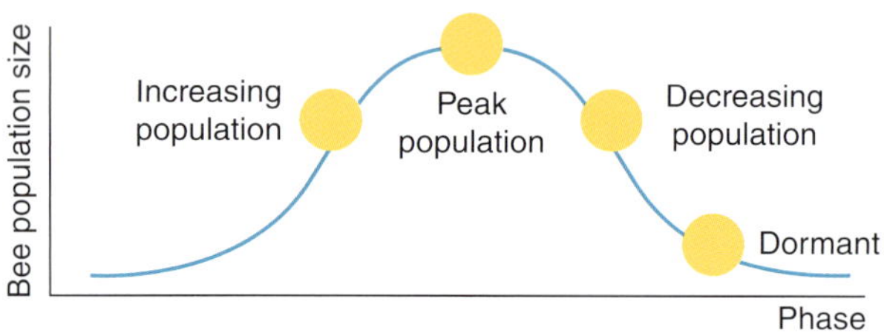

Figure 27.2 Four annual bee colony phases. *Source:* Modified from Honey Bee Health Coalition Tools for Varroa Management.

The peak population phase may or may not coincide with the peak nectar flow. The nectar flow for a region may be very short or extended, depending on colony success in the spring growth phase, as well as weather, location, and season. Potential honey harvest by the colony beekeeper is improved with synchronization with local nectar flow dynamics. Peak colony population may occur as early as April/May or as late as August/September, depending upon location and floral resources.

A population decrease phase occurs during the fall, extending over a couple of months. As with spring increase phase and peak population, colony decrease may be very rapid or occur over a more extended period. During the decrease phase diutinus worker bees (winter bees) are raised; such individuals will survive months, versus survival of only a half dozen weeks during the active season. Control is physiological with glyco-lipoprotein vitellogenin and brood pheromone interaction. (Amdam and Omholt 2002; AliDoke et al. 2015).

The fourth phase, the overwintering or dormant period, will vary considerably depending upon location. This phase may be essentially non-existent in the southern US and as long as six months in northern regions. During this phase, brood rearing is reduced or even halted as external temperatures dip below 45 °F, resulting in the bees clustering together within the hive for warmth. This fourth, overwintering, phase is the most critical phase for survival and when colony death rates are the highest.

Performing a Hive Necropsy

A dead colony examination starts outside in the apiary. Prepare for the visit with the assembly of a Diagnostic kit. If there are living colonies in proximity and weather is conducive to bee flight, don personal protective gear prior to entering the apiary (Figure 27.3).

First, examine the entrance area to the hive and immediately around the front of the colony for dead bees or brood and note any entrance activity. Colonies may have more than a single entrance. Some dead adult bees and/or dead brood will be normal in front of the colony and sometimes on the landing board at the entrance. Look for signs of mammal activity (skunks or bears principally) or large livestock that might disturb colonies.

The hive base (bottom board) should be checked but this may not be possible with some types of hives and will mean removing boxes to access. If the bottom has a debris board examine it or if it has a screened bottom board, look into "garbage pit" beneath the hive where debris will have fallen through. You must become familiar with ordinary hive debris to determine if the material observed is abnormal.

Figure 27.3 Personal protective equipment – bee suit with veil and gloves. *Source:* Photo courtesy of Terry Ryan Kane.

The Diagnostic Kit

- Personal protection equipment – bee veil, sturdy footwear, long sleeve shirt/pants or coveralls (for bee colony inspection), wrist and ankle closures, disposable nitrile/latex gloves
- Apiary inspection tools – smoker and hive tool
- Sample equipment basics – flashlight, toothpicks, magnifying glass, sample containers, forceps, hive marking pencil, notebook

Personal Protective Equipment

- Veil: Always wear a veil, even if you are approaching a hive for simple, quick tasks.
- Clothing: Wear clothing that covers all skin. Periodically inspect bee clothing for tears or openings.
- Footwear: Boots or work shoes are recommended when working with bees to protect your legs and ankles.
- Tuck coveralls or pants into footwear or close pant legs with strapping to keep crawling bees out.
- Gloves: Wear gloves to protect your hands and wrists to avoid stings. Tight fitting gloves are best because they allow you to move nimbly within the hive and avoid crushing bees.
- Body Odor: Scents in perfumes, shampoos, soap residues, cologne, etc. can attract or irritate bees, which are highly sensitive to scents. Do not apply anything with a scent.

Scavenger insect presence (wax moth, small hive beetle, yellow jackets, ants, various beetles or flies) should be noted; if these insects are occupying comb, they may complicate this inspection. Although beekeepers often implicate these scavengers as a reason for colony death, in most instances such populations only increase following colony weakening or demise. Scavenger presence may indicate another issue that weakened the colony.

In most instances it will be necessary to open a colony and look inside to diagnose loss. Start with removal and examination of the hive covers. Be familiar with normal hive odors (pleasant) and note if malodorous, sign of a problem. Determine if covers/frames are stained, have signs of water condensation or bee fecal matter. Look for dead adult bee bodies on the comb face, workers head first inside cells or on the bottom board, and non-emerged brood within combs. Look for fecal waste staining. Observe evidence of honey stores but be aware that any remaining honey may have been robbed by bees from other colonies in the vicinity.

Continue your examination by moving to the central part of the colony where bees are/were rearing their brood (brood = eggs, larvae, and capped/sealed pupae). Remove and examine frames holding comb; non-traditional hive comb may limit such inspection (top bar hive, Warré hive, skep). The brood pattern is a key diagnostic for the size of the colony and healthy bees. In healthy colonies, similarly aged brood should be grouped together in a more or less spherical pattern extending across several frames (depending upon season). In a dead colony, brood remains can be used to estimate the former size of the colony (Figure 27.4).

Brood in dead colonies will likely have been scavenged and will not likely be useful to examine for disease. For failing colonies, examine the capped brood which should occupy nearly every cell, with perhaps less than 5% of cells unoccupied. A scattered (spotty) capped brood pattern or lack of consistent age grouping, requires closer inspection to determine the reason for the lack of uniformity (Figure 27.5). In healthy colonies, the adult bee population should be adcquate to cover the brood. In very populous colonies, you may need to shake the bees from the frame or gently blow on them to move them aside so you can see the brood pattern.

Look for signs of active or remains of queen rearing – capped queen cells are distinctive. Roughly peanut-shaped, queen cells hang vertically, in contrast to horizontal worker and drone brood. Queens may develop from queen cups, constructed by workers at the margins of comb in either a planned division of the colony (swarming) or queen replacement (supersedure). Queen cells started from modified worker cells indicate emergency queen rearing and imply a catastrophic loss of the queen. The remains of a queen cell used to rear a new queen often remains on comb following use for a couple of weeks.

Figure 27.4 Dead colony (note remains of cluster top center-right) showing formerly extensive brood rearing pattern and cannibalized brood cells. Honey remains in the upper right and upper left-center still present). *Source:* Photo courtesy of Dewey M. Caron.

Figure 27.5 An example of a spotty brood pattern. *Source:* Photo courtesy of Dewey M. Caron.

Figure 27.6 Photo of drone cells occupying worker cells. *Source:* Photo courtesy of Ana Heck.

Evaluate the amount of drone brood; it too should be grouped. Larger, dome-shaped capped drones occupying worker cells signals a queen issue (Figure 27.6) and that the colony is on the path toward demise. Both queen cells and excessive numbers of developing drones can signal major problems in a colony, a colony stress that will modify the normal colony brood appearance.

Seasonal Loss Analysis

A bee colony necropsy will vary by seasons and diagnosis must account for previous colony events. If at all possible, analysis of the hive owner's experience/skills should be included in the history. Note that some of the following "reasons" for colony death may occur during other bee seasons than the one under which they are listed.

Overwinter (Dormant Phase) Colony Loss

The majority of bee colonies are lost during the overwintering phase. Fall losses (loss prior to winter solstice) may be the consequence of an event that occurred during the active bee season, such as pesticide poisoning or a colony queen event. However, fall losses are more likely to be due to virus epidemics and, as varroa mites are implicated in viral outbreaks, an analysis of varroa mite numbers is indicated.

Early spring losses (post winter solstice) also may be related to a previous seasonal event that weakened the colony population. Overwintering losses of spring usually appear as colony starvation or colonies too small to survive. Smaller colonies, especially newly established colonies with insufficient drawn comb, may have accumulated too few honey stores to enable dormant period survival. The population may not have been sufficiently large to cluster successfully for warmth (Figure 27.7).

Weak Colony

Bee colonies require a critical mass to keep adults and brood at proper survival temperatures. (See Box weak vs strong). Colonies with a reduced adult population may not be able to rear sufficient diutinus bees or form a sufficiently large cluster to handle winter weather extremes. Viruses, or less commonly another disease or pest situation, may shorten the life expectancy of spring, summer, or fall bees, negatively affecting population size and/or the bees' ability to store sufficient resources for the overwintering period. For the critical overwintering period, too few adults result in difficulty maintaining the temperature needed to rear brood during the early spring expansion phase to replace the aging adults or, alternately, does not permit adults to leave their cluster to access stored honey reserves when temperatures dip overnight. (Figure 27.7)

Figure 27.7 Photo of tiny dead spring cluster (smaller than softball in size). *Source:* Photo courtesy of Dewey M. Caron.

Determine if the adult bee population is sufficient to cover the brood. With variable spring temperatures, including cold spells extending over more than a couple of days, brood at the margins may die due to an insufficient adult population. When conditions improve, the brood area will expand rapidly and the adult population, experiencing heavy death rates of the older fall-raised bees but not yet rearing a sufficient number of replacement adults, will be imbalanced. Late springs of cool, wet weather with limited forage will accentuate this imbalance.

Is It a Strong or a Weak Spring Colony?

Colony adult population size and the amount of brood in a colony are relative. A weaker colony will have a brood rearing area perhaps not larger than a softball in early spring on a single or 1–2 combs whereas a larger colony may have a brood area of basketball size or larger, with brood on 4–5 parallel combs. In normal development the brood area, in a spherical pattern, will expand to adjacent combs, growing to the size of a beach ball or exercise ball within a generation or two. Weather permitting the collection of resources, the genetics of the queen, amount of resources stored the previous fall, and health of the bees are all factors that will impact colony growth and subsequent size.

Small colonies, along with newly established colonies, such as from packages or nucs, can die from lack of proper initial care. Signs include a handful or more of dead bees on the bottom board and/or a small dead cluster with bees head-first ("tails up") in cells around a small patch of dead capped brood located on one, or at most two, brood combs. Mold may be evident and if the bees are "wet" fly maggots can be common. Determination of "too weak" is relative and external weather conditions are often a factor, along with lack of attention by the beekeeper.

Starvation

Honey bees, as social organisms, depend upon external food sources, though beekeepers can supplement natural foods. Bees are "junk-food junkies"; they only eat processed food. Their dietary carbohydrate source is nectar, collected primarily from flowering plants, processed into honey. Other natural sugar sources, as, for example, extrafloral nectaries, the sugar exudate of some insects (i.e. honey dew from aphids/scale insects, etc.), or plant sugars (stalks of sugar cane at harvest) and human-sourced sugar sources (sucrose or sugar blends like high fructose corn syrup [HFCS], candy/syrup manufacturing sources, human garbage sources, etc.) can also be processed into honey. Once fully processed, the honey-filled cells are capped with beeswax.

The dietary needs of vitamins, minerals, amino acids, cholesterol, fatty acids, etc. are supplied by pollen, collected from plants; it is processed into "bee-bread" for bee consumption. Cells of bee bread vary in color depending upon floral source; cells of bee bread are never covered with a capping, though a glaze of honey may reflect light when bee bread cells are examined. In the normal condition, brood is central, bee brood cells ring the top and sides and honey is stored above and to the sides beyond the brood (Figure 27.8).

Beekeepers should evaluate the amount of stored honey and bee bread during fall population decrease phase prior to the dormant (over-wintering) period. Beekeepers can feed sugar syrup, dry sugar and/or pollen patties to strengthen their colony. Sufficient colony stores for successful overwintering are dependent upon the extent of beekeeper honey harvest, the availability of resources during the post-peak (summer) population phase, and rapidity of reduction in population size during the fall phase.

Seemingly contradictory, both large colonies and smaller colonies may starve overwinter. Starvation is a common reason for colony death in the spring, especially of the weaker colonies, but it may occur also in stronger colonies during spring expansion. (See Box what does starvation look like) Spring is a critical time for bee colonies as they are "living on the edge." They need large amounts of fresh pollen to expand brood rearing and nectar for their energy needs. Spring weather and amount of honey and bee bread stores remaining from the previous fall are critical to colony spring expansion.

Figure 27.8 Bee nest organization in Fall; capped brood central, bee bread stores and honey to sides and top. *Source:* Photo courtesy of Dewey M. Caron.

What Does Starvation Look like?

Starvation can be diagnosed by a total lack of stored honey in comb of a dead colony. A distinct cluster formation, around some capped brood will likely be evident. Dead bees will be tightly packed; brushing the dead aside will show dead bees head-first in cells – all in an effort to form as tight a cluster as possible. Capped brood may be cannibalized with cell perforations with no evidence of larvae or eggs. Some of the adults will have fallen to the bottom of the hive. If honey stores remain, it will likely be crystallized and remote for the dead cluster of bees. Mold will be present on the dead bees and growth in bee bread cells will likely be evident (Figure 27.9).

If the dead cluster has been removed diagnosis becomes more difficult. However, a circle of mold may persist as will dead bees head first inside the cells and widely scattered capped brood remains will still be evident.

Starvation of larger, stronger colonies will have some of the same characteristics but a dead bee cluster might be lacking. The bottom board will have dead adult bodies even to the point of their blocking the colony entrance/exit. Many of these bees will have extended tongues. Evidence of larvae attempting to escape their cells should be evident. There will be a lack of capped honey (though distance frames might still have some) and no or few cells of bee bread will be present. Uncapped brood may or may not be evident; normally it will have been cannibalized so will not be not present.

Check weather records to determine if a two–four day wet, cold weather event occurred before reported colony loss. Fly maggots or scavenger beetle larvae may be active among the dead bees, especially if there is a higher moisture level. If the colony is examined shortly after the point of collapse, some of the bees may be revived by warming them up and/or feeding of sugar syrup.

Other reasons for overwintering losses.

Nosema, the adult bee intestinal disease caused by a microsporidian fungus, may be a cause of spring colony death. Worker adults with Nosema infections may become precocious foragers and have reduced lifespans, upsetting the normal age-related sequence of duties in colonies (Higes et al. 2008; Clint et al. 2015). Colonies with small groups of adult bees and a queen clustered over capped brood cells will show high Nosema counts. Fecal spotting outside the hive or fecal remains on the comb may or may not be related to Nosema infection.

Figure 27.9 An example of starvation. Dead bees in a tightly packed cluster and dead bees head first in cells. *Source:* Photo courtesy of Jenny Warner.

Queen failure might be an alternate diagnosis if remains of queen cells or capped drone cells in worker-sized cells are observed with a small dead cluster. The brood area will not be organized in colonies that lack a queen. A spotty brood pattern, thought to be an indicator of poor colony heath, might not be useful in diagnosis of unhealthiness in bees (Lee et al. 2019).

Cold, by itself, is not responsible for killing a bee colony, provided the bees have enough food and population to cluster. Cold weather may, however, reduce brood rearing and spring colony expansion and thus indirectly result in weak colonies that will be stressed and susceptible to other factors negatively impacting bee health.

Newly established colonies from package bees may not survive if wetted extensively or if cold weather follows hiving. Alternately, their caged queen may die of exposure if abandoned without sufficient covering cluster. These occur more frequently in bees hived on foundation or when installed into alternative top-bar hives. The dead bodies will be massed on the bottom.

Climatic events such as flooding or wind related events that overturn or expose the hive to sunlight/moisture may result in colony death. Soggy ground may undermine the hive base, tipping colonies over and exposing the combs to wind, rain, and sunlight. Water usually leaves mud evidence and or water stains of hive equipment. Hives tipped over from livestock or animal activity and not immediately restored, likewise, might not survive exposure.

Colony fecal spotting is sometimes evident in early spring. Normally bees fly away from the hive entrance to void their semi-solid, yellow-brownish colored feces. Look for spotting around the entry/exit area, on box sides or on the colony top. It will clear up as soon as weather conditions improve. If fecal matter is evident inside the colony (such as on top bars or comb) it signals highly stressed bees. Fecal spotting is generally attributed to inadequate nutrition; i.e. a colony with "upset stomachs" and was formerly associated with Nosema.

Spring Increase Phase Colony Loss

Bee colonies can recover from early spring problems such as starvation (provided it does not kill the entire colony) and colony population weakness once weather conditions improve. Starvation is still possible but becomes less likely a reason for colony loss. Colony fecal spotting may continue for a period but lessen as weather permits foraging. When spring is slow to develop with wet, cooler-than normal weather, losses due to weak populations or starvation may increase.

Disease Loss

Colonies will often show signs of European foul brood (EFB) disease in the spring and other "spring stress diseases" such as chalkbrood (a fungal disease) and sacbrood (a larval virus). With spring inspection, beekeepers should begin to look for American foul brood (AFB) disease. Both foulbroods are bacterial. AFB is a disease caused by the spore-forming bacterium *Paenibacillus larvae*. This disease infects developing larvae leading to death in late larval, prepupal, and pupal development stages. Thus, the signs of AFB infestation are most easily distinguished in the brood cell cappings and in the dying prepupal/pupal stages of a developing honey bee (Figure 27.10). Dehydrated remains of pupae killed by the bacteria dry into a scale that melds tightly onto the bottom of the brood cell. Such scales cannot be removed without damage to the cell wall, even with forceps.

EFB is a disease caused by the non-spore-forming bacterium, *Melissococcus plutonius*. It affects young larvae, killing this brood stage before brood cell capping is initiated. It is thought to be related to poor nutrition. When dead larvae dehydrate, they form a scale that is easily removed. Thus, open brood cells need to be examined to see the signs for this disease (Figure 27.11).

The spread of both diseases occurs via scales not removed by the bees, bees "drifting" from infected colonies to healthy ones, bees robbing an infected colony that is collapsing, with beekeeper exchange of infected

Figure 27.10 American foul brood. *Source:* Photo courtesy of Cynthia Faux.

Figure 27.11 European foul brood. *Source:* Photo courtesy of Ana Heck.

equipment, and from spores in honey. Unlike AFB, a colony can recover from an EFB infection as foraging conditions improve. EFB is thought to be stress related and is evident in poor colony nutritional situations. Colonies with AFB die rapidly as the bees are unable to rear healthy replacement adults and the usual outcome is colony death.

How to Confirm AFB

1) Look for characteristic signs of AFB (Figure 27.10)
   a) Capped cells that are perforated, sunken, and greasy looking.

b) late larvae/pupae look sickly, discolored dark-brown to black.
c) Distinctive foul odor (not merely of death but unique odor)
d) Scale is present – desiccated remains of pupa that sticks tightly to the lower cell wall and is not removable without damaging the cell wall.
e) Extended pupal proboscis ("false" tongue) stretching from one cell wall to opposite cell wall (only see if death occurred in early pupal stage) – very positive sign of AFB but not always present.
f) Distinctive "foul" odor – not the typical sour smell of dead brood.

2) Conduct the "ropiness" test
   - This test relies on the unique characteristic of AFB-infected honey bee brood.
   - Select a brood cell that looks infected, but which is not dehydrated (the prepupa/pupa structure is still evident and gooey). Take a stick or toothpick to swirl the contents of the cell and slowly withdraw it. If the contents draw out up to an inch in length (2.5 cm) then snaps back, the cell is most likely infected with AFB (Figure 27.12).
3) Use an AFB diagnosis kit.
   - These are available from several bee supply companies. Be cautious to interpret the results correctly because the test may yield false results. Ideally, the test results should be considered with other lines of evidence (such as 1 and 2 above).
4) Have the colony examined by an individual trained in disease identification, such as a state apiary inspector.
5) Send a sample to a diagnostic lab
   - Collect a sample of the suspected brood by either cutting out a piece of the comb (2 by 2 in.- 5 cm × 5 cm) or coating a stick or toothpick with contents of one or more suspect cells. Wrap the sample in paper (not plastic or aluminum) and send it to a lab (several state Department of Agriculture labs have lab and trained personnel to do this examination) or USDA Bee Lab in Beltsville, MD. Make sure to include return contact information.

Figure 27.12 American Foulbrood "ropiness test." *Source:* Photo courtesy of Cynthia Faux.

How to Confirm EFB (Figure 27.11)

1) Use characteristic signs of dead/dying larvae:
   a) Larva twisted in cell;
   b) Larva with yellowish color or yellow color streaks;
   c) Rubbery scale is easily removed from cell without damaging cell walls;
   d) Only larvae appear sickly – capped cell appearances are normal;
   e) Dead larvae does not "rope out," nor show "false tongue"
   f) Odor mildly sour, not distinctive AFB foul odor;
2) Use an EFB diagnosis kit available from several beekeeping suppliers (see above under AFB).
3) Have the colony, suspect comb examined by an individual trained in disease identification, such as a state apiary inspector or send a sample to a diagnostic lab (as described above for AFB).

Queen Replacement

Spotty brood pattern and excessive numbers of drones (or drones being reared in worker cells Figure 27.6) are signs of queen replacement. Successful replacement will weaken a colony and if something goes wrong, the colony can be left queenless and not strengthen normally. Queen replacement behaviors begin with workers rearing replacement queens. Queen cells are larger, peanut shaped when capped and hang vertically downward from the margin of the comb or from the comb face itself. This occurs in over one-half of spring colonies but some colonies, for largely unknown reasons, abort the process. Swarming, division of the colony, occurs largely in the spring and colonies usually possess several developing/capped cells whereas in supersedure only one or a few replacement queen cells will be present.

If a queen is injured, or even killed, in a colony examination, the colony will seek to rear new queens by modifying the cells of several workers. This is termed emergency queen rearing. Emergency queen cells are reared from modified worker cells, not started from a queen cup. It occurs when a queen is suddenly lost, killed (during colony examination for example), or when the beekeeper removes the queen from a colony, such as when a colony is split (divided) by the beekeeper. It may rarely happen however, that a queen may fail to return to her proper hive following a mating flight or she could be eaten by a predator during flight (Figures 27.13 and 27.14).

Drone Layer or Laying Workers?

Drones in worker cells might be the result of a queen who is only laying unfertilized (drone) eggs in the worker cells or from a queenless condition where, in the absence of a queen worker, ovaries develop and several workers put unfertilized eggs into worker cells. Both will eventually result in colony death. The key observation is drone brood in worker cells and a weakening colony with a high proportion of developing (and adult) drones.

In the case of a drone layer, the queen exhibits her normal egg laying pattern, putting a single egg in a worker cell but the capped stage is the dome-shaped cells of drone pupae. Uncapping a few cells will reveal pupae with the larger eyes of the drone. The reason for the lack of fertilized eggs may be the result of improper mating of the queen, age of the queen, a genetic abnormality, cold/heat injury to the queen herself or exposure to a pesticide. Removal of the existing queen and introduction, via a cage, of a new, properly mated queen will correct the condition, if the bees accept the new queen.

A colony that has no young brood to raise an emergency queen will soon show indications of several workers developing the ability to lay eggs. Worker bees are not capable of mating with drones so all the eggs they produce will be haploid and develop only into drones. Cells will have eggs positioned at odd angles and multiple eggs may be seen within the cells. The brood pattern will be highly spotty, irregular, and capped cells will have drone pupae. Laying worker colonies are more difficult to requeen than drone laying queen colonies. In most instances, such colonies should be discarded.

Figure 27.13 Emergency queen cell. *Source:* Photo courtesy of Ana Heck.

Figure 27.14 Large capped supersedure queen cell. *Source:* Photo courtesy of Randy Oliver.

If too early in the season, virgin queens may not be able to successfully mate. Conventional wisdom says that emergency reared queens are not as desirable as those reared in swarming or supersedure. Large scale commercial rearing of queens uses a controlled emergency (queenless) condition; this management practice leads to excellent new queen stock.

Summer (Peak Population) Loss

Summer events in bee colonies may result in weaker adult populations, and/or reduced honey stores. Losses over the summer are not as frequent as over the winter period. (see BeeInformed Partnership website www.beeinformed.org for annual loss statistics; both winter and summer [active season] losses are tallied).

Sudden appearance of large numbers of dead bees on the ground in front of colonies or piled up within the colony, many with tongues extended, signals a pesticide kill. Scavenger activity, such as ants and yellow jackets, may quickly remove the dead bodies. Not all pesticide damage has the same field signs and severity of a loss will vary among colonies even in the same apiary. Some pesticides kill forager bees in the field before they can return home. A sudden drop in foragers might be noted, but this is difficult to detect.

Some pesticides may kill brood when pesticide contaminated pollen is carried into the hive and fed to larvae. Such colonies may lose a generation or more of developing brood. Once again this is very difficult to detect. Other pesticides weaken a colony, killing some of the foraging population and, if contaminated pollen is converted into bee bread, some brood. (Hooven et al. 2016).

A study of bees moved to California almond orchards (an early spring event) show 40% of colonies with some pesticide damage from dead adults, deformed brood or entire dead colonies (Wade et al. 2019). Incorporation of information into Best Management Practices guide for almond growers (*Honey bee Best Management Practices for California almonds* http://www.almonds.com/pollination) now recommends that no insecticide be mixed into the tank of other pesticides for application to almonds in bloom.

Some pesticides applied by beekeepers to kill varroa mites may also harm adult bees or result in temporary halt of queen egg laying and/or kill the colony queen. Beekeeper applied miticide damage may show up as adult loss and/or a weakened colony. Adult bees may become early "precocious" foragers and die two or three weeks earlier than normal, leading to an imbalance in the ratios of age-duty behaviors of a normal colony (Clint et al. 2015). This, too, is very difficult to diagnose. Pesticides applied to kill mites that halt queen egg laying for a period of time (a few days up to two–three weeks) results in the adult population aging and eventual decrease in hive population. Once again this is difficult to diagnose.

Queen replacement behaviors that do not result in a mated queen leads to a queenless colony that will eventually only raise drones and thus not survive. However even in normal queen replacement, a colony may be several weeks without evidence of a mated, egg-laying queen and subsequently experience a colony population decrease. Non-performing colonies, spotty brood pattern, interval without a laying queen, etc. all may lead to heavy consumption of previously stored honey, resulting in insufficient stores in the fall to raise the wintering bees and/or too few stores to survive the winter.

A curious behavior, common in Africanized bees, is absconding. Absconding is sudden abandonment of a hive without rearing of replacement queens as occurs when bees swarm. Following an abscond only a few straggler bees may be left. Factors that cause absconding are not well known. In Africa and in the American Africanized bees, ants may cause bees to abscond as well as lack of honey stores or excessive examination by the beekeeper (Schneider and McNally 1992; Caron 2001). With European bees, it is postulated that absconding in late summer into the fall might be related to heavy mite populations in the colony.

Heat, as in instances with excessive high temperatures or prolonged heat periods in colonies without an adequate adult bee population may weaken colonies, especially if water is not readily available for hive air conditioning. Foraging workers switch from nectar collection to water collection and if water sources are not close by the colony expends considerable energy trying to keep cool, reducing honey stores needed for the overwintering period.

If there is a robbing event, the bees of a colony under attack expend considerable energy seeking to repel the robbers, inside hive temperatures become elevated and bees, both those of the colony and some of the robbing population, may die from heat exhaustion. (Figure 27.15) Brood needs to be reared in a narrow temperature range and exceeding this range leads to their seeking to "wander" from their cells and their death. Rarely, in instances without adequate adult bee population, the beeswax comb can actually melt.

Bee colonies will usually become clear of EFB, which might be extensive during early spring buildup. However, infestations of EFB, chalkbrood, sacbrood, American foulbrood or adult disease Nosema may extend into summer, resulting in death of developing brood, needed

(a) (b)

Figure 27.15 (a) Bees are robbing the hive on the left. (b) Close-up of Figure 27.15. *Source:* Photo courtesy of Ann Heck.

to replace aging adults. AFB will eventually kill an entire colony, and subsequently likely spread to other colonies; the other brood/adult diseases are not likely to kill a colony, though they may weaken colonies leading to inadequate honey stores/reduced population for the overwintering period.

Some other, relatively uncommon reasons for dead colonies might be bear or human vandalism.

Fall Population Decrease Phase

The major factor in fall bee colony demise is varroa mites. Heavy mite populations, uncontrolled by the beekeeper, may lead to what is now labeled PMBS (Parasitic Mite Brood Syndrome). Look for evidence of spotty brood pattern and non-viable brood, termed snot or cruddy brood (the alternative term Idiopathic brood disease has not been widely adopted) (vanEngelsdorp et al. 2013). Some of the dying larvae will have symptoms of EFB, although lab analysis may not find the EFB pathogen. Capped brood may have perforations and be sunken but otherwise not have the symptoms of AFB and lab examination does not find the AFB pathogen. Looking into brood cells, you might note a bright white "stain" on the upper cell wall (mite guano) and one or more mites on dying capped pupae. You may or may not see dead bees or brood on the bottom board or in front of the colony.

What Does PMBS Look like?

Colonies with Bee PMBS present several signs (Snyder 2013). Initially it can be detected with a spotty, scattered brood pattern and reduced adult bee population. Depending upon degree of hygenicity of the workers, pupae with mites within capped cells are removed so partially removed, chewed pupae can be seen and/or opened brood cells with white, apparently healthy pupae in the cell visible. Mites may be evident on pupae where cell cappings have been removed. Adult sampling will reveal high mite numbers. Mite guano in opened brood cells is often evident. The colony may be reluctant to remove sugar syrup from feeders (Figure 27.16).

Other factors of fall colony loss are accumulation of disease effects, accidental exposure of colonies (weather, animals), unsuccessful queen replacement during the season, pesticide exposure, colony abandonment (absconding) and even starvation (although that is rare).

Parasitic Mite Brood Syndrome

Eventually colonies with PMBS demonstrate decreased adult bee numbers as fewer replacement bees are reared. Viral diseases will cause a sudden collapse of the colony (initially termed CCD – Colony Collapse Disorder) and an otherwise "strong" colony may collapse within three–four weeks. The empty hive will have scattered brood, with many of the cappings with perforations. Honey may remain and a small cluster of bees with the queen might be seen if weather has been cold at night. In some instances the remaining bees and queen abandon the brood area and move to a new portion of the comb and seek to rear new brood in an area without the mite-infested brood. In other instances they completely abandon (abscond) their nest.

Mite control treatments may be too late to rescue such highly stressed colonies, particularly if they have a virus epidemic. If the colony does not collapse, the critical mass of diutinus bees is not raised and the colony with a dwindling adult and brood population seek to cluster into the dormant winter phase but are dead by mid-winter, very spring. Mite treatment should be started earlier in the season after sampling to determine if mite numbers exceed 2% of the adult bee population. (HBHC 2018).

Figure 27.16 Advanced case of PMBS – spotty capped brood, disorganized brood pattern and opened pupal cells. Often small adult population present. *Source:* Photo courtesy of Ana Heck.

Conclusion

Whatever the season, beekeepers are continuing to suffer extensive documented losses. Factors of bee health are complexly interrelated and individual colony demise is likely due to multiple factors. A bee necropsy may help eliminate some possible reasons but may not ultimately lead to a single definitive diagnosis of reason for loss. However, for the veterinarian, this is a good first step.

References

AliDoke, M., Frazier, M., and Grozinger, C. (2015). Overwintering honey bees: biology and management. *Current Opinion in Insect Science* 10, August 2015.

Amdam, G.V. and Omholt, S.W. (2002). *The regulatory anatomy of honeybee lifespan. Journal Theoretical Biology* 216: 209–228.

Caron, Dewey M. (2001). *Africanized Honey Bees in the Americas*. Medina OH: A.I. Root, Co.

Caron, Dewey M. (2018). *Dead colony forensics. Bee Culture* June 2018.

Caron, Dewey M. and Fitzpatrick, J. (2019). *Winter bee losses of Oregon backyard beekeepers, 2018–19*. Access: http://pnwhoneybeesurvey.com/survey-results/2018-19-survey-reports and previous survey reports

Clint, J.P., Søvik, E., Myerscough, M.R., and Barron, A.B. (2015). *Rapid behavioral maturation accelerates failure of stressed honey bee colonies. Proceedings of National Academy of Sciences* https://doi.org/10.1073/pnas.1422089112.

Higes, M., Martín-Hernández, R., Botías, C. et al. (2008). *How natural infection by Nosema ceranae causes honeybee colony collapse. Environmental Microbiology* 10 (10): 2659–2669.

Honey Bee Health Coalition (HBHC) (2018). *Tools for Varroa Management* http://www.honeybeehealthcoalition.org/varroa.

Hooven, L., Sagili, R.R., and Johansen, E. (2016) rev. *How to Reduce Bee Poisoning from Pesticides*. PNW 591.

Lee, K.V., Goblirsch, M., McDermott, E. et al. (2019). *Is the brood pattern within a honey bee Colony a reliable indicator of queen quality? Insects* 10: 12. https://doi.org/10.3390/insects10010012.

Schneider, S.S. and McNally, L.C. (1992). Factors influencing seasonal absconding in colonies of the African honey bee, *Apis mellifera* scutellata. *Insectes Sociaux* 39 (4): 403–423.

Snyder, R. (2013). *Parasitic Mite Syndrome (PMS)*. BeeInformed Partnership blog Oct 15: https://beeinformed.org/2013/10/15/parasitic-mite-syndrome-pms.

Underwood, R. and vanEngelsdorp, D. (2007). *Colony collapse disorder: have we seen this before? Bee Culture* 35: 13–18.

vanEngelsdorp, D., Hayes, J. Jr., Underwood, R.M., and Pettis, J. (2008). *A survey of honey bee Colony losses in the U.S., fall 2007 to spring 2008. PLoS One* https://doi.org/10.1371/journal.pone.0004071.

vanEngelsdorp, D., Tarpy, D.R., Lengerich, E.J., and Pettis, J.S. (2013). *Idiopathic brood disease syndrome and queen events as precursors of colony mortality in migratory beekeeping operations in the eastern United States. Preventive Veterinary Medicine* 108 (2–3): 225–233.

Wade, A., Lin, C.H., Kurkel, C. et al. (2019). *Combined toxicity of insecticides and fungicides applied to California almond orchards to honey bee larvae and adults. Insects* 10: 20. https://doi.org/10.3390/insects10010020.

28

Common Husbandry Issues

Charlotte Hubbard

Schoolcraft, MI, USA www.hubbardhive.com

Honey bee husbandry requires the complicated orchestration of appropriate nutrition, protection from disease, and suitable shelter to promote colony health and minimize stress. Some specific husbandry topics, such as nutrition, disease, and structures, are covered elsewhere in this publication.

Husbandry benchmarks used for other species – such as a review of feed records and evaluation of the facility – may not readily apply to honey bees. However, there are some overall aspects that may be indicative of appropriate husbandry. Please consider that beekeeping operations may be managed appropriately even if they do not follow all the guidelines in this chapter.

Upon approaching and entering an apiary, evaluate:

Location Considerations

- Resources: Can the colony efficiently obtain sufficient, appropriate water and food? If not, has the beekeeper analyzed and anticipated the impact and appropriately supported the colony? (For example, feeding in times of dearth and bee-friendly watering stations in times of drought.)
- Environmental risks:
  - Is the location relatively safe from flooding and wildfire?
  - Are there barriers to prevent large mammals from knocking over hives? (Figure 28.1)
  - Does the apiary location require electric fencing to protect it from bears?
  - Are hives located where they may be subject to structural damage? For example, are they located under a ramshackle overhang or trees with apparent weakened branches?
  - Is there relentless heat, like on an extensive, dark surface where heat extremes are common? (Many colonies do well in full sun when they are located on a surface of brush or grass.)
  - Is the location predominantly sunny with good air drainage? Moderate to full sun exposure is helpful in abating Small Hive Beetles and providing beneficial ultraviolet light to exposed surfaces.
- Safety:
  - Is the apiary away from well-traveled pedestrian and recreational areas?
  - Is there appropriate signage cautioning passersby that honey bees are in the area?
  - Is the location so remote that contacting emergency personnel is impossible or impractical? Does the beekeeper have the knowledge and tools to deal with emergency situations? Can the location be found by others and accessed without special vehicles?
  - Can a vehicle be used for relocating colonies, removing heavy loads of honey, etc.?

Management Considerations

- Are the colonies inspectable and manageable? Colonies located in trees or other natural cavities, structures not intended for bee inhabitation (walls of buildings) and risky locations restrict the routine evaluation required for health.
- Are there too many colonies for the location?
  - While the number of colonies an area can support is a function of forage, a beekeeper may not know how many colonies (feral or managed) are in the area, or not realize if/when colonies will be moved into the area to provide pollination services.
  - Even if available forage supports a great number of colonies, diseases, and pesticide poisoning can occur. A beekeeper may want to consider limiting risk by geographically dispersing colonies, if possible.

Honey Bee Medicine for the Veterinary Practitioner, First Edition. Edited by Terry Ryan Kane and Cynthia M. Faux.

Figure 28.1 Curious cows. *Source:* Photo courtesy of Charlotte Hubbard.

Figure 28.2 This colony was particularly defensive, perhaps because of raccoons' interest in it. *Source:* Photo courtesy of Charlotte Hubbard.

- Are drift (bees returning to a colony not their own) reduction techniques deployed? Honey bees can drift to other colonies, providing opportunities for disease or mite transmission. Drifting may be mitigated with discernable landmarks (trees or other natural features, farm implements, traffic cones, etc.), entrances offset 10–15° from adjacent colonies, and differing exterior appearances (paint color, patterns, decals, or other visual distinguishers) (Wyns 2018). Robbing screens also aid in drift reduction.
- Are grass and weeds trimmed?
  - Excessive vegetation can hide ground wasps, ant mounds, uneven ground and trip hazards, and harbor ticks and rodents.
  - Plants in contact with hives provide bridges for easier insect egress.
  - Dry vegetation may catch fire if in contact with a smoker or a vehicle's exhaust system.
- Do hive stands discourage snakes, rodents, etc. from sheltering in them or ants from building mounds under them? (Anderson 2020).
- Is the yard clean of items, such as discarded comb, chunks of brood, and open feeders that invite small mammals and robbing from bees or insects?
- Are mammal footprints observable on the ground or hives? (Figure 28.2)
- Are there signs of threats from other insects? These signs include headless bees, yellow jackets, hornets, or other wasps hovering at the entrance or dead outside the colony, and excessive ant activity.
- Is the equipment in appropriate condition? Ideally, there should be no major gaps allowing driving rain, wind, cold, snow.
- In areas with possible high winds, are hive tops secured and measures taken to keep hives from toppling over?
- Is equipment utilized properly? Examples of problems: covers skewed or not seated properly, boxes misaligned causing inadvertent openings. Hive entrances should be appropriate for the season in both size and number. (As examples: new colonies need an entrance small enough so that only a few bees can enter at a time until they gain sufficient population to more easily defend the entrance; multiple, upper entrances may be helpful during nectar flows for production efficiency or in times of extreme heat.)

Upon closer observation of hives:

- Is the general impression that activity at entrances is appropriate for the season, time of day, and weather?
- Are there unusual or offensive odors?
  - Unusual odors could be an indication of European and/or American foulbrood, although bee death from other causes also emits foul odors.
  - Some nectars are characterized as smelling a bit offensive. Goldenrod nectar is often described as smelling like dirty gym socks.

Best Practices Considerations

- Are key activities tracked? (Examples: colony demeanor, queen age, colony's queen status, feeding, health, treatments, etc.)
- Are biosecurity practices followed? (Hive tools sterilized, gloves [if used] restricted to an apiary, etc.) It is common that many beekeeping operations do not change or sterilize equipment between apiaries.

Some Common Questions Regarding Starting Beekeeping

Number of colonies: While a significant investment, beginner beekeepers should start with more than one colony for several reasons. First, comparing and contrasting helps the beginning beekeeper identify any differences that may warrant intervention, such as assisting a colony that is building up slower. Secondly, colonies can support each other, and having multiple colonies can help a beekeeper independently support a struggling colony using resources from within their own yard. Finally, having multiple colonies also increases the rate of education, and when the power of multiple colonies is understood and utilized, the apiary moves toward sustainability. Since many new beekeepers underestimate the challenges of beekeeping and/or lose all of their colonies within the first year, it is often recommended that beekeepers do not begin with more than four colonies.

Using old equipment: Unless the history and the reasons that the equipment is no longer in use are known and acceptable, old equipment should not be used. American Foulbrood contamination lasts for decades and is not easily detectable.

Woodenware may be painted or otherwise protected: if desired, but only the outside, nothing within the hive, with the exception of the landing board. Stains, wax, tung oil and linseed oil may also be used for protection.

Obtaining bees: Bees are typically purchased as a package, or a nuc (nucleus colony). Some beekeepers purchase colonies in full-sized hives.

A package is typically three pounds (about 10 000) bees and a laying queen in a shipping container, complete with syrup for feed during transport. The queen is caged within the shipping container so that the workforce, usually taken from multiple colonies, has time to develop loyalty to her. The bees must be installed into their hive as soon as practical, ideally within a few days of when received.

A nuc is a small colony, complete with everything found in a production colony – a workforce of appropriately varying ages, a queen laying predominantly worker brood, brood of all ages, and some level of honey and pollen stores.

A nuc purchased for starting beekeeping is typically five-frame Langstroth type, although three- and four-frame nucs are also available. As a nuc will outgrow its box, it is easily (and gently) relocated into a larger hive fairly soon after acquisition, although more experienced beekeepers may acquire and manage nucs as resource banks for production colonies.

Some beginners want to start keeping bees by removing a colony from a structure or cavity, or by capturing a swarm. While possible, these are less-ideal options. Colony removal from a structure or cavity, called a "cutout" is fraught with issues. Most beginners lack the knowledge to assess the health of the bees they hope to remove and the specialty equipment and advanced carpentry knowledge the job might demand. Upon removal, inexperienced beekeepers may also be unable to accurately assess what they have removed (was the queen captured? injured?) and appropriately support the highly stressed, relocated colony.

Swarm captures present similar challenges. The unknown health status is a concern for any swarm capture, and an inexperienced beekeeper may not accurately assess whether they have captured the queen or, for example, if they do find a queen, whether she is a virgin with an afterswarm, and will thus require additional monitoring.

Whether to start with a package or a nuc is debatable, but not an important decision. Both nucs and packages can grow into thriving perennial colonies with appropriate management. Considerations include:

- Cost: Packages are *generally* less expensive than nucs.
- Availability: Packages are typically more available. However, both packages and nucs are seasonally produced in limited quantities and may need to be reserved early in the year for delivery later, when the season commences.
- Timeliness: In northern climates, packages are generally available earlier relative to nucs. And, while a package (or nuc) may be available, an earlier start may not provide any advantage. Adverse weather may impact forage and colony build-up, subsequently stressing the colony, and presenting additional feeding challenges to the beekeeper. Concurrently, delivery of nucs and packages may be delayed due to weather conditions in the regions where they are produced.
- Chances of success: There are many impactful variables, including:
  - Packages are initially broodless. The brood break may help manage Varroa and provide the package supplier and/or customer an opportunity to apply miticides.
  - Package queens are often superseded. Monitoring for and remediating that failure, if it occurs, can be challenging for new beekeepers (Reese 1951).
- Locally adapted genetics: Purchasers often prefer bees with genetics that survived in their area – obtaining nucs from local overwintered stock and acquiring local queens as the season advances, to replace the queens supplied with packages. Unfortunately, demand often outstrips supply.
- Quality: Concerns accompany both packages and nucs, ranging from improperly mated queens to chemical contamination of bees and combs, to diseased bees, to nuc suppliers selling off broken, contaminated and/or excessively used frames from their apiary, etc. Use reputable

suppliers and monitor honey bee health throughout the season.

- Hive configuration: Nucs are generally composed of Langstroth-style deep frames of bees, brood, and stores. The receiving equipment must be able to accommodate frames; use of a top bar hive configuration is best accomplished with a package instead of a nuc, for example. The receiving equipment must also match the frame size delivered (deep frames into deep hive bodies, for example) or will require less-than-ideal workarounds.
- Getting started: Bees from a package must be "installed," which can be an intimidating exercise for a new beekeeper, as can be the required queen release and verifying appropriate laying. (A nuc is more easily transferred into its full-sized hive and does not require queen release.) (Burlew 2017).

Subspecies: Beginners may wish to start with the less expensive, commercially available stock, likely Italian, Carniolan, or Caucasian; hopefully, also offering beneficial characteristics such as mite tolerance in addition to gentleness. Too often, beginners attribute failure to the subspecies rather than determining root causes. Beginners might consider gaining experience and success before making acquiring less common subspecies.

Figure 28.3 Some beekeepers, upon spotting the queen, carefully contain her in a queen clip so they may further inspect without worrying about injuring her or where she is. This queen's retinue is very attentive to her. *Source:* Photo courtesy of Charlotte Hubbard.

Common Mistakes

Successful beekeeping is largely an art, based upon unforgiving application of a large amount of science. The variety and creativity of missteps is equally large. Following are areas where mistakes are most commonly made – areas not necessarily unique to beginners, and in no particular order:

Trying to find the queen at every inspection: There is no need to find her every time, although following each inspection, a beekeeper should be reasonably confident the colony is queenright. Assuming it is the time of year when the queen should be laying, key indicators include finding one egg per cell in the bottom of the cell, larvae (indicates a mated queen present within last nine days), or capped brood (indicates a queen present within last 21 days for worker pupae). Experienced beekeepers may also note the colony's demeanor. While colonies can sound and behave differently due to genetics and other stressors such as dearth, queenless colonies often have a "roar" that experience can detect. Sometimes the demeanor will seem unduly restless or highly reactive to the slightest provocation.

Finding only larvae or pupae warrants consideration of why the queen might not be laying (season, lack of pollen, recent treatment, etc.) and rechecking in a week or so to affirm the colony is queenright.

Mishandling the queen: It is wonderful to find her, and she is worthy of veneration . . . but the frame she occupies should be held over the open hive while she is briefly admired and then returned to the colony where she is reasonably secure. A "queen clip" may be helpful (Figure 28.3).

Managing Varroa: Varroa is indeed an issue for any colony but especially for new beekeepers. Early in the learning curve, new beekeepers may need education on the life cycle of Varroa as it relates to the life cycle of the honey bees. Varroa is the primary vector for a multitude of viral diseases. (See Chapter 20 for a complete discussion on Varroa.)

Veterinarians can:

- Teach and encourage beekeepers to monitor Varroa and to treat proactively throughout the season, from spring to late fall (when mite counts can spike).
- Explain that winter losses are primarily due to high mite levels.
- Make sure beekeeper clients recognize when mite levels are nearing critical thresholds.

- Encourage participation in Citizen Science by reporting mite counts to Bee Informed Partnership's annual Mitecheck.
- Explain management tools that help control mites, such as brood breaks, drone comb, requeening with hygienic stock (all part of Integrated Pest Management) and proper biosecurity methods.
- Guide beekeepers in using treatments as labeled, and appropriate for the time of year, weather conditions, and the level of infestation.

Feeding issues: New beekeepers may see an abundance of flowers and assume their bees will be well fed. Bloom is essential, but bees also require flowers producing nectar (not challenged by drought or washed out by rain) and an appropriate workforce and weather to gather it.

To get new colonies off to the best possible start, sugar syrup should be provided until bees no longer consume it or the brood nest comb is completely drawn out (typically two deep or three medium hive bodies) (Milbrath 2016) and again whenever conditions warrant supplementing nutrition. A few reminders:

- Sugar syrup should be a 1 : 1 concentration of white sugar to water, by weight, for spring feeding; 2 : 1 for fall feeding.
- Avoid using external feeders to prevent attracting robbing bees and other insects to the apiary.

Inspection frequency/time: Unless there are time parameters (applying/removing a treatment, releasing a queen, etc.), inspections should be performed no more frequently than weekly to every 10 days. Over time and with experience, a beekeeper will learn to couple exterior observations with seasonal knowledge to guide efficient in-hive inspections. To avoid issues (like queen absence) advancing too far to be correctable, inspecting about every two to three weeks is advisable.

New beekeepers often keep the hive open too long. Efficiency comes with experience. Beekeepers may minimize inspection time by:

1) Having a purpose for disrupting the colony: For a new beekeeper, "just seeing how they're doing" is legitimate, but add specificity, such as ". . .and if they're ready for the next box," or ". . . and if the queen released last week is laying."
2) Staging what is needed: Anticipate what might be needed (the next hive body and frames, the tub for pulling honey), and have it in the apiary.
3) Accomplishing the purpose as smoothly and quickly as possible. If you do not know what actions to take at that moment, close the hive to review your notes or consult, and make decisions when you are not rushed.

Smoker use: *Never open a hive unless a smoker is lit and functioning.*

Use gentle puffs, several inches away from bees, to herd bees where they should move – off the edges of the box so you can replace the lid with minimal squishing of bees, down into the colony, off the honey frame, etc. Too often beginning beekeepers flood the area with smoke, leaving the bees confused and keepers teary-eyed and coughing.

Moving frames: Wax building requires heat. Occasionally an undrawn frame may be relocated to above or adjacent to the brood to enhance wax building, replacing a completed frame of nectar or honey from the brood nest area (which is then relocated to the sides of the box where bees tend not to work). Keep brood frames together for colony efficiency.

Splitting the brood nest: In spring, reversing hive bodies by relocating the top brood chamber to the bottom (bees will have moved up through the winter) is a reasonable practice; it gives the colony room to store nectar above the brood nest, assisting in swarm management. However, the beekeeper must take care to keep the brood nest contiguous. Often, by the time weather allows for a thorough spring inspection, the nest will occupy the top portion of the bottom box and extend upward into the next box. In this situation, reversing hive bodies splits the brood nest into non-contiguous locations and may lead to chilled brood and colony inefficiency.

Orientation flights versus swarming or absconding: The sudden emergence of bees from a colony, swirling crazily and noisily about, is commonly mistaken for swarming or absconding. It occurs when bees, now aged into the role of foragers, conduct their initial flights to familiarize with their location and hone flight skills. They will make small, sometimes seemingly haphazard, increasing circles about the face of the hive. This hovering-like activity lasts several minutes, with individual bees departing over time, as opposed to swarming or absconding, when the exit is massive and generally over within minutes. Absconding and swarming are discussed below.

Too much/too little space: *Too little* space in the hive may initiate a swarm response, even in a first-year colony. Concurrently, *too much* space requires bees to expend more effort patrolling and protecting, and moderating hive temperature. The time of year and the size of a colony is key for understanding space needs. Beekeepers must anticipate the activities and needs of the colony until the next inspection.

Beginners, not considering the dynamics of population build-up and key honey bee biology (like it will take a few days and perhaps even a week before the queen lays, and then 21 days from egg to worker brood emergence), may stack on all the hive bodies purchased with their beginner kit before they are needed. A starter colony from a package needs only one hive body for at least the first month, assuming 8- or 10-frame equipment, as an initial population decline for about a month is expected.

When colonies are small (struggling, or just starting), a rule of thumb is to add the next box when "plenty of bees" (not just a dozen or so wandering about) are active on about 70% of the frames – drawing out comb, polishing cells, etc. A large colony during a honey flow, however, may need many boxes, and it is appropriate to add multiple boxes at one time. Consider:

- Are bees working drawn comb or do they first need to build comb? If comb is already drawn (typically not the case for a beginning beekeeper's new package), the colony will need more space, added more frequently, as its efforts are applied to comb preparation and use (for hopefully a strong nectar flow). Drawn comb is a very valuable resource.
- Is the population ready to greatly increase? A single deep frame may yield close to 9000 bees. A couple of deep, capped frames of worker brood indicate a significant workforce emerging within a few days. (Figure 28.4) Waiting until 70% of the frames are being actively worked may be too long for a rapidly expanding workforce, especially if a nectar flow is pending or underway. (This assumes use of Langstroth equipment.)
- Is there a nectar flow coming? More space, added more frequently above the brood nest, might be required to ensure nectar and pollen do not clog the brood nest. Referred to as "honey-bound" or "pollen-bound," these conditions may initiate swarming. During a nectar flow, there should always be space (empty, drawn comb) above the brood nest.

Figure 28.4 This single deep frame will yield about as many bees as arrived in a package. *Source:* Photo courtesy of Charlotte Hubbard.

- Is the colony population and forage seasonally reducing? If so, even if there is not much room available, it may be the wrong season to add space, and may even be time to reduce it.

Disrespecting the 3/8″ bee space: Too much space (wrong-sized frames, eight frames in 10-frame equipment, an empty hive body, etc.) and bees will fill it with comb. Too little space, bees will propolize the gap, also wasting bee efforts and causing inspection inefficiencies.

Understanding what is in the hive: Beyond identifying diseases and pests, inexperienced beekeepers will probably make mistakes and take inappropriate actions until they can identify worker versus drone brood; swarm versus emergency versus supersedure queen cells; capped brood versus capped honey . . . to name a few. Beekeeping requires year around management. New ideas, new solutions, new research is rapidly developing. Continuing education is important for both beekeepers and bee doctors.

Euthanizing a Colony

There are times when euthanasia is appropriate, such as in the face of disease or when the colony presents a safety issue. Euthanizing for disease requires a specific approach to prevent further contamination; euthanasia methods include:

- Alcohol sprayed directly onto bees: This method does not allow for reuse of comb by bees, harvesting of honey, etc. (Westernwilson 2017).
- In the case of American Foulbrood, burning of the hive parts may be required in some areas: See Chapter 22 on Bacterial Diseases. It is recommended that the colony be humanely euthanized before burning the hive, if possible.
- Dish soap: Considered a more humane approach because of its quick efficacy when applied in sufficient amounts. Using a 1 : 4 ratio of liquid soap to water, (Terminix 2015) pour onto all bees in a colony. This method does not allow for reuse of comb or honey. Alternatively, bees may be shaken into a container of soapy water when bees are not flying (end of the day, cooler weather, poor health, etc.)
- Dry ice: This method requires sealing all openings, placing eight to 10 pounds of dry ice on the top bars, and closing the hive. Bees are killed by CO_2.
- Gasoline or kerosene are not recommended as these are flammable and can be a danger around a smoker and in arid regions where there is risk of fire.

(The American Veterinary Medical Association (AVMA) Guidelines for Euthanasia of Animals, 2020 Edition, did not include honey bees.)

Spring Issues and Considerations

Beyond the challenges affiliated with beginning beekeeping, spring reveals two general states for overwintered colonies: either stressed or strong. The first will likely need supportive feeding, space reduction, disease/mite intervention; the second – management for swarm prevention and production.

It is often surprising how quickly a strong, overwintered colony will grow, especially with ideal weather, and more so for a beekeeper enjoying first-time overwintering success. A starter colony could not commence building until arriving in their new hive in April or May; an overwintered colony's population increase begins a few months earlier. The starter colony spent much of their first active season focused on comb production; the survivor colony launches into preparing and filling that comb with brood and stores. The beekeeper's challenges are now swarm and nectar management.

Swarming: In *swarming*, just over half the colony departs in a frequently hasty, often very noisy exodus. (When witnessed, the sheer majesty suggests it was the entire colony.) Swarming more often occurs in the spring. It is how honey bee colonies/superorganisms naturally propagate when healthy and robust (called a reproductive swarm), or it is their response to a lack of space (a crowded swarm).

Swarming is natural behavior. However, when a swarm issues, the beekeeper just lost part of their investment, and what was left behind is at risk until they successfully requeen, with some intermittent loss of production. Nearby structures may gain unwelcomed, potentially costly tenants. (Figure 28.5)

Figure 28.5 Two beekeepers worked two days relocating this colony, constructed across three floor joists of a bedroom. *Source:* Photo courtesy of Charlotte Hubbard.

Effective swarm management involves identifying the regional cues that initiate swarming and proactively managing them (splits, appropriate space).

Nectar management: Nectar management – moving frames and adding space so the queen always has room to lay – diminishes the chances of swarming and optimizes the honey crop (Koss 2011; Milbrath 2018).

Honey harvest: Honey harvest is highly variable by area forage, weather, and geographic region. Some areas have two honey harvests – after the spring nectar flow, and again late summer. Regional practices should be applied along with three general considerations:

1) Honey cannot typically be obtained from starter colonies; commonly they may need all of the honey they produce their first year to survive winter.
2) Depending upon mite management strategies, it may be ideal to remove the honey for human consumption, and then apply a treatment. Removing honey supers reduces the beekeeper's effort if in-hive access is needed for treating, and it is required if the treatment cannot be applied to honey for human consumption.
3) It is best to learn about when and how much honey other beekeepers keep in their hives for winter and what is an appropriate configuration for overwintering in your area. The best strategy may not necessarily be to leave all the honey for winter, for three reasons. First, all the honey may not be enough for the upcoming winter. Second, it may actually be too much. In spring, the bees will quickly shift to needing space. Having too much capped honey can restrict their space in the spring and make swarm management more difficult. Third, the honey might not be arranged for easy access by a winter cluster. In cold temperatures, it can be harder for a cluster to access honey in a box of partly drawn or empty frames than in a box of full honey frames.

(Largely) Summer-Early Fall Issues and Considerations

Absconding: When all bees – except those too ill, young, or out foraging – abandon the hive *en masse* to find a new home, it is called absconding. It occurs when conditions in the hive are so stressful that the colony leaves to find a new location.

Causes for absconding include Varroa or other infestation, disease, excessive disruptions from mammals (including beekeepers), an inability to defend or cool the location, and nutritional deficits. Bees of African origin may be naturally more prone to absconding than bees of European origin.

Absconding may occur any time of year when temperature allows flight, perhaps more often in fall (Burlew 2010b). Newly introduced packages and recently captured swarms are also more prone to absconding, possibly in response to stress or a dislike of the home that was not their selection. Overcrowded nucs are also more likely to abscond or swarm (Burlew 2016).

Robbing: This mob-like, violent behavior occurs when a colony, unable to adequately defend itself, is attacked by bees seeking its nectar and honey. Robbing typically occurs in late summer or early fall, but particularly during times of nectar dearth. Robbers fight the colony's defenders (with losses on both sides) and rip open capped honey to plunder.

Signs include:

- Handfuls (or more) of bees investigating any parts of the hive where the scent of honey is detected – cracks between hive bodies, sealed entrances, covers, etc.
- Fighting bees – tumbling and wrestling on the landing board and outside the hive. Robber bees are often shiny and darker in appearance because they have lost their hair through fighting.
- Increased buzzing during robbing.
- Bees continually coming and going – as opposed to swarming or absconding, where they will dissipate typically within a matter of minutes.
- Dead and dying bees on the landing board and/or in front of the hive, or perhaps all deceased in a robbed-out colony.
- If bees are closely observed, you might also see bees dipping when flying out the hive; they are loaded with honey. Often robbers will climb up the side of the hive to take off, seeking more clearance due to the weight they are carrying.
- Decimated honey stores with jagged edged comb.
- Wax comb fragments from the destruction of sealed comb on the bottom board and in front of the hive.

Minimize robbing opportunities by:

- Understanding when it will likely occur and taking proactive measures such as removing external feeders, no open feeding, sealing multiple entrances (i.e. upper entrances often added during nectar flows), reducing entrances to sizes more easily defended, and installing robbing screens. Robbing screens also aid with drifting, and thus reduce pathogen transmission and the spread of contaminants.
- Keeping colonies strong and healthy.
- Delaying inspections until the threat of robbing is over, if possible. If inspections during heightened robbing conditions are essential, use smoke, minimize inspection time, and keep hive components as covered as possible and practical, using a sheet, hive covers, etc. (Burlew 2012).

Bearding: This fascinating behavior occurs typically during warm, late afternoons and early evenings when there is much honey to be ripened, as foragers gather outside the colony to help the internal temperatures stay cool. Often the gathered bees will also fan, pushing air inside. The large gathering is not a preamble to swarming or absconding, but rather a response to interior heat. (Figure 28.6)

Figure 28.6 Bearding in the late afternoon of a very steamy summer day. *Source:* Photo courtesy of Charlotte Hubbard.

Queen replacement: A queen should be replaced if found improperly mated. Too often though, queen replacement is advised for seemingly any issue – a colony not building up as expected, a spotty laying pattern, etc. Replacing the queen may be part of the solution, but the root(s) of the problem should first be determined by evaluating:

- Is it the time of year for the queen to be laying?
- Are clean, drawn cells available in the brood nest?
- Is the colony nutritionally stressed?
- Was a mite treatment recently applied that may have temporarily disrupted laying and brood?

Other reasons for queen replacement include:

- Promoting favorable characteristics in a colony, such as gentleness, Varroa mitigation, low swarming tendencies, disease resistance, or increased honey production.
- Replacing package queens with locally raised queens to gain locally adapted genetics.
- Ensuring colonies overwinter with a young, vigorous healthy queen, eager to build up in spring, as opposed to risking that an older queen will be superseded or swarm during the critical spring season. Routinely replacing queens after summer solstice is typical in commercial operations and often considered best practice.

Depending upon apiary management goals, timing, and beekeeper knowledge, requeening may be accomplished by letting the colony attempt to requeen itself, or with a replacement queen (mated or virgin), or with a queen cell (Connor 2015).

Weak hive/combination: Overwintering is often accomplished, even in northern regions, in nucs with special management consideration of warmth and stores (Milbrath 2017). Another approach may be to combine a struggling colony with a strong colony, with stipulations:

- The struggling colony must be weak for known, acceptable reasons, like it was started as a nuc late in the season. Avoid putting a healthy colony at risk by combining it with unhealthy bees.
- Combine with sufficient time (and hopefully weather) to allow the bees to rearrange their brood nest and stores as they deem appropriate.
- Debate continues whether to kill one of the queens or simply combine the colonies (generally recommended with a sheet of newspaper giving them time to sort out a few things) and let them figure it out.

Fall Issues and Considerations

Hopefully a colony entering the crucial fall season is strong and well on its way to being prepared for winter. Beekeeper considerations include:

Sufficient stores: What stores are needed is a function of the geographic area. If the estimated stores seem lacking, feeding with 2 : 1 sugar syrup is good insurance. An emergency food ceiling, discussed under overwintering, may also be added late fall.

Space reduction: Remove unused frames and hive bodies, although some unused frames may need to remain to respect bee space. (If so, position them toward the outside of the hive body.) Consolidate filled honey frames in the same hive bodies, positioned adjacent to and overhead of the brood nest, which is naturally forced downward throughout fall (reproduction slows and the brood nest is backfilled).

Critter guards: When bees cluster, small mammals seeking a dry, warm shelter with food may move into the hive. Barriers prohibiting entry by anything larger than a bee should be installed *prior* to when small mammals move into hives.

Unneeded components: Remove queen excluders, shims that may have been used for treatment application, and internal feeders.

Treatments: Perform final mite monitoring and management strategies.

Overwintering

The largest predictor of overwintering success is colony health. Early season health stumbles may be overcome, but a healthy workforce is critical mid- to late-summer. Late-summer bees have the responsibilities of raising specially adapted winter bees in sufficient numbers for the colony to deal with cold, all while collecting and processing the nectar to achieve necessary winter stores (as much as 100 pounds in some climates) and often in less than favorable foraging conditions.

Beyond colony health, there are some practices that may promote successful overwintering. These practices are used in climates where winter is characterized by months of consistently below-freezing temperatures, a lack of forage, and bees generally clustered for weeks or months. Local practices vary, but offer a reasonable guide for what is helpful and include:

Emergency food ceiling: Sugar blocks, candy boards, and the mountain camp method are common practices to ensure bees have carbohydrates if they run low on stores and cold or wet weather conditions persist through the spring. (Figure 28.7) If the cluster is too small to move in the cold and/or is covering brood away from the supplemental stores, the colony may starve to death regardless of provisions. Bees will starve before abandoning brood.

Moisture remediation: Bees in cluster shiver to generate heat, releasing rising, moist air. Various tactics can

Figure 28.7 Bees working sugar blocks on an unusually warm Michigan day. The beekeeper added more blocks to the shim before covering with a quilt box. *Source:* Photo courtesy of Todd Smith.

Figure 28.8 Various wraps and windbreaks for a midwest winter. *Source:* Photo courtesy of Todd Smith.

wick away moisture from the colony, such as a quilt box, a moisture board placed below the inner cover, and/or sugar (dry or candy form) on the top bars of the colony. An upper entrance is also often deployed and can serve as an important egress when snow height prohibits cleansing flights.

Wrapping/wind protection: A large, healthy cluster with sufficient stores can endure lengthy winters and extreme cold. Some beekeepers wrap their hives with the intent of mitigating stress. (Figure 28.8) Options are numerous, including commercially available polypropylene wraps, black house wrap, and foam. Some wraps negate the warming advantage of sun.

Wrapping does increase wind stability and provides wind protection as do windbreaks, ranging from straw bales to sideways lawn chairs to sophisticated, reinforced barriers.

Prevent wind gusts up into the hive, either by skirting the bottom with a wrap, straw bales, etc., and/or using a slider board (if there is not a solid bottom board).

Monitoring: When there is an unusually warm winter day and bees are flying outside of the hive – beyond a hasty cleansing flight – beekeepers are delighted. This thrill is quickly offset by increased worrying as bees are burning crucial stores in an attempt to find non-existent forage.

If that joyous event is missed, occasional light tapping on the hive, in hopes of detecting a buzz, is permissible (Burlew 2010a). But, a reassuring buzz in response does not mean the colony will still be alive when it counts – in the spring. It only assures there is that possibility. Unfortunately, tapping disrupts torpor and increases colony energy use. Increasingly, affordable infrared cameras provide valuable information including:

- Life: Knowing of a dead colony (deadout) allows for replacement strategizing. Moving deadouts should wait for warmer weather though; cold wax is very brittle.
- Cluster size: A small cluster may warrant additional protection (perhaps wrapping) to support them through the remaining winter.
- Location: Bees start toward the bottom of the hive and work up throughout the winter; a cluster nearing the end of stored food could likely benefit from supplementation. (Figure 28.9)

Opening a hive in cold weather to address issues is justified, although warmer, low-wind days are preferable, and intervention should be efficient.

As in most livestock operations, management issues are often the cause of poor performance or lead to less than desirable production results. Veterinarians can familiarize themselves with common husbandry practices and pitfalls to help beekeeping clients attain their goals, whether it is the enjoyment of having bees, or commercial honey production, or supplying pollinator services to agriculture.

Figure 28.9 This infrared camera photo reveals differences in cluster size and location of overwintering hives. All three colonies are still working below their ample stores (beekeeper ensured the top hive bodies full of honey in fall). *Source:* Photo courtesy of Todd Smith.

References

Anderson, C. (2020). *Keep-ants-out-of-your-beehive*. [Blog] http://Carolinahoneybees.com. Available at: www.Carolinahoneybees.com [Accessed 20 Jan 2020].

Burlew, R. (2010a). *The-winter-hive-to-tap-or-not-to-tap*. [Blog] http://Honeybeesuite.com. Available at: www.honeybeesuite.com [Accessed 20 Jan 2020].

Burlew, R. (2010b). *Why-do-honey-bees-abscond-in-the-fall*. [Blog] http://Honeybeesuite.com. Available at: www.honeybeesuite.com [Accessed 20 Jan 2020].

Burlew, R. (2012). *Robbing-bees-questions-and-answers*. [Blog] http://Honeybeesuite.com. Available at: www.honeybeesuite.com [Accessed 20 Jan 2020].

Burlew, R. (2016). *Did-they-abscond-or-die-from-varroa*. [Blog] http://Honeybeesuite.com. Available at: www.honeybeesuite.com [Accessed 20 Jan 2020].

Burlew, R. (2017). *Nuc-or-package-how-to-buy-honey-bees*. [Blog] http://Honeybeesuite.com. Available at: www.honeybeesuite.com [Accessed 20 Jan 2020].

Connor, L. (2015). Queen Rearing Essentials, 2e. Kalamazoo, MI: Wicwas Press.

Koss, R. (2011). *Nectar Management Works!* [Blog] http://Beesource.com. Available at https://beesource.com/point-of-view/walt-wright/nectar-management-works. [Accessed 20 Jan 2020].

Milbrath, M (2016), *Establishing A Honey Bee Colony*, http://sandhillbees.com, viewed 20 February 2020, https://static1.squarespace.com/static/56818659c21b86470317d96e/t/5900c95f1b631b3a5b23b9c1/1493223777073/Getting+your+first+year+honey+bee+colony+established.pdf.

Milbrath, M 2017, *Sustainable Northern Beekeeping A Method to Improve Survival and Reduce Replacement Costs*, http://sandhillbees.com, viewed 20 February 2020, https://static1.squarespace.com/static/56818659c21b86470317d96e/t/5a00cd449140b785dda50a4a/1510002000326/Milbrath_Sustainable FallNucs.pdf.

Milbrath, M 2018, *Swarm Biology and Control in Northern States*, http://sandhillbees.com, viewed 20 February 2020, https://static1.squarespace.com/static/56818659c21b86470317d96e/t/5c0acac0758d4684e95caf23/1544211142569/Swarms_Milbrath2018.pdf.

Reese, C. (1951). *Package Bees for Honey Production and Pollination*, Agricultural Extension Service. The Ohio State University, Ohio.

Terminix (2015). http://terminix.com. Available at https://www.terminix.com/pest-control/bees/stings/symptoms/spray [Accessed 23 Jan 2020].

Westernwilson (2017). *When Needs Must: Euthanizing Honey Bee Colonies* [Blog] http://herewebee.wordpress.com. Available at https://herewebee.wordpress.com/2017/08/18/when-needs-must-euthanizing-honey-bee-colonies [Accessed 20 Jan 2020].

Wyns, D. (2018). *Drift* [Blog] http://beeinformed.org. Available at https://beeinformed.org/2018/06/11/drift [Accessed 20 Jan 2020].

Further Reading

AVMA (2020) AVMA Guidelines for the Euthanasia of Animals, 2020 Edition. https://www.avma.org/sites/default/files/2020-01/2020-Euthanasia-Final-1-17-20.pdf.

Caron, D. and Connor, L. (2013). Honeybee Biology and Beekeeping. Kalamazoo, MI: Wicwas Press.

Milbrath, M 2018, *Help I Need a Queen*. http://sandhillbees.com, https://static1.squarespace.com/static/56818659c21b86470317d96e/t/5a00cd449140b785dda50a4a/1510002000326/Milbrath_SustainableFallNucs.pdf.

Plonski, S. (2017). *Beekeeping 101: Learn how to use the bee smoker*. [Blog] http://Thehoneybeeconservancy.org.

29

Queen Rearing and Bee Breeding

Krispn Given

Department of Entomology, Purdue University, West Lafayette, IN, USA

US Commercial Queen Rearing Industry

Each year tens of thousands of queens are produced in the US with the biggest producers located in California and Hawaii. Queen producer businesses produce 6000–16 000 queens annually.

Some universities are breeding mite resistant stocks. The success of these breeding programs is dependent on proper breeder selection as well as encouraging beekeepers to use these honey bee strains to help reduce the use of miticides and ameliorate overall honey bee health.

The veterinarian should have a basic understanding of the process of queen rearing as part of the overall management of honey bees. Remember, however, that most hobby beekeepers will produce new queens through splits, nuclei, and swarms.

Benefits of Queen Rearing

Indeed, queen rearing saves the apiarist from having to purchase queens and provides the opportunity to develop locally improved colonies of bees adapted to a particular area. Selecting from overwintered colonies from your local area can increase disease resistance in addition to increasing honey crops.

Queen Cell Building

The ability to produce queen cells is dictated by weather conditions and the colony population, along with abundant nectar and pollen sources. Queen production is often most successful when the colony has an impulse to swarm, which varies with locality.

Grafting

Gilbert M. Doolittle is the father of modern commercial queen rearing, He first described grafting in his book *Scientific Queen Rearing* (1889), accidently using the word grafting to describe queen rearing. The term stuck even though grafting typically refers to plant grafts. Grafting is the transfer of larvae from a brood comb to the prepared artificial cups using a grafting tool.

Equipment Needed

- *Grafting tool*. Used to transfer the larvae from the cell to the cups
- *Cell cups*. Most queen breeders use plastic cups
- *Grafting frames*. These typically have three removable bars that easily accommodate 15 cell cups each.
- *Internal feeder*. Used for syrup

There are several types of grafting tools on the market today (Figure 29.1). The Chinese grafting tool with a wooden handle and flexible quill tip is one of the best for beginners. Lifting the larvae from the cell is easily accomplished with some practice. The Chinese grafting tool is particularly useful because it allows you to lift some royal jelly with the larvae (like a spoon), insuring more acceptance from the cell builder. Slide the flexible quill of the tool along the cell wall to the bottom of the cell and lift the back side of the larvae. However, you need larvae that are 24 hours old or less (Laidlaw 1977). Older larvae, three and four days old, can result in an "inter-cast" queens,

Honey Bee Medicine for the Veterinary Practitioner, First Edition. Edited by Terry Ryan Kane and Cynthia M. Faux.

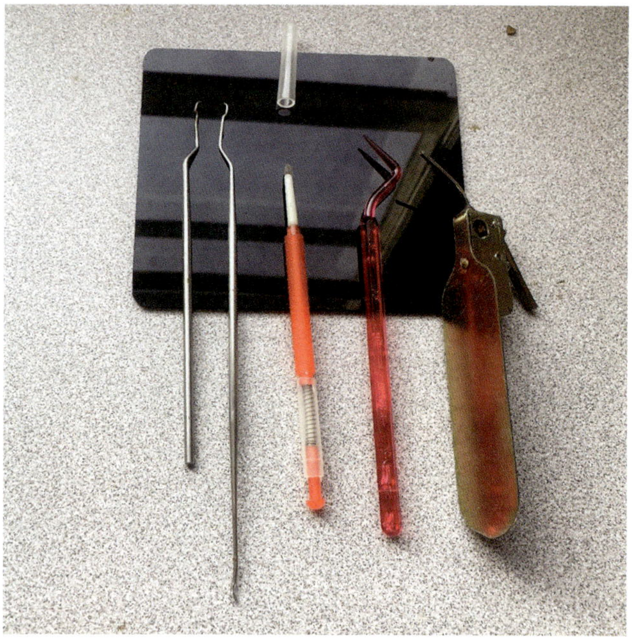

Figure 29.1 A variety of grafting tools.

Table 29.1 Simple table for queen rearing.

Simple table for queen rearing	
Day 1	Selected breeder queen lays eggs
Day 3	Eggs hatch
Day 4	Transfer the larvae to the prepared cups
Days 6–9	Queen cells are sealed
Days 12–14	Transfer cells to mating nuclei or queenless hives
Day 16	Virgin queens emerge
Day 21	Virgin queens fly to DCA to mate with 15–20 drones
Day 30	Mated queen should start egg laying
Day 35	Look at queen's brood pattern for uniformity

where she is part queen, part worker, much smaller in size, but behaviorally a queen. These are inferior queens and will eventually result in supersedure by the workers.

Grafting Process

A well-designed grafting station will have a lighted magnifying glass along with a comfortable place to sit. It is recommended to have a cell bar holder and a water well to dip and rinse the grafting tool periodically.

At the grafting station, cover the frame of selected larvae with a clean moist towel to prevent the larvae from

Figure 29.2 Grafting larvae into queen cups.

desiccating during the grafting process. Look for the right age larvae to graft; if you can barely see them, they are the correct larvae to use (Figure 29.2).

With the grafting tool, slide down the base of the cell wall until you are under the larvae (they look much like the letter C) and then gently lift and place into the cup.

Producing queens from natural swarm cells is also an easy way to raise a new queen. One can remove frames containing queen cells and place them in "nuc" (nucleus) colonies. These cells produce excellent queens because the larvae in these cells were likely very well fed and chosen at the optimum stage of development.

It is striking how quickly the workers start feeding the larvae in the cell cups. The nurse bees are producing large amounts of proteinaceous "royal jelly" from their paired hypopharyngeal and mandibular glands located in the head, these secretions are synthesized with nectar to the incipient queens. They will feed the larvae more than they can eat up to the fourth day, when the cell is sealed (Laidlaw and Page 1987).

After 10 days there should be well sculpted large cells (Figure 29.3).

Virgin queens will emerge 13 days after the egg hatches (day 16 after laying) (Figure 29.4).

Banking Queens

A queen bank is simply a de-queened colony with many young bees and emerging brood to take care of the young queens.

Queen Piping

Piping is most common when there is more than one queen in a colony. These so-called battle calls, which are sonic beeps lasting about two seconds, are produced by the queen

Figure 29.3 Queen cells on the frame.

Figure 29.4 Queens emerge from the end of the cell. Queen cells with a hole chewed in the side have indicate the occupant was killed by another queen.

rapidly moving her thoracic wing muscles. Piping is not a buzz. Piping is very common in the spring when many swarm cells may be present and usually just before the colony swarms.

Keeping Good Records

It is essential to keep good records. Below is an example of the information to capture.

Selective Breeding

Beekeepers will often choose a superior breeder queen to produce more queens with desirable characteristics such as to:

- Increase honey production
- Improve overwintering ability
- Decease defensive behavior
- Reduce swarming
- Change color
- Decrease incidence of diseases
- Resistance to varroa
- Other personal reasons

Producing superior bees with desirable traits, like resistance to the obligate parasite *Varroa destructor*, is the goal that most honey bee breeders wish to accomplish. The great pioneering honey bee breeder Harry H. Laidlaw once said *"80% of the colonies overall fitness is governed by simply having a superior queen"* the other 20% we can influence through selective breeding.

The fundamental principle of any breeding program is that the offspring resemble their parents. In our lab, we synthesize our rigorous selection with selected daughter queens and drones produced from parents expressing desirable traits, with hierarchical behavioral phenotypes for grooming and mite-biting. We want bees that display phenotypic behaviors and bees that overwinter well here in the northern part of the US

Drones

Drones develop from unfertilized eggs produced by the queen, so they do not have a father but a grandfather. Although drones have the reputation of being the laziest members of the colony, it is important to produce high quality drones from the best queens and to try to flood the mating yards with these drones. One drone comb in each breeder colony will yield about 6000 drones. Drones are sexually mature in 12–15 days post emergence. It is a life of leisure; sexually mature drones are attracted to light in the afternoon between 2 and 6 p.m. Under optimum conditions, they will take several mating flights to various "drone congregation areas" (DCA's) in search of a virgin queen. Only about 1% of drones will successfully mate with a virgin queen, and die for their effort, if successful. Once a drone achieves success at mating, it instantly becomes paralyzed and falls back. The endophallus break also occurs at this time, resulting in mortality due to rapid blood "hemolymph" loss.

Basic Bee Genetics

One does not need to be a skilled geneticist to breed honey bees with success. Honey bees, like other *Hymenoptera* (ants and wasps), have haploidy diploidy which, at its most simple explanation, is a sex determination system in which males develop from unfertilized eggs (haploid) and females develop from fertilized eggs (diploid). Female workers and queens have 16 paired or 32 total chromosomes (Laidlaw and Page 1987). Drones only have half, 16 chromosomes (8 paired). Alleles are different phenotypic characters, like color or behavior characteristics. For example, pollen collection is influenced by specific related alleles.

If a queen is mated to her brother, the progeny would share many of the same alleles, and we would call them homozygous. If a queen is mated to unrelated drones, these progenies would have many diverse unrelated alleles and are referred to as heterozygous. These polygenic-colonies would express more traits, possibly reducing the impact of diseases and parasites.

Breeding is a systematic process where one selects for certain phenotypes.

Instrumental Insemination

As in other species, instrumental insemination (artificial insemination) is essential in any honey bee breeding program. One of the challenges in honey bee breeding is controlled mating. As mentioned above, honey bees have extreme polyandry, meaning the queen mates with many individuals in a (DCA) from various colonies resulting in a "subfamily" superorganism. You cannot place a colony in an enclosed structure and achieve mating success.

Figure 29.5 Photograph of an artificially inseminated queen laying an egg surrounded by her retinue.

Dr. Lloyd Watson first demonstrated a successful instrumental insemination of queens in 1927. Modern bee breeding programs use instrumental insemination to speed up stock selection and purify desirable traits like grooming behavior or resistance to pest and pathogens (Figure 29.5).

References

Doolittle, G.M. (1889). Scientific Queen Rearing. Chicago, IL: Thomas G. Newman and Son. 186 p.

Laidlaw, H.H. (1977). Instrumental Insemination of Honey Bee Queens: Pictorial Instructional Manual. Hamilton, IL: Dadant and Sons. 144 p.

Laidlaw, H.H. and Page, R.E. (1987). Queen Rearing and Bee Breeding. Cheshire, CT: Wickwas Press. 224 pp.

30

The Future Direction of Honey Bee Veterinary Medicine

Jeffrey R. Applegate, Jr.[1,2]

[1] *Nautilus Avian and Exotics Veterinary Specialists, Brick, NJ, USA*
[2] *North Carolina State University College of Veterinary Medicine, Raleigh, NC, USA*

The future of honey bee veterinary medicine is yet to be determined. At the time of this writing, relatively recent federal regulatory changes have caused US veterinarians to play a bit of "catch-up" in order to become proficient in honey bee veterinary medicine and to understand the needs of various factions of apiarists (i.e. backyard hobbyists, small "farmer's market-level" producers, and large commercial operations). This new discipline led to the creation of the Honey Bee Veterinary Consortium (HBVC; www.hbvc.org), and drove a need for continuing education through this and numerous other CE-focused veterinary conferences. Many US veterinary schools are now collaborating with on-campus bee research labs to plan curricula for veterinary students and to offer hands-on "bee labs" to both students and practitioners.

To help this trend, this book, Honey Bee Medicine, is offered as the first collaborative effort among veterinarians, entomologists, toxicologists, pharmacologists, laboratorians, and the beekeeping community, to provide a foundational scholarly and credible approach to honey bee veterinary medicine.

US veterinarians became involved in honey bee veterinary medicine to help control medically important antibiotic use in honey bees and thus reduce antibiotic residues in the related human consumables. The concern over the epidemic of antimicrobial resistance has led many countries outside the US and Canada to ban or restrict the use of antibiotics in food producing animals, including honey bees. Europe, Australia, and New Zealand have more stringent rules on reporting and mitigating contagious foulbrood diseases.

The future for honey bee veterinarians will be driven by education and training. A number of veterinary schools in Europe, Asia, and other parts of the planet, have already incorporated a honey bee curriculum. Beekeepers in those countries depend much more on veterinarian's expertise for disease prevention and management, parasite and pest control, and honey bee husbandry. In France, veterinarians are the apiary inspectors.

Inclusion of honey bee veterinary medicine in US and Canadian veterinary colleges will be the first step to building a foundation of knowledge for those who want to pursue this sector of veterinary medicine. A basic knowledge of honey bee biology/behavior and ecology in veterinary curricula would benefit the veterinary community twofold: bringing awareness to the fact that honey bees are food animals in need of veterinary care and addressing the environmental and economic impact of honey bees on global food security.

Following the academic framework established through didactic and hands-on labs, further opportunities can be found through collaboration with entomologists, conference offerings, and mentored training. In the future, post-veterinary degree training programs specific to honey bees, in the form of a residency similar to that of other food animal-focused veterinarians, or simply a certification program, like the World Aquatic Veterinary Medical Association (WAVMA) CertAqV program, may coalesce and become valuable training opportunities. As new research and new technologies continue to develop, industry entities may help fund the training, in concert with an entomology program, so that honey bee veterinarians become more knowledgeable and thereby be increasingly incorporated into the practices of small and commercial operations.

The veterinary profession is an inquisitive one and the growing medical science in honey bee and pollinator health/management provide career research opportunities. New veterinary research will enable conferences to call for abstracts, present poster-worthy information, and present

Honey Bee Medicine for the Veterinary Practitioner, First Edition. Edited by Terry Ryan Kane and Cynthia M. Faux.

novel veterinary-focused information to interested parties to further their education beyond the basics. New clinical testing methods that model other species with which veterinarians work, may become available. Lateral flow testing for more than just bacterial diseases may become the norm. Dissemination of the new information is paramount. Depending on how vigorous the veterinary research becomes it may lead to a veterinary honey bee focused journal.

The mandated antibiotic oversight – the judicious use of medically important antibiotics in all US food producing livestock-required by the US Food and Drug Administration's 2017 Veterinary Feed Directive rule, requires a legal Veterinary–Client-Patient-Relationship with the apiarist. This is a significant change for the US beekeeping industry as well as for veterinarians. We are all responding to this transition, but as of yet, there are not enough trained "bee-doctors" to meet the needs of the US beekeeping industry.

American foulbrood (AFB) and European foulbrood (EFB) hive-side field tests and diagnostic test kits are readily available. As with other sectors of veterinary medicine, this technology will advance and result in broadened testing possibilities. Hive-side test kits or advanced testing methods may become available for other diseases of concern in the future. This testing will easily parlay into disease surveillance, as is done with other food animals. Disease surveillance may become increasingly important due to hive migration, natural foraging behaviors, mating interactions as well as the impact of global climate change on agriculture. Early detection will help to contain outbreaks and foreign animal disease, pests, and parasites.

Veterinarians will be at the forefront of upcoming diagnostic tests and disease diagnosis in hives. Veterinary input will be necessary as the industry advances into managing the honey bee microbiota, gene expression, and genomics. Collaboration among veterinarians, entomologists, regulatory officials, and of course, apiarists, will be necessary to build a smooth interdisciplinary working relationship for the advancement of honey bee health and the industry.

Ethical concerns of humane management of invertebrates, including euthanasia, are important issues. The veterinary perspective on these topics, along with industry and regulatory officials and beekeepers, will be necessary for a comprehensive understanding of the future of honey bee management and veterinary medicine. These future practices will require a variety of management strategies; honey bees are not only a commercial entity, single hive owners are known to develop a significant human animal bond with the hive and its queen.

Integrating veterinarians into the regulatory framework may become necessary and beneficial as veterinarians in the US and Canada work more closely with state and provincial bee inspectors and regulatory authorities. A team approach will help ensure smooth transitions and changes in regulatory plans and, hopefully, ease the minds of concerned apiarists. Beyond domestic honey bee veterinary involvement, international opportunities will present themselves. For instance, in 2019, the HBVC represented US veterinary honey bee regulatory updates at Apimondia Roma and attended an international veterinarian-only honey bee medicine meeting focused on Worldwide honey bee health. The 46th Apimondia International Apicultural Congress Montréal, 2019 was attended by many US and Canadian bee veterinarians. Apimondia International 2021 will be held in Russia. Honey bee and pollinator health is a global concern.

Future honey bee veterinarians will be challenged in multiple ways:

- With managing hive biosecurity in the face of the bees' natural behaviors.
- With needing to be versatile enough to use their skill set to guide hobbyist hive owners and still be able meet the expectations of a commercial environment – as veterinarians already do with commercial mammalian and avian livestock, as well as hobbyist owners of the same livestock.
- With integrating into integrated crop pollination using a combination of honey bees and native bees, and to use digital tools, not only to monitor hives, but to monitor crops and resources. The use of drone pollinators is on the horizon.
- With working with the myriad of honey bee-focused small to large businesses directing their efforts toward new urban beekeeping, suburban beekeeping, and beekeeping to benefit charitable causes.
- And with addressing research opportunities and career options that will likely arise, in both domestic and international venues.

The future of honey bee veterinary medicine is wide open and bright. It is up to the current professional veterinarians with bee expertise, as well as the veterinarians entering this fascinating and important part of medicine, to determine our profession's overall eventual roles. It is up to each of us to help create the future we want.

Honey Bee Medicine

Resources

Websites

Agriculture and Agri-Food Canada, https://www5.agr.gc.ca
American Association of Professional Apiarists, https://aapa.cyberbee.net/
American Beekeeping Federation, www.abfnet.org
Animal and Plant Health Inspection Service, APHIS, https://www.aphis.usda.gov/aphis/ourfocus/planthealth/plant-pest-and-diseaseprograms/honey-bees/outreach-videos
Apiary Inspectors, https://apiaryinspectors.org/, https://apiaryinspectors.org/provincialinspection-services/
Apimondia, https://www.apimondia.com/en/
American Veterinary Medical Association, AVMA,
https://www.avma.org/honey-bees-101-veterinarians
Bee Informed Partnership, BIP, https://beeinformed.org/
Bee_Health Extension, https://bee-health.extension.org/
Bee People, Get Stung, https://bee-people.com/
2 Million Blossoms, https://www.2millionblossoms.com/
Canadian Association of Professional Apiculturists, CAPA, http://www.capabees.com/
Canadian Food Inspection Agency (CFIA), https://inspection.gc.ca
Canadian Honey Council, http://honeycouncil.ca/
Canadian Veterinary Medical Association, CVMA, https://www.canadianveterinarians.net/documents/treating-honey-bees-andpollinators-what-veterinary-medical-professionals-need-to-know
Deschambault Animal Science Research Center, http://crsad.qc.ca/
Environmental Protection Agency, EPA, https://www.epa.gov/pollinator-protection
Food Animal Residue Avoidance Databank (FARAD), http://www.farad.org/vetgram/honeybees.asp
Food and Drug Association, FDA,
https://www.fda.gov/animal-veterinary/development-approval-process/using-medically-important-antimicrobials-bees-questions-and-answers
Government of Canada Honey Bee Producer Guide to the National Bee Farm-level Biosecurity Standard, https://www.inspection.gc.ca/animal-health/terrestrial-animals/biosecurity/standards-and-principles/honey-bee-producer-guide/eng/1378390483360/1378390541968
Honey Bee Veterinary Consortium, www.hbvc.org
Honey Bee Health Coalition, https://honeybeehealthcoalition.org/
Honey Bee Suites, https://www.honeybeesuite.com/
Canadian National Bee Farm Level Biosecurity,
https://www.inspection.gc.ca/animal-health/terrestrial-animals/biosecurity/standards-and-principles/bee-industry/eng/1365794112591/1365794221593?chap=0
Government of Canada Bee Health Managment, https://www.inspection.gc.ca/animal-health/terrestrial-animals/biosecurity/standards-and-principles/honey-bee-producer-guide/eng/1378390483360/1378390541968?chap=4
National Academy of Sciences, Engineering, Medicine ,https://sites.nationalacademies.org/sites/climate/SITES_190724
National Bee Diagnostic Centre, https://www.gprc.ab.ca/research/nbdc/
National Honey Board, https://www.honey.com
National Veterinary Accreditation Program, NVAP Modules 23, 29 and 30, https://www.aphis.usda.gov/aphis/ourfocus/animalhealth/nvap/ct_aast
Pollinator Partnership, https://www.pollinator.org/
Pollinator Partnership Protection Plan, Pollinator Health Task Force, https://www.whitehouse.gov/sites/whitehouse.gov/files/images/Blog/PPAP_2016.pdf
Scientific Beekeeping, http://scientificbeekeeping.com/
The Sand Hill, www.sandhillbees.com

Honey Bee Medicine for the Veterinary Practitioner, First Edition. Edited by Terry Ryan Kane and Cynthia M. Faux.

The United States Climate Alliance, http://www.usclimatealliance.org/
United States Department of Agriculture, ARS, Bee Research Labs, https://www.ars.usda.gov/
USDA National Agriculture Statistical Service:
https://www.nass.usda.gov/Surveys/Guide_to_NASS_Surveys/Bee_and_Honey/
The World Bee Project, http://worldbeeproject.org/
Veterinary Information Network, VIN, www.vin.com, courses on honey bees

University Bee Labs

Auburn, https://auburnbees.com/
Colorado State, https://pollinationbiologylab.agsci.colostate.edu/, https://ramlink.campuslabs.com/engage/organization/csuapiculture
Cornell, https://pollinator.cals.cornell.edu/resources/, https://pollinator.cals.cornell.edu/
Dalhousie University, https://www.dal.ca/faculty/agriculture/plant-food-env/research/centres-and-labs.html
Univeristy of Florida, http://entnemdept.ufl.edu/honey-bee/ https://ag.tennessee.edu/EPP/Pages/Bees%20and%20Beekeeping/eXtension.aspx https://www.mdac.ms.gov/bureaus-departments/plant-industry/honeybee-program/
Iowa State, https://www.ent.iastate.edu/pollinators/
Laval University, https://www.fsg.ulaval.ca/en/research/natural-resources/
Louisiana State University, https://entomology.lsu.edu/beeresearch.html, https://www.ars.usda.gov/southeast-area/baton-rouge-la/honeybeelab/
Mississippi State, http://extension.msstate.edu/agriculture/livestock/beekeeping
Minnesota, https://www.beelab.umn.edu/
Michigan State University, https://www.canr.msu.edu/home_gardening/pollinators/
North Carolina State, https://entomology.ces.ncsu.edu/apiculture/bees/, https://www.ncsuapiculture.net/
University of Nebraska, https://entomology.unl.edu/bee-lab
Ohio State, https://u.osu.edu/beelab/
Oregon State, https://honeybeelab.oregonstate.edu/
Penn State, https://ento.psu.edu/pollinators
Purdue, https://extension.entm.purdue.edu/beehive/
Simon Fraser University, https://www.sfu.ca/people/eelle.html
Texas A&M, https://honeybeelab.tamu.edu/
Tufts, http://ase.tufts.edu/biology/labs/starks/Default.htm
University of California, Davis, https://honey.ucdavis.edu/, https://elninobeelab.ucdavis.edu/
University of British Columbia, https://beehive.ubc.ca/, https://ubcfarm.ubc.ca/csfs-research/honey-bees-pathogen-response/
University of Calgary, https://www.ucalgary.ca/
University of Georgia, https://bees.caes.uga.edu/
University of Guelph, https://www.uoguelph.ca/honeybee/
University of Illinois, https://publish.illinois.edu/dolezalbeelab/research/
University of Manitoba, https://umanitoba.ca/faculties/afs/dept/entomology/index.html
University of Maryland, https://www.vanengelsdorpbeelab.com/
Universidad Nacional Autónoma de México, UNAM, https://www.pollinator.org/mexico; https://ourworld.unu.edu/en/protecting-native-bee-populations-in-mexico
University of Missouri, https://extension2.missouri.edu/m403
University of Puerto Rico, http://www.giraylab.org/
University of Saskatchewan, Western College of Veterinary Medicine, https://wcvm.usask.ca/departments/pathology/pathology-people/elemir-simko.php#AcademicCredentials
University of Tennessee, Knoxville, https://ag.tennessee.edu/EPP/Pages/Bees%20and%20Beekeeping/home.aspx
University of Wisconsin, https://pollinators.wisc.edu/honey-bees/
University of Vermont, https://www.uvm.edu/gund/bees
Virginia Tech, https://www.freelyflyingbees.com/, https://www.ento.vt.edu/Projects/vap.html
Washington State University, http://bees.wsu.edu/research-lab/
York University, http://zayedlab.apps01.yorku.ca/wordpress/

Beekeeping Courses and Mentoring Are Offered at State (or provincial) and Local Bee Clubs

https://www.abfnet.org/page/states
https://americanbeejournal.com/tiposlinks/beekeeping-associations/
https://www.mannlakeltd.com/beekeeping-education/beekeeping-directory

State Dept. of Agriculture Beekeeper sites

https://www.nasda.org/states/state-directory

Provincial Government Apiculture Sites

British Columbia: https://www2.gov.bc.ca/gov/content/industry/agriculture-seafood/animals-and-crops/animal-production/bees
Alberta: https://www.alberta.ca/apiculture.aspx
Saskatchewan: https://www.saskatchewan.ca
Manitoba: https://www.gov.mb.ca
Ontario: http://www.omafra.gov.on.ca/english/food/inspection/bees/apicultu.html
Quebec: https://www.mapaq.gouv.qc.ca/fr/Productions/santeanimale/maladies/RAIZO/Pages/reseauapicole.aspx

Nova Scotia: https://novascotia.ca/sns/paal/agric/paal020.asp
New Brunswick: https://www2.gnb.ca/content/gnb/en/departments/10/agriculture/content/bees.html
Newfoundland and Labrador: https://www.faa.gov.nl.ca/agrifoods/plants/apiculture.html
Prince Edward Island: https://www.princeedwardisland.ca/en/topic/small-fruit-and-bees
Northwest Territories: https://www.enr.gov.nt.ca/en/services/insects-and-spiders/bees

Laws and Legal Information on Bees and Honey

https://www.lawserver.com/bees-and-honey
https://pollinatorstewardship.org/index.php/state-beekeeping-laws/

Books

The ABC and XYZ of bee culture; a cyclopedia of everything pertaining to the care of the honey-bee; bees, hives, honey, implements, honey-plants, etc. By A.I. Root and E.R. Root. 41st Edition 2007
Africanized Honey Bees in the Americas, Dewey M. Caron, 2001
Attracting Native Pollinators: The Xerces Society Guide, Protecting North America's Bees and Butterflies, the Xerces Society, 2011
Bee, photographs by Rose-Lynn Fisher, Princeton Architectural Press, 2012
Bee Health and Veterinarians, OIE, World Organization for Animal Health, 2014
Beekeeping For Dummies. Howland Blackiston, John Wiley & Sons, 2016
Bee Time, Lessons From the Hive, Mark L. Winston, Harvard University Press, 2014
Beginning Beekeeping, Everything You Need to Make Your Hive Thrive, Tanya Phillips, Penguin Random House, 2017
Buzz, The Nature and Necessity of Bees, Thor Hanson, Basic Books, 2018
Diagnosis Of Honey Bee Diseases: United States. Department of AgricultureHoney Bee Biology and Beekeeping, Revised Edition. Dewey M. Caron, Lawrence John Connor, Robert G. Muir, Ann Harman, David Heskes, Jon Zawislak
Elimination of American Foulbrood Disease Without the Use of Drugs, a Practical Manual for Beekeepers, Apiculture New Zealand, by Mark Goodwin 2018
Honey Bee Biology and Beekeeping, Revised Edition, Dewey Caron et al. 2013
Honey Bee Democracy, Thomas D. Seeley, Princeton University Press, 2010
Honey Bee Veterinary Medicine, *Apis mellifera,* Nicolas Vidal-Naquet
Our Native Bees, North America's Endangered Pollinators and the Fight to Save Them, Paige Embry, Timber Press, 2018
Pollinator Conservation Handbook: A Guide to Understanding, Protecting, and Providing Habitat for Native Pollinator Insects, Xerces Society with The Bee Works, Matthew Shepard et al. 2003
Simple Smart Beekeeping, Kirsten S. Traynor, PhD and Michael J. Traynor, 2015
The Anatomy of the Honey Bee. R.E. Snodgrass
The Backyard Beekeeper, An Absolute Guide to Keeping Bees in Your Yard and Garden, Kim Flottum, 4th edition. Quarto Publishing Group, 2018
The Bee: A Natural History. Noah Wilson Rich
The Bee Book, Fergus Chadwick et al.,Penguin Random House, 2016
The Bees in Your Backyard, A Guide to North American Bees, Joseph S. Wilson & Olivia Messinger Carril, Princeton University Press, 2016
The Beekeeper's Bible, Richard A. Jones and Sharon Sweeney-Lynch, Stewart, Tabori & Chang, 2011
The Beekeeper's Lament: How One Man and Half a Billion Honey Bees Help Feed America. Hannah Nordhaus, Harper Perennial, 2011
The Classroom, Beekeeping Questions and Answers, Jerry Hayes, Dadant and Sons, 1998
The Dance Language and Orientation of Bees, Karl Von Frisch, Harvard University Press, 1971
The Lives of Bees: The Untold Story of the Honey Bee in the Wild. Thomas D. Seeley,Princeton University Press, 2019
The Sixth Extinction: An Unnatural History, Henry Holt Publishers, by Elizabeth Kolbert, 2014
100 Plants to Feed the Bees, The Xerces Society, Story Publishing, 2016
Understanding Bee Anatomy, a full colour guide, Ian Stell, The Catford Press, 2012

Suppliers

ApiHex, https://apihex.ca/
Stores located in Quebec, Ontario and Alberta
Toll Free: 855-666-3233

BeeMaid Bee Supplies, http://www.beemaidbeestore.com/
625 Rosemary Street
Winnipeg, Manitoba
R3H 0T4
Canada
Toll Free: 1-866-783-2240 ext. 228

Better Bee, www.betterbee.com
8 Meader Road
Greenwich, NY 12834
Toll Free: 1-800-632-3379
Fax: 518-314-0576

Dadant & Sons, www.dadant.com
51 South 2nd St.
Hamilton, IL 62341
Toll Free: 888-922-1293
217-847-3324 Fax: 217- 847-3660 email: dadant@dadant.com
Multiple branch offices

Dancing Bee Equipment, https://dancingbeeequipment.com/
3384 Loyalist Road
Port Hope, ON
L1A 3V7
Canada
Ph: 1-905-753-2623

Mann Lake, https://www.mannlakeltd.com,
Main Headquarters - Mann Lake
501 1st St S
Hackensack, MN 56452 US
Phone: 800-880-7694
Fax: 218-675-6156
Email: beekeeper@mannlakeltd.com

There are many local and state bee clubs/organizations and suppliers as well. Check local listings.

Notes on Editors and Contributors

Editors

Terry Ryan Kane, DVM, MS, earned her MS in Ecology from the University of Illinois, Chicago. As a licensed pilot, she was inspired by the waggle dance and the navigation skills of the honey bee to take beekeeping while in veterinary school at UI–Champaign-Urbana. She founded Michigan's first feline-only hospital in 1981. After decades of practice she pivoted to policy having received an AVMA-AAAS Congressional Science and Technology Policy Fellowship serving as a science policy advisor in the US Senate. The FDA ruling on antibiotic stewardship was being debated while she was in Congress. Returning home to Michigan she started up her bee hives again and founded A2 Bee Vet, a practice devoted to honey bee medicine (www.a2beevet.com). She serves on the AVMA Committee on Environmental Issues (CEI) and is on the Board of the Honey Bee Veterinary Consortium, an organization for honey bee veterinarians.

Cynthia M. Faux, DVM, PhD, DACVIM-LA, earned her DVM from Iowa State University and her PhD (Geology) from Yale University. She became interested in bees from her grandfather who kept bees when she was a child. She has her own bees and has taught honey bee continuing education for veterinarians as well as the honey bee lab at the AVMA convention with Dr. Kane. Cynthia is currently a Professor of Veterinary Medicine at the University of Arizona.

Contributors

Esmaeil Amiri, PhD, used to be a migratory professional beekeeper in Iran along with his undergraduate studies. He then traveled to several European countries to further broaden his knowledge, and earned his PhD from Aarhus University, Denmark. Thereafter, he continued his research activities by joining the honey bee research collaborative team in NC, USA. Since 2011, he has been contributing to the field of honey bee science and has published several scientific articles. His academic interests lie at the interaction of the fields of honey bee health and virology, specifically virus–host interactions and virus transmission pathways with the aim to identify alternative ways to combat viral diseases in honey bees.

Jeffrey R. Applegate, Jr., DVM, DACZM, was born and raised on the coast of New Jersey where he still enjoys spending free time surrounded by salty air. He attended Virginia Tech for undergraduate studies in animal science and biology and Kansas State University for veterinary school. Following a small animal rotating internship at the Animal Medical Center in New York City, he worked as an exotic animal clinician in a busy New Jersey private practice. He then served as faculty at North Carolina State University, College of Veterinary Medicine in the exotic animal medicine service for over five years where he remains involved as adjunct faculty. Dr. Applegate has lectured on honey bee veterinary medicine both domestically and internationally. He is a diplomate in the American College of Zoological Medicine and is a founding member of the Honey Bee Veterinary Consortium (HBVC), in which he served terms as secretary, conference chair, and president. Dr. Applegate has also worked as a small animal veterinary technician, and a volunteer firefighter. In his free time he enjoys the outdoors and tending his own apiary.

Dewey M. Caron, Emeritus Professor of Entomology & Wildlife Ecology, University of Delaware, & Affiliate Professor, Department Horticulture, Oregon State University. Bachelor's degree Zoology University of Vermont 1964, MS 1966 University of Tennessee Ecology, PhD 1970 Entomology (Apiculture) Cornell University. Faculty member UMD 1970–1981 and University of Delaware 1981 until retirement in 2009. He remains active writing for bee magazines/newsletters and giving Bee Short Courses on both West and East coasts. He annually gives over 100 presentations to bee clubs and state bee organizations. He is the author of *Honey Bee Biology and Beekeeping*, seven additional books and over 20 book

Honey Bee Medicine for the Veterinary Practitioner, First Edition. Edited by Terry Ryan Kane and Cynthia M. Faux.

chapters. He is active in Master Beekeeper training in OR, CA and with EAS on the East coast. He represents WAS on Honey Bee Health Coalition and is the principal author of *Tools for Varroa Management and BMPs.* He conducts annual Bee loss survey of Pacific Northwest backyard and commercial beekeepers and annual Pollination survey of PNW beekeepers.

Christopher J. Cripps, DVM, started keeping bees while earning his Boy Scout Merit Badge in Beekeeping. He took several beekeeping classes including labs at Cornell University and worked for Professor Roger Morse as his lecture room assistant. While earning his DVM at Ohio State University, he worked as a bee inspector for Franklin and Delaware Counties. After graduation, he moved to Greenwich NY where he worked as a food animal veterinarian for 17 years while keeping bees as a hobby. In 2012, he bought Betterbee, a beekeeping supply business, with his partner from the veterinary clinic, Dr. Joseph Cali. He manages the business, teaches classes, and works with customers to help diagnose and correct problems that come up with their bees. He is a Past President of the Southern Adirondack Beekeepers Association and helped organize the Honey Bee Veterinary Consortium. Chris has presented at many veterinary meetings about honey bee diseases.

Yanping (Judy) Chen, PhD, is Research Scientist at the United States Department of Agriculture (USDA) Bee Research Laboratory (BRL) in Beltsville, MD. Judy Chen received her BA from Hunan Agriculture University, P.R. China, her MS from Brigham Young University, and her PhD from Texas A&M University. After completing her postdoctoral research at the University of Maryland Medical School, National Institutes of Health, USDA–ARS Insect Biocontrol Laboratory, she joined the USDA BRL in 2002. Her current research includes investigation of the epidemiology, transmission, and pathogenesis of honey bee viruses and the microsporidian Nosema, development of in vivo and in vitro systems for honey bee virus and Nosema propagation, identification of new emerging viruses in honeybees, characterization of the genomic structures of viruses and Nosema, and development of RNAi-based therapeutics for the treatment of honey bee diseases.

Gigi Davidson, BSP, DICVP, is the former Director of Clinical Pharmacy Services at the NC State University College of Veterinary Medicine where she practiced veterinary pharmacy for 35 years. Ms. Davidson received a pharmacy degree from UNC Chapel Hill in 1983 and was awarded Diplomate status in the International College of Veterinary Pharmacy in 2001. She was inducted into Phi Zeta in 2006 for distinguished contributions to the advancement of veterinary science. Gigi is a past President of the Society of Veterinary Hospital Pharmacists and is the current President of the American College of Veterinary Pharmacists. She has many publications in peer-reviewed scientific journals, and her primary area of research interest is stability, safety, and efficacy of compounded therapies in non-human species. Her research interests in retirement include drug disposition in *Apis mellifera* and pharmacological action of plant pollens and nectars on honey bee colonies. Ms. Davidson is the owner of Vetpharm Consulting and is a conservationist beekeeper committed to the survival of honey bee colonies.

Jay D. Evans, PhD, is Research Leader at the United States Department of Agriculture (USDA) Bee Research Laboratory in Beltsville, MD, where he has worked for 20 years. The BRL is focused on the development of management strategies to help honey bees thrive in the face of disease, chemical stress, and inadequate forage. Jay's own research uses genetic techniques and controlled challenge experiments to address the impacts of parasites and pathogens. Current projects involve honey bee immunity, interactions among stress factors, and the development of novel, safe, controls for mites and viruses. Jay received his AB in Biology from Princeton University in 1988 and his PhD in Biology from the University of Utah in 1995, where he studied ants before developing an interest in bees and beekeeping.

Krispn Given, is recognized as one of the leading authorities on honey bee instrumental insemination and honey bee breeding. He is the Apiculture Specialist at Purdue University's Department of Entomology in West Lafayette Indiana where he teaches an annual queen rearing short course and instrumental insemination class. He also lectures to beekeepers and researchers around the country. His current research focus is on selecting for behavioral resistance to varroa destructor by selecting for bees that groom themselves free of mites and bite them. Grooming behavior is an important mechanism for resistance. Krispn was instrumental in developing the "Mite-biter" honey bee strain. This hygienic behavior is of much interest to beekeepers and researchers of honey bees. Krispn is author and co-author of numerous publications, including scientific and trade journal articles. He also helped design innovative instrumental insemination devices that national and international breeders and researchers are using (https://apisengineering.com).

Jerry Hayes started out in life as a High School teacher but later learned he liked working with a beekeeper much better. Jerry became more interested and fascinated and started reading everything he could get his hands on about honey bees. He turned into the consummate backyard beekeeper and wondered if he could get into the beekeeping world and support a young family. So, with the support of his family he went back to school under the tutelage of Dr. Jim Tew, at Ohio State University. Jerry said it was the "best thing he ever did". Jerry has authored the "Classroom Q&A" column of the American Bee Journal for almost 40 years, the "Classroom" book and is author/co-author on 23 research papers and a variety of honey bee related articles in a wide variety of publications. Jerry has served as Past President of Apiary Inspectors of America, Heartland Apiculture Association, Colony Collapse Working Group, and has shared his wisdom through many professional presentations internationally and through media opportunities. Looking back on his opportunities as a Research Technician at the USDA/ARS Baton Rouge Bee Lab, Dadant and Sons Regional Mgr., Dadant And Sons New Product Dev., and AR Mgr., Chief of the Apiary Section for the Florida Dept. of Agriculture and Consumer Services, Monsanto Honey Bee Lead, VP. of Vita Bee health North America and now Editor of Bee Culture magazine, Jerry is filled with awe and amazement.

Brandon Hopkins, PhD, is an Assistant Research Professor at Washington State University in the Department of Entomology. Initially working on the development of cryopreservation of honey bee germplasm for breeding and conservation, work that enabled the establishment of the world's first honey bee germplasm repository at WSU and inclusion of honey bee semen in the USDA National Animal Germplasm Program. His research efforts have been focused on developing practical solutions for the beekeeping industry ranging from bee breeding to varroa control.

Charlotte Hubbard, Michigan's 2018 Beekeeper of the Year, mentors beekeepers and teaches beekeeping for several colleges, clubs, and conferences – almost anywhere there's a group seeking knowledge wrapped with humor. She writes and edits bee-related works, is the author of *Dronings from a Queen Bee; the First Five Years,* a collection of humorous essays about beekeeping, and *If I Could Sit on a Bee's Knees,* a children's book answering questions about bees. She is passionate about bee education and helping bee experts share their knowledge to most effectively help pollinators. She is very involved with the Kalamazoo Bee Club and the Michigan Beekeepers Association. She and her husband manage about three dozen colonies.

More information about her work is found at www.hubbardhive.com and on her facebook page, Charlotte Hubbard, Beekeeper.

Adam Ingrao, PhD, is an US Army veteran and Agricultural Entomologist at Michigan State University Extension (MSUE). He holds a BS in Agriculture and Environmental Plant Science from California Polytechnic State University, San Luis Obispo, and a PhD in Entomology from Michigan State University. His research and professional backgrounds include expertise in apiculture, agricultural pest management, chemical ecology, predator–prey interactions, invertebrate taxonomy, and production horticulture. He is the founder and lead instructor for Michigan State University's Heroes to Hives program (https://www.canr.msu.edu/veterans/),a beekeeping education program for military veterans and their dependents. Dr. Ingrao owns and operates Bee Wise Farms LLC with his partner Lacey Ingrao.

Reed M. Johnson, PhD, got his start in research beekeeping while looking for a summer job in his hometown, Missoula, Montana. He knocked on the door of Dr. Jerry Bromenshenk at the University of Montana, was offered employment, and was quickly drawn into the world of bees and bee toxicology. Reed went on to receive a PhD in Entomology from the University of Illinois at Urbana-Champaign working with Dr. May Berenbaum where he was involved in the honey bee genome project. Reed is currently an Associate Professor in the Department of Entomology at The Ohio State University in Wooster, Ohio. He teaches two courses at Ohio State: one on the biology and practical aspects of beekeeping and another on pesticide toxicology and application. His research focuses on determining how bees are exposed to pesticides and measuring the effect that toxic exposure has on the health of honey bees, with the goal of promoting bee health in the context of modern agriculture.

Britteny Kyle, DVM, graduated from the Ontario Veterinary College in 2009. Upon graduation she entered practice at a busy 24 hour small animal hospital in Toronto, Ontario. She worked as an associate for several years before leaving practice to focus on her young family. During her time away from practice, Dr. Kyle began to study honey bee medicine. She was elected to the board of the Honey Bee Veterinary Consortium in 2019 and serves as President of the organization in 2020. In 2019 she established a pollinator garden at her local elementary school where she works to engage the children, as well as the community at large, in conservation efforts to protect pollinators. In 2020 Dr. Kyle returned to the Ontario Veterinary College to earn her Master's degree in Epidemiology with a focus on honey bee diseases. Dr. Kyle enjoys lecturing to the veterinary community and writing about her experiences as a new beekeeper and as a honey bee veterinarian. When not studying bees she is busy with her three boys, two cats, one dog, and her honey bee colonies.

Katie Lee, PhD, received her MS and PhD from the University of Minnesota with Dr. Marla Spivak. She is the Apiculture Extension Educator at the University of Minnesota. Her research focuses on measures that indicate queen bee and colony health, and the benefits of pollinator plantings on honey bee health. She developed a parasitic mite sampling protocol that is now a nationwide standard. For the non-profit organization the Bee Informed Partnership, Katie founded the Northern California and Minnesota-based TechTransfer Teams that provide services for commercial beekeepers. She served on the board of the American Beekeeping Federation and co-chairs the research and education committees.

Marcie Logsdon, DVM, graduated from Washington State University College of Veterinary Medicine in 2012 and has worked in the WSU Veterinary Teaching Hospital Exotics and Wildlife Department since 2014. Since 2017 she has collaborated with the University of Minnesota based Partners for Wildlife initiative – a program aimed at improving animal welfare in wildlife rehabilitation. She recently moved to a farm in rural Palouse and was thrilled to find a thriving colony of honeybees in one of the outbuildings. Since then she has inherited another bee hive and now considers herself a beekeeper.

Margarita M. López-Uribe, PhD, is an Assistant Professor of Entomology and the Lorenzo L. Lansgstroth Early Career Professor at Penn State University. She is broadly interested in understanding responses of bee populations to environmental change. She received her bachelor's degree in biology from Universidad de los Andes in Colombia, her master's degree in genetics and evolution from Universidade Federal de São Carlos in Brazil, and her doctoral degree in entomology from Cornell University. López-Uribe was an NSF Postdoctoral Fellow at North Carolina State University before moving to Penn State. Her research integrates population genetics, comparative phylogenetics, landscape ecology, and field experiments to address fundamental questions in bee ecology and evolution.

Scott McArt, PhD, an Assistant Professor of Pollinator Health, helps run the Dyce Lab for Honey Bee Studies at Cornell University in Ithaca, New York. He is particularly interested in scientific research that can inform management decisions by beekeepers, growers, regulatory agencies, and the public. Research in the McArt lab focuses on the impact of pesticides, pathogens, habitat, and management practices on the health of honey bees and wild pollinators.

Meghan Milbrath, PhD, has been a beekeeper for over 25 years, and runs The Sand Hill Apiaries (http://www.sandhillbees.com/), a small queen rearing and honey farm in South Central Michigan. She works as an Assistant Professor in the Department of Entomology at Michigan State University where she is the coordinator of the Michigan Pollinator Initiative. Dr. Milbrath's research and extension work is focused on protecting honey bees and other pollinators from environmental health risks. She teaches Apiculture and Pollination at MSU, as well as rotation at MSU college of veterinary medicine, and frequently writes for Bee Culture and American Bee Journal. Dr. Milbrath has a master's in Public Health, and a PhD in Environmental Health Sciences, and her public health background strongly drives her perspectives on pollinator health.

Kristen K. Obbink, DVM, MPH, DACVPM, is a public health veterinarian and hobbyist beekeeper dedicated to the promotion of animal, human, and environmental health. She holds her BS (Zoology) and DVM from Iowa State University, her MPH from the University of Minnesota, and is a Diplomate of the American College of Veterinary Preventive Medicine. In addition to working in private practice, she previously served as the State of Iowa's enteric disease epidemiologist and coordinator for food/feed emergencies. Currently, she is a lead public health veterinarian with the Center for Food Security ad Public Health at the Iowa State University College of Veterinary Medicine where she co-managed the development of materials for the USDA's National Veterinary Accreditation Program, including a module on honey bee veterinary medicine. Dr. Obbink also serves as co-chair of the Honey Bee Veterinary Consortium Education Committee, as an Executive Board member and Public Health Committee Chair to the Iowa Veterinary Medical Association, and as Delegate to the American Veterinary Medical Association representing the American Association of Food Safety and Public Health Veterinarians.

Randy Oliver earned BS and MS degrees in biological sciences, specializing in insect culture, and is an independent researcher and professional beekeeper. He has kept honey bees for over 50 years, and writes for the American Bee Journal, posting his articles on all aspects of bee biology and beekeeping to his website, Scientific Beekeeping (http://scientificbeekeeping.com/). The success of his 1500-hive operation in arid California is highly dependent upon assessing the nutritional status of his colonies, and providing supplemental feeding when indicated.

David T. Peck, PhD, studies the behavioral interactions between *Varroa destructor* and honey bees. His doctoral work at Cornell University involved studying mite-resistance traits of unmanaged tree-living bees in New York's Fingerlakes region. His ongoing research examines mite-resistance traits in mite-naïve managed bees in Newfoundland, Canada, as well as native honey bees in Anosy, Madagascar that have only faced mites since varroa arrived on the island around 2010. He also studies how mites spread between honey bee colonies, examining the impacts of bee drift, honey robbing, and even mite behavior on flowers in an attempt to understand when and how mites manage to disperse from a dying colony to its neighbors.

Robin W. Radcliffe, DVM, DACZM, is a wildlife veterinarian at the Cornell University College of Veterinary Medicine where he recently completed the Cornell Master Beekeeping Program. He directs a unique research and training program that connects health and conservation of endangered species with experiential training opportunities for students together with Jane Goodall. His fascination with honey bees began when his grandfather built him a silver bee box and together they followed wild bees to his first bee tree in the forests of Wisconsin. Robin and Tom Seeley recently teamed-up to explore the health of wild honey bees by comparing a man-made bee tree cavity to a traditional thin-walled hive box. Robin lives in an 1890 homestead on the edge of Shindagin Forest in Brooktondale, New York.

Rolfe M. Radcliffe, DVM, DACVS, DACVECC, is a veterinarian specializing in large animal surgery and emergency critical care. He works mostly with critically sick horses, cattle, and other large animals at the Cornell University College of Veterinary Medicine. Dr. Radcliffe attended veterinary school at the University of Minnesota, was in private veterinary practice in the Midwest for several years prior to returning for a residency in Large Animal Surgery at Minnesota. Following his residency, he traveled to the Ontario Veterinary College working as a research associate and emergency faculty. Dr. Radcliffe next completed advanced training in Large Animal Emergency and Critical Care Medicine at Cornell University. Together with his twin brother, Robin, Rolfe has started a honey bee health course for veterinary students, and is active in continuing education training for veterinarians interested in learning about the fascinating biology, diseases, and management of the honey bee.

Kasie Raymann, PhD, is a microbial ecologist using the honey bee as a model system to study the evolution and dynamics of host-associated microbial communities. Her PhD is in Microbiology and Genomics from Pasteur Institute (Paris, France) and she was a USDA NIFA postdoctoral research fellow at the University of Texas at Austin. She is currently an Assistant Professor at the University of North Carolina at Greensboro (UNCG), and one of the Directors of the UNCG Plant and Pollinator Center. Her previous research has investigated the impacts of chemical exposure on the honey bee gut microbiome and honey bee health having shown that both antibiotics and pesticides can severely alter the gut microbiome which leads to increased pathogen infection and mortality. Recent research is on factors that shape fine-scale microbial community structure within the gut microbiome of honey bees. Dr. Raymann has published many research articles on the impact of chemicals on the bee microbiome and has been invited to speak at national and international meetings including, the American Veterinary Medical Association (AVMA) National meeting, the Honey Bee Veterinary Consortium Conference, the International Pollinator Conference, and the Presidential Advisory Council Meeting on Combating Antibiotic-Resistant Bacteria (PACCARB).

Gary S. Reuter has a BS in education from the University of Minnesota and has been a beekeeper since 1982. He has worked as a technician for Dr. Marla Spivak at the University of Minnesota bee lab since 1992 where he maintains the research colonies, trains and works with students in the field, designs and builds specialty equipment, and speaks to beekeeping, student and civic groups. He plans the Extension short courses and, together with Marla, teaches beekeeping to all experience levels from beginner to expert.

James A. Roth, DVM, PhD, is a Distinguished Professor in the College of Veterinary Medicine at Iowa State University and a member of the National Academy of Medicine. He received the DVM (1975) and PhD (1981, immunology) degrees from ISU. He is the Director of the Center for Food Security and Public Health. Dr. Roth's primary area of research expertise is immunity to infectious diseases of food producing animals. He has worked with federal, state, and industry officials to develop the Secure Food Supply plans in the event of a foreign animal disease outbreak for poultry, beef, dairy, pork, and sheep. These plans include biosecurity recommendations. He has testified before Congressional committees on biosecurity preparedness, on efforts to address bioterrorism and agroterrorism, and on the need for vaccines for foreign animal diseases.

Olav Rueppell, PhD, is a Professor in the Department of Biological Sciences at the University of Alberta Canada. His academic interests bridge fundamental research on the causes and consequences of social evolution in insects and applied research to improve honey bee health. He is the author of over 90 scientific publications and has mentored numerous students. His integrative research includes bioinformatics, genetic analyses, studies of cells, behavioral and physiological observations and experiments, and demographic and ecological approaches. He has studied honey bees since 2001 and his honey bee health research is particularly focused on understanding how viruses, parasitic mites, and stress contribute to the ongoing honey bee health crisis with the goal of identifying sustainable apicultural solutions.

Thomas D. Seeley, PhD, is the Horace White Professor Emeritus in Biology at Cornell University. His research focuses on the behavior, social life, and ecology of honey bees. He has been an avid beekeeper since he was 16, hence for more than 50 years. He is the author of several books on bees, including *Honeybee Democracy* (2010), *Following the Wild Bees* (2016), and *Honey Bees in the Wild* (2019). For fun, besides beekeeping he enjoys bee trapping (catching swarms in bait hives) and bee hunting (finding wild colonies by bee lining). He lives in Ithaca, New York.

David Tarpy, PhD, is a professor of Entomology and the Extension Apiculturist at the University of North Carolina State University since 2003. He has been contributing to the field of honey bee science and is the author of more than 100 scientific publications. He has supervised numerous graduated students and mentored several postdoctoral fellows. His research interests focus on the biology and behavior of honey bee queens, using techniques including field manipulations, behavioral observation, and molecular genetics, in order to better improve the overall health of queens and their colonies.

Dan Wyns has accumulated experience as a commercial beekeeper and bee inspector in addition to work as a Field Specialist with the Bee Informed Partnership (BIP) Tech Transfer Team. BIP is a non-profit organization facilitating collaboration between research institutions and beekeepers with an objective of improving honey bee colony health. The Tech Transfer Team works with commercial beekeepers to provide in-field and laboratory diagnostic services to help evaluate and improve colony management practices. In addition to the Tech Transfer Team, BIP also manages citizen science projects including the Sentinel Apiary program and conducts the annual colony loss and management surveys. More information about all BIP programs and research findings are publicly available at BeeInformed.Org

Index

Honey Bee Medicine for the Veterinary Practitioner, First Edition. Edited by Terry Ryan Kane and Cynthia M. Faux.

b

c